NATIONAL ACADEMIES Sciences Engineering Medicine

AF772429

NATIONAL ACADEMIES PRESS
Washington, DC

Thriving in Space

Ensuring the Future of Biological and Physical Sciences Research
A Decadal Survey for 2023–2032

Committee on Biological and Physical Sciences Research in Space 2023–2032

Space Studies Board

Aeronautics and Space Engineering Board

Division on Engineering and Physical Sciences

Consensus Study Report

NATIONAL ACADEMIES PRESS 500 Fifth Street, NW Washington, DC 20001

This activity was supported by Contracts NNH17CB02B and 80HQTR22DA001 with the National Aeronautics and Space Administration. Any opinions, findings, conclusions, or recommendations expressed in this publication do not necessarily reflect the views of any organization or agency that provided support for the project.

International Standard Book Number-13: 978-0-309-69498-8
International Standard Book Number-10: 0-309-69498-1
Digital Object Identifier: https://doi.org/10.17226/26750
Library of Congress Control Number: 2023948182

This publication is available from the National Academies Press, 500 Fifth Street, NW, Keck 360, Washington, DC 20001; (800) 624-6242 or (202) 334-3313; http://www.nap.edu.

Printed in the United States of America.

Suggested citation: National Academies of Sciences, Engineering, and Medicine. 2023. *Thriving in Space: Ensuring the Future of Biological and Physical Sciences Research: A Decadal Survey for 2023–2032*. Washington, DC: The National Academies Press. https://doi.org/10.17226/26750.

The **National Academy of Sciences** was established in 1863 by an Act of Congress, signed by President Lincoln, as a private, nongovernmental institution to advise the nation on issues related to science and technology. Members are elected by their peers for outstanding contributions to research. Dr. Marcia McNutt is president.

The **National Academy of Engineering** was established in 1964 under the charter of the National Academy of Sciences to bring the practices of engineering to advising the nation. Members are elected by their peers for extraordinary contributions to engineering. Dr. John L. Anderson is president.

The **National Academy of Medicine** (formerly the Institute of Medicine) was established in 1970 under the charter of the National Academy of Sciences to advise the nation on medical and health issues. Members are elected by their peers for distinguished contributions to medicine and health. Dr. Victor J. Dzau is president.

The three Academies work together as the **National Academies of Sciences, Engineering, and Medicine** to provide independent, objective analysis and advice to the nation and conduct other activities to solve complex problems and inform public policy decisions. The National Academies also encourage education and research, recognize outstanding contributions to knowledge, and increase public understanding in matters of science, engineering, and medicine.

Learn more about the National Academies of Sciences, Engineering, and Medicine at **www.nationalacademies.org**.

Consensus Study Reports published by the National Academies of Sciences, Engineering, and Medicine document the evidence-based consensus on the study's statement of task by an authoring committee of experts. Reports typically include findings, conclusions, and recommendations based on information gathered by the committee and the committee's deliberations. Each report has been subjected to a rigorous and independent peer-review process and it represents the position of the National Academies on the statement of task.

Proceedings published by the National Academies of Sciences, Engineering, and Medicine chronicle the presentations and discussions at a workshop, symposium, or other event convened by the National Academies. The statements and opinions contained in proceedings are those of the participants and are not endorsed by other participants, the planning committee, or the National Academies.

Rapid Expert Consultations published by the National Academies of Sciences, Engineering, and Medicine are authored by subject-matter experts on narrowly focused topics that can be supported by a body of evidence. The discussions contained in rapid expert consultations are considered those of the authors and do not contain policy recommendations. Rapid expert consultations are reviewed by the institution before release.

For information about other products and activities of the National Academies, please visit www.nationalacademies.org/about/whatwedo.

COMMITTEE ON BIOLOGICAL AND PHYSICAL SCIENCES RESEARCH IN SPACE 2023–2032

ROBERT J. FERL, University of Florida, *Co-Chair*
KRYSTYN J. VAN VLIET, Cornell University, *Co-Chair*
ADAM P. ARKIN, University of California, Berkeley
SUSAN M. BAILEY, Colorado State University
DEBJYOTI BANERJEE, Texas A&M University
PAUL M. CHAIKIN (NAS), New York University
KATHLEEN E. CULLEN, Johns Hopkins University
DANIEL H. GESCHWIND (NAM), University of California, Los Angeles
ROBERT W. HYERS, Worcester Polytechnic Institute
YIGUANG JU, Princeton University
CHRISTOPHER E. MASON, Weill Cornell Medicine
MICHAEL J. PECAUT, Loma Linda University
WILLIE S. ROCKWARD, Morgan State University
ELBA E. SERRANO, New Mexico State University
PETER VOROBIEFF, University of New Mexico
RONALD L. WALSWORTH, University of Maryland, College Park
SARAH WYATT, Ohio University
LUIS ZEA,[1] University of Colorado Boulder
ZHUOMIN ZHANG, Georgia Institute of Technology

Staff

DWAYNE A. DAY, Senior Program Officer, *Study Director* (from February 2022)
ARUL MOZHI, Associate Board Director, *Study Director* (from August 2022)
GREGORY MACK, Senior Program Officer, *Study Director* (February 2022 to August 2022)
SANDRA J. GRAHAM, Senior Program Officer, *Study Director* (January 2021 to February 2022)
JULIE ANNE SCHUCK, Senior Program Officer
ERIK SVEDBERG, Scholar
STEVEN MOSS, Senior Program Officer
GAYBRIELLE HOLBERT, Senior Program Assistant
DIONNA WISE, Program Associate
COLLEEN N. HARTMAN, Director, Aeronautics, Astronomy, Physics, and Space Science

Panel on the Biological Sciences

SUSAN A. BLOOMFIELD, Texas A&M University, *Co-Chair*
CHRIS WOLVERTON, Ohio Wesleyan University, *Co-Chair*
DEBRA T. AUGUSTE, Northeastern University
JENNIFER A. BARRILA, Arizona State University
ENRIQUE M. DE LA CRUZ, Yale University
AMANDA C. EVANS, Los Alamos National Laboratory
MONIKA FLESHNER, University of Colorado Boulder
PETER W. GRABHAM, Columbia University
RASHA HAMMAMIEH, Walter Reed Army Institute of Research
PANKAJ JAISWAL, Oregon State University
JANET K. JANSSON, Pacific Northwest National Laboratory

[1]Resigned on June 28, 2022.

EDWARD J. KELLY, University of Washington
ELLIOT M. MEYEROWITZ (NAS), California Institute of Technology and Howard Hughes Medical Institute
KAREN OCORR, Sanford Burnham Prebys Medical Discovery Institute
RONKE M. OLABISI, University of California, Irvine
AUBRIE O'ROURKE, NASA Kennedy Space Center
IMARA Y. PERERA, North Carolina State University
MANU O. PLATT, National Institutes of Health
ROBERT J. PLOUTZ-SNYDER, University of Michigan School of Nursing
BRINDA K. RANA, University of California, San Diego
SCOTT TRAPPE, Human Performance Laboratory
MARTHA H. VITATERNA, Northwestern University
JOSEPH C. WU (NAM), Stanford Cardiovascular Institute

Staff

STEVEN MOSS, Senior Program Officer
LINDA WALKER, Program Coordinator
DIONNA WISE, Program Associate

Panel on the Physical Sciences

BEVERLEY J. McKEON, Stanford University, *Co-Chair*
ALEXANDRA NAVROTSKY (NAS), Arizona State University, *Co-Chair*
JAMEL ALI, Florida A&M University-Florida State University College of Engineering
EDEN FIGUEROA BAGGAGAN, Stony Brook University
DANIEL L. DIETRICH, NASA Glenn Research Center
ADRIENNE R. DOVE, University of Central Florida
TANVIR FAROUK, University of South Carolina, Columbia
AMBER GENAU, University of Alabama at Birmingham
TEJA GUDA, The University of Texas at San Antonio
JOHN I. HOCHSTEIN, The University of Memphis
MICHELE V. MANUEL, University of Florida
ROBERT J.C. McLEAN, Texas State University
JOHN B. McQUILLEN, NASA Glenn Research Center
JEFFREY F. MORRIS, The City College of New York
ISSAM MUDAWAR, Purdue University
HOLGER MUELLER, University of California, Berkeley
ALEKSANDAR G. OSTROGORSKY, Illinois Institute of Technology
AMY M. PETERSON, University of Massachusetts Lowell
WILLIAM D. PHILLIPS (NAS), National Institute of Standards and Technology
MARIANNA S. SAFRONOVA, University of Delaware
PETER B. SUNDERLAND, University of Maryland, College Park

Staff

ERIK SVEDBERG, Scholar
LINDA WALKER, Program Coordinator
DIONNA WISE, Program Associate

Panel on Engineering and Science Interface

MICHAEL G. RYSCHKEWITSCH, Johns Hopkins University Applied Physics Laboratory, *Chair*
GALE J. ALLEN, G-Allen Strategies
ROBERT V. DUNCAN, Texas Tech University
ANTONIO L. ELIAS (NAE), Orbital ATK (Retired)
HANS KOENIGSMANN, Space Exploration Technologies (SpaceX) (Retired)
BRUCE M. LINK, Southeastern Universities Research Association
LISA W. SMITH, NASA George C. Marshall Space Flight Center
KATHRYN C. THORNTON, University of Virginia
KASTHURI VENKATESWARAN, Jet Propulsion Laboratory
DEBORAH S. WELLS, Leidos
DANIELLE WOOD, Massachusetts Institute of Technology

Staff

ARUL MOZHI, Associate Board Director, *Study Director*
GAYBRIELLE HOLBERT, Senior Program Assistant
DIONNA WISE, Program Associate

SPACE STUDIES BOARD

MARGARET G. KIVELSON (NAS), University of California, Los Angeles, *Chair*
JAMES H. CROCKER (NAE), Lockheed Martin (Retired), *Vice Chair*
GREGORY P. ASNER (NAS), Carnegie Institution for Science
JEFF M. BINGHAM, Consultant
ADAM BURROWS (NAS), Princeton University
JEFF DOZIER, University of California, Santa Barbara
VICTORIA E. HAMILTON, Southwest Research Institute
CHRYSSA KOUVELIOTOU (NAS), The George Washington University
DENNIS P. LETTENMAIER (NAE), University of California, Los Angeles
ROSALY M. LOPES, Jet Propulsion Laboratory
STEPHEN J. MACKWELL, American Institute of Physics
DAVID J. McCOMAS, Princeton University
LARRY PAXTON, Johns Hopkins University Applied Physics Laboratory
ELIOT QUATAERT, University of California, Berkeley
MARK SAUNDERS, NASA (Retired)
BARBARA SHERWOOD LOLLAR (NAE), University of Toronto
HOWARD SINGER, National Oceanic and Atmospheric Administration
HARLAN E. SPENCE, University of New Hampshire
MARK H. THIEMENS (NAS), University of California, San Diego
ERIKA WAGNER, Blue Origin
PAUL WOOSTER, Space Exploration Technologies (SpaceX)
EDWARD L. WRIGHT (NAS), University of California, Los Angeles

Staff

COLLEEN N. HARTMAN, Director, Aeronautics, Astronomy, Physics, and Space Science
ARUL MOZHI, Associate Director, Aeronautics, Astronomy, Physics, and Space Science
TANJA PILZAK, Manager, Program Operations
AHMED ZAKI QASIM, Financial Officer

Reviewers

This Consensus Study Report was reviewed in draft form by individuals chosen for their diverse perspectives and technical expertise. The purpose of this independent review is to provide candid and critical comments that will assist the National Academies of Sciences, Engineering, and Medicine in making each published report as sound as possible and to ensure that it meets the institutional standards for quality, objectivity, evidence, and responsiveness to the study charge. The review comments and draft manuscript remain confidential to protect the integrity of the deliberative process.

We thank the following individuals for their review of this report:

ELIZABETH AINSWORTH (NAS), University of Illinois at Urbana-Champaign
JULIA BAILEY-SERRES, University of California, Riverside
RODOLPHE BARRANGOU (NAS/NAE), North Carolina State University
KAREN DANIELS, North Carolina State University
HENRY J. DONAHUE, Virginia Commonwealth University
DAN DUMBACHER, American Institute of Aeronautics and Astronautics
DEREK DUNN-RANKIN, University of California, Irvine
CATRIONA JAMIESON, University of California, San Diego
JUNGHO KIM, University of Maryland
PETER LEE, Brown University
KAREN McDONALD, University of California, Davis
MICHAEL L. MEYER, NASA Glenn Research Center
GBADEBO OWOLABI, Howard University
MARSHALL PORTERFIELD, Purdue University
SHANBIN SHI, Rensselaer Polytechnic Institute
DAN M. STAMPER-KURN, University of California, Berkeley
ERIKA WAGNER, Blue Origin
JAMAL YAGOOBI, Worcester Polytechnic Institute
Y. SHRIKE ZHANG, Harvard University

Although the reviewers listed above provided many constructive comments and suggestions, they were not asked to endorse the conclusions or recommendations of this report nor did they see the final draft before its release. The review of this report was overseen by **ELIZABETH R. CANTWELL**, Utah State University, and **GABRIELA GONZALEZ (NAS)**, Louisiana State University. They were responsible for making certain that an independent examination of this report was carried out in accordance with the standards of the National Academies and that all review comments were carefully considered. Responsibility for the final content rests entirely with the authoring committee and the National Academies.

Contents

Preface

This decadal survey is the second report on biological and physical sciences (BPS) research in space, following the 2011 decadal survey report *Recapturing a Future for Space Exploration: Life and Physical Sciences Research for a New Era* (NRC 2011). This study was requested by the National Aeronautics and Space Administration (NASA). The statement of task was developed in direct consultation with NASA and the National Academies of Sciences, Engineering, and Medicine. While developing the statement of task, the National Academies sought input from its Space Science Board chair and members of its Committee on Biological and Physical Sciences Research in Space.

The National Academies appointed an ad hoc committee to carry out a decadal survey of biological and physical sciences research in space. This study generates consensus recommendations to implement a comprehensive strategy and vision for a decade of transformative science at the frontiers of biological and physical sciences research in space. The results of the study will assist NASA in defining and aligning biological and physical sciences research to uniquely advance scientific knowledge, meet the needs of human and robotic exploration missions, and provide terrestrial benefits. (The full statement of task is reprinted in Appendix A.)

The Committee on Biological and Physical Sciences Research in Space 2023–2032 was appointed by the National Academies in summer 2021. The steering committee obtained community input from the biological and physical sciences research community via two types of solicited input papers: scientific topical input and research campaign input, from June to December 2021, in the form of written papers submitted via an Internet solicitation and outreach and publicly available at a National Academies' site dedicated to this decadal survey.[1] The work of the steering committee was supported by three panels that benefited from this community input as well as dedicated working group meetings among the panelists. These panels were the Panel on the Biological Sciences, the Panel on the Physical Sciences, and the Panel on Engineering and Science Interface. These panels worked under the direction of the steering committee to assess the current state of knowledge in relevant fields and to identify and articulate potential space research objectives; working groups were organized around the development of Chapters 1 through 7 to intentionally draw from all three panels of distinct expertise rather than contributing input as isolated, research expertise–narrowed panels.

---

[1] See resources on community input, including links to view topical and research campaign papers submitted to the committee, on the decadal survey project page at https://www.nationalacademies.org/our-work/decadal-survey-on-life-and-physical-sciences-research-in-space-2023-2032.

For the first time, this decadal survey included a technical risk and cost evaluation (TRACE) of research campaigns that were considered as priorities as additional funding became available. The application of TRACE within this BPS decadal survey differs from its application on other decadal surveys, where those surveys are regularly charged with evaluating distinct preestablished mission concepts. Such mission-level concepts are new to BPS. The decadal survey panels worked with the TRACE contractor, The Aerospace Corporation, along with the steering committee, to produce recommendations for research campaign prioritization. The TRACE process served as input, along with many other inputs, to assist the steering committee with its deliberations. Just as other decadal surveys have learned and adjusted to this process, the BPS community will also learn from it in coming decades.

The steering committee held its first meeting in August 2021. This meeting was followed by a series of open- and closed-session meetings on a roughly bi-monthly basis through February 2023. Early meetings of the steering committee and most of the panel meetings were held virtually owing to the COVID-19 pandemic, but by summer 2022, the steering committee and the panels began holding hybrid virtual/in-person meetings, which greatly aided deliberations. A goal for this decadal survey was to start to make permeable some of the artificial barriers between the biological and physical sciences. In an effort to accomplish this, the panels often met as combined working groups. The working groups began drafting chapters in summer 2022. This report was then finalized by the steering committee and submitted to the National Academies for independent peer review in March 2023. Upon completion of this review, the report was prepared for publication and release.

Robert J. Ferl, *Co-Chair*
Krystyn J. Van Vliet, *Co-Chair*
Committee on Biological and Physical Sciences Research in Space 2023–2032

Summary

Not since the Apollo era have U.S. capabilities and ambitions for space exploration grown in intensity or breadth of interest as is seen in the present. More astronauts conduct more experiments in space now than ever before. Commercial trips to space expand participation in science and spawn increased desires for personal space travel. Industrial processes are being developed for space, and NASA is headed back to the Moon with the Artemis program, and then on to Mars. All of this exists because the United States has invested in the science and technology to further develop space-based research and applications. Research in the biological and physical sciences has been critical to those advances being humanly possible, safer than ever before, and inspiring. For the United States to continue to lead among the pioneering nations that embrace space exploration for both national security and global sustainability interests, it is necessary to resolve several scientific challenges that leverage or require the space environment in the coming decade. For the United States to thrive in that international competition and to benefit the majority of citizens who will not themselves travel to space, it is imperative to foster a biological and physical sciences (BPS) research and technology community that includes scholars, practitioners, explorers, and enthusiasts. *This goal can be accomplished by substantially increasing national investment in BPS research infrastructure and investigators, concentrating U.S. effort on ambitious yet focused key scientific questions that strategically advance space exploration and transform knowledge of how the world works, and considering new investment in large-scale research campaigns.*

The past decade saw the launching of new capabilities and scientific experiments into space, providing a deeper understanding of our planet, solar system, and universe. Human presence in space enables complex operations and observations. Additionally, space experiments have changed the understanding of the physical nature of the universe, led to the development of unique materials, and given insight into plant, animal, and microbial health and productivity in the extreme environment of space. Progress in understanding the effects of extended space travel on human health has led to new findings on aging; musculoskeletal, respiratory, and neurological health; and host–pathogen interactions. These successes have fueled further exploration ambitions, with more humans spending time in space and reaching out beyond low Earth orbit (LEO), separated from the rich and protective resources of Earth. The Committee on the Biological and Physical Sciences Research in Space 2023–2032 of the National Academies of Sciences, Engineering, and Medicine envisions a future that is encapsulated in the name of this report: *Thriving in Space: Ensuring the Future of Biological and Physical Sciences Research: A Decadal Survey for 2023–2032*. This report sets forth the priorities and challenges in scientific pursuits that position the United States not just to participate and survive in that expanding sphere of discovery *but also to lead and flourish in the coming decade*, as the nation returns humans to the Moon and reaches on toward Mars. The recommendations, key scientific questions (KSQs), overarching themes, and research campaigns in this report guide the country toward that future.

Current excitement in the benefits and urgency of space exploration is driven by several key developments. The return of humans to the Moon through the NASA Artemis program is occurring during a time of tremendous public and commercial interest in traveling into deep space. The commercial development of space, especially in LEO, is occurring as science on the International Space Station (ISS) invites both discovery and application. The burgeoning launch capacity from the United States, combining capacity provided by government and private entities, has fueled public interest, including the advent of civilian spaceflight participants, while diversifying the opportunities to study and exploit the space environment.

The world is entering a new era fueled by humanity's connection to space and marked by greater opportunities, interest, resources, and innovation than ever before, rivaling and indeed exceeding the magnitude and relevance of the Apollo era. At the same time, there are many significant challenges in Earth-based societies, such as impact of population growth and climate change on the environment and civil organization. The success of NASA and the United States in fully developing this new era depends on the nation's ability to effectively address the wide range of engineering, biological, and physical sciences needed to explore space safely and productively. That the nation now stands on this threshold is a direct result of implementations from the first and pivotal decadal survey of this community, *Recapturing a Future for Space Exploration: Life and Physical Sciences Research for a New Era* (NRC 2011). That survey in 2011 set about the task of rebuilding the science necessary for space exploration and enabling science to be done from space to benefit Earth. The science community has indeed recaptured that vision; it is now critical for the nation to secure the path forward.

This report presents a high-priority research strategy to enable and derive knowledge from the exploration of space in the decade just begun, 2023–2032. Federal investment in this science occurs primarily through the NASA Division of Biological and Physical Sciences (BPS Division), now within the Science Mission Directorate (SMD). However, space exploration missions—and especially the exploitation of space environments for science and technology—are increasingly supported by other government agencies and the private sector. Therefore, in developing its recommendations, the committee was keenly aware of the lead role of NASA in forming a hub, linking its mission to the interests of other government agencies and the increasingly important commercial entities supporting exploration and development. Constituent panels of the committee identified critical fundamental science opportunities, including those that support and are supported by human presence in space and that are enabled by this potentially synergistic public/private network. Those science opportunities are expressed in the form of KSQs that are the priorities for the next decade. This summary highlights the committee's top findings and recommendations based on those priorities.

STATE OF THE SCIENCE

The past decade experienced a very large expansion in new routes to space and new uses for space exploration, with concomitant increased capacities and utilization of space science platforms. One example of the results of this expansion is the science output from the ISS, driven by international academia, government researchers, and the commercial sector. More than 3,000 experiments have been run on the ISS over the past 2 decades, with nearly twice the publications in the past decade compared to the previous one (Witze 2020). More than 2,000 science publications emanated from ISS research since 1998 (Guzman 2023; Johnson 2022). The private investment in space has grown from $300 million in the United States in 2012 to over $10 billion today (Bland et al. 2022). These investments seeded new ways to move experiments and people into space and fueled ambitions to create commercial space stations to further enable science and manufacturing, while also inserting a certain tension of priorities between program offices. The United States stood up the Space Force as a branch of the armed services[1] and developed a national strategy for LEO research and development (NSTC 2023). The rise of industry and expanded space programs in the United States and other nations also increased the size and diversity of the space oriented workforce on Earth and in space.

[1] United States Space Force, "United States Space Force History," https://www.spaceforce.mil/About-Us/About-Space-Force/History.

Additional parts of the expansion of new routes to space include the transition of the ISS to commercial LEO, commercial suborbital vehicles, orbital vehicles from several countries and agencies, and the Artemis lunar exploration program. All of these routes provide potential science venues for BPS research, as well as emergent science needs that BPS researchers are well-suited to address.

STATE OF THE PROFESSION

In part, the 2011 decadal survey was a call to action for the space exploration science community, NASA, and the nation. The committee finds that NASA and the science community answered that call. NASA stood up the Division of Space Life and Physical Sciences Research and Application, which later became BPS and moved to SMD. The creation of the BPS Division and its subsequent placement in SMD resulted in a scientifically engaged division that is perceived to be moving to accommodate the science intensive needs of space exploration. The science community has developed a level of accomplishment and stability; however, the community has been severely underfunded to meet the large portfolio of science deemed important in the 2011 decadal survey. This conclusion is based on surveying the size of the community relative to the work needed, as well as the funding available relative to the amount of activity in space. During the space shuttle era, funding for the equivalent NASA division was approximately eightfold higher than current funding within NASA (NASEM 2018). Yet, as this report will argue, the opportunity and demand for science served by human exploration and use of space are more numerous and ambitious than the past eras, and science now operates continuously in LEO compared to the episodic orbital sorties that characterized those past eras. There is therefore an increasingly apparent and critical risk of underfunding this foundational area of science, the science that is an engine for innovation in space and on Earth. Last, research enabled by unique access to the space environment confers Earth-based benefits, including national security as well as human and environmental health.

Finding: NASA responded to the 2011 decadal survey by standing up what is now the Division of Biological and Physical Sciences within the Science Mission Directorate. As a result, the scientific community in these important and diverse fields has begun to be rebuilt. Importantly, however, much work remains to establish a healthy and sustainable biological and physical sciences community with resources commensurate with its long-term scientific and technological mission. Much of this work lies in the area of funding beyond simple inflationary adjustments toward large investments that scale to the work needed, similar to the levels that existed during the space shuttle era. (Chapter 7, Finding 1)

Finding: A robust and resilient BPS program requires

- A healthy and regular cadence of proposal calls and grant dollar awards that are consistent with sustaining a diverse and productive BPS community over the course of the next decade, including the necessity of training a diverse scientific workforce of sufficient size and caliber to maintain the BPS community over a generational timescale;
- Broadened and more inclusive participation in the U.S. BPS community, including diversity of both scientific expertise and lived socioeconomic experience, recognizing the slow progress in attracting and retaining women and persons of color into graduate and post-graduate research roles;
- A total science budget sufficient to meet current national needs and international competitor/collaborator challenges;
- Interactions with other U.S. government and non-U.S. space agencies necessary for optimal BPS community productivity in science and technology development; and
- Significant awareness and collaboration with the emerging commercial space science, platforms, and activities, as appropriate for BPS program goals. (Chapter 7, Finding 2)

Finding: The BPS program is severely underfunded relative to current need, essentially preventing the development of a truly robust and resilient program that can meet the space exploration science needs of the nation. (Chapter 7, Finding 10)

The recommended funding increase is an inescapable conclusion based on three driving forces—cost analysis, political reality, and historical precedence—that all converge to strongly suggest that *the BPS budget must rise by a factor of 10 well before the end of the decade*. (See Chapter 7, Box 7-2.) Research campaigns can be considered only if the recommended increase in funding occurs. This funding recommendation is also based on the assumption that transportation costs and mission integration and operations costs will continue to reside outside of the BPS budget, in other units of NASA.

Recommendation: To retire many of the key scientific questions by the end of the decade, NASA should establish support for the Biological and Physical Sciences Program to levels that reflect the current national need and to build the science community in size, diversity of technical expertise and lived experience, and capability to reach the science goals of the nation, toward levels that are an order of magnitude above the current funding and well before the end of the decade. (Chapter 7, Recommendation 9)

Space is no longer a realm solely explored and inhabited by agencies of nations. Significant commercial interests now characterize virtually all space venues, from suborbital space through LEO and on to the Moon and Mars. Appropriately, NASA seeks to enable those commercial developments while simultaneously becoming a user of those developments for national science needs. A prime example lies in those companies currently engaged with NASA in the Commercial LEO Destinations (CLD) program, who in collaboration with NASA are working to ensure that capacity and capability is available to meet research needs after the ISS end of life that is anticipated by the end of this decade. The recognized goal is impactful science, both national and commercial, conducted under conditions of certainty and reliability regarding the availability of one or more multi-user platforms in LEO.

Finding: The private sector is engaged in development of commercial LEO destinations, on which the nation's research in BPS will depend. However, science-design requirements have yet to be published. This delay may result in an unintended consequence that CLD companies develop revenue sources to focus on commercial markets, deemphasizing government-funded or fundamental research for public benefit. (Chapter 7, Finding 3)

Recommendation: Because the nation benefits from global leadership in space science and technology, and given the emergence of commercial platforms that can be tasked to the nation's science, NASA should:

- **Seek significant funding increases for biological and physical sciences with new monies or through rebalancing the portfolio across the Science Mission Directorate, and in coordination with other U.S. government agencies, as the community needs to grow significantly in size to reach the science goals of the nation;**
- **Actively engage commercial spaceflight firms, using science funding as a driver and with all due haste, to ensure that science needs are met with clear priority, guaranteeing that national science needs are enabled along with those of potential commercial customers using those platforms; and**
- **Ensure that the funded science community fully engages diversity and inclusivity in the pursuit of the nation's space exploration science priorities.**

(Chapter 7, Recommendation 1)

ESTABLISHING SCIENCE PRIORITIES

Starting from an initial set of hundreds of community-submitted input papers and the decade of published articles, three interdisciplinary panels of U.S. experts worked with the committee to establish the current state of knowledge and capabilities in BPS (Chapters 1 and 2) as a baseline of opportunity for research in the coming decade. From this diverse and wide-ranging input arose three themes and a set of key scientific questions (KSQs) that prioritize science that enables safe and ambitious space exploration (see Chapter 4), and science that is enabled by access to space (see Chapter 5). These themes and KSQs form the core science recommended for BPS and need to be supported by the entire BPS core funding within NASA.

The first theme, *Adapting to Space*, concerns how the fundamental physics of space environments impact the ability of living systems to survive transition to and extended stays in space. The second theme, *Living and Traveling in Space*, explores living systems and supportive environments over long durations in space, while deriving resources in space under the logistical and physical constraints of space. The third theme, *Probing Phenomena Hidden by Gravity or Terrestrial Limitations*, seeks scientific insights that can be found only in space.

These KSQs were, by design and where possible, intended to be biologically motivated yet physics aware or physics motivated yet biologically aware, but recognizing that there could be subquestions framed as purely biological or physical. Importantly, and as a result of the maturity and increasing interdisciplinarity of space science research over the past decade, many KSQs are broader than any one scientific discipline—spanning multiple biological species, material classes, or physical principles. These concepts of interdisciplinarity and comprehensiveness informed all of the recommendations.

Recommendation: NASA should direct its research resources toward the key scientific questions identified in this study (Table S-1 and Chapters 3, 4, and 5). (Chapter 3, Recommendation 1)

Recommendation: NASA should work with other U.S. government agencies and other nations' space agencies to coordinate research resources toward the key scientific questions, as relevant to multiple agency missions. (Chapter 3, Recommendation 2)

TABLE S-1 Key Scientific Questions in BPS Space Research Over the Decade 2023–2032

Themes	Key Scientific Questions
Adapting to Space (Chapter 4)	• How does the space environment influence biological mechanisms required for organisms to survive the transitions to and from space, and thrive while off Earth? • How do genetic diversity and life history influence physiological adaptation to the space environment? • How does the space environment alter interactions between organisms?
Living and Traveling in Space (Chapter 4)	• What are the important multi-generational effects of the space environment on growth, development, and reproduction? • What principles guide the integration of biological and abiotic systems to create sustainable and functional extraterrestrial habitats? • What principles enable identification, extraction, processing, and use of materials found in extraterrestrial environments to enable long-term, sustained human and robotic space exploration? • What are the relevant chemical and physical properties and phenomena that govern the behavior of fluids in space environments?
Probing Phenomena Hidden by Gravity or Terrestrial Limitations (Chapter 5)	• What are the mechanisms by which organisms sense and respond to physical properties of surroundings and to applied mechanical forces, including gravitational force? • What are the fundamental principles that organize the structure and functionality of materials, including but not limited to soft and active matter? • What are the fundamental laws that govern the behavior of systems that are far from equilibrium? • What new physics, including particle physics, general relativity, and quantum mechanics, can be discovered with experiments that can only be carried out in space?

Recommendation: Because key questions identified in this study benefit from access to multiple spaceflight-related platforms, the Biological and Physical Sciences Program should

- **Coordinate funding opportunities with the Space Technology Mission Directorate such that access to the range of spaceflight and spaceflight-related platforms is efficiently employed to answer key science questions, especially those questions that inform technology development for space exploration; and**
- **Maintain a foundational approach to science, building through a strong, vibrant program of ground-based, suborbital, orbital, lunar, martian, and beyond missions.**

(Chapter 7, Recommendation 3)

RESEARCH CAMPAIGNS

The KSQs present imperatives that demand a level of activity—research community growth, replicate experiments, and increased access to space environments and crew time—that is currently insufficiently resourced. Therefore, this study explored the concept of research campaigns to drive synergistic sets of KSQs toward resolution and impactful societal milestones within the next decade. The research campaigns are highly directed efforts on a scale that is large, scientifically robust and dedicated to specific mission queries in a manner common within other divisions of SMD. They were chosen by prioritizing significance, feasibility of scientific goals, and potential for capability building.

Research campaigns would be a new approach at the BPS Division. The BPS Division currently directs budget to advancement of scientific areas defined in the 2011 decadal survey. Research campaigns, however, are designed to directly address specific science goals that could be driven to completion. BPS research campaigns conceptually align with SMD spacecraft missions in scope of inquiry, mission duration, acceptable risk, and cost. For this reason and for the first time in this decadal survey series, the committee conducted technical risk and cost evaluation (TRACE) for the research campaigns. This exercise informed an analysis of multiple potential campaigns and resulted in two research campaigns to be considered for new funding and two futuristic concepts to be considered only with multi-agency inputs.

Research campaigns of the next decade are not presented as priorities within the current BPS budget model. Specifically, campaigns are not meant to supplant the current granting process that is recommended herein to focus on KSQs. Rather they are presented as directed, large-scale efforts to be pursued only with the addition of budget specifically directed to the full execution of the campaign. They are intended to be additional components of the BPS science portfolio beyond the budget currently dedicated to BPS science.

Recommendation: NASA should pursue dedicated research campaigns that, through the coming decade, will drive solutions to specific groups of key scientific questions. Coordination beyond NASA, including other federal agencies and the private sector as well as public–private partnerships, should be considered for the dedicated new funding and materials to support these research campaigns.

- **BLiSS (Bioregenerative Life Support Systems) to build and understand the systems that would provide high-quality food, refresh air and water, process wastes, and enable the creation of space environments sustainable for long periods of time independent of Earth.**
- **MATRICES (Manufacturing Materials and Processes for Sustainability in Space) to understand and harness the physical processes by which materials and complex fluids can be repeatably utilized in space, to enable sustainable exploration and circular lifecycles for the built environment on Earth and in space.**

(Chapter 6, Recommendation 1)

Campaign: Bioregenerative Life Support Systems

BLiSS is targeted to build sufficient knowledge of the biological systems, interactions, and systems of systems to provide high-quality food, refresh air and water, process and valorize wastes, and enable the creation of space environments sustainable for long periods of time independent of Earth. Sustainable bioregenerative life support has been a science goal for many decades and is encompassed within NASA's technology roadmap, which states that self-sufficient life support systems are crucial for sustaining life and mitigating negative physiological effects on long-duration missions. This campaign seeks to understand the multiple biological phenomena at play while providing a distinct technology gain for space exploration and also presenting high return-on-investment for development of sustainable technologies for Earth.

Campaign: Manufacturing Materials and Processes for Sustainability in Space

MATRICES maps advances in materials science and manufacturing via in-space research both to advance fundamental knowledge of physical principles enabled by space and to create approaches that will enable future human space exploration in sustainably built environments. This campaign addresses two challenges in the journey and destination of space travel and habitation: (1) limited mass of resources launched from Earth for long journeys, and (2) limited knowledge of how to repeatably use Earth-origin and space-origin resources to manufacture and repair the world around us, with minimal impact to that world. This campaign envisions the types of materials science, complex fluid dynamics, and manufacturing, near and far from equilibrium conditions, that will be enabled over the next decade in an expected ecosystem that includes ISS, commercial space destinations in LEO, and planetary space experimental platforms. This vision enables both public- and private-sector resources for new science and engineering outcomes that advance science while demonstrating that the concept of circular material and process design will also benefit Earth-based processing and products.

Multi-Agency Initiatives and Stretch Concepts

The study identified an ambitious and scientifically transformative effort described here as an initiative rather than a campaign. It is framed as multi-decadal and requiring whole-of-government investment and coordination. The Probing the Fabric of Spacetime (PFaST) initiative is centered on deploying an advanced quantum sensing network whose performance will be many orders of magnitude better than previously thought possible, enabled by scientific discoveries that occurred only recently and that will be continuously refined throughout the decades to come. Establishing space-based optical lattice clocks (OLCs) for ultraprecise quantum sensing will enable breakthrough science for probing the fabric of spacetime within the solar system and beyond the solar system, while testing whether gravitation fields have quantum aspects. This initiative would be at the forefront of developing quantum technologies with a huge indirect Earth benefit, notably quantum sensing, computing, and quantum information processing.

The study also identified a novel BPS research infrastructure concept—the Biological and Physical Science Free Flyer (BiPS-Free) vehicle for polar-orbital, untended space science research venues addresses current constraints of crew-tended, LEO-limited BPS research, including months-long access by non-human living organisms and engineered materials to the gravitational forces and cosmic ray radiation exposure beyond LEO.

THRIVING IN THE FACE OF UNCERTAINTY

The KSQs and research campaigns were prioritized in the context of the rapid spaceflight evolution and Earth-based knowledge advancement anticipated over the coming decade. In mounting resources to answer these scientific questions and mission-style research campaigns, the "so what?" of successful research outcomes must be compelling even when the practical utility is not immediate. The uncertainties span the pace of technological change, available funding levels and economic stability in the United States, geopolitical tensions and cooperation, ISS and CLDs and service provider stability, launch cost and frequency for space science research, and competition to recruit and retain the nation's best minds to address BPS challenges. In addition, the underlying pace of BPS portfolio fields is among the most rapid in the world, and this rapid change is unlike that faced by other space sciences. Thus, while this decadal survey intentionally identified KSQs and notional research campaigns, a process needs to exist to adapt, evaluate, and pivot as new opportunities and threats emerge. Decision rules that guide the potential for major changes—in public funding availability, access to space assets for BPS research, commercial destination and payload services, NASA missions, or international cooperation—over the decade will help to stabilize BPS research following negative changes and capitalize quickly on opportunities following positive changes (Chapter 7). Chief among the decision rules is the concept of pursuing research campaigns only with new funding beyond typical BPS research efforts, returning to the themes and KSQs when new platforms or opportunities are presented.

Recommendation: To maintain research campaign momentum, NASA should require external advisory committees to evaluate research campaign team progress and emergent technologies annually. (Chapter 7, Recommendation 2)

SUSTAINING A THRIVING, NATIONAL BPS RESEARCH COMMUNITY

For space science to thrive, the community of research needs to be diverse, inclusive, and accessible to the full range of U.S. talent. Ensuring broad access and participation is essential to maximizing excellence in an environment of fierce competition for limited human resources, and to ensuring continued U.S. leadership in space exploration. The committee applauds the hard-earned progress that has been made—most notably with respect to the entry and prominence of women in the field—as well as the exemplary goals and intentions of NASA science leadership. However, much work remains to be done to address the particularly persistent and troubling issues of historically underrepresented groups in BPS research within academia, industry, and government. A related critical issue in retaining talent within the BPS community is the difficulties in ensuring funding over the course of a person's training endeavors.

Finding: The BPS program and the community it supports are increasingly diverse and inclusive, yet work still needs to be done to ensure momentum and broader participation at all levels of the research enterprise. (Chapter 7, Finding 15)

In 2022, the National Academies' Space Studies Board produced two reports addressing diversity, equity, and inclusion in principal investigator–led space science missions and the health and vitality of the space science communities. Both studies were requested by NASA. Titled *Advancing Diversity, Equity, Inclusion, and Accessibility in the Leadership of Competed Space Missions*, and *Foundations of a Healthy and Vital Research Community for NASA Science*, these reports covered NASA's SMD. The *Advancing* report outlined near- and long-term actions that NASA can take to make opportunities for leadership and involvement in competed space missions more accessible, inclusive, and equitable. Recommendations included changes to the mission proposal process as well as investments in science, technology, engineering, and mathematics (STEM) education and career pathways. Many of the report's recommendations can also be applied broadly to research at NASA and other federal agencies and institutions, leading to a more diverse research workforce. The *Foundations* report identified the characteristics of a healthy and vital research community, defined implementable measures for assessing the health and vitality of a research community, described the types of data that NASA should be collecting to enable future assessments of the health and vitality of the scientific work force, and recommended best practices to improve the health and vitality of NASA's research communities. Several of that report's recommendations, as well as its "tenets of a healthy and vital research community," can also be applied to other federal agencies and institutions.

Notably, the BPS Division was not included in the above NASA studies because the BPS Division does not have competed space science missions as do the other four SMD divisions. However, shared responsibility to foster BPS research excellence in the United States—by NASA as one of the primary funders of the space science research community and by sponsored investigators who build and mentor their research teams—is included as part of the decision rules that were part of this decadal survey's statement of task. (See Chapter 7, Box 7-1.) This explicitly comprises expectations and suggested mechanisms to annually increase the inclusive participation of BPS researchers including students, postdoctoral researchers and technical staff, and research team leaders of broader lived experience over the coming decade.

BPS science on fundamental biological processes, both within and among organisms in the animal, plant, and microbial kingdoms, complements and informs NASA's Human Research Program (HRP), which is focused directly on human astronaut safety and health. (See Appendix D.) HRP and the BPS Division co-develop roadmaps that articulate roles and responsibilities for specific tasks; however, the BPS–HRP relationship is based on the working-level arrangements rather than on formal coordination. Vitally important research in and for human space exploration would move forward and the outcomes could be shared and coordinated, not be circumscribed artificially by programmatic funding boundaries. Thus, given the important opportunity for synergy among these research communities and NASA resources to improve human well-being, it will be imperative in the coming decade that scientific exchange (of plans, data, and communication of important results) and coordination (of opportunities and funding) are prioritized between the BPS Division and HRP.

Recommendation: NASA should continue to strengthen the science exchange between the Biological and Physical Sciences Program and the Human Research Program. Such effort may include establishing a coordinating body and shared research initiatives as well as the two-way exchange of technologies, data, mission science, specimen banking, and plans. (Chapter 4, Recommendation 1)

Additionally, as this decadal survey is concluded in the U.S. Year of Open Science (White House 2023), it is timely to note the important historic investments and advancements in the sharing of data and physical artifacts among the BPS community. Investment by NASA and the whole of government in curated, maintained databases and physical repositories has been critical to the ability to gain more insight and value from each research project than could be achieved by only the original investigator team. Findable, accessible, interoperable, and reusable (FAIR) access principles (Wilkinson et al. 2016) for research data and artifacts will also become increasingly important. Continued emphasis on open science encourages the research community to use existing data for improved (or reduced) design of experiments, and offers the potential to build the space science research community beyond those investigators who are directly funded to generate such data.

Recommendation: NASA should continue to expand the investment in open and shared computational infrastructure (CI) to support storage, analysis, and dissemination of its biological and physical data, while ensuring linkage to the original and archived samples.

- **For biological sciences, GeneLab should be continued and efforts made to ensure findable, accessible, interoperable, and reusable access from other critical international biological resource CIs.**
- **NASA should recognize the need for long-term investment to maintain, update, and improve such community-serving CI and physical repositories over time.**

(Chapter 7, Recommendation 6)

Quite distinct from the prior decade, the increased participation of the private sector in space exploration and destinations has increased both the potential capacity for and complexity of BPS research in space environments. With the anticipated sunsetting of the ISS within the coming decade, NASA investment priorities will need to drive commercial supplier support through the transition to commercial platforms. The substantial investment that NASA has made in BPS research on behalf of U.S. taxpayers, including the costs of launching key experiments to the ISS, has been critical to the current progress of BPS research but has been less visible to the research community itself. (See Figure S-1.) Private-sector development of BPS research platforms and destinations can advance national priorities, but NASA's leadership and support of fundamental research with societal benefit will remain critical as costs and access approaches remain uncharted territory for BPS. NASA thus plays an important role in coordinating this new space science ecosystem.

Recommendation: NASA should work with the other appropriate U.S. government agencies with a goal to establish an office or a mechanism for commercial sponsorship and collaboration with non-profit organizations, including academia and government research agencies. That office/mechanism should have the primary focus of

- **Coordinating the work between these commercial sectors and government agencies;**
- **Providing guidance on or facilitating research compliance, data security, and material transfer agreements, including prototype agreements;**

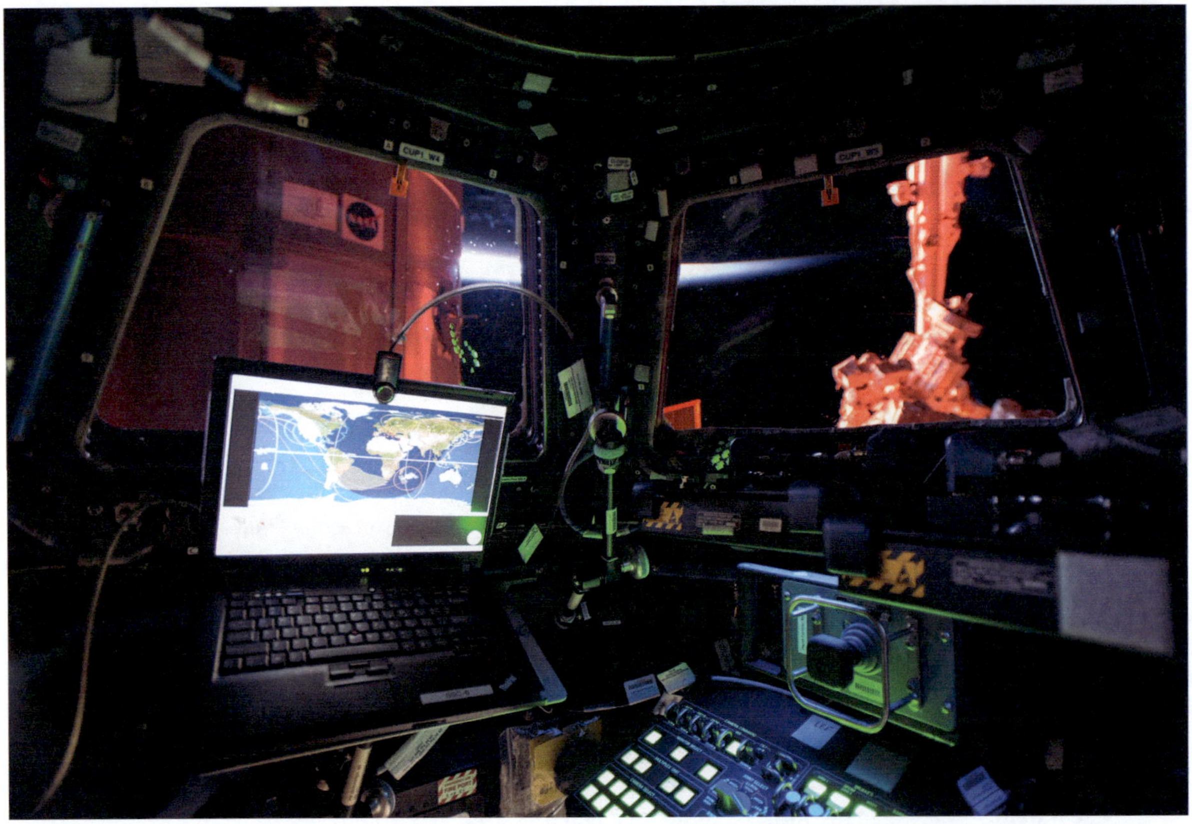

FIGURE S-1 The International Space Station has been critical to the current progress of biological and physical sciences research. Within the decade it will be replaced by new capabilities. SOURCE: Courtesy of NASA, Expedition 42.

- **Representing multiple space environments and destinations (e.g., not only the International Space Station in low Earth orbit); and**
- **Communicating these opportunities to the research community.**

(Chapter 7, Recommendation 7)

The challenges of space exploration to the Moon and on the way to Mars will be realized in the coming decade. Adapting to, living and traveling in, and leveraging the space environment to maximum responsible U.S. advantage comprise a charge to which the BPS space research community including the U.S. public and private sectors can respond with focused enthusiasm. *Thriving in Space: Ensuring the Future of Biological and Physical Sciences Research* charts the progress, potential, and pitfalls that are anticipated, as well as approaches to pivoting as technological successes or surprises unfold over the coming decade. The following chapters describe the findings and recommendations that will together ensure the future of space exploration and the well-being of society on Earth, through the resilient and collaborative efforts of the nation's BPS research community. This is indeed an awesome era in which science can contribute to a brighter future on our home planet and beyond.

1

Introduction

Imagine what might be accomplished if science could effectively grow plants for life support away from Earth, manufacture better materials from lunar or martian resources, and reduce risks of extended space exploration by humans or by automated machines. The ability to escape the confines of Earth has been achieved in only the most recent moments of life on this planet. This exciting capability makes it possible to ask both new and persistent questions about how the Earth environment has shaped the development and functioning of organisms throughout time, as well as how organisms will adapt and respond to novel space environments in the future. Despite the many advancements made on Earth with present understanding of physical laws, the vast majority of the universe remains unexplored by human beings and is unknown to humans firsthand. Exploration of that universe over greater distances and durations and by more humans requires scientific understanding of the impact of that space on fundamental biological and physical systems, systems that until now were confined to Earth. It is the role of the NASA Division of Biological and Physical Sciences (BPS Division) and its science community to conduct this research. That division manages and pursues biological and physical sciences (BPS) that enable the exploration of space, and also pursues unique opportunities to advance BPS generally by utilizing unique features of the space environment.

Many important scientific questions can be explored and answered only in the space environment, including questions that interrogate the nature of dark matter and dark energy, or that probe the behavior of engineered and living materials when critical interactions are no longer concealed or constrained by Earth's gravitational force. Studying the nature of gravity itself through experiments on Earth's surface is limited. In contrast, experiments performed in space present opportunities to examine a wide section of variable gravitational potentials. Additionally, experiments conducted in the space environment can leverage much larger distances to reveal fundamental science, which is often constrained by the diameter of Earth and by seismic activity. Extreme features like these and others, including different and significant radiation and chemical exposures, create conditions not attainable on Earth.

Thus, research conducted in space can significantly open the observation window, allowing scientists and engineers to measure fundamental biological and physical processes and mechanisms that can only be theorized today. In addition to providing inspiration and perspective (Figure 1-1), the capacity to expand scientific discovery to the Moon and on the way to Mars tangibly benefits life on Earth as well as exploration of space.

FIGURE 1-1 Realm of low Earth orbit (LEO). Views like this one define the iconic perspectives of the realm of LEO, where the International Space Station firmly established the need for and the value of science in space near Earth. SOURCE: Courtesy of NASA, https://www.issnationallab.org/2022-issnationallab-science-highlights.

As an illustrative example, advancements in space science have the potential to address very down-to-Earth issues. Degenerative pathologies associated with aging range from dementias and frailty to cardiovascular disease and cancer. Currently, the annual cost of dealing with age-related medical issues owing to muscle wasting and bone loss alone is in the billions of dollars for the United States (Van Houtven et al. 2008). Muscle wasting and bone loss problems are encountered by biological organisms living in the space environment at a significantly accelerated timescale, making spaceflight an interesting model for aging. Engineered materials can also undergo accelerated aging in the space environment, although that same term is taken to mean physical or chemical changes in the material structure with attendant reduction in properties, on exposure to other features of the space environment such as ultraviolet (UV) radiation or wide cycles in temperature (ESA 2020). Therefore, studies conducted in space to better understand the underlying biological and physical processes will have profound implications and impacts on the health and well-being of societies on Earth.

Biological and physical sciences also benefit space exploration. For example, a sustained lunar presence is just one of NASA's exploration objectives. The Artemis program (Figure 1-2) is a series of complex missions to provide a foundation for research and knowledge accumulation and to expand capabilities to extend human exploration to the Moon and beyond to Mars. Sustained human exploration beyond low Earth orbit (LEO) presents science objectives and challenges not yet approached through the era of the International Space Station (ISS). Human crews of varied age and background will grow older during transit and perhaps while living and working in lunar or martian stations. Other biological organisms including microbes and plants and animals on those missions will eventually reproduce over several generations, thus elevating the need for science to consider longer durations in space and advancing biophysical approaches to create or monitor nutrients. Likewise, precise clocks

FIGURE 1-2 Earth and the Moon viewed from the Orion spacecraft. The Artemis 1 mission redefines the envelope for human exploration of space in the next decade, which again includes cislunar (lying between Earth and the Moon or the Moon's orbit) space and now extends to distances well beyond the Moon. This singular image creates a point of view never seen by humans firsthand, yet one that will be a part of the explorations of the coming decades. SOURCE: Courtesy of NASA, https://www.nasa.gov/feature/view-the-best-images-from-nasa-s-artemis-i-mission.

and positioning systems and communications for space navigation all stem from physical science advances. Many of the underlying science studies underpinning those exploration necessities may have seemed esoteric before the use cases of the Global Positioning System (GPS) and the like became so ubiquitous as to seem ever present.

NASA is key to U.S. leadership of and participation in the diverse space research community, including access to the space environment for research discoveries and translation through many flight platforms. With the sunset of the U.S. space shuttle program, the ISS took over as the major science platform in LEO. (See Box 1-1.) The ISS directly serves NASA research as well as the research needs of international partners. To expand access, the NASA Authorization Act of 2005 created the ISS National Laboratory (ISSNL). The ISSNL enables access to space research and development (R&D) access to a broad range of commercial, academic, and government users. The ISSNL—chiefly as a manager, promoter, and broker of research onboard the ISS—has engaged with other U.S. and international government agencies, industrial sectors, and financial markets to fuel the future growth of space-based research. These public, private, and public–private partnerships have resulted in impactful new scientific discoveries, and use-inspired innovation in terrestrial applications (ISSNL 2018). For example, optical fibers are used to transmit data, and tiny defects formed during processing reduce the fibers' performance to slow transmission speed and degrade information. In 2018, microgravity conditions at the ISS made it possible

BOX 1-1
The International Space Station

In 1984, President Ronald Reagan announced the beginning of the space station project. Throughout the remainder of the 1980s, the station progressed slowly as the United States sought to build an international coalition of station partners. It faced cancellation in the early 1990s and underwent major changes as Russia was added to the partnership in 1993. The first space station component was launched in 1998, and it was finally completed in 2010. Even before the International Space Station (ISS) was substantially complete, NASA declared its intention to discontinue its use by 2015, a decision that was later reversed by 2010. The ISS's lifetime has been extended multiple times, into the 2020s, and now to 2030. The ISS is the primary focus and location of U.S. biological and physical science (BPS) in space research. What its long and bumpy history demonstrates is that timelines for development of new capabilities are long; major decisions affecting these capabilities are made at high political levels (i.e., by the U.S. executive and legislative branches, and among international partners); and space science has not been a major driver in establishing priorities for BPS capabilities. In addition, NASA's BPS program, which underwent substantial restructuring and budget cuts in the 2000s, was also joined by a new entity mandated by Congress, the ISS National Laboratory (ISSNL), which placed demands on ISS resources.

Center for the Advancement of Science in Space and the International Space Station National Laboratory

In 2005, U.S. congressional legislation created the ISSNL. From 2005 through 2010, NASA managed the ISSNL through a NASA Research Announcement, "Opportunity for the Use of the International Space Station by Domestic Entities Other Than U.S. Federal Government Agencies." In 2010, Congress directed NASA to select an entity to manage the ISSNL, and the following year the agency signed a 10-year agreement with the Center for the Advancement of Science in Space (CASIS), a non-profit organization. In 2017, NASA extended its contract with CASIS through 2024, and in 2022 it further extended the contract to 2027. In 2018, a NASA Inspector General report concluded that CASIS had not met many of the expectations for research pathways and education programs. It also determined that NASA had been insufficiently involved in overseeing CASIS activities. In 2019, CASIS formally changed its name to the ISS National Laboratory, although this was a branding change and CASIS is still the organization's legal name. The name change was intended to raise awareness about the laboratory and attract more users.

In February 2020, an independent review panel delivered a report to NASA that criticized the management of the ISSNL. In response, NASA and CASIS promised to make changes, including establishment of an advisory committee. In June 2022, the Government Accountability Office issued a report that stated that CASIS was not making full use of a user advisory group that had been established in response to the earlier independent review team report.

Substantial ISS resources have been taken up by an organization that has come under significant criticism by the NASA Office of the Inspector General, an independent review team, and the Government Accountability Office. This situation has undoubtedly affected the ability of the United States to efficiently and effectively use the capabilities of the ISS for impactful research, including BPS research in space.

Implications of the NASA ISS Management Approach

Half of all resources in the U.S. segment of the ISS is allocated to the ISSNL. By early 2023, NASA had effectively reached full utilization of the ISS given crew time limitations and transportation resources. The NASA Liaison given the responsibility to oversee and manage the entity managing the ISSNL resides in the International Space Station Program Office (ISSPO) at NASA headquarters, in the Space Operations Mission Directorate. This management arrangement places the operations of the major orbital resource for BPS execution outside BPS Division control and essentially places BPS in competition with all users of the ISS, including the Human Research Program and perhaps especially the ISSNL.

Moreover, the ISSPO pays for a very large share of the total cost of doing science on the ISS. The ISSPO pays what are called the mission integration and operations costs (MI&O). These costs are significant and play a major role in budget considerations for BPS (see Figures 6-5 and 6-9), especially if these costs are ever envisioned be transferred out of ISSPO or borne by the BPS Division.

The Future of the ISS, ISSNL, and Commercial LEO Development

Alternatives to the ISS as a location for research in LEO, and the relative priority and cost they will offer for fundamental research, remain unclear. This uncertain future affects the development pace of commercial LEO destinations (CLDs) for research, as well as national competitiveness and collaborations in BPS research. During the development of this report, one nation (China) completed construction of a separate, crewed space station in LEO (Tiangong, or Sky Palace), to which U.S. researchers do not anticipate having access. In contrast, NASA sponsorship of BPS in space, including at the ISS, has largely been subject to other political and management decisions made by the agency and NASA leadership, such as the decision on when to retire the ISS. Unlike other areas of science managed by the NASA Science Mission Directorate, science priorities for BPS have not driven the major funding for the program. While NASA seeks orderly transition of ISS functions to CLDs in the coming decade, it remains unclear how fundamental research that lacks near-term commercial potential will be weighted or costed in LEO. The committee believes science priorities to be a more effective driver of ISS and follow-on capabilities such as CLDs for science research.

to produce optical fibers from a special fluoride glass, avoiding the formation of microstructural defects that currently plague such processing on Earth.

Applied research such as this optical fiber processing example is often described as supporting a practical application, where the research outcome is tangible and includes a new material, component, or device. Fundamental research, in contrast, can be described as generating new knowledge of how the universe works. Fundamental biological and physical sciences in space identify the pieces necessary to eventually mitigate risks, reduce costs, and improve efficiencies of future developments. Over the past decade, fundamental biological science has made great strides in understanding how microbes and plant systems grow and replicate differently in microgravity environments, and such knowledge can advance long-term life support systems for space exploration. Basic physical science research on understanding fluid behavior in low gravity has been extended to applications of three-dimensional (3D) printing, metal alloy solidification, novel cooling systems, and maintenance of plants on spacecraft. Increasingly, the ISS has also housed a number of innovative laboratories for critical fundamental investigations. For example, in 2018, NASA launched the Cold Atom Laboratory (CAL) to enable macroscopic observation of quantum properties of atoms.

Science onboard the ISS has been dependent on international and commercial missions to send cargo to the ISS (upmass) and return cargo to Earth for any analysis (downmass). The ISS is collaboratively managed by a group of international partners and physically consists of two main sections: the Russian Orbital Segment (ROS), which is operated by Roscosmos (the Russian federal space agency), and the U.S. Orbital Segment (USOS), which is operated by NASA, the Canadian Space Agency (CSA), the European Space Agency (ESA), and the Japan Aerospace Exploration Agency (JAXA). In addition, ground facilities in the United States, France, Netherlands, Germany, Russia, Kazakhstan, and Japan support launch, operations, and payload services for the ISS (Joseph and Wood 2019). Visiting crew and cargo vehicles (governmental and commercial) originate from Kazakhstan (Russian Soyuz and Progress spacecraft), Japan, and the United States. As of October 2022, at least 263 individuals from 20 countries have visited the ISS, with the United States having the largest number of unique visitors, at 161.[1]

[1] NASA, "Visitors to the Station by Country," https://www.nasa.gov/feature/visitors-to-the-station-by-country, accessed October 2022.

Hardware and laboratories inside the ISS support life and experiments in a highly controlled and habitable built space environment, and facilities outside the station provide platforms for testing equipment and performance in the extreme conditions of space.

On December 31, 2021, the White House announced a decision to extend operations of the ISS through 2030, and then announced the National Low Earth Orbit Research and Development Strategy. The ISS international partners are working with their respective governments to extend agreements through 2030. This extension will allow continued science on the ISS while new CLDs come to fruition (Gatens 2022). However, because this extension does not span the targeted duration of this decadal survey through 2032, and BPS transitions will need to be planned and prioritized ahead of ISS retirement, such non-ISS capabilities are a major concern within the period of this survey. (See Box 1-1.)

BIOLOGICAL AND PHYSICAL SCIENCES IN SPACE

Investigating biological and physical systems and related phenomena under the extreme conditions associated with the space environment improves understanding of how such systems function, both in the space environment and in the more familiar conditions on Earth. Figure 1-3 illustrates the conceptually broad swath of scientific expertise within biological and physical sciences, collectively referred to as BPS, in space science research. From an Earth-focused research organization perspective such as a university or industry laboratory, each of these hexagons could represent a single department or division—defined by shared terminology, experimental practices, and research infrastructure. As Figure 1-3 moves from left to right, the scientific underpinnings traverse from the purely biological systems and empirical data-rich studies to the purely physical concepts described by Newtonian laws

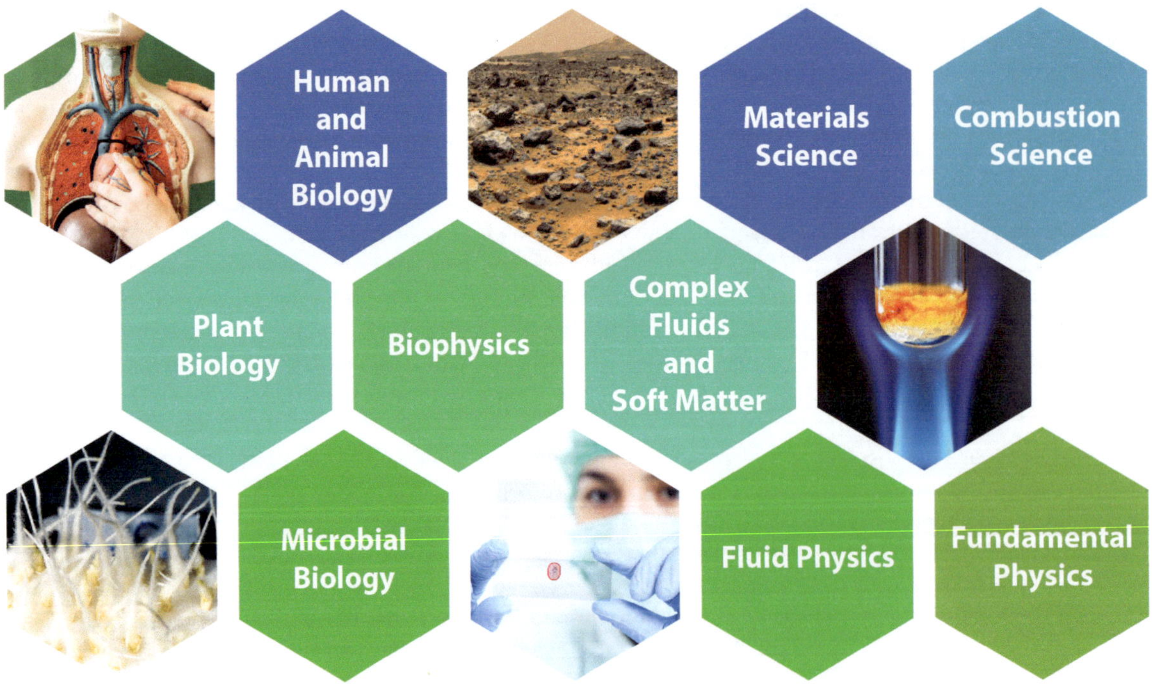

FIGURE 1-3 Biological and physical sciences in space include a breadth of disciplines, research communities, associated research infrastructure, and societal impact. SOURCES: Composed by Tim Warchocki. Image credits: *Learning anatomy*: Abo Photography/Shutterstock; *Mars surface*: NASA/JPL; *Combustion reaction*: ggw/Shutterstock; *Plant study*: NASA, https://www.flickr.com/photos/nasamarshall/15102363689/in/album-72157667439026641, CC BY-NC 2.0; *Viewing microscope slide*: Alex_Traksel/Shutterstock.

or Maxwellian equations or quantum theories. On Earth, these research communities and their funding sponsors may not interact much or share a common framework for scientific inquiry, beyond the expectation of reproducible and ethically reported results. Increasingly, however, interdisciplinary research blurs these boundaries to achieve impressive impact, using physical concepts to predict biological cells' motion or using complex biological fluids to illustrate a physical phase separation. Indeed, these scientific communities are coupled specifically in NASA's strategic plan and BPS Division, as well as in the first decadal survey by this intellectually diverse community (NRC 2011).

Development of innovative technology and tools is essential for conducting experiments in extreme space environments across BPS disciplines. The constraints associated with limited available resources, crew time, and accommodation on a spacecraft or space station in turn demand considerable attention to automated and intelligent systems that can also provide the ability for remote monitoring and real-time control of experiments. There is also a need for integrated modules for easily interchanging experiments in existing systems. Resource limitations also motivate a well-coordinated ground-based research program for experiment development, and analysis of samples returned to Earth when possible. Resource limitations also drive a need for systems to be made in space to reduce the need for transporting equipment and samples, as well as to enable longer mission durations without resupply. Similarly, needs drive improvements in space analysis and monitoring capabilities, including the development of sensors, cameras, fluorescent detectors, and microscopy, as well as the ability to control experiments from Earth and improving computational capabilities through enhanced data storage and transmission.

It is essential that the technology and tools perform as expected or needed in the space environment and that space laboratory equipment interact well with human scientists where appropriate. Simply put, technologies need to generate reliable and reproducible data and be manipulable for human interactions. With increasing science in space, and increasing numbers of astronauts in space, further iterative technology designs are likely to be needed. (See further discussion of particular technologies developed for BPS experiments over the past decade in Chapter 3.)

SPACE, SPACEFLIGHT, AND PLANETARY ENVIRONMENTS

The ISS has been a premier destination for conducting BPS research for more than 20 years, perhaps most notably examining the effects of gravity on organisms, combustion, fluids, and materials. However, the influence of microgravity is only one of many extraordinary features of the extreme environment onboard spaceflight vehicles, including the ISS. (See Box 1-2.) In addition to being its own system of closed life support and attendant vehicular mechanisms, the ISS travels at a speed of approximately 17,000 mph in LEO at about 250 miles above the surface of Earth. The ISS experiences 16 light/dark cycles every 24 hours, as well as a host of chronic and combined stressors, including space radiation exposure,[2] confinement and isolation, and a closed environment (Afshinnekoo et al. 2020). Although the ISS and its inhabitants are partially protected by Earth's magnetosphere, they are still exposed to an average dose of approximately 100–200 milliSievert (mSv) per year (Cucinotta and Durante 2006), a dose range 2–4 times higher than the annual limit for radiation workers on Earth (50 mSv/yr). Once completely outside the protection of Earth's magnetosphere on exploration missions to the Moon and Mars, radiation doses will be higher still. For example, astronauts on future missions of about 3 years to and from Mars will experience effective doses approaching and even exceeding 1,000 mSv (1.0 Sv) (Zeitlin et al. 2013).

Although the ISS has been the premier platform for significant advancement and success of space science, it has several unique and unavoidable constraints for research execution and progress, including limited crew time and constraints and logistics complexities that limit science delivery to the ISS and sample return from the ISS. Ground-based experiments and non-ISS flight platforms remain appropriate and necessary for many questions and preliminary testing prior to spaceflight. Planned vehicle developments will also provide new

[2] The space radiation environment includes high-energy protons produced from sporadic solar particle events (SPEs) and galactic cosmic rays (GCRs), which consist of electrons and positrons (2 percent), protons (85 percent), helium nuclei (12 percent), and heavier, more densely ionizing and damaging particles known as high atomic number and energy (HZE)-charged particles (1 percent) (Durante and Cucinotta 2008).

BOX 1-2
What Exactly Is the Space Environment?

The space environment is indeed very different from the terrestrial environment, the conditions on or near the surface of Earth. Microgravity is often used as a proxy term for the space environment. However, microgravity as a lone descriptor fails to encompass the variety of environmental factors encountered during spaceflight and exploration of other planetary body surfaces. In addition, the definition of microgravity is quite vague in general usage. NASA defines microgravity as "the condition in which people or objects appear to be weightless" (May 2017), whereas science generally defines microgravity as acceleration forces below 10^{-6} *g*. The unique features of the space environment that define it as different from the terrestrial environment include variations in gravity, radiation exposure, temperature, atmospheric pressure, atmospheric and surface compositions, magnetic fields, and the effects of extremely large distances, as illustrated in Figure 1-5. Moreover, many of these features have very complex properties and unknown interactions. For example, the temperature of the solar wind may be more than 1 million K, but its density is so low that negligible amounts of heat are transferred from this hot plasma. Such features of the space environment create remarkable opportunities for scientific discovery regarding the effects of these novel environments on animals, plants, microbes, materials, and physical laws of space, time, and matter. These same features also create complications and hazards that need to be considered not only when conducting research in space, but also when sending equipment and living organisms to explore space.

- *Variable gravity:* During spaceflight, objects and materials do not experience the effects of gravity, such as buoyancy or sedimentation. This "microgravity" environment is useful for studying a range of phenomena, including fluid shifts, body forces, cellular gravity perception, plant growth and mechanical response, fundamental combustion and fire safety, two-phase flow, solidification, processes during additive manufacturing, and soft matter structure and dynamics. Variable gravitational potential in elliptical orbits and other complex geometries enables tests of gravity and other fundamental physics studies. On future lunar or Mars missions, the difference in the gravity of the lunar (1/6 *g*) or Mars (approximately 1/3 *g*) surfaces will provide new opportunities for discovery of additional novel effects. In addition, transitions among gravitational environments will be a factor as biological systems and technologies move from launch through transit to landing.
- *Radiation:* In the spaceflight environments, organisms and physical materials will be exposed to higher doses of different and more damaging types of ionizing radiations than experienced on Earth. Cosmic radiation results from particles trapped within Earth's magnetic field, particles ejected into space from the Sun during solar flares (solar particle events, or SPEs), and from galactic cosmic rays (GCRs). GCRs are high-energy protons and heavy ions from outside our solar system that also interact with spacecraft materials to produce secondary particles (neutrons). Thus, radiation exposure represents a major hazard for long-duration space missions, as it increases health risks not only for astronauts but also for other biological organisms such as plants, and risk of physical damage to equipment.
- *Temperature:* In direct sunlight, the temperature of a vehicle in the space environment could reach temperatures of 150°C (302°F), while in the shadows it can be as low as –270°C (–454°F). Extreme temperature changes cause thermal expansion and contraction that can degrade life support systems and precision instruments and reduce stability of constructed materials and welds.
- *Atmospheric compositions:* Biological systems of animals and plants and physical properties of materials respond differently in atmospheres of altered gas compositions. Inside spacecraft, even with efforts to create safe, viable, "Earth-like" atmospheres in terms of gas composition and pressures, there are effects and even hazards owing to volatile organics and elevated CO_2, lower O_2, biofilm production, biocorrosion, and uncertainties regarding fire safety, as well as the intrinsic

metabolism consequences to astronauts, microbiomes, and plants. Outside of spacecraft, hazards and damage are magnified in combination with radiation exposure and altered temperatures.

- *Low atmospheric pressures:* Atmospheric pressure, as well as gas composition, is among the engineering variables that significantly impact the trade spaces considered in designing spacecraft and lunar and Mars surface habitats, making the physical and biological impacts of atmospheric pressure an important space science consideration. The hard vacuum of space (10^{-6}–10^{-9} Torr) will cause outgassing, releasing of volatiles from materials.
- *Magnetic fields and ionosphere:* The electromagnetic field produces convection of ionic flow in the ionosphere. The ionic flow affects heat and mass transfer as well as charges, conductivity, and reaction dynamics. In addition, magnetic fields have known effects on quantum devices, fluids, and biology, including bacteria that are sensitive to electromagnetic fields. Magnetic fields also interfere with communication systems and alter radiation, complicating the understanding of radiation effects to exploration systems.
- *Extraterrestrial surface environment:* Research is advancing the prospect of mining lunar resources such as noble gases, ice, and minerals from lunar regolith, as well as creating products from lunar resources and constructing them on the Moon (in situ resource utilization, or ISRU)—for example, cement/concrete production, thermal energy storage, and metal and/or oxygen extraction from the regolith of the lunar surface. However, lunar regolith and dust present known and anticipated toxicity hazards, as well as thermal/physical/electrical complications that need to be better understood. Furthermore, low seismicity and availability of permanent cryogenic regions on the Moon provide a unique environment for gravitational wave detection and other experiments.

opportunities and platforms for conducting BPS research. (See Figure 1-4.) BPS science will therefore engage commercial and government ground-based facilities, including commercial sub-orbital vehicles, CLDs, the lunar orbit (Gateway), the lunar surface (Commercial Lunar Payload Services, Human Landing System), and Mars transit vehicles.

OVERVIEW OF RELEVANT DISCIPLINES

NASA's Division of Biological and Physical Sciences (BPS Division) engages a broad range of research expertise and practices. This disciplinary breadth fosters interrelationships on a mechanistic level that inform scientific needs. In particular, there is strong overlap between many physical and biological phenomena, especially in the area of interfaces, fluids, phase separation, self-assembly, and colloids.

BPS disciplines support and interact with other major biological programs at NASA, including the Human Research Program (HRP). The relationships between HRP and BPS communities is discussed throughout this report: BPS provides fundamental science that impacts humans as well as other life-forms and natural or engineered systems, and HRP deals directly with astronaut health and safety, including so-called human factors research.

Biological Sciences in Space and Planetary Environments

Science has made great strides in understanding how life evolves and persists on Earth while under the constant influence of, selective pressure resulting from, and protection of its physical environment. Much research in space-related biological sciences pursues identifying the health risks and possible mechanisms for reducing such risks of the space environment on biological systems of animals, plants, and microbes. It also examines the hazards and benefits of interactions within and between organisms in the closed environments of space vehicles. An important objective is to understand, at a fundamental level, how the extreme environment of space influences biological systems. Biological sciences in space can be classified as human and animal biology, plant biology, and microbial biology. This nomenclature is useful for organized discussion and programmatic efficiency. However, it is important to recognize that many of the underlying biological processes

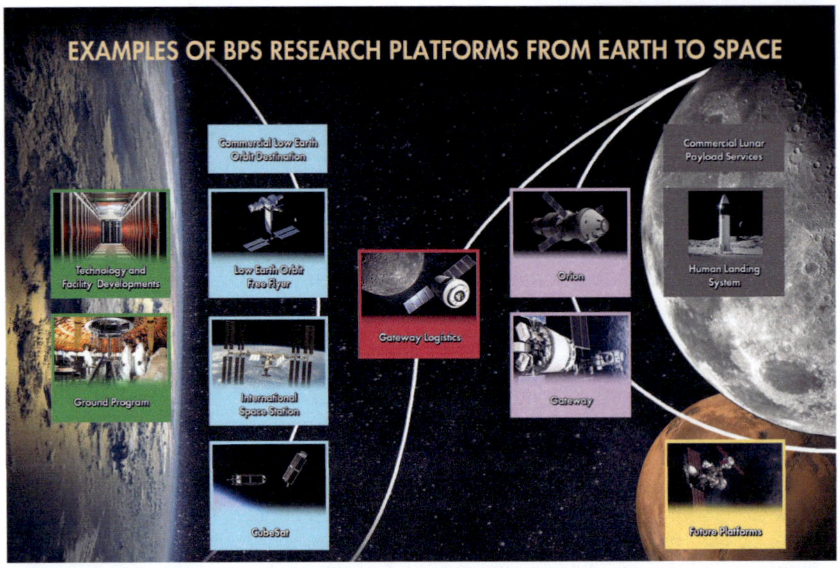

A

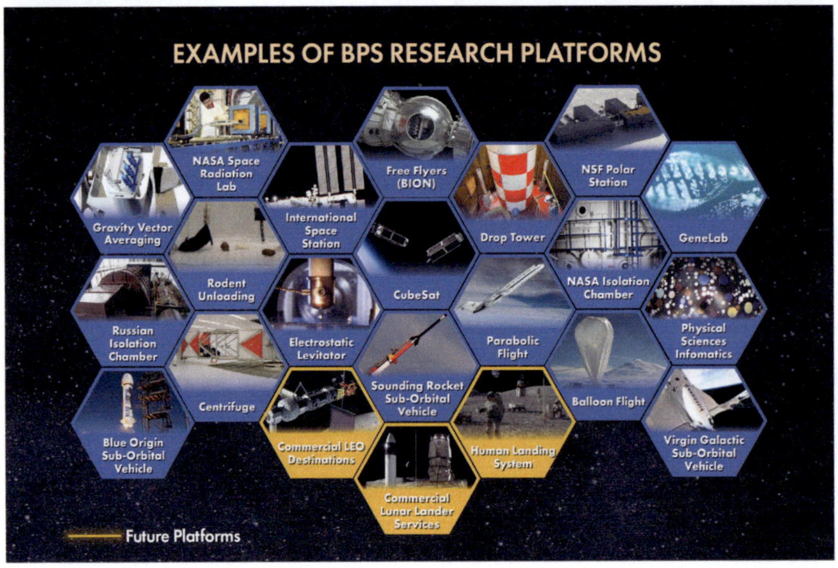

B

FIGURE 1-4 Biological and physical sciences (BPS) research platforms from Earth to space. (A) The space environment as operationally described for the next decade includes an increasingly large number of platforms in suborbital space and in low Earth orbit, as well as platforms in lunar orbits, on the surface of the Moon, and considered vehicles on the way to Mars. (B) Example BPS research platforms as summarized by NASA, current or in short-term planning for known missions. SOURCES: Composed by Tim Warchocki. Image credits for (A): *Technology and Facility Developments*: NASA Neil A. Armstrong Flight Research Center; NASA Photo/Tom Tschida. *Ground Twins Program*: NASA/Langley Research Center. *Low Earth Orbit Free Flyer*: Orbital Reef/Blue Origin. *International Space Station*: NASA/ISS. *CubeSat*: NASA Goddard Space Flight Center. *Gateway Logistics*: Courtesy of NASA, https://www.flickr.com/photos/nasa2explore/51046733772/in/album-72157716027881092, CC BY-NC-ND 2.0. *Orion* and *Future Platforms*: NASA. *Gateway*: NASA/Alberto Bertolin. *Human Landing System*: NASA. Background images—*Earth* and *Mars*: NASA. *Waxing Gibbous Moon 160321*: R. Pettengill, http://astronomy.robpettengill.org, CC BY-NC-ND 2.0. Image credits for (B): Most images courtesy of NASA, including *Rodent Uploading*: NASA, Tahimic et al. (2010), CC BY-NC 4.0; *Sounding Rocket*: By NASA/Wallops, https://sites.wff.nasa.gov/mpl/w_blackbrantxii.html, *Balloon Flight:* NASA/BPO. Other image credits: *Gravity Vector Averaging*: As One International; *Blue Origin Sub-Orbital*: Blue Origin; *Space Radiation Lab*: Brookhaven National Laboratory; *Electrostatic*: NASA/MSFC/Emmett Given; *GeneLab*: Shutterstock/Vector/Garry Killian; *Sub-Orbital Vehicle*: Virgin Galactic.

FIGURE 1-5 Features of the space environment relevant to space, spaceflight, and planetary surfaces. NOTE: Arrows indicate relative magnitude with respect to Earth surface (e.g., microgravity in low Earth orbit annulus). SOURCE: Composed by Tim Warchocki with images courtesy of NASA.

and the physical phenomena driving these processes are shared among all organisms. Also, humans, animals, plants, and microbes will interact with each other both intentionally and inadvertently in spaceflight environments—as they do on Earth.

Human and Animal Biology

Spaceflight poses numerous health challenges to human and animal systems, including alterations in cardiovascular, muscle, bone, vestibular, vision, brain, and likely all organ systems, which have far-reaching effects from the level of whole body to individual genes (Bloomfield et al. 2016; English et al. 2019; Garrett-Bakelman et al. 2019; Van Ombergen et al. 2017). Deleterious health effects are impacted by several factors, including duration of spaceflight, radiation exposure, behavioral aspects, living in confined and highly controlled spaces, and temporal alterations in hormonal and circadian rhythms (Goel et al. 2014; Kennedy et al. 2014; Mark et al. 2014). These effects are further complicated by heterogenous responses among individual crew members to spaceflight and implemented countermeasures (Fitts et al. 2010; Mark et al. 2014; Moore et al. 2014; Trappe et al. 2009).

Current knowledge about the deleterious health effects of extended exposure to the space environment comes primarily from a limited number of studies on astronauts on the ISS (Delp et al. 2016; Elgart et al.

2018; Garrett-Bakelman et al. 2019; Reynolds et al. 2019), and a number of studies done on animal and cell model systems while aboard the ISS (and earlier on the space shuttle), or through ground-based analogues of simulated microgravity, hypergravity, and space radiation. The NASA Twins Study represented the most comprehensive evaluation of human health effects associated with long-duration spaceflight ever conducted (Garrett-Bakelman et al. 2019). These findings highlight potential implications for improved understanding of aging and appropriate countermeasures—both in space and on Earth. Furthermore, individual differences in response were observed, highlighting the importance of personalized, genotype-specific approaches for evaluating effects of spaceflight, particularly as the number and diversity of space travelers increases (Xu et al. 2022; Willey et al. 2021).

Owing to the limitations associated with obtaining astronaut samples, a key research focus of space health in the past decade was investigating biological effects of space environmental factors using in vivo animal models or in vitro/ex vivo human cell models. The majority of studies in animal cells and model systems in altered gravity environments focused on function of bone (Sarkar and Pampaloni 2022), muscle (Juhl et al. 2021; Moosavi et al. 2021), nervous tissue (Kohn and Ritzmann 2018) including the eye (Khossravi and Hargens 2021; Paez et al. 2020), and the immune system (Akiyama et al. 2020; Ludtka et al. 2021).

Although a number of studies have been performed using mammalian models, including humans, the numbers of individual animals in these studies are relatively low and relatively few have been precisely replicated. Thus, it will be important to establish facilities for larger-scale studies that include sufficient replicate experiments for statistically robust conclusions and that include combined features of the space environment. These approaches may also include greater reliance on so-called tissue chips that replicate some features of the physiological micro-environment and require contributions from BPS to harness in the space environment. In the next decade, human space research is expected to emphasize a systems biology, multi-omics approach to understand how the space environment impacts the interplay between complex biochemical networks that regulate human biological systems as a whole, and the interspecific interactions between humans and microorganisms.

Plant Biology

The space environment provides unique opportunities for fundamental plant sciences, as distinct from applied studies that may be concerned with cultivation of plants for life support in space. The study of green plants in microgravity and partial gravity environments encountered in space enables the interrogation of scientific questions that cannot be investigated on Earth, such as the isolation and study of mechanical effects on plant cells (Sampathkumar et al. 2014), environmental responses, and plant growth. Because plants are rooted in place and depend on sunlight for photosynthesis, environmental cues play a relatively larger role in the regulation of plant developmental processes compared to other organisms. Experiments conducted in microgravity allow for disentangling the interaction between light and gravity (Millar et al. 2010; Vandenbrink et al. 2016), which are primary cues directing plant growth on Earth (Hangarter 1997).

The space environment also poses unique challenges to plant growth and development, with issues related to water delivery, air composition and circulation, lighting and temperature, as well as altered pressure and radiation. Because plants are vital components of bioregenerative life support systems, the successful propagation of plants in space is essential for the success of future space missions (Wheeler 2017). This requires a fundamental understanding of plant adaptations to spaceflight and extraterrestrial environments.

Moving beyond LEO, the need to grow plants for lunar and Mars missions will require new types of plant growth facilities, optimizing plant growth, and selecting or engineering traits optimized for growth on supplemented lunar or martian regolith and under altered gravitational conditions and within the complex microbiome of space vehicles and habitats.

Microbial Biology

Microorganisms and their associated viruses have the potential to be present in all areas of a spacecraft or extraterrestrial habitat where plants, humans, and animal may coexist, in addition to the human gut microbiome.

As such, understanding of space microbial biology will impact the success of current and future missions. Ongoing culture-based isolations and characterizations are being performed to monitor microbiological conditions on spacecraft (Castro et al. 2004), and the establishment of culture collections of these isolates remains a key resource for studies related to microbial control and microbial evolution (Mora et al. 2016; Yang et al. 2016). There is ongoing development of monitoring technology to enhance analysis of organisms during spaceflight and reduce the need for samples to be collected and analyzed on Earth (Khodadad et al. 2021). Culture-independent studies are becoming more widely used to describe microbial communities associated with spacecraft (Be et al. 2017; Haines et al. 2019). In clinical settings, sequenced-based identification has an advantage in that identification is quite rapid, with results available within a few hours, as compared to 2–3 days for culture-based identification (Schmoch et al. 2021). DNA sequencing has been demonstrated on the ISS (Castro-Wallace et al. 2017), and ongoing developments in sequencing and other technologies enable some underlying biological mechanisms to be identified. While -omics and other technological developments enable considerable experimental data to be acquired, it is vitally important that these data become available and are used to identify underlying biological processes and other scientific principles, rather than being merely descriptive catalogs of data (Biteen et al. 2016; Thompson et al. 2017).

Microorganisms can also impact spacecraft materials. The water recovery system recycles crew urine and humidity from breath condensate (Yang et al. 2018; Zea et al. 2020). As such, it would receive a regular introduction of microorganisms. Bacteria readily attach to surfaces and form adherent microbial populations encased in an extracellular polymer layer, biofilms, that can potentially coat, clog, and foul surfaces. When growing as biofilms, such microorganism communities become highly tolerant to antimicrobial agents and are thus very difficult to control (Orazi and O'Toole 2019).

On Earth, microbial communities carry out functions essential for life, including cycling of carbon and other nutrients, degradation of pollutants, biosynthesis of valuable materials, and support of plant growth (NASEM 2019). It is important to know how conditions in space influence microbial communities and the beneficial services they provide. At a fundamental level, studies are needed to determine the impact of reduced gravity and increased radiation on microbial community interactions. Systems biology and multi-omics approaches are applicable to decipher mechanisms underlying metabolic and other interactions between community members. This research would eventually entail increasing system complexity: from model or synthetic microbial communities (SynComs) in liquid cultures, toward SynComs in and on solid human-made materials, and eventually to soil-associated and plant-associated microbiomes.

Biophysics in Space and Planetary Environments

Biophysics is a rapidly developing area of space science research for the BPS community. It is an inherently interdisciplinary, bridging field that seeks to understand and describe biological phenomena through application of the theories and approaches of physics (Limbert 2017; Rapin et al. 2021).

Biophysics is sometimes treated as a discipline within the physical sciences, and other times treated as a discipline within quantitative, applied life sciences. From the latter perspective, alternative terms may be used including physical biology, chemomechanics, and mechanobiology. This decadal survey does not dwell on those distinctions in terminology, but instead aims to use the accumulated expertise of this community to address KSQs at the interface of biological systems and physical principles.

Many new developments in biophysics are deeply connected to progress in other experimental and computational fields. In the past decade, expansion of experimental tools such as single-molecule fluorescence techniques and atomic force microscopy, both products of advancements in biophysics, has made biological systems far more accessible to quantitative experiments. Engineering advances contribute to biophysics, and provide the current capacity to weigh, deform, and sort individual cells without biochemically (e.g., fluorescently) labeling them; such biophysical attributes can be correlated with cell function (Lee et al. 2014; Stockslager et al. 2021) through advanced data analytics. Thus, all of this progress on Earth and in space-based research is driven by a complex interplay of biology, chemistry, engineering, data science, and physics (NASEM 2022).

In space, biological systems ranging from unicellular microbes to individual mammalian tissue cells to entire humans adapt in multiple ways to the altered influences or cues of the space environment. Biophysics aims to bring insight into underlying cellular and molecular mechanisms, or signals and responses, that contribute to well-documented phenomena, including the formation of biofilms, mechanical force generation via a cell's motor proteins along its cytoskeleton such as muscle cell contraction, migration of cells during tissue development or cancer metastasis, and a range of reduced-gravity-related physiological effects on biofluid flow in a plant stem or a human artery. As discussed further in Chapter 2, gravitational forces have increasingly dominant effects on biological assemblies and organized systems that are typically greater than 1 mm in length; systems of smaller length and mass scales are dominated typically by viscous and electrostatic forces rather than gravitational forces.

Physical Sciences in Space and Planetary Environments

Some important questions about the physical world can be answered only in the space environment, leveraging unique conditions such as low gravity and low pressure. For example, a wide range of physical sciences experiments, from investigating the nature of dark matter and dark energy to the behavior of complex materials, are affected by Earth's gravity. Feeble but key interactions in many experiments, ranging from those conducted on soft matter to those on ultracold atoms, are hidden by effects of Earth's gravity. Access to the space environment enables not just elegant but otherwise impossible research that opens up new insights on how to steer light, control magnetic effects at the nanoscale, and process quantum information (Ballantine and Ruostekoski 2020).

Physical sciences also concern the search for precise understanding of what happens to physical systems and processes in the space environment, and how to use both Earth-based and off-Earth materials effectively in space. Such knowledge of how to design, process, and use the solid, liquid, and gaseous materials that define the built environment of space travel and habitation is critical for sustaining long-term space exploration. Physical sciences in space also advance understanding of physical systems to improve life on Earth, including how to adhere or separate or recycle materials. Fundamental research across a wide range of disciplines and settings contributes to the robust physical sciences program. The primary relevant disciplines within the physical sciences include materials sciences such as physical metallurgy, complex fluids and soft matter, fluid physics, combustion science, and fundamental physics. In the United States, it is as common to find researchers in these disciplines who are trained as engineers as those who are educated as physicists or applied mathematicians. As noted for the biological science kingdoms, these physical science discipline references are useful for organized discussion and programmatic efficiency, even though the separation between domains is seldom stark and many phenomena in each domain are driven by common physical processes across some of these domains.

Materials

Materials science and engineering is a multi-disciplinary field that connects the processing, internal structure, and properties of materials to predict and control key functional properties of those materials. By understanding the thermodynamics and kinetics by which changes in material composition, phase, and structure can be engineered, materials scientists have developed the aluminum alloys and polymer composites of aircraft fuselages as well as the semiconducting thin films in the chips that fill the aircraft computers and the specialty textiles that comprise wearable sensors. Practically, materials in the space environment need to be able to function well when exposed for long times to different temperatures and pressures, altered gravity and other potentials, and radiation. Materials need to protect the payload of devices, equipment, and life-forms from harmful effects. At the same time, the space environment is a unique laboratory for making new materials and studying materials properties unhampered by typical terrestrial conditions. Both advancing materials for future space exploration and manufacturing materials within the space environment require knowledge of the behavior, microstructures, and fundamental properties of materials in extreme environments and the mechanisms governing their formation, transformation, damage, and destruction.

Advances in materials science in the past decade have highlighted the synthesis of new materials comprising metals, ceramics, and/or polymers fabricated as thin films, composites, and nanomaterials. Concurrent advances in materials engineering experiments and theory have afforded more accurate control and measurement of their thermodynamic, structural, and functional properties. There have been tremendous advances in computation, including density functional theory, molecular dynamics, and, more recently, machine learning and artificial intelligence to predict properties and processing approaches to make such materials. The theoretical prediction of materials' properties could continually be linked to experiment. Experiments in space, where one can separate the effect of gravity from that of other drivers, offers new opportunities for fundamental understanding of the driving forces for chemical reactions, melting, crystallization, phase separation, and microstructure development. Furthermore, there are opportunities to synthesize new high-purity materials with controlled topologies and microstructures, including high-value semiconductors and pharmaceuticals. Last, additive manufacturing approaches including 3D and 4D (time-variable) printing of materials and material composites affords new possibilities in processing and prototyping.

For sustainable operation on the Moon or Mars, local resources could be utilized to create the built environment; it is prohibitively expensive and time consuming to bring all needed engineering materials from Earth. Mining, transport, processing, and manufacturing could be adapted to the local space environment. To do this successfully, science will need to be integrated with engineering to ensure that appropriate processes are proposed, developed, and tested. Likewise, there needs to be an active interface between engineered materials and biological materials and systems, whether in considering materials that interface with humans in medical devices or materials that augment human potential in soft robotics.

In addition to the mechanically stiff and hard crystalline materials comprising most metallic and ceramic load-bearing structures on Earth and in the ISS, the space environment includes the familiar glassy or amorphous stiff materials (e.g., borosilicate glass) and the more easily deformable materials including colloidal suspensions and gels that comprise our adhesives, pharmaceutical capsules, and personal hygiene products. These materials, alternatively called complex fluids or soft matter, are potentially useful in space; their formation in low-gravity environments may lead to new microstructures and properties. While study of materials naturally includes the study of soft matter, this subset is described next because the communities that conduct research on this topic can be described variously as materials scientists, physicists, and plasma scientists.

Complex Fluids and Soft Matter

A class of materials that has characteristics of both solids and fluids is referred to as soft condensed matter, or simply soft matter. Soft matter has been described as "everything squishy," and is a term used to denote materials that flow readily under low applied forces, including gravitational forces on Earth; they are also termed complex fluids. Soft matter includes research areas such as polymer flow, colloids, gels, foams, liquid crystals, complex/dusty plasmas, and granular media.

The lack of buoyancy in the microgravity environment eliminates natural convection and sedimentation, which allows for study of 3D systems, as opposed to 2D and planar systems. Slower chemical reactions as well as reorientation of molecular arrangements that are not possible in a normal gravity environment where density-driven flows disrupt these processes can be studied in the space environment. For example, the liquid content in foams in a terrestrial environment drain from within the intercellular network (Langevin and Vignes-Adler 2014), but in microgravity, the coarsening mechanism is substantially different, because surface tension forces dominate the redistribution of liquid. Extensional rheological measurements of polymers without the fluid "sagging" become possible in a microgravity environment (Hall et al. 2009). In the absence of gravity, attractive forces in granular media may become dominant and allow for other packing arrangements of the material as well as easier dispersion. (See Figure 1-6.)

Knowledge gained about soft matter in the space environment can improve many Earth-bound applications ranging from stability of colloidal solutions, gels, and foams in the consumer market; extrusion processes for polymers; the use of liquid crystals in electronics; and the extraction, transport, and reacting of various raw minerals into usable commodities.

Utilizing soft matter in productive ways will be important as space exploration on other planetary bodies develops, especially with regard to handling granular media. Several areas of research will facilitate these activities

FIGURE 1-6 Commander Barry Wilmore installing a 3D printer on the International Space Station. SOURCE: Courtesy of NASA TV, https://www.flickr.com/photos/nasamarshall/15628687807/in/album-72157667439026641, CC BY-NC 2.0.

and mitigate the risks. In situ resource utilization (ISRU) will also rely on extraction, transport, and processing of various types of lunar soil or regolith (Nangle et al. 2020). Designing safe and effective means of collecting, handling, and processing these materials in this environment will require a thorough understanding of handling granular media. Conversely, mitigating the effect of fine particles of granular media, or dust, to prevent it from adhering to photovoltaic solar cells and thermal radiators and diminishing their performance is necessary for missions that last weeks to months. Last, understanding the processes associated with 3D printing to manufacture delicate biocompatible structures or replacement parts from soft matter is also being investigated in both the microgravity and partial gravity environments (Nangle et al. 2020).

Fluid Physics

The study of single-phase fluids and their behaviors in space is critical to a wide range of functions in the space environment. These include controlling liquid fuels in ways that enable spaceflight and keep crew members safe (Fester et al. 1975; Hansen et al. 2020), recycling air and water and supporting food production within spacecraft (Meyer and Schneider 2018); power production (Bennett et al. 2020) and temperature control (Sarraf and Anderson 2007); and understanding fluid-like behaviors of lunar (Agui and Wilkinson 2010) and martian surface materials.

The initial, and continuing, purpose of such research is to develop and validate techniques for controlling the position and motion of fluids in the space environment based on their physical properties, with particular emphasis on understanding the effects of reduced gravity on fluid behaviors.

Fluid research has been conducted using drop towers (less than 5 seconds) (Masica and Salzman 1967; Wollman et al. 2016) and parabolic flight aircraft (less than 20 seconds) (Dukler et al. 1988; Konishi et al. 2015b; Raj et al. 2009) for behavior that quickly transitions from normal gravity to the reduced gravity environment. Suborbital capabilities such as sounding rockets (Zell et al. 1984) and commercially available reusable rocket systems (Collicott and Alexeenko 2019) extend the possible duration of experiments (up to about 2–3 minutes). Fluid research aboard orbital spacecraft occurred as early as Scott Carpenter's Mercury 7 flight (Nussle et al. 1963). Fluids experiments were conducted on many space shuttle flights (Lee et al. 1997) and continue to be conducted aboard the ISS (Chatterjee et al. 2013; Motil et al. 2021). Most of these investigations have been conducted in the low-gravity environment ranging from 10^{-6} to 10^{-2} *g*'s. Research in a partial gravity, similar to lunar or martian gravity, has been conducted aboard aircraft flying through a modified parabolic arc (Hurlbert et al. 2004). Centrifuges in drop towers and the ISS also have been used to generate a partial gravity environment, but compensation for spinning fluid motion generated by the centrifuge is required for these environments (Ferkul 2017). Fluid and gas behaviors in a high-gravity rotating environment (giant planets) also have their peculiarities (Bouchet and Sommeria 2002; Dowling 2020).

Gravity affects both the shape of the vapor–liquid interface in bubbles and droplets and the flows of gas and liquid along that interface, thus impacting mixing within each phase and between phases as well as the mass and heat transfer through the interface. As a result, the length, velocity, and timescales of fluid behavior are dramatically altered by moving to a reduced gravity environment. These alterations make detailed scientific investigation of fluid behavior possible. On Earth, heavier objects such as solid crystals fall through liquids, while lighter phases like gas bubbles rise. This motion can affect the shape and size of the bubble and crystal, as well as the rate at which the crystal forms (or dissolves). In the space environment, the gravitational pull causing this motion is reduced. However, other physical mechanisms become more prominent and can induce motion and consequently affect the shape of bubbles, droplets, and crystals in different ways. For example, in space, the dominance of capillary effects is significantly increased, resulting in larger bubbles during boiling. Temperature or concentration differences along a bubble surface can cause bubbles to "swim" toward regions of warmer liquids, whereas the motion of growing crystals allows mixing that increases the rate of growth and can lead to shapes and internal structures that are different than if there was no flow. In addition, as a crystal falls to the bottom of a terrestrial vessel, it stops its growth along the bottom surface and has limited growth along its sides until it merges with neighboring crystals. Different crystal orientations and dimensions are possible in microgravity, as the crystal growth is not blocked by any surfaces or support structure, and the mixing is non-existent.

The fundamental science of fluids impacts other science in space. Fluids provide cooling to science instruments aboard space telescopes and spacecraft to prevent overheating of electronics and motors. The ability to prepare chemical compounds in space by transferring reagents to sample chambers such as pipetting reagents into sample vials for analysis or storage safely without releasing stray droplets is required for crew safety. Material science investigations require controlling bubbles for solidification or crystallization studies. In addition to conducting successful plant biology research in space, plants need the correct balance of water, nutrients, and air to their roots to prevent them from succumbing to situations of drought or flooding.

Combustion

Combustion science involves fluid mechanics, heat transfer, phase change, transport phenomena, thermodynamics and chemistry. The major products of combustion are carbon dioxide and water, but that is an oversimplification of a much more complex physicochemical process. For even the simplest fuels there are a large number of intermediate reactions and species. In addition, flow motion, molecular diffusion, and thermal radiation are coupled with the reaction processes (Turns and Haworth 2021). Gravity has profound effects on most flames (Law and Faeth 1994).

Altered gravity environments, including those experienced in spaceflight or on the Moon or Mars, create significant challenges in combustion, whether in an unwanted fire in a spacecraft or an extra-terrestrial combustor.

The lack of a buoyant flow, however, also creates a unique opportunity for researchers—the ability to perform fundamental studies of combustion phenomena without the complicating and dominating influence of the buoyancy-induced flow. Over the past 3 decades, combustion and fire researchers have exploited the reduced-gravity environment in ground-based and space-based facilities to study fundamental combustion problems to unlock and discover aspects of combustion that have eluded researchers owing to the pervasive influence of buoyancy-induced flows in typical terrestrial laboratory studies. Entirely new combustion systems, such as so-called sofballs (Ronney 1998) and cool diffusion flames (Ju 2021), have been developed. Such understanding may extend to improved prediction and management of terrestrial wildfires, in that the large length scales and temperature differences suggest that some aspects of these fires are suited to more fundamental study in environments such as space, where buoyant forces are reduced.

Analysis of combustion in microgravity can improve the understanding of fundamental combustion physics, leading to more efficient and less polluting terrestrial combustion devices such as engines and burners (Ju 2021). Currently, 86 percent of the world's energy demand is met through combustion (Ritchie et al. 2022). With movement toward hydrocarbon fuels from greener or carbon-neutral feedstocks (as opposed to conventional petroleum-based fuels), future engines will be required to simultaneously increase efficiency and reduce emissions. These objectives can be met only through a fundamental, predictive understanding of combustion physics that does not currently exist (Ju 2021).

Combustion science within the space environment also informs spacecraft fire safety, leading to improved material qualification, fire detection, and fire suppression in spacecraft and outposts (Guibaud et al. 2022). NASA's future exploration goals are just beginning with the start of the Artemis program. The cabin atmospheres in the Artemis program will have lower cabin pressure with an increased ambient oxygen concentration (to minimize the required prebreathing time for extravehicular activities). Because fires are very sensitive to ambient oxygen mole fraction, reactants and materials, diluents, pressure, and gravitational forces, fire safety practices deemed safe in the microgravity environment of the ISS (with a cabin atmosphere identical to air on Earth) may no longer offer the same factor of safety (or be completely inapplicable) for future exploration missions. Last, combustion within space environments may be utilized beyond propulsion, such as in supercritical water oxidation for waste disposal, or for ISRU processes including the synthesis of cementitious materials.

While there have been dozens of microgravity combustion and fire studies in drop facilities, parabolic aircraft, and spacecraft (Kono et al. 1996; Sun et al. 2020), as well as the ISS (Dietrich et al. 2014), those studies have been limited in terms of flame and fire sizes, diagnostics, pressures, and the effects of lunar and martian gravity.

Fundamental and Quantum Physics

The goals of fundamental physics research are to discover and explore the physical laws governing matter, space, and time. Such understanding can help predict physical behaviors in space and on Earth. Present physical theories (e.g., general relativity, the standard model of elementary particles and fields, and quantum mechanics) provide an excellent description as well as prediction of many physical phenomena. However, most (more than 95 percent) of the universe's composition is unexplained—a "dark matter" and "dark energy" of unknown origin. Many other puzzles, such as obvious matter–antimatter asymmetry, the near-horizon geometry of black holes, and the physics of the early universe prior to recombination, remain unanswered.

Some of the most important questions of modern physics can be answered only in the unique environment of space. Space provides a unique platform of exploration away from Earth's gravity and access to a variety of variable gravitational potentials and enables experiments needing baseline distances beyond Earth scale. The Moon provides a unique combination of very low seismic backgrounds, vacuum, and a permanent cryogenic environment. Other solar system objects, including asteroids, could provide access to unique fundamental physics explorations.

Deployment of high-precision optical clocks in space and demonstration of space-to-Earth and space-to-space optical time transfer are the next steps toward new physics discoveries enabled by space. Space optical clocks will also greatly advance relativistic geodesy (measurements of Earth's gravitational potential for Earth science applications), enable autonomous spacecraft navigation throughout the solar system, provide links for high-precision optical clocks on Earth, and dramatically expand the capabilities of very long baseline interferometry (VLBI). Applications of other quantum technologies and corresponding R&D, including atomic interferometers, solid-state-defect magnetometry, quantum gases, quantum memories, entanglement sources, and others, will leverage fundamental and applied advances enabled by quantum devices in space. In particular, optical atomic clocks and atom interferometers have been proposed for novel future gravitational wave detectors in wavelength ranges not accessible by either Earth-based detectors or the Laser Interferometer Space Antenna (LISA), providing a powerful new window on diverse phenomena, including ultra-massive black holes at the center of most galaxies and the physics of the early universe.

RESEARCH LANDSCAPE

The current BPS research landscape is a highly complex system of government space agencies, federal research agencies, commercial entities, private funders, and a diverse array of researchers across the biological and physical sciences. The complexity of this system over the next 10 years is likely to increase. The government agencies operating the ISS—NASA, ESA, Roscosmos, JAXA, and CSA—are among the many space agencies striving to make substantial progress in space exploration in the coming years. (See Figure 1-7.) With ISS operations expected to expire in 2030 (NASA 2022b) and goals focused on deep-space exploration, government agencies are examining how they can maintain and surpass the research progress made thus far on the ISS. How the launch costs associated

FIGURE 1-7 NASA astronaut and Expedition 68 Flight Engineer Nicole Mann poses with a pair of free-flying, cube-shaped Astrobee robotic helpers inside the Kibo laboratory module. The International Space Station lifetime has been extended to 2030, but NASA plans to replace it with a commercial option. SOURCE: Courtesy of NASA, https://www.flickr.com/photos/nasa2explore/52536103444, CC BY-NC-ND 2.0.

with BPS research are shared in the coming decade among U.S. government agencies and research performers also remains unknown at present. Clarity on that point will affect the pace of research that is not also prioritized by the private sector, including corporate research and philanthropy. For example, will facilities for BPS experiments be made available on private spaceflights? The description of the research landscape here is a summary of the changing opportunities that will affect U.S. leadership of science priorities and avenues for research over the next 10 years. This is not an exhaustive review of all the changes affecting the BPS research community, but it reflects the types of developments under way.

NASA has participated in many international partnerships that have included the bartering of experiment resources such as crew time and facility usage, launch mass, and so on, and enabled sharing the data among principal investigators from many nations. Now, commercial entities have entered the suborbital and LEO environments. NASA is supporting the growth of a robust LEO economy. A number of companies are engaged in crew and cargo transport (e.g., to the ISS), free-flyer operation, and development of new stations to continue the ability for humans to live and work in space. Such companies include Axiom Space, Blue Origin, Boeing, Nanoracks, Northrop Grumman, Sierra Space, and SpaceX. These companies are also likely to operate ground-based replica science facilities (in addition to space platforms) in order to conduct ground controls for space experiments and thus compete for the business of BPS.

The possibilities of expanded LEO destinations and laboratory resources are exciting, but there will be many challenges to consider in terms of the orchestration of research over the next decades. These include uncertainty of the frequency and extent of government-funded research (by NASA and other federal agencies) on these platforms, how unique hardware will be provided on various platforms, data sharing requirements, crew and data protections, broadening of the BPS research talent pipeline, intellectual property rights, export control restrictions, and policy considerations when at least one CLD is in operation (Gatens 2022).

Databases for BPS Research

Development of public databases in the past decade have expanded the opportunities for researchers to mine and utilize data from prior flight investigations to further scientific inquiry. Efforts within NASA to establish and promote use of open databases were supported by a 2013 Executive Order that directed the default practice for government data. These data are used to identify underlying biological and other scientific principles (Biteen et al. 2016; Thompson et al. 2017). There are also biospecimen repositories such as the NASA Biological Institutional Scientific Collection (NBISC) and the Physical Science Informatics (PSI) data repository for physical science experiments performed on the ISS and other flight or space analogs.[3] The growth of data repositories has been an extremely positive outcome in support of open science within the BPS community, and is in line with the national open science initiative (NASA 2023b; White House 2023)[4] and attendant expectations that NASA champion FAIR access principles (Wilkinson et al. 2016) for the BPS research community. Table 1-1 assembles the major data repositories of relevance to BPS research.

Research Platforms and Laboratories

BPS researchers utilize a range of platforms and laboratories to carry out experiments and study space-related phenomena. These include the ISS (see Figure 1-8), drop towers, parabolic flight, balloons, suborbital vehicles, and free-flyers, as well as a number of terrestrial analogs and ground-based laboratories. (See Figure 1-4 and Box 1-3.) A number of these platforms produce microgravity conditions through free fall of an experiment for different time durations: less than 5 seconds in vertical drop towers, less than 30 seconds at the apex of aircraft parabolic flight, up to 3 minutes in sub-orbital vehicles, and 5 to 20 minutes in sounding rockets (Joseph and Wood 2021). Space-like radiation environments can be accessed with high-altitude balloons, with the conditions varying with location in Earth's magnetic field. These facilities are used for concept development, testing matrix refinement,

---

[3] This text was modified after the release to sponsor to clarify what is included in the PSI data repository.

[4] NASA, "2023: Year of Open Science," https://nasa.github.io/Transform-to-Open-Science/year-of-open-science.

TABLE 1-1 Summary of Data Repositories for Biological and Physical Sciences Research

Database	Data Type	Payloads	Species or Physical Areas of Investigation
NASA Life Sciences Data Archive (LSDA)	Mission or experiment or biospecimen metadata, experiment assay data, hardware specifications, environmental telemetry, mission science reports, imagery	Apollo, BION, Biosatellite, Cosmos, Gemini, Mercury, MIR, Skylab, space shuttle, the International Space Station (ISS), SpaceX, bed rest, Human Research Facility, NEEMO, ground analogs	Human, monkey, rat, mouse, fish, frog, fly, nematode, quail, plant, microbe, others
NASA GeneLab	-omics	Space shuttle, ISS, parabolic flight, NASA Space Radiation Laboratory (NSRL)	Animal, microbial, plant, human
JAXA-ToMMo Integrated Biobank for Space Life Science (ibSLS)	-omics	ISS	Mouse
NASA Biological Institutional Scientific Collection (NBISC) and Biospecimen Sharing Program (BSP)	Animal, microbial	Cosmos, space shuttle, ISS, rodent research, ground analog studies	Mouse, rat, quail, microbes
ISS Microbial Analysis	Microbes monitoring within the ISS	ISS	Microbes
JAXA Biorepository	Animal, microbial	Mouse habitat unit, aquatic animal experiment facility	mouse
Physical Science Informatics (PSI)	Machine-readable textual or numerical form documents, digital images and videos; also contains analyzed or reduced data and any supporting data, including science requirements, experiment design and engineering data, analytical or numerical models, publications, reports, and patents, and description of commercial products developed; also beginning to archive flight samples from past experiments	Space shuttle, ISS, free-flyers, ground-based	Biophysics, combustion science, complex fluids, fluid physics, fundamental physics, and materials science
ESA Erasmus Experiment Archive (EEA)	Database of European Space Agency (ESA)-funded or co-funded experiments covering a wide range of scientific areas that were performed during missions and campaigns on/in various space platforms and microgravity ground-based facilities, starting from 1972; supersedes the former ESA Microgravity Database (MGDB)	Space shuttle, ISS, free-flyers, sounding rockets, parabolic flights, ground-based	Wide range of biological and physical sciences and technology

SOURCE: Adapted from E. Afshinnekoo, R.T. Scott, M.J. MacKay, E. Pariset, E. Cekanaviciute, R. Barker, S. Gilroy, et al., 2021, "Fundamental Biological Features of Spaceflight: Advancing the Field to Enable Deep-Space Exploration," *Cell* 184(24):6002, Copyright 2021, with permission from Elsevier.

FIGURE 1-8 Earth observation taken during a day pass by NASA astronaut Scott Kelly aboard the International Space Station (ISS). The ISS is due to be retired within the period covered by this decadal survey. SOURCE: Courtesy of NASA, https://www.flickr.com/photos/nasa2explore/24392399919, CC BY-NC 2.0.

BOX 1-3
Space Experiment Platform Options

- *The International Space Station and commercial low Earth orbit (LEO) destinations (CLDs)*—Microgravity, long-duration experiments (months), lower ambient radiation doses than in deep space, use of available laboratory resources (e.g., freezers, microscopes, glove boxes/biosafety cabinets, incubators, bone densitometers, PCR machines, etc.).
- *Suborbital*—Short exposure to microgravity (minutes), manual or automated experiments, low ambient radiation exposure owing to duration and orbital height.
- *Free-flyers/small satellites*—Microgravity, very long duration experiments (months to years), automated and miniaturized equipment, data telemetry, high doses of ambient radiation for beyond LEO free-flyers.
- *Balloon flights*—Long-duration experiments possible (weeks to months), radiation exposure can be elevated by adjusting altitude, location, and duration of flight.
- *Cislunar (e.g., NASA Gateway)*—Microgravity, long-duration experiments (months to years), automated and miniaturized equipment, data telemetry, high doses of ambient radiation.
- *Lunar surface experiments*—Partial gravity (1/6 *g*), long-duration experiments (long-term goal of sustained lunar habitation), automated and miniaturized equipment, data telemetry, high doses of ambient radiation.

SOURCE: Adapted from S. Bhattacharya, NASA, presentation to the steering committee, August 16, 2021.

and hardware verification. The committee anticipates that this range of platforms and laboratories will continue to be used and even expand as the BPS community seeks to leverage commercial LEO destinations, developments in free-flyers and ground-based simulators, Artemis mission capabilities, and other platforms beyond LEO. Box 1-3 briefly describes the environment and testing conditions for different platforms in space.

Chapter 2 next summarizes key advances in BPS science over the prior decade. These discoveries and developments—made by a resilient BPS research community that still has much room to broaden its talent pipeline—establish a baseline of knowledge and capabilities that motivate the framework for BPS research themes of the coming decade as outlined in Chapter 3. This baseline of scientific insights highlighted in Chapter 2 also serves as a reference for the key scientific questions of the coming decade described in Chapters 3–5, as well as the research campaign concepts outlined in Chapter 6 and overall strategy described in Chapter 7. These chapters, as well as the associated findings and recommendations, are responsive to the statement of task in Appendix A. Acronyms and abbreviations are listed in Appendix B and defined at the first instance within each chapter. References are listed in a separate section preceding the appendixes.

2

Current State of Knowledge in the Biological and Physical Sciences

The prior decade of research in the space environment is highlighted by remarkable accomplishments in the biological and physical sciences (BPS), from serendipitous discoveries to painstakingly planned demonstrations of concepts. This baseline of knowledge and capabilities is to be celebrated as the product of government support including from NASA, private-sector engagement, and resilient space science research communities. This baseline also highlights gaps in knowledge, and sets the context for future research that will access the space environment for maximum societal impact in the coming decade. Chapter 2 overviews the current state of knowledge in BPS disciplines gained since the last decadal survey (NRC 2011) was initiated, as well as the advances in engineering, technology, and data analytics that have facilitated both progressive research advances and transformative pivots in research trajectories. This baseline is organized by disciplines, described in Chapter 1, serving as a jumping-off point to identify key scientific questions (KSQs) and research campaign opportunities for the coming decade in Chapters 3–6. Reference to the platforms and technologies used by the BPS research community to gain that knowledge is included throughout Chapter 6, underscoring the interdependence of specialized experimental capabilities and successful scientific inquiries. Here, success is not gauged only by impact within the research community as quantified by publications or patents or wider adoption of methods. Rather, BPS research is also defined by the demonstrated or anticipated positive societal impact of research findings.

The interplay between technology and science is highly visible over the past decade of BPS in space. Technological advances have led to giant leaps at the level of platform infrastructure, research campaigns, experiments, and data analysis. Indeed, the past decade of the International Space Station (ISS) has welcomed new facilities for plant, animal, and microbial studies; protein crystallization; sequencing; solidification and properties of materials; microscopy; hyperspectral imaging; combustion science; quantum fluids; and fluid dynamics studies. Simultaneously, the number of commercial suppliers providing access to space has increased dramatically, along with advances in novel platforms like cube-satellites and commercial space vehicles. With the increase of platform availability has come an increasing number of experiments, research focus areas, and technological developments.

Fundamental science underlying the impact of space environments on biological systems from cells to ecosystems has advanced owing to innovation in molecular and cellular biology; platforms for biological research in space; increased focus on biological data science; and an impressive array of studies spanning humans, animals, plants, microbes, and their communities. (See Figure 2-1.) These have enabled significant discoveries of the challenges

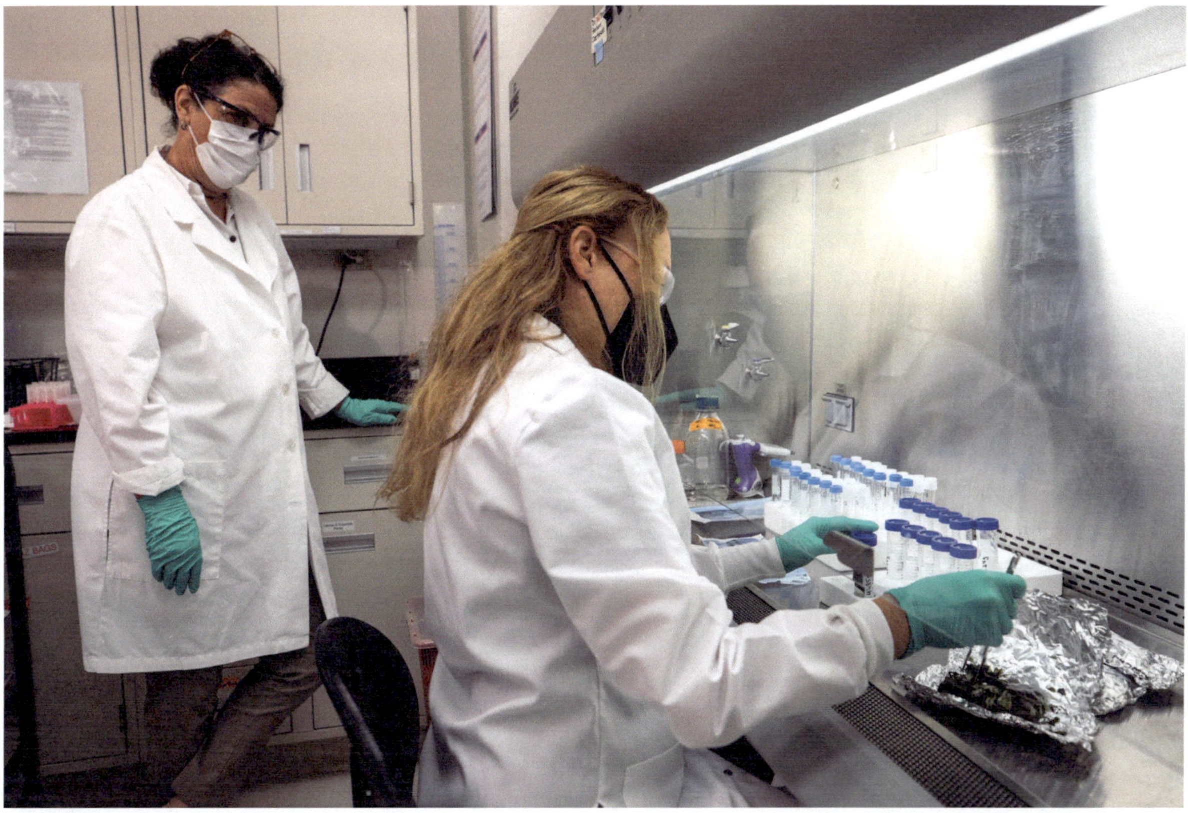

FIGURE 2-1 The Microbiology Laboratory at NASA's Kennedy Space Center, where microbiologists are working with frozen lettuce samples that recently returned from the International Space Station. NASA is studying how to effectively grow crops in space so that plants can provide supplemental nutrients to astronaut crews on long-duration missions, such as a mission to Mars. SOURCE: Courtesy of NASA/Ben Smegelsky, https://www.flickr.com/photos/nasakennedy/51585196635, CC BY-NC-ND 2.0.

and adaptations of organisms to the spaceflight environment, leading to recognition and improved understanding of unique space-induced effects. Across all life-forms, gene expression and epigenetic changes are observed, as well as stress effects in response to special space conditions. For example, in plants, red light responses that cannot be observed on Earth have been discovered, as well as the limits of root gravity sensing. In animals, the past decade has revealed a greater understanding of spaceflight-induced musculoskeletal, cardiovascular, and immune system changes. This new understanding has led to the identification of pathways that may improve mitigation strategies for bone loss and the successful growth of multiple food crops. The effects of space environment on host–microbe interaction have been probed, showing new routes to possibly pathological interactions. The microbial ecology of the built environments in space has been measured, and how microbes can disperse, form films, and possibly interact with other organisms and other mission-critical components under the novel conditions found in low Earth orbit (LEO) has been probed. The new physics encountered by life in these environments entwines this research with that of the physical sciences in soft and condensed matter physics, as new routes to high-quality crystallization of otherwise recalcitrant proteins have been discovered and how the impact of hydrodynamic forces that are different in space affect biofilms, organ printing, and phase transitions in newly discovered membraneless organelles in cells has been studied. The uses of naturally selected or engineered microbes and plants to provide other services—including in situ resource utilization (ISRU), waste recycling, and manufacturing of pharmaceuticals, chemical components, and materials—have been explored as means of providing services to explorers logistically separated from Earth. These examples suggest that living organisms can not only survive off Earth but can also thrive and, in some cases, can provide critical resources to humanity wherever located between the surfaces of Earth and Mars in the coming decade.

Findings in the biological or life sciences are deeply supported by discoveries and achievements in the physical sciences, both fundamental and applied. For example, the emerging understanding of membraneless intracellular structures as different phases that exist in a non-equilibrium environment is a finding that builds on years of effort supported by NASA to expose behavior of non-biological colloidal materials and protein solutions undergoing phase transitions. This bridging of biology and physics is critical to understanding how cells work, but also has practical implications—for example, in the thermostabilization of viruses to enable transport and storage of vaccines outside deep refrigerated environments. (See Figure 2-2.)

BPS has been served by the development of core infrastructures, ensuring predictable resources for carrying out experiments and analyzing and disseminating data. For example, the development of the NASA GeneLab database in the past decade has allowed meta-analyses of molecular biology data from spaceflight samples, revealing common recurring space effects across experiments and across species. As an example of platforms and scientific programs advancing the physical sciences, the creation of a Cold Atom Laboratory (CAL) in the ISS has enabled the creation of ultra-cold gas bubbles, a key step on the way to making and studying Bose-Einstein condensates (the fifth state of matter, as gas at nanokelvin temperatures) in space. Atom interferometry has been demonstrated on the ISS, opening the door to a variety of experiments, including long-flight-time atom interferometry, which promises higher-precision fundamental measurements as well as better inertial navigation.

FIGURE 2-2 NASA astronaut Peggy Whitson storing blood samples in the International Space Station's (ISS's) ultra-cold freezer for eventual return to Earth. Whitson returned to the ISS in 2023 as a private astronaut. SOURCE: Courtesy of NASA, https://www.flickr.com/photos/nasamarshall/36005395331/in/album-72157667439026641, CC BY-NC 2.0.

Research from two new facilities for containerless processing of materials, the European Space Agency's (ESA's) ISS electromagnetic levitator (ISS-EML) and the Japan Aerospace Exploration Agency's (JAXA's) Electrostatic Levitation Furnace (ELF), on the ISS is providing new insights into the initial formation and growth of solids without the distortion and convection caused by gravity, particularly rapid solidification from liquids cooled far below their equilibrium melting temperatures. The reduced gravity extends the range of time, temperature, and mixing that can be investigated before other phenomena intervene. Such containerless processing in microgravity also allows scientists to determine temperature-dependent material properties like viscosity and surface tension, which are necessary to accurately model and predict material behavior. Furthermore, the microgravity environment enables or facilitates production of materials that cannot be produced on Earth, or only with significant difficulty. Basic manufacturing processes, sometimes from materials derived from both abiotic and biotic ISRU, can now be conducted in space, especially important for in-space manufacturing and repair given the long timescales and limited mass of objects transported to the space environment from Earth. These practical advances are made possible by fundamental if more abstract advances in the theory and practice of non-equilibrium kinetics and transport in high pressure, reacting and non-reacting systems, and plasmas. Those principles, once understood, dictate how in situ material synthesis can be optimized and tailored for lunar or Mars conditions as well as microgravity spaceflight conditions. Combustion is a non-equilibrium energy conversion process; combustion and plasma can be used to generate non-equilibrium material states, open new non-equilibrium reaction pathways, and give rise to new combustion regimes. That knowledge gained from combustion far from equilibrium can in turn be applied to enable cold fire combustion, reignition in microgravity, and plasma-assisted combustion and material synthesis that affect feasibility and technical risk of future space exploration modes.

ENGINEERING AND TECHNOLOGY ADVANCING BIOLOGICAL AND PHYSICAL SCIENCE RESEARCH

New technologies have led to improvements in testing and techniques, without which scientific progress is slowed for any research community or nation. The relative ease of sequencing human DNA or of computing the structure of engineered materials in batteries today compared with a decade ago is owed to that parallel investment in research-purposed technology. Translation of those seemingly abstract research practices to societal impact hinges on the accuracy, speed, and accessibility of such tools—including adaptation of first-on-Earth research workhorses into industrialized machines, such as gene sequencers now operating in U.S. hospitals or machine learning–based weather prediction. Major advances in engineering and technology enable progress in space-relevant BPS, including potential technologies beyond LEO that can further enable fundamental physics.

Manufacturing Technologies

Advances in manufacturing over the past decade have facilitated development of science and technology applied to new research capabilities on Earth and in space environments. These same manufacturing innovations have also opened new avenues of research and production enabled by unique space environment attributes such as microgravity. Digital manufacturing, robotics-assisted manufacturing, and additive manufacturing are three examples in which the United States has created public–private partnerships to foster research in manufacturing processes on Earth among academic researchers, industry, start-up companies, and non-profit organizations along the manufacturing pipeline (Vickers 2021).[1] The ISS National Laboratory (ISSNL) has focused recent efforts in funding opportunities to leverage the enabling features of LEO for manufacturing in space (ISSNL 2023).

While space-based research and exploration are anticipated to rely increasingly on robotics to augment crew safety and productivity—and include crewless missions and Earth-based production of space-destined platforms and vehicles—whether and how to automate production is a decidedly human challenge related to production

[1] This network and website (https://www.manufacturingusa.com) is coordinated by the U.S. National Institute of Standards and Technology (NIST) with specific institutes co-sponsored by U.S. agencies including NIST, the Department of Defense (DoD), and the Department of Energy (DOE) in partnership with U.S. public and private sectors.

complexity and product life cycle. Initial efforts in servicing, assembly, and manufacturing in space environments have focused on replacement of astronaut extravehicular activities (EVA) (Papadopoulos et al. 2021) and robotics-assisted production of crops—with concepts initiated by NASA-sponsored research or college student-focused research participation (Dee 2018) and innovation prizes (Martinovich 2016) and translated to start-up companies. Indeed, underscoring the reciprocal relationship between biological and physical science research, researchers have explored the connection between plant biology and robots as inspiration for how to "grow" robots based on principles of plant growth (Mazzolai et al. 2020).

Bioprinting is a subset of additive manufacturing, implying printing of biologically compatible materials and/or inclusion of living biological cells in the printed structure. The BioFabrication Facility (BFF), flown to ISS in 2019, is the first U.S. additive manufacturing system capable of manufacturing human tissue in the microgravity of space, using bio-inks—which have natural or synthetic proteins, nutrients, and other growth factors blended with living cells (mammalian, plant, and/or microbial)—to construct living material that can be conditioned to grow (Vialva 2019). (See Figure 2-3.) Current research is under way to enable the creation of bone implants for astronaut transplantation during long-term interplanetary expeditions. While bioprinting can be performed on Earth, the effects of gravity cause mechanically compliant (i.e., low stiffness), additively manufactured polymers and composites relevant to so-called soft tissues comprising organs such as brain and kidneys to sag and deform under their own weight (Espinosa-Hoyos et al. 2017). Bioprinting in reduced gravity space and/or development of programmable biomaterials has the potential to mitigate these limitations.

FIGURE 2-3 NASA astronaut and Expedition 69 Flight Engineer Frank Rubio uses a glovebag to service the BioFabrication Facility, replacing and installing components inside the research device designed to print organ-like tissues in microgravity and learn how to manufacture whole, fully functioning human organs in space. SOURCE: Courtesy of NASA, https://www.flickr.com/photos/nasa2explore/52886560902, CC BY-NC-ND-2.0.

There are also efforts to adapt the more standard 3D materials printing infrastructure by engineering microbes that can produce, for example, diverse recyclable plastics on demand (Dissanayake and Jayakody 2021) from in situ resources. Research into how to engineer these organisms to provide materials with better printing and mechanical properties compatible with these space-based 3D printing platforms and that optimally utilize mission-available resources are ongoing (Cestellos-Blanco et al. 2021). Other early research accomplishments linked to additive manufacturing across mission needs include work that has explored how microbes can produce biomined metals (Santomartino et al. 2022) and biosynthesized plastics with mission-defined performance characteristics (Averesch et al. 2021, 2022) and compatible with the 3D manufacturing approaches. Recent progress in additive manufacturing of nonbiological materials such as metals is equally impressive, using lasers and electron beams to melt and deposit metal into structures that carry significant loads and can withstand high pressures and temperatures (Herzog et al. 2016) and thin-layer deposition of polymers, including in LEO conditions. With potential to impact production of propulsion system components or to repair space-deployed components faster and with lower upmass, safe adaptation of additive manufacturing for metals in space environments will require innovation in the coming decade. Such anticipated advances will leverage the physical science data repositories and insights gained from prior years of research on metal solidification in microgravity (Fredriksson 2022) and other features of space environments.

Informatics Platforms and Data Repositories

The Physical Sciences Informatics (PSI) data repository was opened in 2015 and contains most of the original data (including images and videos) from experiments in reduced-gravity environments such as the ISS, space shuttle flights, and free-flyers. PSI also includes data from some related ground-based studies. This open access repository allows for investigators to reanalyze data in new ways. To support these efforts, NASA has an annual funding call for 2-year projects that use the PSI. To date, 42 awards have been funded in the area of materials science, combustion, fluid physics, complex/fluids and soft matter, and informatics.

GeneLab is the first comprehensive space-related -omics database in which users can upload, download, share, store, and analyze spaceflight and corresponding model organism data.[2] It is a tool that has demonstrated the capability to improve information sharing and increase the pace of scientific discovery from space biology experiments. It is a multi-year, multi-phase project that includes a unique data repository and partnerships with investigators performing spaceflight and spaceflight-related experiments, including human data from commercial providers. Four grants have been awarded to GeneLab Analysis Working Group (AWG) members. See Chapter 1 for further information on these and other informatics platforms.

Finding 2-1: Open data and software enable both funded scientists and the broader community.

New Technologies in Biological Research

Massive improvements in sequencing of nucleic acids, together with the emergence of AI and advanced computation tools, now capture whole genomes, associated transcriptomes down to that of individual cells, and epigenomes. The exploration of the meta-genomic and meta-transciptomic space provides pictures of biological communities, informing who is in them, and how they might be interacting. Improvements in gene editing—changing that sequence in non-germ cells—permit rapid testing of gene functions, alterations of expression, or rewriting of genetic elements to ask complex scientific questions previously limited to inefficient iteration or inference.

Simultaneous advancements in instrumentation, sensing, and measurement are powering advances in biological and physical sciences and creating challenges for "big data." Data sets and especially the libraries required to analyze them need to be housed in the cloud because it is too difficult to move terabytes and petabytes over the Internet—or over increasingly long distances associated with space travel. Furthermore, multiple computer

[2] See GeneLab, "Publications," https://genelab.nasa.gov/publications.

processing units (CPUs) are needed to analyze them. Large data sets enable machine learning and artificial intelligence (AI), where AI algorithms can be trained on one data set and tested on another. This has led to the development of new technologies like AI systems that predict, for example, the structure of all proteins based on sequence alone, a "holy grail" for protein science (Baek et al. 2021; Jumper et al. 2021; Tunyasuvunakool et al. 2021) that has led to even newer AI to design new proteins (Watson et al. 2022). AI, especially in combination with multiple sensors, multi- or hyperspectral imaging, and knowledge bases is increasingly used to analyze physical system, plant, or biological performance, with an ultimate promise of enabling machine decision making on specific experiment steps or processes.

Finding 2-2: The increased sophistication of required measurement, sample process, and cellular manipulation in space suggests that the future will require support for space-based large-scale synthesis, molecular/cellular biology, and biological measurement backed by significant data systems.

Finding 2-3: Technology developed to advance scientific inquiry on Earth has advanced rapidly in the past decade. Lagging incorporation into research infrastructure for space environments has the potential to reduce scientific and technical impact of space-based research in the coming decade.

Space-Based Research Platforms

Figure 2-4 illustrates existing and envisioned space-based research platforms, extending from suborbital and LEO to future missions to the lunar surface and Mars. Each platform has constraints and requires both engineering science and technology development, iteratively involving the scientific researchers who will later be users, for years before the first scientific discoveries can be reported. Such platforms developed prior to this decade have enabled remarkable advances in multiple disciplines, as highlighted below.

Biological Sciences

Development of a microbial observatory on the ISS was a high priority in the 2011 decadal survey, *Recapturing a Future for Space Exploration: Life and Physical Sciences Research for a New Era* (NRC 2011). This area of research continues to provide valuable insight into not only the health of the crew but also the health of the vehicle environment. A series of three microbial monitoring investigations (Microbial Tracking-1, -2, and -3) was designed to investigate cataloging and characterizing of potential disease-causing microorganisms onboard the ISS. Traditional methods of monitoring are culture-based, and results are limited to microorganisms that can be grown in the laboratory. The series of experiments employ two methods: culture-based and DNA/RNA-based analysis to measure microorganisms that cannot be cultured in vitro, creating a larger inventory of data to understand the biome in space. In 2019, the Spectrum hardware reached the ISS. Spectrum is a multi-spectral fluorescence imaging system designed for capturing in vivo genetic expression in the low-gravity environment. This hardware is designed to work with Petri plates and can be used for studying any organism (microbes, plants, insects) expressing common fluorescent tags (NASA 2019). Examples of others include specialized reactors that have been developed by the international community, including the ESA's BioRock biomining reactor that allowed exploration of how microbes could extract rare Earth minerals from lunar and martian regolith (Cockell et al. 2020). (See Figure 2-5.)

Finding 2-4: Even modest increases in hardware capability have advanced for characterizing gene expression in organisms, including microbes. However, the next phase requiring more advanced culturing bioprocess control and monitoring will also require new invention and experimental protocols for microbiology.

Plants

Several hardware packages have been designed and are operational on the ISS, including the Apex chamber plant growth facilities designed to be used in the ISS Vegetable Production System (Veggie 2014)

FIGURE 2-4 U.S. development of space research platforms and capabilities. NOTES: Figure is an artistic illustration and generalization of existing and envisioned space research platforms from ground simulation to deep space with capabilities of examining conditions from Earth on way to Mars. It is not intended to be all inclusive. Development of platforms and capabilities is very dynamic. CLPS, commercial lunar payload services; DLR, German Aerospace Center; HALO, Habitation and Logistics Outpost; HLS, Human Landing System; iHab, international habitation module; ISS, International Space Station; MELiSSA, Micro-Ecological Life Support System Alternative; MSSF/KSC, Microgravity Simulation Support Facility at Kennedy Space Center; PPA/KSC, Plant Processing Area at Kennedy Space Center. SOURCE: Composed by Tim Warchocki with planet images courtesy of NASA.

FIGURE 2-5 The Spectrum prototype unit is here being used to expose organisms in a Petri plate to blue excitation lighting. SOURCE: Courtesy of NASA, https://www.flickr.com/photos/nasamarshall/37897775936/in/album-72157667439026641, CC BY-NC 2.0.

for open crop production, updated versions of Biological Research in Canisters (BRIC) including the Petri plate fixation unit BRIC-PDFU and the BRIC-LED, which includes customizable discrete LED lighting with four wavelengths. The most advanced plant chamber is the Advanced Plant Habitat (APH 2017), which is a nearly closed crop system that controls almost all growth parameters. A number of pick-and-eat crops including lettuce, kale, and chili peppers have been grown in space using these capabilities with the culmination of harvest and human consumption. (See Figure 2-6.) These are the first steps in growing crops for long-duration missions.

The BRIC hardware provided profound insights into the mechanism of action of the effects of microgravity on root hair and root tips (Kwon et al. 2015), confirmation of the fundamental hypothesis that survival of plants in the spaceflight environment requires adaptive changes that are both governed and displayed by alterations in gene expression (Paul et al. 2012), insights into the existence of post-transcriptional and post-translational modifications in response to spaceflight (Kruse et al. 2020), and full mapping of spaceflight-induced hypoxic signaling, response, and transcriptional reprogramming in *Arabidopsis* (Angelos et al. 2021; John et al. 2021).

The past decade witnessed the retirement of the European Modular Cultivation System (EMCS) in 2018. EMCS provided an on-orbit centrifuge to allow for an on-orbit 1 *g* control, multiple colored light regiments, rotation at variable speeds to generate multiple fractional *g* levels and imaging of seedlings on orbit. The hardware supported multiple TROPI experiments and Plant Gravity Perception (PGP) experiments assessing phototropic

FIGURE 2-6 The Materials International Space Station Experiment (MISSE) has collected data on the exposure of materials to the space environment. SOURCE: Courtesy of NASA, https://www.flickr.com/photos/nasamarshall/15834368051/in/album-72157667439026641, CC BY-NC 2.0.

responses, photo-/gravi-tropic interactions and mechanisms, and sensitivity of gravity perception (Vandenbrink et al. 2016). The last experiment to fly on EMCS before its decommission was PGP, to examine a range of *g* levels to further explore gravity perception in plants (Wolverton 2022).

Vertebrates

The Rodent Research Facility is a hardware system to carry rodents safely from Earth to the ISS and provide long-term accommodation aboard the station. Previous shuttle missions housed rodents in microgravity for about 10 days. The facility is designed to allow rodents to spend up to 90 days in space, greatly improving the ability to use animal models to study the effects of spaceflight on human health. Rodent spaceflight experiments have contributed significantly to understanding the effects of microgravity on biological processes that are directly relevant to humans in space. Rodent studies provide information of the whole biological system, including the effects of microgravity on the structure and function of the sensory-motor, musculoskeletal, nervous, cardiovascular, reproductive, and immune systems.

Tissue chips or organs on chips are small devices containing living cells in 3D compartmentalization and/or 3D matrices as well as physiological-like features of the microenvironment such as fluid flow, but they are not functional organisms. These formats enable testing of the cells for response stresses, drugs, radiation, and genetic changes—including human tissue cells derived from and specific to each person as needed. Indeed, development of tissue chip–based research in space environments has made remarkable progress in the past decade, and included coordination between NASA and the National Institutes of Health (NIH) National Center for Advancing Translational Sciences (NCATS) for both funding and study design for Tissue Chips in Space that relied in part on ISS access for academic research and also on National Science Foundation (NSF) contributions.[3] Five basic research investigations were funded in 2018 to look at microgravity impacts on diverse topics including immunology, the blood brain barrier, kidney function, musculoskeletal disease, and stem cell biology (ISSNL 2017). Tissue chip technologies allow basic research studies on human tissue-like constructs in microgravity that other systems do not. When validated against existing methods and protocols on Earth and in space environments, tissue chips may also offer an alternative (non-animal, non-astronaut) platform for testing drug efficacy and toxicity. In addition, they help solve the problem of collecting samples from astronaut's bodies, allowing for research with greater potential for design-of-experiment and replicate trials that establish reliable principles under the severe constraints of space-based research time and mass.

Invertebrates/Small Organisms

The Fruit Fly Laboratory is a research facility deployed to space for long-duration studies of fruit flies. Experiments involving the fruit fly can allow scientists to determine how microgravity and other aspects of the space environment affect these insects, delivering valuable information for long-term human spaceflight with special focus on illnesses because about 77 percent of human disease genes have a close match in the fruit fly genome. The Vented Fly Box (VFB) platform houses 15 standard fly vials that contain flies and fly food. The VFB has mesh-covered vent holes that allow airflow to the samples and clear windows on the sides to allow for evaluation. In addition, the VFB contains temperature/humidity data loggers inside the unit.

Physical Sciences

Key resources that enable materials science and soft matter research have been brought online since the last decadal survey (NRC 2011). The ESA Materials Science Laboratory Electromagnetic Levitator (MSL-EML; name later changed to ISS-EML) was launched and brought online in 2014. The ISS-EML enables contactless processing of conductive samples (metals, alloys, and semiconductors) in a high-temperature, ultra-clean environment. This allows for study of heating, melting, (under)cooling, and solidification and concomitant high-quality measurement of thermophysical properties via a pyrometer and two high-speed cameras to capture surface images. Properties that can be so measured include viscosity, surface tension, heat capacity, heat of fusion, emissivity, and electrical conductivity, in addition to studies of nucleation, crystal growth velocity, and phase selection. A flight-identical ground model was installed at DLR in Cologne, Germany, for experimental preparation and comparison of terrestrial versus space measurements (Glaubitz et al. 2015). The JAXA Electrostatic Levitation Furnace (ELF) was also launched to the ISS the following year (Tamaru et al. 2018).

These levitation resources have been heavily used since then, with more than 100 publications published based on ISS-EML and ISS-ELF experiments to date. The Materials International Space Station Experiment (MISSE), which began in 2001, continues to collect data about how exposure to the LEO space environment affects material properties and performance. MISSE 1–8 missions were placed outside the ISS for 1–4 years in a Passive Experiment Container (PEC), while MISSE-9 and higher have been installed on the MISSE flight facility (MISSE-FF), a permanent materials science platform. MISSE-FF was launched in 2018 (De Groh and Banks 2021). The Light Microscopy Module (LMM) was being tested on station in 2011 and was upgraded during its on-orbit life to include 3D views. NASA used it for both biological and physical sciences (Giannone 2017). (See Figure 2-6.) LMM returned to Earth in 2023.[4]

---

[3] National Institutes of Health. "2017 Tissue Chips in Space Projects," updated July 18, 2022, https://ncats.nih.gov/tissuechip/projects/space2017.

[4] The text was modified after the release to sponsor to update the description of the LMM.

Multipurpose Facilities

Facilities also exist for multiple research purposes. Some, like the LMM and Spectrum, have been mentioned above. In addition to these, the BioCell (flown 2014) was developed to allow researchers to perform complex laboratory-fidelity cell cultures on ISS. The plate habitat (PHAB) complements the BioCell by providing a secondary containment capability. This system supports research on mammalian cells/tissues, small organisms, yeast, bacteria, fungi, algae, and biofilm formation studies. The Advanced Space Experiment Processor (ASEP) is a culture-based system with cassettes that can be operated independently and range in temperature from 4°C to 40°C. This system is designed for cell cultures but can be expanded to other applications. The Space Automated Bioproduct Laboratory (SABL) is another advanced incubator facility that can operate from –5°C to 43°C.

The Multi-Use Variable-G Platform (MVP) was delivered to the ISS in 2018. It is a commercial testbed for centrifuge-based science. Because gravity determines so much of a live organism's behavior and growth, centrifuge-based experiments have long been a part of biological investigations in space. While the pull of Earth's gravity makes this type of investigation difficult at home, the ISS's microgravity environment makes it the perfect place for fractional gravity experimentation. MVP greatly expands that testing capability for the ISS. (See Figure 2-7.) Owing to its size and capability, it can accommodate a wide variety of samples, including cultured cells, *Drosophila*, plants, *C. elegans*, aquatics, tissue chips, bacteria, and organoids. JAXA's Cell Biology Equipment Facility (CBEF)

FIGURE 2-7 NASA astronaut and Expedition 66 Flight Engineer Mark Vande Hei sets up components for the MVP-Plant-01 space botany study and nourishes *Arabidopsis* plants grown on Petri plates. That investigation is exploring how plant molecular mechanisms and regulatory networks adapt to the weightless environment of space. SOURCE: Courtesy of NASA, https://www.flickr.com/photos/nasa2explore/51858491771, CC BY-NC-ND 2.0.

was instrumental in providing 1 *g* controlled and partial gravity experiments over the past decade. This facility was much larger than the MVP, requiring nearly an entire express rack. Other small systems are being developed, including Mobile Spacelab and Space Taxi, a centrifuge facility that can be activated minutes after launch.

As cube-satellites (CubeSats) have become more popular, the concept has morphed into "cube labs." A standard cube is 10 cm^3 (1 U). Standard sizes are 1, 2, 4, 6, and 8 U and can be configured to meet the experimenter's design. The cube(s) are then installed in another piece of standard hardware supplied by the commercial vendor. These systems include Tangolab, Nanode, and YURI. The small size and configurability of these systems work well with the smaller launch vehicles in the post-space shuttle era. Cube designs may also be transferred to true CubeSats (free-flyers) if the samples do not need to be returned to Earth.

Freezer facilities have improved dramatically over the past decade. Originally, there were no freezer or cold stowage spaces on the ISS. Currently, there are three MELFI units (–80°C, –26°C, +4°C), four GLACIER units (–160°C to +4°C), four POLAR units (–80°C to +4°C), and four MERLIN units that can also be used as incubators (–20°C to +48.5°C). The MELFI, GLACIER, and MERLIN units can be used in an express rack or on launch/return vehicles. There is also a glove box freezer for rapidly freezing samples with a contact temperature of –185°C. In addition to these assets, there are many passive freezer and cold storage options. (See Figure 2-8.)

As more commercial partners and operators have moved into space, the number of hardware designs and concepts has increased dramatically; not all of them have been mentioned here, owing to the challenges of tracking such private sector changes comprehensively. This is a trend that is expected to accelerate over the next decade as more commercial partners are being solicited to provide ISS services, develop new space stations, and provide lunar surface or Mars transport access in addition to the launch, return, and astronaut services that they are providing now. At the time of this decadal survey report writing, two entirely commercial, orbital astronaut (i.e., crewed) missions have launched from the United States with no space agency astronauts aboard.

FIGURE 2-8 European Space Agency astronaut Andre Kuipers, Expedition 30 flight engineer, prepares to insert biological samples in the Minus Eighty Laboratory Freezer for ISS (MELFI-1) in the Kibo laboratory of the International Space Station. SOURCE: Courtesy of NASA, https://www.flickr.com/photos/nasa2explore/6819319112, CC BY-NC 2.0.

Quantum Science

In July 2018, the Cold Atom Laboratory (CAL) became the first facility to produce the "fifth state of matter," called a Bose-Einstein condensate (BEC), in Earth orbit. Applications of BECs include atom lasers, atomic clocks, and sensors with high sensitivity. This fifth state of matter has been studied intensely for more than 25 years, resulting in Nobel prize-awarded understanding of the fundamentals of atomic interactions and applications that may enable communications and travel in space environments (Georgescu 2020). CAL, a fundamental physics facility, cools atoms down to ultracold temperatures in order to study their basic physical properties in ways that would not be possible on Earth, advancing basic understanding as well as such applications. Currently, upgrades are in progress, including the use of augmented reality.

Artificial Intelligence and Data Science

Artificial intelligence (AI) algorithms allow processing of large streams of data and can enable decision making in situ with reduced communications to Earth. AI-based approaches can extract and transmit only the valuable information from a large data stream, reducing the demand on limited or delayed communications. These are valuable capabilities as distance from Earth increases and crew availability decreases. On Earth, plans are under way for GeneLab to utilize machine learning algorithms and AI to maximize the usage of the data. Focus will be on modeling and extrapolation to human risks. The initial step will be to provide a portal proposing mechanistic models describing known health risks (cancer, muscle atrophy, bone loss, etc.). Development of tools like these enable assessment of how well GeneLab -omics data and data from other databases can inform these models and test or develop alternative models.

These engineering and technology capabilities and the persistence of the space research community form the basis for considering the biological and physical sciences to be conducted as part of the BPS research portfolio in the coming decade.

BIOLOGICAL SCIENCES

The major motivations for understanding biological processes in space is the fact that humans are exploring space and the fundamental question of whether life is limited to Earth. Any effect of space environment on molecular biology, cellular biology, physiology, or biochemistry has the potential for influencing astronaut health and safety. In addition, as humans travel to and live in space, terrestrial and human microbiomes travel with them. Biology is also part of effective life support systems in space travel. Therefore, the biological sciences in space span a wide range of microbes, animals, and plants—all in service of effective space exploration and habitation. Non-human biology also provides excellent experimental systems to define and understand fundamental biological processes that are affected by spaceflight. Such space-based research in non-human systems and/or with human cells in engineered constructs provide keen insights for terrestrial applications. Over time, these Earth-based benefits of space-based research may include spaceflight as an accelerated model of aging, a unique stressor for agricultural research or lever for agricultural yield, and environments with specific advantages for creating and studying 3D biofidelic constructs for discovery of better therapies.

Biological sciences in space are also motivated by the recognition that many signals and processes within biology are affected by gravity in ways that cannot be understood while gravity is influencing the system. Experiments in space allow deeper understanding of the roles of gravity and weak physical signals in biological systems.

The interplay between the fundamental biological processes revealed by spaceflight and the applications of biological processes to enhance space exploration fuels the BPS portfolio, deriving data from appropriate model systems to inform those applications. The roles and responsibilities of BPS within the spaceflight sciences at NASA have evolved, especially with the move of BPS from the Human Exploration and Operations Mission Directorate to the Science Mission Directorate. (See Chapter 7.) Throughout that evolution, BPS has maintained a strong interaction with the biological sciences at the Human Research Program (HRP) within NASA. That relationship continues to evolve, with HRP focusing ever more directly on the science of human health in space, and with

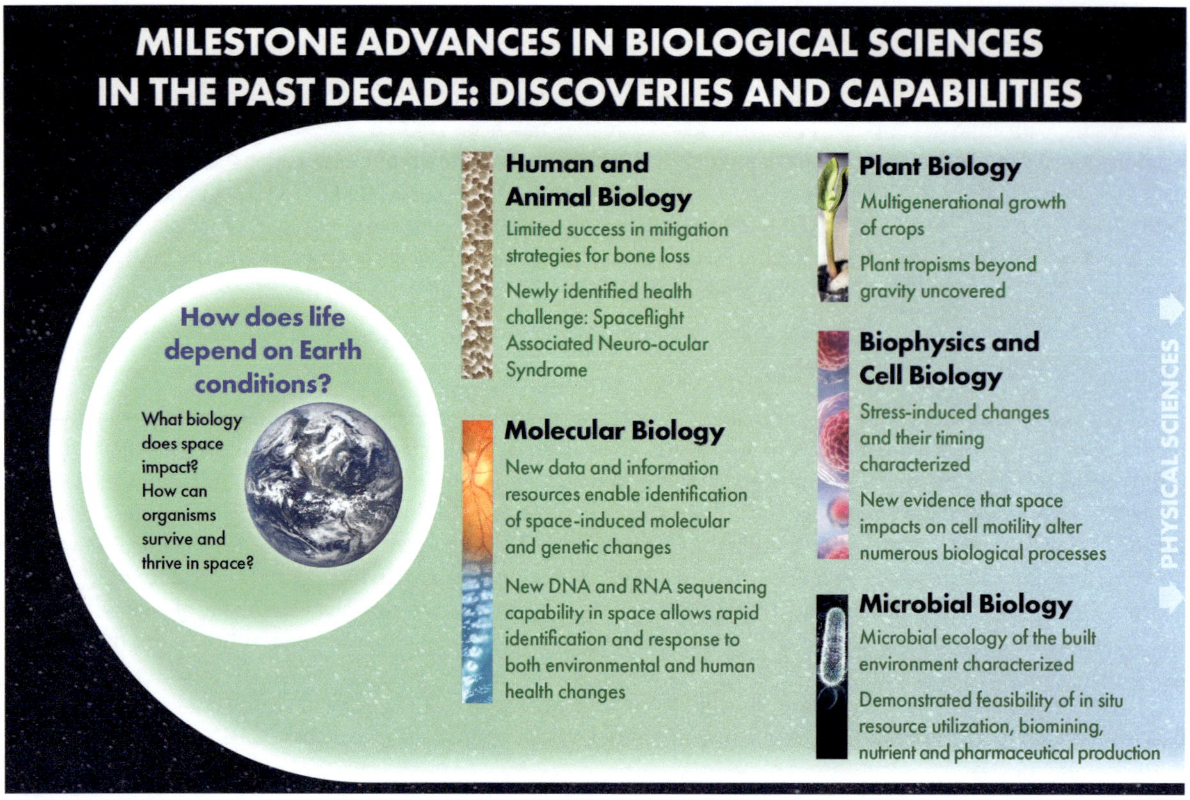

FIGURE 2-9 Milestone advances in biological sciences in the past decade. SOURCES: Composed by Tim Warchocki. Image credits: *Earth*: NASA/DSCOVR EPIC. *Osteoporosis*: Shutterstock/Crevis. *Human eye*: Shutterstock/Left Handed Photography. *Double helix*: Shutterstock/Vector/Garry Killian. *Plant*: NASA/ISS. *Cells*: Shutterstock/Anusorn Nakdee. *Virus*: Shutterstock/Crystal Light.

BPS focusing more on the fundamental biology of model systems to inform the basic science that impacts human health and the biology of all systems that are part of traveling in space.

The current state of biological sciences in space therefore includes strong references to human biology in addition to the other biology that is part of the exploration portfolio. (See Figure 2-9.)

Human and Animal Biology

This section is organized by physiological systems for clarity; however, multi-system research is a growing area that could be incorporated to an even greater degree in the coming decade. Some key topics that are also relevant to NASA's Human Research Program are presented to frame some of the fundamental biology functions and advances in knowledge over the prior decade.

Musculoskeletal Integrity

By 2011, the rapid and compartment-specific bone loss incurred during longer-duration LEO missions was fairly well defined (Vico and Hargens 2018). Yet undefined were the impacts on skeletal integrity of multiple missions, partial gravity environments, and the addition of space-relevant radiation. Much new information has emerged over the past decade on specific mechanisms for the elevated resorptive activity observed in human crew members, which may provide targets for specific countermeasures (Juhl et al. 2021). (See Box 2-1.)

BOX 2-1
Is the Moon's Gravitational Field Strong Enough to Maintain Muscle Strength and Bone Mass?

After over a decade of study, the answer is no. The 1/6 *g* of the lunar environment does not provide enough mechanical loading to mitigate the bone loss seen in simulated microgravity. Indeed, graded reductions of muscle and bone mass and muscle strength are observed with stepwise reductions in weight bearing (Mortreux et al. 2022). In vivo animal research will be critical to predicting adaptations to partial gravity environments on lunar or planetary surfaces, because the one attempt at modeling this in humans using head-up-tilt bed rest (Cavanaugh et al. 2013) did not lead to further investigations. A major advance in bone loss countermeasures was achieved in an ISS study that increased nutritional intake of energy and vitamin D while astronauts regularly performed high-intensity resistance training; however, bone resorption biomarkers remained elevated (Smith et al. 2012). Notably, a trial testing the addition of an anti-resorptive medication (alendronate) to the high-intensity resistance training regime revealed better mitigation of bone mass density loss and suppression of resorptive activity (Leblanc et al. 2013). Longer-acting anti-resorptive agents (e.g., zolendronate) and newer anabolic pharmaceutical agents are under investigation in rodent models.

Pharmacological inhibition of myostatin, which promotes muscle breakdown, demonstrated prevention of muscle mass and grip strength declines in mice exposed to 6 weeks of LEO (Smith et al. 2020). If activin A signaling is also blocked along with myostatin signaling, both bone and skeletal muscle loss was prevented in mice flown onboard the ISS for 33 days (Lee et al. 2020), and there is some evidence for gender- or sex-specific differences in skeletal muscle adaptations to the space environment (Shackelford et al. 2004; Trappe et al. 2009) and during recovery back on Earth (Clark and Manini 2012). However, only minor differences have been observed between male and female rodents exposed to hindlimb unloading and no definitive differences between male and female human space flyers (Ploutz-Snyder et al. 2014).

Morphological changes occur in muscle fibers of mice 9 weeks after exposure to ~0.3 Gy 56Fe ions, indicative of muscle remodeling (fewer smaller fibers) and regeneration (threefold more centrally located nuclei) (Bandstra et al. 2009), and exposing rodents to space-relevant doses of radiation illustrated a rapid acceleration of bone resorption promoting bone loss (Vico and Hargens 2018). Acute exposures to high-energy ion species available at NASA's Space Radiation Laboratory accelerates the bone loss observed with simulated microgravity (Alwood et al. 2010) and partial gravity.

Cardiovascular

The major findings and research tools to help define these cardiac and cardiovascular pathologies that have been developed since the last decadal survey (NRC 2011) are the NASA Twins Study (Garrett-Bakelman et al. 2019), use of tissue organoids and iPSC-derived tissues in long-term spaceflight on the ISS (Wnorowski et al. 2019), and computational modeling approaches based on patient specific metrics (Gallo et al. 2020). (See Box 2-2.)

The ability to identify at-risk individuals who are currently asymptomatic is relevant for both NASA and public health. This opens compelling questions about accumulation of epigenetic modifications to DNA. The Twins Study showed that there were changes, but most reverted on return to Earth. Development of insulin resistance in spaceflight has been demonstrated in astronauts and in mouse models, and this may cause cardiovascular diseases while in-flight (Hughson et al. 2016). Sex-specific differences were also measured in mice flown in space, with an increase in arterial stiffness in male mice, but reduced stiffness in female mice. These hemodynamic concerns have been observed in astronauts on the ISS, with stagnant or reverse flow in the internal jugular vein and an occlusive

BOX 2-2
From Twins to Tissue Chips, the Past Decade Increased Knowledge About the Effects of Prolonged Flight on Organ Systems and Fluid Dynamics

The NASA Twins Study included 10 U.S. research teams and the Kelly brothers. While admittedly limited to analysis of only one male astronaut's extended spaceflight (Scott) in comparison to his male twin's Earth-based co-existence (Mark), this ambitious study of human subjects in space both confirmed some preliminary findings developed from prior studies and also uncovered surprising observations motivating future studies at the space science research bench. Key results reported in 2019 included the following (Garrett-Bakelman et al. 2019):

1. Gene expression (e.g., expression of proteins encoded by DNA) changed in Scott while in space (low Earth orbit [LEO] on the International Space Station [ISS]), and a small percentage did not return to baseline within 6 months of his return to Earth. Why and whether this can be repeated in other humans or in tissue/organ chip studies remain open questions.
2. Biomarkers of aging (telomeres in white blood cells) in Scott increased in length while in space (LEO of the ISS), decreased upon his return to Earth, and then returned to average length half a year after his return. These surprising dynamics are important to better understand if repeatable, as they relate to stability of the genome and may differ among individuals.
3. An Earth-manufactured vaccine (against influenza) functioned in Scott when administered in space as would be expected on Earth. This implies that immune system response was preserved during extended spaceflight for this particular individual, vaccine, and space environment.

Concurrent with the NASA Twins Study, rapid progress was achieved in re-creating aspects of tissue or organ function in engineered, miniaturized devices called tissue chips. These devices typically include microfluidics and biological cells and facilitate the key feature of repeated scientific observations (technical and biological replicate experiments under controlled conditions) that would be impractical with humans or animals. Tissue chips in space over that period, as coordinated by NASA and the National Institutes of Health (NIH 2018), quantified effects of microgravity on drug responses in heart-like tissue and assessed dysfunctions in heart-like tissue grown from human-induced pluripotent stem cells. Such advances also support the NIH position that, when sufficiently developed and proven, non-animal model approaches can reduce the need for animals in research related to basic science and therapeutic drug development.

and a partial thrombus identified in other crew members (Marshall-Goebel et al. 2019), sparking the concern for undetectable cardiovascular disease or for the acceleration of disease while in-flight.

Furthermore, it is now well established that exposure to high-dose radiation accelerates atherosclerosis on Earth, and is thus anticipated as a risk during long-duration spaceflight. Low-dose, high-energy, charged-particle irradiation reduced ejection fraction and fractional shortening in mice exposed to ^{16}O particles,[5] but these effects were not seen after preexposure to γ-rays or protons (Seawright et al. 2019). Aortic stiffness (Soucy et al. 2011) and accelerated atherosclerosis (Yu et al. 2011) was also observed in mice space radiation studies. Other factors affected were DNA oxidation, myocardial fibrosis, and modified cardiac function (Yan et al. 2014). Reactive oxygen species (ROS) generation and activity in space are also elevated during spaceflight, and these molecular mechanisms of damage to cardiac and vascular cells and structures can be determined molecularly.

[5] Radiation damage is caused by exposure to high-energy particles. ^{16}O or 16O is the most abundant particle in galactic cosmic rays, as oxygen's most common isotope with mass number of 16. See Kiffer et al. (2020).

Pulmonary

Delivery of ambient air and of arterial blood to various zones of the lung changes in microgravity, but the critical ventilation-to-perfusion ratios remain constant, ensuring efficient gas exchange (Prisk 2014). Earlier concerns about a reduced ventilatory response to hypoxia while in microgravity were resolved when preflight testing on crew members revealed that the same response is produced in 1 *g* in the supine posture (Prisk 2014). Important progress has been made on testing the toxicity of inhaled lunar dust from the Moon. Adult rats exposed to lunar dust (generated from lunar regolith) exhibited dose-dependent increases in pulmonary lesions and biomarkers of toxicity but to a lesser degree than those exposed to two standard toxicity-reference dusts. Importantly, dust toxicity was not related to the oxidative stress-generating capacity produced by those dusts in vitro but rather to the persistent accumulation of neutrophils in exposed lung tissue (Lam et al. 2022).

Fluid Shifts

The NASA Fluid Shifts Study[6] is assessing effects of microgravity on ocular or eye function/physiology and intracranial pressure (ICP). Given that ~60 percent of astronauts experience adverse changes in visual acuity after 6 months in microgravity (Mader et al. 2011), this is a major concern. Work by NASA's Human Research Program (HRP) does not address changes at the molecular and cellular level; instead, such studies focus on functional risks to humans with noninvasive techniques to quantify blood flow, ocular pressure, and intracranial pressure changes. (See Appendix D for discussion of NASA's delineation and the BPS community's perceived distinctions between these two research funding divisions of the agency.)

To investigate changes in eye physiology, structure, and function at cellular level, 12 male mice were exposed to microgravity in the JAXA Habitat Cage Unit (HCU) for 35 days on the ISS (Mao et al. 2018). The study included centrifugal 1 *g* in addition to HCU and vivarium ground controls. Microgravity resulted in significant apoptosis in the retina vascular endothelial cells (Mao et al. 2018).

While all of these deleterious effects point out obvious concerns for the long-term health ocular effects induced by microgravity, interspecies differences at the molecular and cellular level remain unanswered (Volland et al. 2015). Compared to other human organs, the eye is quite neglected in the research funded by the NIH NCATS tissue chips initiatives (Onyak et al. 2022). (See Box 2-2.) Given the criticality of healthy ocular function to all aspects of astronaut mission effectiveness in missions of extended duration in both LEO and deep space, understanding the basic biology is necessitated to devise interventions and therapeutic options to mitigate acute and long-term adverse effects.

Neuroscience

The space environment poses challenges to all mammalian biological systems, including the nervous system. Because the absence of gravity unloads the vestibular otolith organs, they are no longer stimulated as they would be on Earth. The brain needs to adapt to this altered sensory input during spaceflight, and then readapt to the presence of gravity on return. As a result, approximately 70 percent of astronauts experience impaired balance, locomotion, gaze control, dynamic visual acuity, eye–head–hand coordination, and/or motion sickness within the first days of spaceflight (Lackner and Dizio 2006; Souvestre et al. 2008). Post-flight disturbances in perception, spatial orientation, posture, gait, and eye–head are also commonly reported and provided early evidence that the nervous system can adapt to some aspects of the space environment. These earlier reports involved relatively short-duration missions. Elevated cerebrospinal fluid pressure has been reported in some astronauts following spaceflight, owing to increased intracranial pressure. The neurophysiological effects of this fluid shift, while not fully understood, are known to affect visual performance by contributing to Spaceflight Associated Neuro-ocular Syndrome (SANS) (Otto 2016).

Space-based experiments have further established that altered gravity produces structural and functional changes at multiple stages of vestibular processing, spanning from the otoconia and hair cells of the otolith sensory

---

[6] NASA, "Fluid Shifts Before, During and After Prolonged Space Flight and Their Association with Intracranial Pressure and Visual Impairment," https://www.nasa.gov/mission_pages/station/research/experiments/explorer/Investigation.html?#id=1126.

organs to the Purkinje cells of the vestibular cerebellum (Carriot et al. 2021). The brain accounts for these changes by reweighting extra-vestibular information to substitute for the altered and less reliable vestibular input (Hupfeld et al. 2022). Ground-based studies have shown that extra-vestibular (i.e., somatosensory and motor) signals will substitute for unreliable vestibular input (Carriot et al. 2015). Overall, significant sensory reweighting occurs within 24 hours and becomes relatively steady within 4–5 days. Notably, it is during this reweighting-related intensive time window that astronauts display motion sickness (Mulavara et al. 2012).

Molecular, Cellular, and Blood–bBrain Barrier Impacts of the Space Environment

Many neural changes are key to human ability to function in space (Hupfeld et al. 2021). Rodent studies show that prolonged exposure to space produces changes in the structure of vestibular apparatus and changes in the vestibular cortex (Kharlamova et al. 2021) as well as changes in gene expression of neurotransmitter systems and neurotrophic factors in specific regions of the brain (Popova et al. 2020). Male mice that spent 35 days in the space environment had histological evidence of damage to the blood–brain barrier (BBB) and increased expression of aquaporin4 compared to ground-control housing-matched mice (Mao et al. 2020).

Stress and Immunology

The mammalian immune system is molded by environment, gut microbiome, and pathogen exposure; it develops throughout a lifetime and is unique to every individual. One of the most consistent observations in space biology is that launch, landing, and the space environment create changes in the immune system resulting in suboptimal or imbalanced function that can persist after returning to Earth, including the reactivation of latent viruses (Pierson et al. 2005; Mehta et al. 2014, 2018). Although rarely associated with clinical symptoms, latent viral reactivation is associated with other immune and stress hormonal measures and could be considered a benchmark of space-associated immune dysregulation (Crucian et al. 2014; Mann et al. 2019). Viral shedding can be detected in saliva samples collected from astronauts during their mission on the ISS, suggesting that changes to in vivo immunity are not owing to reentry, and viral reactivation can occur for up to 6 months (Crucian et al. 2015; Mehta et al. 2014; Voorhies et al. 2019).

Both long- and short-duration missions dysregulate many features of innate and acquired immunity, some of which persist after return to Earth (Crucian et al. 2015). Currently, stressors associated with the space environment are difficult to model (Britten et al. 2021). Rodent studies, using ground-based models, revealed that single stressor versus complex stressor exposures and radiation produce complex patterns of changes in in vivo immune function (Tesei et al. 2022; Turner et al. 2020; Villacampa et al. 2021).

> Finding 2-5: Ground-based experiments are important components of impactful biological sciences studies ranging from vascular to neuro- to immune systems, as comparisons of conditions and more reproducible baselines of genetic diversity.

The risk of the emergence of neoplastic cells and potential cancer is also increased when immune system surveillance is diminished (Bigley et al. 2019). The impairment of NK cell (a specific immune cell type) function was more prominent in astronauts who underwent their first flight compared to experienced astronauts. Simulated microgravity changes tumor cell behavior and metabolism, leading to the acquisition of an aggressive and metastatic stem cell–like phenotype (Masini et al. 2022). Although both NK cell function and potential tumor cell behavior are changed by space-associated stressors, there is no evidence that astronauts have a high risk of developing cancer.

Skin and Wound Healing

Wound healing is a complex process that consists of hemostasis, proinflammation and inflammation, angiogenesis, proliferation and migration, contraction, and remodeling. This process involves intricate interactions of immune cells, soluble factors, and skin cells, the coordination of which are disrupted by microgravity, which negatively impacts skin wound healing. During simulated microgravity, skin cells showed enhanced proliferation and viability in epidermal stem cells and enhanced growth rate and migration in keratinocytes (Bacci and Bani 2022). Conversely, in vivo simulated microgravity showed that prolonged mechanical unloading results in delayed and impaired dermal wound healing. These conflicting results have few on-orbit studies to resolve the conflict or identify mechanisms. Microgravity-induced hypovolemia has been associated with thrombocytopenia, a deficiency of sufficient platelets to adequately initiate the coagulation cascade, and a decrease in platelets contributes to poor wound-healing outcomes in microgravity. In leeches, platelet-rich plasma was able to counteract wound-healing delays induced by simulated microgravity (Cialdai et al. 2020).

In addition to dermal cells, immune cells play a role in wound healing, and it has been established that T cells are functionally impaired by microgravity (Morrow 2006; Pippia et al. 1996; Simons et al. 2006; Walther et al. 1998). Epidermal T cells in acute wounds have been shown to participate in wound healing, while epidermal T cells in chronic wounds are in an unresponsive state (Toulon et al. 2009). Microgravity-induced disruption of epidermal T cells may contribute to the delays in wound healing that are observed and are also an avenue of investigation for wound-healing studies in space. Considering the many factors that contribute to wound healing, it seems likely that the most relevant data will come from wound-healing studies in a whole organism.

Metabolic Energy Balance

Activity, thermoregulation, and metabolism, major elements of energy balance in animals, are all impacted by the spaceflight environment. Analysis of mouse behavior in the rodent research habitats has revealed dramatic differences in behavior, including running behavior in-flight versus on the ground (Ronca et al. 2019). Recent studies in rodents have highlighted similarities between spaceflight and cold exposure (Wong et al. 2021), which can have implications for bone and muscle as well as metabolic homeostasis. A recurrent effect of spaceflight has been insulin resistance, including prediabetes (Hughson et al. 2016). Mitochondrial dysfunction may represent a hub connecting all of these alterations in energy balance (da Silveira et al. 2020).

The short-lived *D. Melanogaster* and *C. elegans* models are well suited to understanding the role of genes in aging and longevity (Kenyon 2010). Both are excellent models for elucidating the role of metabolism in aging (Dabrowska et al. 2022; Parkhitko et al. 2020; Piper et al. 2018) and have been used to study the possible benefits of food-delivered antioxidants as measures for improving longevity and aging on Earth (Yi et al. 2021). Such studies have the potential to identify countermeasures to prevent space-related cellular damage and explore the "aging" effects of the space environment, but to date such studies have been very limited in number and scope.

Studies on roundworms in microgravity suggest that context is important. Some tissues, such as muscle tissue, have shown early aging effects reflected in changed gene expression in microgravity. However, the systemic down-regulation of metabolic genes observed in space has been shown to increase lifespan of worms on Earth (Honda et al. 2014). These relationships are incompletely understood, and studies in microgravity in these simpler model systems will help to identify the fundamental cellular components involved both in space and back on Earth.

Reproduction

Longer-duration missions with more humans and other organisms motivate understanding of not just aging and genetic variation but also reproduction and thence the multi-generational genetic and epigenetic effects of extended space missions. For over a century, Drosophila has been used as a model organism to understand how genes work together to create and maintain all life-forms, chiefly because the short life span of the fly (months) facilitates studies on the genes involved in reproduction and aging (Holtze et al. 2021). Now that the fly genome has been sequenced, it is known that the fruit fly shares roughly 75 percent of disease-causing genes with humans, permitting us to use this model to identify fundamental cellular processes that contribute to disease. A key challenge to space exploration is the effects of radiation on genes that maintain life-forms and ultimately on those that are passed on to progeny. Drosophila was the first organism to be flown in space (in a 1947 non-orbital rocket mission) (Royal Museums Greenwich 2023), in part thanks to its small size, and is well suited to studies on the mutational effects of radiation. There are short-term reductions in sperm motility in flies reared onboard the ISS (Ogneva et al. 2022). In simulated microgravity, increased motility has been observed (Ogneva et al. 2020) along with increases in respiration, cytoskeletal mRNA, and proteins, likely the result of epigenetic gene regulation (Ogneva and Usik 2021; Usik et al. 2021).

In mammals, reproductive research in the past decade has examined the effects of the space environment on fertility, but not other aspects of reproduction (Mishra and Luderer 2019), but there is a paucity of studies examining the effects of cosmic radiation, microgravity, or other stressors of spaceflight on gonadal function and fertility. Not only is this of critical concern for female astronauts who often delay parity until post-flight and have increasing maternal age at first flight, but it is also important to a complete understanding of fecundity in the cosmic environment in general (Steller et al. 2020). Female mice on-orbit for 37 days experience estrous cycles, essential for fertility (Nguyen et al. 2021). Sperm harvested from male mice, after 35 days on orbit (two spaceflight groups: one experiencing microgravity and another artificial gravity at 1 *g* via centrifugation while on the ISS) and subsequent return to Earth, were successful in fertilizing and siring healthy offspring and, further, that offspring appeared fecund or fertile (Matsumura et al. 2019).

Finding 2-6: Complex biological processes from wound healing to reproduction are known to be moderated by the space environment, in organisms from fruit flies to human. Current limitations in milestone studies are associated with access to observation timelines commensurate with those longer biological processes.

Plant Biology

A strong foundational understanding of the underlying cellular and molecular mechanisms that shape the spaceflight phenotypes of plants now exists, from the model organism *Arabidopsis thaliana* to various crop species. While personnel onboard the ISS enjoyed the fresh greens provided by Veggie, NASA's Vegetable Production System, model plant experiments continued to decipher the microgravity response data to aid in modification of plants and growth practices to enable long-term plant production to support human exploration. (See Figure 2-10.)

To achieve their research objectives, scientists have used combinations of -omics technologies with wild-type, mutant, and transgenic lines of plant species. Experiments have been conducted in flight (ISS, suborbital, balloon) and exploiting SMG (clinorotation) and centrifugal devices that can provide fractional levels of gravity (variable-*g* platform). Furthermore, the spaceflight environment has provided a way to study phototropic responses unconfounded by the gravitropic response. This decoupling of tropisms (movements) has allowed for the novel observation that red-light sensing, which was known in older plant lineages, is masked by normal 1 *g* conditions in more recently evolved lineages.

Transcriptomics/Proteomics of Spaceflight

The molecular landscape of plants is altered dramatically in spaceflight. The impacts are apparent in both the transcriptome and proteome (Kruse et al. 2020) and vary according to tissue (Paul et al. 2013), ecotype (Choi et al. 2019), genotype (Paul et al. 2017; Zupanska et al. 2017), and developmental age (Beisel et al. 2019). Post-transcriptional regulation and post-translational modification patterns are also altered (Beisel et al. 2019; Kruse et al. 2020), and heritable epigenetic changes have been observed as well (Xu et al. 2021). Among the most common trends observed in spaceflight is an enrichment in response to reactive oxygen species (ROS) and oxidative stress (Choi et al. 2019; Kruse et al. 2020; Paul et al. 2017; Sugimoto et al. 2014). Proteomic data revealed that nearly 80 percent of the membrane proteins differentially abundant in space showed greater prevalence in their oxidized form, and 90 percent of the differentially abundant soluble proteins showed decreased oxidation in response to spaceflight (Kruse et al. 2020). Pathogen response is another recurring ontology term in -omics data analyses (Correll et al. 2013; Paul et al. 2012, 2013; Zhang et al. 2015). This may be an effect of disruption of pathogen-sensing mechanisms involving the cell wall, plasma membrane, and cytoskeleton (Paul et al. 2012).

Other major molecular signal players in plant spaceflight responses include genes/proteins related to auxin biology.[7] Space-flown plants have thinner, less rigid cell walls (Hoson and Wakabayashi 2015). So, it is not

[7] See Ferl et al. (2015), Kamada et al. (2020), Kruse et al. (2020), Mazars et al. (2014), and Zupanska et al. (2017).

FIGURE 2-10 A zinnia plant "pillow" floats through the U.S. Destiny Laboratory onboard the International Space Station (ISS). The zinnias are part of the flowering crop experiment that began on November 16, 2015, when NASA astronaut Kjell Lindgren activated the ISS Vegetable Production System (Veggie) and its rooting pillows containing zinnia seeds. SOURCE: Courtesy of NASA, https://www.flickr.com/photos/nasa2explore/24566046491, CC BY-NC 2.0.

surprising that evidence of cell wall biosynthesis and remodeling/reorganization has been a clear trend among many studies.[8] Some class III peroxidases mediate cell wall dynamics through ROS concentrations (Francoz et al. 2015), which raises the possibility of cell wall responses being tied to ROS trends observed in spaceflight (Sugimoto et al. 2014). These trends hold even when considering multiple hardware meta-analyses.

Numerous spaceflight experiments are conducted using etiolated seedlings owing to hardware constraints (Barker et al. 2020; Herranz et al. 2019; Kruse et al. 2020; Villacampa et al. 2021). However, the ROS-specific subcategory of "high early light" (Choi et al. 2019) was similarly observed across several microgravity data sets (Barker et al. 2020). Most notable in this trend is its appearance in the etiolated *Arabidopsis* seedings grown in the BRIC hardware onboard ISS (Kruse et al. 2020), as well as under blue-light treatment (Herranz et al. 2019) and red-light treatment (Villacampa et al. 2021). The presence of light-related differential expression under microgravity (but not under low gravity or 1 *g*) may suggest transcriptional dysregulation when the plant lacks the directional cue of gravity.

Concurrent proteomic studies that evaluated the same plant tissues are generally in concordance with what was suggested by the transcriptomic data sets with additional insights, such as changes in post-translational modifications and a role for calcium ion signaling and phosphorylation in spaceflight. (See Figure 2-11.)

[8] See Choi et al. (2019), Correll et al. (2013), Ferl et al. (2015), Johnson et al. (2017), Kruse et al. (2020), Kwon et al. (2015), and Paul et al. (2013, 2017).

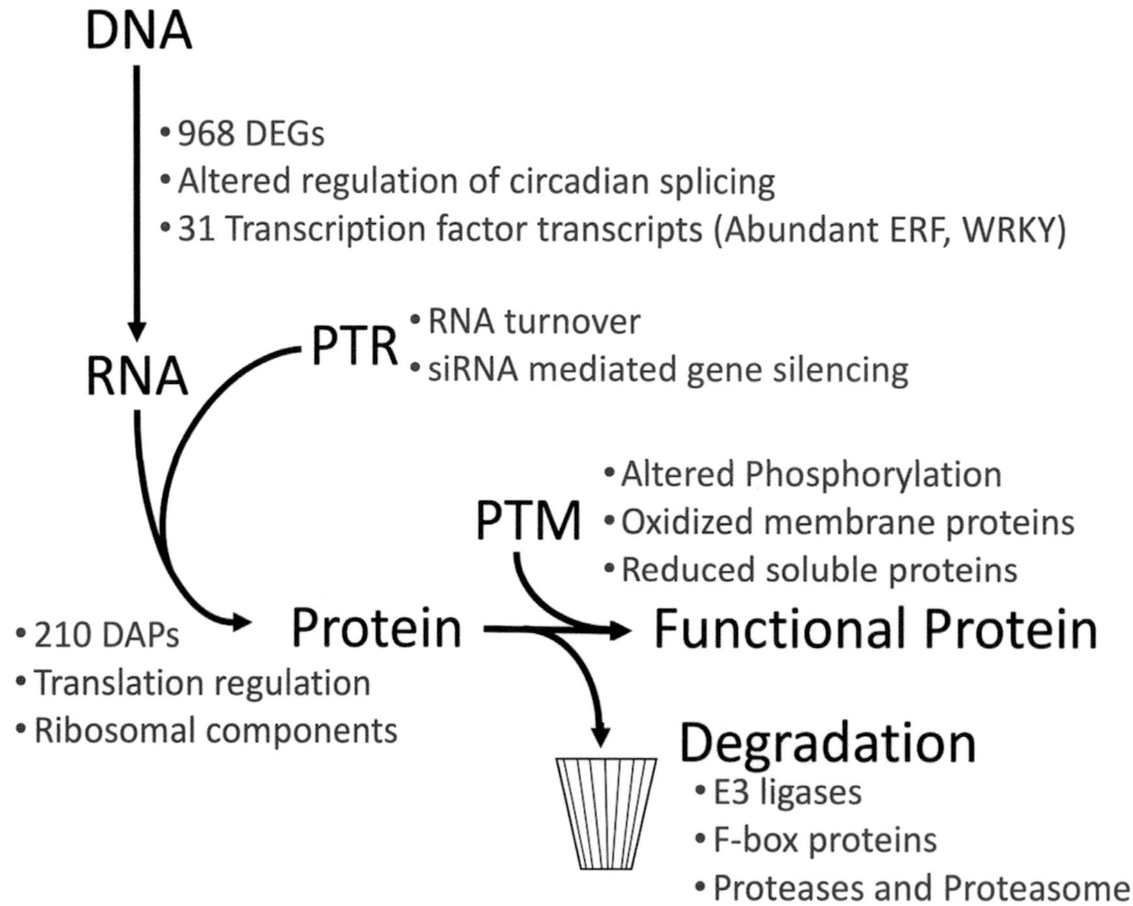

FIGURE 2-11 Regulatory dynamics induced by adaptation to spaceflight. NOTES: Data as indicated by the transcript (RNA), protein and post-translational modification (PTM) data sets. Transcript expression profiles were characterized by RNA-seq and protein expression profiles by iTRAQ LC-MS/MS. Gene ontology analysis identified PTR, PTM, and degradation candidates. DG, differentially expressed gene; DAP, differentially abundant protein; PTR, post-transcriptional regulation. SOURCE: From Kruse et al. (2020), https://doi.org/10.1186/s12870-020-02392-6, CC BY 4.0.

Epigenetics

Whole genome bisulfite sequencing of microgravity-germinated *Arabidopsis* showed that, although spaceflight did not result in much of a change in total methylation, the distribution of methylation was shifted in response to spaceflight, particularly in leaf tissue, and particularly associated with protein-coding genes (Zhou et al. 2019), indicating that the adaptation of plants to microgravity is likely refined by epigenetic modification. Mutants in two different methylation mechanisms (Paul et al. 2021) were differentially impaired in spaceflight and showed different gene expression patterns, suggesting that DNA methylation is important to physiological adaptation of plants to spaceflight.

Signaling

A wide array of signaling components have been implicated in the plant gravity response—calcium, ethylene, ROS, and cyclic AMP, among others, including nitric oxide (NO) (París et al. 2018). (See Box 2-3.) Gravity-stimulated root tips were imaged using NO-specific fluorescent probes, showing a distinct asymmetrical

BOX 2-3
Key Metabolic Players in Plants

Research in fundamental biology of plants in space environments over the past decade has elucidated several key metabolic players that govern plants' root skewing, or bending away from the direction of gravitational force.

1. Nitric oxide is an important early signaling component in plants' microgravity response, including root bending.
2. Phosphorylation of proteins can affect plant cell wall adaptations.
3. Genes involved in circadian clock cycles correlate with plant root bending, although it is not fully understood why or how.
4. Expression of some heat shock proteins and factors is altered to in turn affect cell shape in microgravity.

accumulation of the molecule 30–90 minutes post-reorientation in the lower half of the root tip. This pattern is mirrored by the auxin efflux facilitator PIN2, vital in the asymmetric auxin gradient necessary for root bending in response to gravity. However, in the presence of an NO scavenger, differential PIN2 localization did not occur, leading to agravitropic and even negatively gravitropic responses, strongly implicating NO as vital to the plant gravity response via PIN stabilization and relocalization (Kruse and Wyatt 2022).

Only one study to date has looked at changes in post-translational modifications in plants, finding 16 differential phosphorylation events in spaceflight (Kruse et al. 2020). These proteins were not differentially expressed in spaceflight, only differentially phosphorylated. The *Arabidopsis* plasma membrane proton aptase 2 (AHA2) protein was less phosphorylated (less active) in spaceflight compared to ground controls. Implicated as a link between auxin and acid growth, AHA2 cell wall acidification allows for cellular elongation. AHA2 is inhibited in a FERONIA-dependent manner, which was increased in spaceflight. A cell wall protein involved in cellulose deposition, cellulose synthase 1 (CESA1), also showed increased phosphorylation in spaceflight, as did SHOU4 proteins, which suppress cellulose synthesis by regulating CESA1 trafficking. Tubulin-α 1, 3, 4, and 6 (cellulose patterning) were more phosphorylated in the ground controls at the site that promotes microtubule depolymerization. Insoluble proteins were more oxidized in space and soluble samples had decreased oxidation, and membrane proteins tended to be more oxidized in space.

Circadian cycles are responsible for the regulation of a wide array of plant processes, such as photosynthesis, transcription, and flowering, as well as the *Arabidopsis* gravity response in the root tip (Tolsma et al. 2021). When plated *Arabidopsis* seedlings, grown vertically, were rotated 90 degrees at different times throughout the day, the total growth of the root over the subsequent 24 hours is not affected. However, the degree of root curvature was affected, depending on the time of day of the initial reorientation, with the greatest disparity occurring between plants reoriented at dusk (strongest response) and before dawn (weakest response). These findings indicate the role of the circadian clock as an important upstream regulator of plant gravitropism.

Root skewing has long been thought to be a gravity-dependent growth behavior; however, skewing occurs in microgravity (Paul et al. 2012), suggesting that skewing is independent of both the tropic force of gravity and the gravity-induced contact forces between roots and growth media (Millar et al. 2010; Nakashima et al. 2014; Paul et al. 2012, 2013).

Arabidopsis mutants of SPIRAL1, a skewing-related protein implicated in directional cell expansion, and SKU5, a skewing-related glycosylphosphatidylinositol (GPI)-anchored protein of the plasma membrane and cell wall (Califar et al. 2020), showed different skewing behavior and markedly different patterns of gene expression in the spaceflight environment, indicating unique and opposite contributions to physiological adaptation to spaceflight and suggesting that proper function of both genes is important to spaceflight adaptation.

Cell cultures of *Arabidopsis* perceive and respond to spaceflight, even though they lack the specialized cell structures normally associated with gravity perception in intact plants—in particular, genes for a specific subset of heat shock proteins (HSPs) and heat shock factors (HSFs) are induced (Paul et al. 2012, 2013). HSP-related proteins play a role in maintaining cytoskeletal architecture and cell shape signaling.

NASA GeneLab Initiative as Enabling to Plant Research and Cross-Kingdom Insights

With limited access to experimentation in spaceflight, data sharing is a vital component of maximizing advancement in the field. For example, mitochondrial dysregulation was identified as a fundamental biological feature of spaceflight in humans (da Silveira et al. 2020), but no such finding had been identified in plant -omics data from flight. A reanalysis of RNAseq data in NASA's GeneLab Data System from Arabidopsis Col-0, WS, and *phyD* ecotypes/genotypes (GeneLab Dataset ID GLDS-7, GLDS-120) identified several mitochondrial genes as hub genes via a network analysis (Manian et al. 2021). Another meta-analysis across several studies also identified mitochondrial-related changes, specifically the alternative oxidase (AOX) gene family as common components often altered by spaceflight (Barker et al. 2020). To date, GeneLab hosts more than 40 data sets derived from plant experiments. To encourage community utilization of GeneLab's -omics data, NASA sponsored the development of the Test of Arabidopsis Space Transcriptome (TOAST) software. TOAST is an interactive data visualization environment that was developed to increase the accessibility of -omics data in plants (Barker et al. 2020).

Applied Plant Research

The past decade (2012–2022) witnessed the retirement of the European Modular Cultivation System (EMCS) in 2018 and the introduction of two plant growth platforms: Veggie in 2014 (see Figure 2-12), and the Advanced Plant Habitat (APH) in 2018. Both were developed to support plant research onboard the ISS. These growth chambers allowed for the propagation of larger crop species, those more likely to represent life support species and food source. To date, many space crop production efforts have focused on leafy greens (John et al. 2021;

FIGURE 2-12 International Space Station Vegetable Production System (Veggie) hardware. Veggie hardware was designed as a crop production system and supports both the "pick and eat" cropping system for the astronauts and a variety of plant experiments. (A) 33-day-old VEG-01B lettuce plants in the Veggie unit prior to harvest, and (B) Kjell Lindgren and Scott Kelly sampling the Veggie-grown lettuce. SOURCES: (A) From Massa et al. (2017), https://www.researchgate.net/figure/fig1_313735628, CC BY-NC-ND 4.0. (B) Courtesy of NASA, https://www.flickr.com/photos/nasamarshall/21327479788/in/album-72157667439026641, CC BY-NC 2.0.

Massa et al. 2013, 2017; Mickens et al. 2019). Comparison of nutritional value and microbial/fungal load on lettuce grown in Veggie indicated that key nutrients were not significantly different between ground and space and that food-borne pathogens (*E. coli*, *Salmonella sp.*, *S. aureus*) were negative (Khodadad et al. 2020). The monocot, cereal crop *Brachypodium distachyon* (van der Schuren et al. 2018) and dwarf wheat (Monje et al. 2020) have also been propagated in spaceflight with the goal of characterizing their adaptive responses and comparing them to that of *Arabidopsis*. Additional comparisons for C3 and C4 metabolisms have been conducted using *Brachypodium distachyon* (Masson 2019) and *Setaria viridis* (Handakumbura et al. 2022) as model systems. *Prunus domestica* (plum) has been genetically altered to shorten flowering time to enhance its usefulness as a candidate crop for spaceflight and may be helpful in protecting against the bone loss associated with spaceflight (Graham et al. 2015).

Focused investigations on important plant processes have also been conducted, such as the study of lignification and carbon concentrating mechanisms in spaceflight (Lewis et al. 2022), mechanisms that sustain seed viability, such as branch-chain amino acids (Brandizzi 2022), and the use of polyamines to mitigate stress that often arises as an -omic signature of spaceflight (Masson and Su 2022).

Not only did Veggie increase capability for plant research and crop production, but its development also provided invaluable lessons on the movement of water and nutrients in the microgravity environment (Massa et al. 2017).

The consistent observation that spaceflight responses affect genes associated with pathogens and wound responses suggests the potential for altered host–pathogen relationships in spaceflight (Kruse et al. 2020; Paul et al. 2012). Opinions differ as to the interpretation of these data, which indicate either susceptibility to pathogen colonization in spaceflight or the similarity of the response to spaceflight environmental stress and pathogen infection stress. However, a fungal infection outbreak on *Zinnia hybrida* plants grown in Veggie hardware in 2017 (see Figure 2-13) further confirms the presence and ability of plant pathogens to infect plants in spaceflight experiments (Massa et al. 2017). The diagnosis of this opportunistic infection took about 9 months (Schuerger et al. 2021). This motivated the development of the MinION sequencing platform for use as a rapid disease diagnosis tool in space (Haveman et al. 2022), enhancing the ability to combat in-flight plant infection.

Several strains of *Sphingomonas spp.* (Lombardino et al. 2022) and *Methylobacterium spp.* (Bijlani et al. 2021) have been isolated and identified on the ISS. Whole genome sequencing, analysis, and sequence comparison to

FIGURE 2-13 Diseased zinnia plants on the International Space Station (ISS). (A) Diseased zinnia plants from the ISS returned to NASA Kennedy Space Center on May 18, 2016. Signs of an active disease outbreak on *Zinnia hybrida* tissues during the VEG-01C experiment onboard the ISS. Aerial mycelia (am) of the opportunistic pathogen *Fusarium oxysporum* were observed associated with active or previous water films (wf) on tissues. (B) Plant tissues were cut into longitudinal, transverse, and cross sections and then processed for recovery of the presumptive causal agent of the disease. Fungal cultures on PDA (potato dextrose agar) from ground-Veggie plant pillow materials used in the VEG-01C experiment. *Fusarium sporotrichioides* (Fs) was recovered only from unsterilized gasket foam used in the Veggie plant pillows. Gasket foam also contained the fungi *Alternaria alternata* (Aa), *Cladosporium cladosporioides* (Cc), and *Aspergillus niger* (not shown). Arcillite was mostly free of bacteria and fungi, but *Penicillium olsonii* (Po) was recovered on one of four PDA cultures, each with Arcillite. Last, *Penicillium sumatrense* (Ps) was recovered from Nomex thread. SOURCES: (A) Courtesy of NASA. (B) From Schuerger et al. (2021), http://doi.org/10.1089/ast.2020.2399, CC BY-NC 4.0.

FIGURE 2-14 *Arabidopsis* plants in lunar regolith. *Arabidopsis* plants can grow directly in lunar regolith, suggesting one cultivation method that might support lunar agriculture, as well as lunar resource utilizations. SOURCE: From Paul et al. (2022), https://doi.org/10.1038/s42003-022-03334-8, CC BY 4.0.

well-characterized *Sphingomonas spp*. identified the presence of genes with potential to promote plant and algal resistance to abiotic stress, including heat/cold shock response, heavy metal resistance, and oxidative and osmotic stress resistance (Schuerger et al. 2021). Harnessing the key features of these important microbes may improve plant response to space environment stress.

Plants can grow in lunar regolith (Figure 2-14), suggesting that lunar soils can be used for future lunar farming and biological ISRU. Regolith collected from the Apollo 11, 12, and 17 spaceflights were used to grow plants on Earth. Plant growth was delayed and underdeveloped as plants faced the stresses of the ionic and oxidation-inducing lunar regolith. Growth success was varied across the three different regoliths. Plants developed more poorly in regolith with longer exposure to the lunar surface (mature) as compared to regolith with short exposure (less mature) as determined by sampling location (Paul et al. 2022).

Lunar and martian soils are deficient in all macro and micronutrients derived from decomposing organic material. Therefore, these and regolith simulants cannot produce crops without added fertilizers or organic material. The ISRU approach as part of bioregenerative life support systems looks to use native regolith (Karl et al. 2018) and recycled organic waste for food production (Menezes et al. 2015) as opposed to terrestrial soil or mineral input (Duri et al. 2022). Simulants developed to date are usually less beneficial to plant growth because they have alkaline pH, high sodium ion availability, low cohesion of mineral components, and more macroscale pores than microscale pores with associated poorer holding capacity of water. (See also Figure 2-15.)

The Advanced Plant Habitat (APH) was delivered and installed on the ISS in 2017–2018. Designed for a 10-year mission, APH collects physiological data on plant response to the spaceflight environment using more than 180 calibrated sensors to ensure the autonomous functioning of the system (Monje et al. 2020). Research conducted in the APH represents the first true foray into studies involving space-based agricultural cycles. Utilizing the APH capability, seed-to-seed cultivation of crops such as wheat and radishes (John et al. 2021) has been achieved, with increased opportunity to study plant root zone water stress assessment in spaceflights (Heinse et al. 2022). This multi-generational capacity allows investigations on the complete life cycle of plants. The APH continues to be a choice growth platform for plant biology experiments that focus on

FIGURE 2-15 NASA's lead horticulturist LaShelle Spencer is studying the use of 3D printed materials as media to grow plants at NASA Kennedy Space Center. The activity is taking place inside the Plant Production Area at the Florida spaceport's Space Station Processing Facility. SOURCE: Courtesy of NASA/Isaac Watson, https://www.flickr.com/photos/nasakennedy/52829925346, CC BY-NC-ND 2.0.

crop cultivation in space, and generally instrument development has been required of plant science advances in space environments.

Microbial Biology

Technological advances in recent years have transformed our view of the vast microbial diversity on our planet, with predictions now estimating ~1 trillion species in existence, most of which remain uncharacterized (Locey and Lennon 2016). Micro-organisms are present in most environments, including extreme environments, and are essential for sustaining human, animal, and plant life. In addition to their important roles in regulating the balance between health and disease, microbes can exert beneficial and harmful effects on natural and human-built habitats. Microbes are important for a wide range of processes on Earth with applications to supporting human life in space, including food and medicine production, integrity/function of engineered materials, nutrient cycling, biofuel production, bioremediation, and biomining. Transition of these processes to the spaceflight environment could be impacted by the global alterations in molecular genetic and phenotypic responses observed to occur in microbes. As humans return to the Moon and establish long-term lunar and deep-space outposts, there will be an increased need to harness beneficial microbial capabilities while mitigating harmful effects to the crew and their habitats (e.g., infectious disease, biofouling of life support systems and materials degradation). To this end, space microbiology research conducted over the past decade has significantly broadened our fundamental understanding regarding the effect of spaceflight and spaceflight analog conditions on single-species cultures and made progress in characterizing changes occurring to mixed microbial consortia associated with humans, animals, plants, and space habitats.

Single-Cell Microbial Cultures and Microbial Community Biofilms

Although microbes rarely exist in nature as single species, the isolation and characterization of pure microbial cultures is a cornerstone of microbiology. This approach enables identification of agents responsible for a disease or environmental process, facilitates use of recombinant DNA technologies, and allows targeted evaluation of microbial responses to environmental stimuli (e.g., nutrients, pH, antibiotics). Microbes rapidly sense and respond to changes in their microenvironment, including biological, chemical, and physical cues. During spaceflight, organisms are exposed to a unique combination of stressors, including reduced gravity, lack of convection and sedimentation, and increased radiation. While gravitational forces are not predicted to directly act upon individual microbes owing to their size (see Biophysics section), it has been suggested that the indirect effects of microgravity, such as altered fluid dynamics and extracellular transport, play key roles in regulating microbial responses.[9] The microbiology field has recently begun to better appreciate the importance of physical forces on microbial physiology and virulence, in a subdiscipline known as microbial mechanobiology (Dufrêne and Persat 2020; Fajardo-Cavazos and Nicholson 2021; Nickerson et al. 2004). In addition to the well-known mechanosensitive ion channels (e.g., MscL, MscS), other structures implicated in mechanosensing include the cell envelope, flagella, and pili/fimbrae (Blount and Iscla. 2020; Dufrêne and Persat 2020; Mathelié-Guinlet et al. 2021). In addition to roles in life support systems, microbes have also been explored as possible resources for providing key services for astronauts, including biomining (Santomartino et al. 2022), bionutrient production (Tabor 2022b), and multi-stage carbon fixation to bioplastics using organisms and processing platforms simplified for space (Cestellos-Blanco et al. 2021). The former two have demonstrated how to perform bioprocessing in space and how to mitigate microgravitational effects; the latter system is still ground-based and awaiting tests in flight. The foundational science in support of space bioprocess engineering, however, has really just begun.

Recapturing a Future for Space Exploration suggested rigorous experimentation using model organisms and also recommended the pairing of classic microbiological techniques with robust -omics profiling for deeper mechanistic insight into how spaceflight-associated factors regulate microbial responses. Since then, novel findings have been reported for numerous model microorganisms grown under spaceflight or spaceflight analog conditions, including Gram-negative bacteria, Gram-positive bacteria, and fungi. The collective research to date has repeatedly demonstrated that spaceflight and/or spaceflight analog culture environments modulate a range of microbial characteristics, including growth, metabolism, gene expression, stress responses, biofilm formation and structure, motility, and host-microbe interactions.[10] Given these widespread observations, the 2018 mid-term assessment reiterated high prioritization of microbiological studies to support deep-space exploration.

Most studies to date have been performed on short-duration cultures (hours to days). However, as mission duration and distance from Earth extends, it is critical to evaluate how long-term spaceflight affects microbial function. The knowledge gap in this area was considered significant in the 2011 decadal survey, leading to the suggestion that multi-generational studies be highly prioritized. There is still a lot of work to be performed in this area; however, several studies have evaluated microbial responses to long-term exposure to the microgravity and

[9] See Bijlani et al. (2022), Fajardo-Cavazos and Nicholson (2021), Nickerson et al. (2004, 2016), and Zea et al. (2016).

[10] See Acres et al. (2021), Bijlani et al. (2022), Green et al. (2021), Kim and Mudawar (2013), Nickerson et al. (2016, 2022), and Zea et al. (2020).

simulated microgravity environments relative to ground controls.[11] There are also several ongoing studies with *Saccharomyces*, *Arthrospira*, and *Salmonella spp.* in NASA's research portfolio.

> Finding 2-7: Despite progress over the past decade in understanding short-term effects of microbes and microbial communities, long-term effects (defined as months or years representing many cell division lifetimes) on microbial culture and storage in space environments are currently unclear.

Extending these basic scientific studies on the physiology of microorganisms is work focused on what services these organisms might provide for astronauts on long-duration missions, including ISRU through fixation of carbon and nitrogen from waste or atmosphere (Detrell 2021; Lehner et al. 2019) and through biomining of metals from lunar or martian regolith (Castelein et al. 2021; Cockell et al. 2020, 2021) and biomanufacturing of nutrients (Tabor 2022b), pharmaceuticals (Hilzinger et al. 2022), and bioplastics (Cestellos-Blanco et al. 2021). There are open questions about how to increase the efficiency, scale, and scope of these bioprocesses in space. This includes work that can be accomplished on the ground. For all, it includes better use of in situ resources and mission waste streams (both human and, e.g., plastic waste) to support growth of the microbial cultures. For biomining, it includes improved organic acid production and acid tolerance of the hosts. The spectrum of mission-critical molecules to be biosynthesized, the material properties of the bioplastics, and the biorecycling streams of the same all need to be expanded and barriers to engineering of these organisms for optimal production needs to be explored. The chemical and bioprocessing engineering principles—including scale-up, automation, and online optimization in the space environment—are also a ripe area for research and infrastructural invention (Berliner et al. 2021, 2022; Nangle et al. 2020).

Mixed-Species Microbial Communities

Built environments, being substantially more constrained than completely open systems, provide excellent settings to study interactions between host, habitat, and microbes. The ISS is a unique built environment: occupied by only a limited number of inhabitants at any given time and subject to microgravity and exposed to increased radiation. *Recapturing a Future for Space Exploration* recommended setting the stage for the use of the ISS as a microbial observatory. As a result, the 12 years and 22 missions of routine sampling of surfaces, air, and potable water from the ISS and a collection of 424 bacterial isolates encompassing 34 genera was made available to the research community through funded grants, including a NASA and Sloan Foundation Space Act Agreement postdoctoral call for proposals. This productive research effort led to multiple publications characterizing alterations in genomic and phenotypic profiles of microbial consortia isolated from ISS potable water, air, and surfaces (O'Rourke et al. 2020; Yang et al. 2021). This avenue of research grew to support characterization of microbes of the ISS, focused on ISS microbial population dynamics, microbe–microbe interactions, and microbial systems in long-term life support systems largely sampling water, air, and surfaces. Studies in this area evaluated viral and select bacterial and fungal pathogens and correlated their presence on crew, dust from ISS vacuumed material under varying humidity treatments, the ability to degrade plant compost materials, and even genetic engineering of select strains. Microbial monitoring efforts were also done in collaboration with other space agencies such as the 3-year flight study to examine the changes in the microorganism population on the Kibo-JEM microbial observatory. All investigations are performed with the aim to provide NASA with both a mechanistic understanding of the microbial population dynamics (e.g., cataloging population changes and mapping/linking these to environmental niche and genomic changes) presented over time owing to exposure to spaceflight conditions, as well as insight into practical countermeasures for mitigating risks to humans and environmental systems. Studies in this field are essentially ecological time course evaluations, and largely enabled and evaluated by next-generation sequencing. To gain meaningful comparative assessments of flight samples, the sequencing data sets need to be and are compared to existing terrestrial data sets available in public databases.

[11] See Cassaro et al. (2022), Fernander et al. (2022), Horneck et al. (2012), Napoli et al. (2022), Nicholson and Ricco (2020), and Ott et al. (2020).

Cross-Kingdom and Cross-Disciplinary Biology

During space travel, astronauts and their cargo carry diverse microbial ecosystems that are exchanged with other crew members, their vehicles, or habitats. These communities have the potential to function in beneficial or harmful capacities, depending on the microbes and their microenvironment (Nickerson et al. 2022). Microbes have previously been documented to cause infectious disease in both human and plant hosts during spaceflight (Ott et al. 2020; Schuerger et al. 2021). While relatively little is known about how spaceflight impacts the host–pathogen interaction, key studies using the foodborne pathogen *Salmonella typhimurium* previously showed increased virulence in mice following spaceflight and/or analog culture (Guéguinou et al. 2009; Nickerson et al. 2000; Wilson et al. 2007, 2008). The Hfq protein was identified as a potential regulator of these responses (Wilson et al. 2007). Subsequently, the effects of spaceflight and analog culture on the interaction between *Salmonella* and human 3D intestinal epithelial models were investigated (Barrila et al. 2021, 2022). Results suggested exacerbated infection profiles under these conditions, which aligned with the increased virulence findings in animal models. Another enteric pathogen, *Serratia marcescens*, also exhibited increased virulence under these same conditions using *Drosophila melanogaster* as the model host (Gilbert et al. 2020, 2022). These studies identified asparagine biosynthesis as the underlying pathway involved in regulating the virulence response of the pathogen.

Plant pathogens have also been examined in spaceflight for their ability to cause disease, with some results indicating increased pathogenesis under these conditions (Leach et al. 2001). Recently, the opportunistic fungus *Fusarium oxysporum* was identified as the etiological agent responsible for damaging *Zinnia hybrida* plants grown in the hydroponic Veggie system onboard the ISS (Schuerger et al. 2021).

Spaceflight investigations have evaluated changes in host microbiome and proposed how these alterations might impact humans, animals, plants, and their habitat (Garrett-Bakelman et al. 2019; Turroni et al. 2020; Voorhies et al. 2019). The Astronaut Microbiome project identified longitudinal shifts in microbiome composition at multiple sites in the human body evaluated before, during, and after spaceflight (Voorhies et al. 2019). The Twins Study evaluated longitudinal microbiome changes in fecal consortia in genetically matched subjects, with one twin remaining for almost a year on the ISS while the other twin served as the ground control (Garrett-Bakelman et al. 2019). (See Figure 2-16A,B.) Other studies have investigated the impact of spaceflight and/or analog culture on the microbiome of non-human subjects, such as mice (Jiang et al. 2019; Ritchie et al.

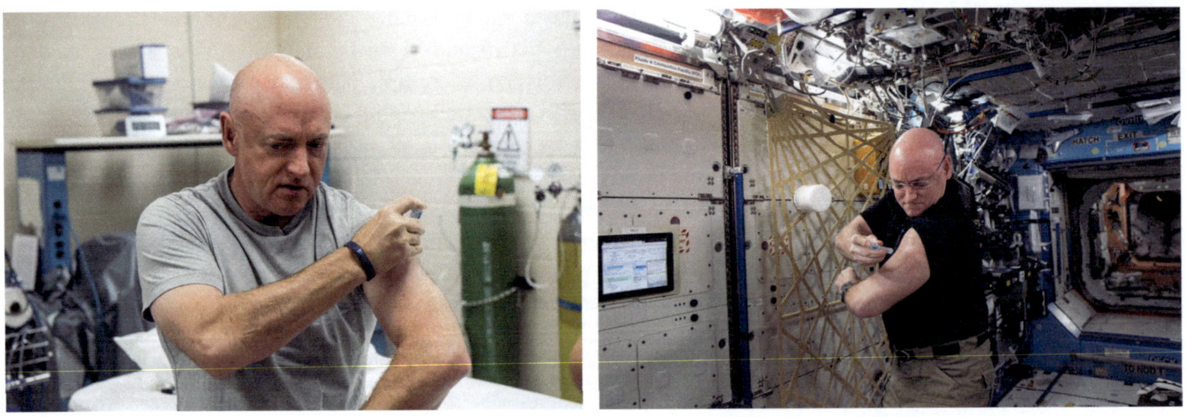

FIGURE 2-16 (A) Former NASA astronaut Mark Kelly gives himself a flu shot for an ongoing study on the human immune system. (B) His brother, NASA astronaut Scott Kelly, gave himself a flu shot while onboard the International Space Station (ISS). The vaccination was part of NASA's Twins Study, a compilation of multiple investigations that took advantage of a unique opportunity to study identical twin astronauts Scott and Mark Kelly, while Scott spent nearly a year onboard the ISS and Mark remained on Earth. SOURCES: Courtesy of NASA, https://www.flickr.com/photos/nasa2explore/22009194102 and https://www.flickr.com/photos/nasa2explore/21792722385, CC BY-NC 2.0.

2022), the symbiotic relationship between the bobtail squid and *Vibrio fischeri* (Casaburi et al. 2017; Foster et al. 2013; Grant et al. 2014), and plants (Haveman et al. 2021). Additionally, microbes isolated from the built environment of the ISS have also been investigated for their ability to infect cell lines and whole organisms (O'Rourke et al. 2020; Urbaniak et al. 2019).

Changes to fundamental molecular biological processes form a common thread connecting effects of spaceflight on single-celled organisms and the diverse physiological effects in multi-cellular organisms. How the space environment induces many of these effects remains poorly understood. Discerning which features of the Earth environment life depends on, and what these dependencies are, will be critical to sustaining and protecting living systems on Earth and in space.

Ground-based studies have examined some individual factors such as radiation (Beheshti et al. 2018). These single-factor studies have identified clear effects on many fundamental biological processes such as DNA damage and oxidative stress. However, how the space environment affects some of these processes, such as telomeres and circadian rhythms, remains unclear. For example, circadian, or daily, rhythms persist in space, with reduced amplitude and increased variability (Flynn-Evans et al. 2016; Sulzman et al. 1984) that can affect clock gene expression in both mammals (Casey et al. 2015; Fujita et al. 2020) and plants (Tolsma et al. 2021). As described in the next section, direct effects of altered gravity levels at the molecular or cellular level are extremely weak. Gravitational influences on molecular circadian rhythms thus may involve additional, indirect effects with mechanisms yet to be discovered.

BIOPHYSICS

Biophysics encompasses a broad interdisciplinary field that uses physical concepts to understand biological phenomena. While still in its infancy in comparison to more established fields of biology and physics, as highlighted by the omission of biophysics in *Recapturing a Future for Space Exploration* (NRC 2011), there have been a number of key findings in microgravity and LEO with respect to biological physics that indicate opportunities for new science and medicine that are both enabled by the space environment and also enable future space exploration.

As highlighted in previous sections, the space environment presents challenges and opportunities to explore many biological systems in the presence of higher radiation, lower pressure, and gravity. However, the laws that govern the relative influence of these physical forces are controlled largely by the scale of biological components. As relocation from terrestrial environments to LEO and deep space are associated with a diverse array of environmental changes, attributing specific physical causes to observed effects in inherently complex biological systems can be challenging. However, certain scaling laws, which hold in space and Earth, allow the elimination of some environmental changes as major driving factors. First, consider the scales at which gravitational forces act on cellular and subcellular systems in space, as context to appreciate the prior decades advances in biophysics-focused space research.

Magnitude of Gravitational Force

In physics, scaling laws and dimensional analysis provide useful insight into how physical quantiles interrelate over broad intervals. The relative importance and magnitudes of many forces depend on size and dimensionality. Of relevance for space-based research are forces that change in the space environment, such as gravity.

Gravitational forces have increasingly dominant effects on biological assemblies and organized systems that are typically greater than 1 mm in effective diameter (Figure 2-17); however, systems of smaller length and mass scales are dominated by other body forces. These small systems are often dominated by other forces such as viscous or electrostatic. Consider the gravitational force on a single swimming bacterium, which is approximately ~10^{-14} N (~0.01 pN). This force magnitude is significantly lower than the viscous drag force ($F = -6\pi\eta rv$, for a spherical bacteria) of an individual swimming bacterium, which is on the order of ~4×10^{-13} N (~0.4 pN), and considerably less than the force generated by adenosine triphosphate (ATP)-driven molecular motor proteins, which

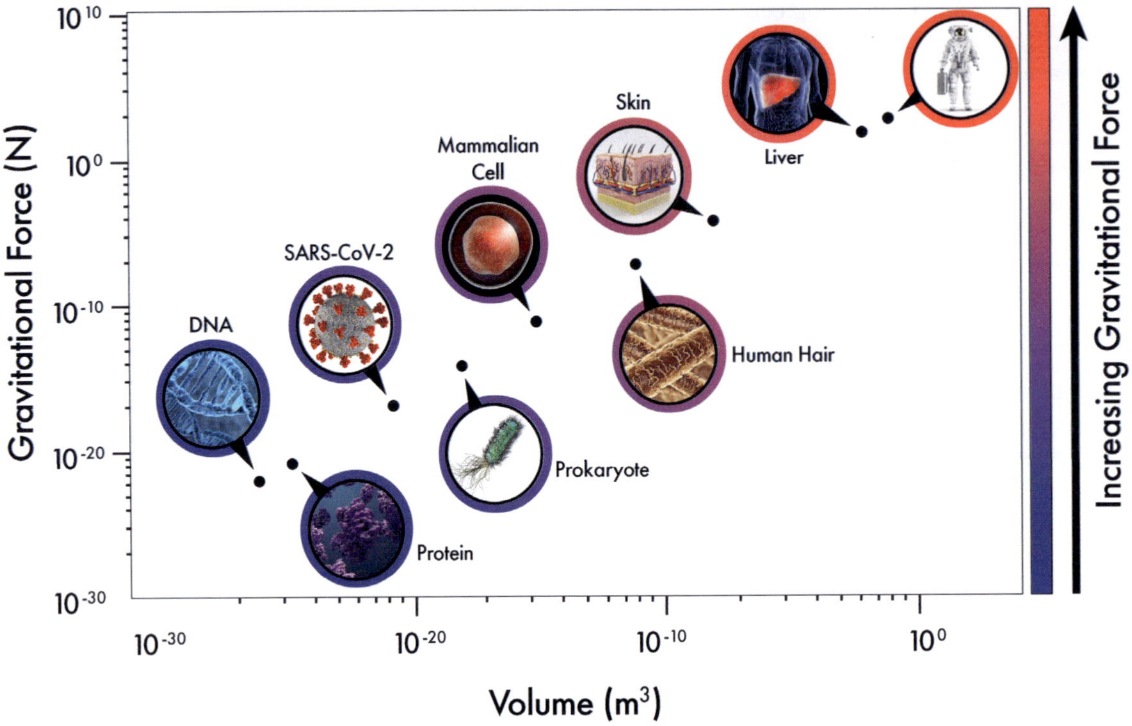

FIGURE 2-17 Gravitational force versus volume in low Earth orbit (LEO; estimates) for common biological systems. SOURCES: Composed by Tim Warchocki. Image credits: *DNA*: vitstudio/Shutterstock. *SARS-CoV-2*: GEMINI PRO STUDIO/Shutterstock. *Mammalian cell*: Kateryna Kon/Shutterstock. *Skin*: Motionblur Studios/Shutterstock. *Liver*: Magic mine/Shutterstock. *Astronaut*: GrandeDuc/Shutterstock. *Protein*: Design_Cells/Shutterstock. *Prokaryote*: Kateryna Kon/Shutterstock. *Human hair*: Kateryna Kon/Shutterstock.

generate a few pN per ATP molecule hydrolyzed. Thus, although gravitational forces act on small-scale systems, the forces that dominate the motion of molecules and cellular life originate primarily from viscous and molecular motor generated forces, as opposed to gravity.

While gravitational forces may have limited direct effect on biological systems below a certain length scale, reduced gravity impacts a number of other phenomena, including reduced bulk fluid flow generated by buoyancy-driven natural convection that can have significant physical effects on biological systems, as can be observed by analysis of non-dimensional numbers (Rayleigh and Grashof numbers that reflect approximate ratios of forces).

Additionally, radiation effects on biological processes are complex but amenable to biophysical analysis. For example, researchers have made progress since the early 2000s in relating biophysical parameters and models to correlate energy deposition by space particles (e.g., ^{16}O oxygen isotope particles originating from cosmic rays) to the probabilities of biological outcomes (Cucinotta et al. 2003). However, that parameterized description is coarse and does not predict patterns of energy deposition occurring at the tissue, cellular, or biomolecular level. More recently, biophysics-based analyses have focused on effects of radiation on specific physiological subsystems such as the nervous system and have also begun to consider the combined effects of radiation and microgravity on DNA damage and repair (Furukawa et al. 2020; Moreno-Villanueva et al. 2017; Onorato et al. 2020). More broadly, biophysics extends to the cellular and molecular scales. At those length scales, knowledge about how macromolecules synthesized by cells can self-assemble, or how mechanical forces exerted by cells on their surroundings can be affected by the space environment, has advanced in the past decade.

Molecular Biophysics

Crystallization of Biomolecules

Determination of the crystallographic structure of protein assemblies is important for answering fundamental biological questions and for developing new therapeutics. Crystal growth of biological macromolecules in the space environment has a rich history (Kundrot et al. 2001; McPherson and DeLucas 2015). Since *Recapturing a Future for Space Exploration* and the mid-term report, there have been several publications indicating that protein crystals grown in the space environment are larger and have fewer defects than those grown on Earth, aligning with the results of past work. In some, but not all, cases, crystals grown in space yielded higher-resolution atomic structures. The observed differences in crystal growth are not directly attributed to lower gravitational forces, but to minimization of convective flows, which favor single nucleation events and the subsequent growth of a single crystal (Reichert et al. 2019). The current reasoning for this behavior is the formation of stable regions of lower protein supersaturation, referred to as depletion zones, that are directly adjacent to crystal surfaces and remain quasi-stable owing to limited gravity-dependent, buoyancy-driven convective transport that typically disrupts crystal growth on Earth (Lin et al. 1995). However, such flows generated from surface tension-driven Marangoni convection can have more significant effects on crystal growth in the space environment. It is also of note that protein crystals of equal quality produced in space and on Earth in gels have been achieved, further indicating that the elimination of convection is key to ideal crystal growth (Artusio et al. 2020; Gavira et al. 2020). Unfortunately, a National Research Council review in 2000 and subsequent independent appraisals since the last decadal survey (NRC 2011) have viewed efforts in the crystallization of biological macromolecules in the space environment as inconclusive and yielding only incremental results that have not had major impact on structural biology (McPherson and DeLucas 2015; NRC 2000; Reichhardt 1998; Scott and Vonortas 2017).

In addition to macromolecular crystallization, small molecule crystallization for drug discovery and optimization has been explored on the ISS (Williamson Smith 2019).

Self-Assembly of Protein Fibrils

Some important biological proteins self-assemble into higher-order structures, such as fibrils and filaments. In some cases, such as amyloid fibrils, these protein aggregates are associated with human pathologies and diseases. A recent study identified significant differences in amyloid fibrils (fibrils associated the brain dysfunction) grown on the ISS compared with those formed on the ground (Yagi-Utsumi et al. 2020). Fibrils assembled more slowly on the ISS than on the ground and displayed multiple morphologies, while those grown on the ground displayed a single morphology. As noted earlier (see Figure 2-17), gravitational forces are negligible at molecular-length scales, including those of the assembled fibrils. Therefore, the observed effects of the space environment on protein assemblies and morphology could originate from additional factors, potentially owing to convective flows and/or surface tension.

Cellular Biophysics

Cell Motility and Taxis

Most mammalian and most microbial cell types possess motility mechanisms that serve as key components in their life cycle. The growth, spread, distribution, virulence, adaptation, and survival of microbes (including single cell eukaryotes, archaea, and bacteria) have been linked to motility, both on Earth and in the space environment (Acres et al. 2021). The motility of mammalian cells, including those of the human immune system, have been observed to have altered motility in space and have been implicated in altered tissue structure (Cervantes and Hong 2016). In addition, taxis or the locomotive response of organisms toward and away from environmental signals, have been observed in space environments, including chemotaxis and gravitaxis

BOX 2-4
Predicting Cell Adhesion and Migration

Biophysics has advanced prediction of how cells respond to extracellular cues, including the local stiffness and pH of the cell surroundings. Studies drawing from experiment, simulation, and modeling set the stage to understand how space environments may augment our knowledge of how cells adhere, migrate, and differentiate into new functions in space environments. The practice of using physical principles to abstract, quantify, and model such biological processes extends down to the level of binding kinetics among individual receptor proteins in the cell membrane that bind or ligate the extracellular materials to the intracellular cytoskeleton in eukaryotic cells (Paradise et al. 2011, 2013) and the analogs in prokaryotes like microbes.

(Häder et al. 2017). In fact, in mammalian cells (ranging from those comprising blood vessels, scar tissues, or stem cells engineered ex vivo for study or application), the cue-signal-response of cell adhesion and migration is now well-described by biophysical principles in terms of the transmembrane integrins and cytoskeletal motor proteins that together engage and exert force against the extracellular material. This biophysics approach enables predictions of cell adhesion and migration as a function of gradients in extracellular stiffness or pH that can be simulated explicitly at the protein binding kinetics level and predicted mathematically at the cellular level. (See Box 2-4.)

In the vast majority of past reports, motility in space environments has been inferred from genomic and transcriptomic analysis of motility genes. Thus far, there have been no direct experiments that have directly explored the biophysical mechanics for motility changes in space that could be associated with the altered dynamics of molecular motors, such as the flagellar motor in bacteria, and the dynamics of the active filaments, such as the bacterial flagellum, that give rise to motion (Acres et al. 2021).

Biofilms

Communities of microbes, known as biofilms, can be beneficial and detrimental in both human health and environmental safety in space. Most biofilm investigations in the space environment have focused on changes in expression of genes involved in metabolisms (Su et al. 2022), virulence (Taylor 2015), and chemotaxis (Acres et al. 2021). However, thus far the direct effects of gravity on biofilms are unclear. What is known is that significantly reduced microbial settling and buoyancy-driven convection alter how biofilms form in space. In comparison to biofilms grown on Earth, it has been shown that in some cases these communities can grow at a faster rate and display distinctly different phenotypes (Senatore et al. 2018). However, from a biophysical perspective, experiments that probe the structural mechanical changes in biofilms and the extracellular polymeric substances (EPSs) that they produce and that encapsulate these microbial communities have yet to be performed. (See Figure 2-18.)

PHYSICAL SCIENCES

The physical sciences comprise a wide array of phenomena and materials properties that are potentially altered by the spaceflight environment. Some of these phenomena are very practical, such as the behaviors of the fluids and materials that support space exploration. Others are more fundamental, such as the physics that is best able to be examined in the absence of gravity or by utilizing the long baseline distances available in space. Figure 2-19 highlights several milestone advances in the physical sciences, emphasizing the intersection with biological sciences for inspiration, shared methodologies, and applications of physical principles. This broad

FIGURE 2-18 Expedition 67 Flight Engineer and European Space Agency astronaut Samantha Cristoforetti packs experiment containers for the Biofilms investigation onboard the International Space Station. The biotechnology study explores ways to protect astronaut health and maintain spacecraft safety from microbes living in the orbiting laboratory's environment. SOURCE: Courtesy of NASA, https://www.flickr.com/photos/nasa2explore/52277282345, CC BY-NC-ND 2.0.

span of the space environment reflects example areas of research need such as ice physics for surfaces of moons, biophysics of interacting species in microgravity, and granular physics relevant to fluid flow on and among planetary surfaces.

Materials Science

Materials science is fundamentally both space-enabling and enabled by space. New materials and materials processing techniques are required for expanded spaceflight and exploration. Additionally, the space environment and its differences from Earth (especially differences in gravity) allow for experiments that advance our understanding of the fundamental physics controlling material structure and properties, in order to better predict and control material behavior for both terrestrial and space applications.

With respect to materials science research, *Recapturing a Future for Space Exploration* focused on the overarching topics of advanced materials for extreme environments (including low-density materials, high-temperature materials, and smart/stimuli-responsive materials), in situ resource utilization, materials science fundamentals, materials synthesis and processing to control microstructure and properties, and computational materials science. Given the complexity of materials and structures used in spaceflight, integrating computational materials science with experimental verification was highlighted as critical, along with database development to reduce cost and

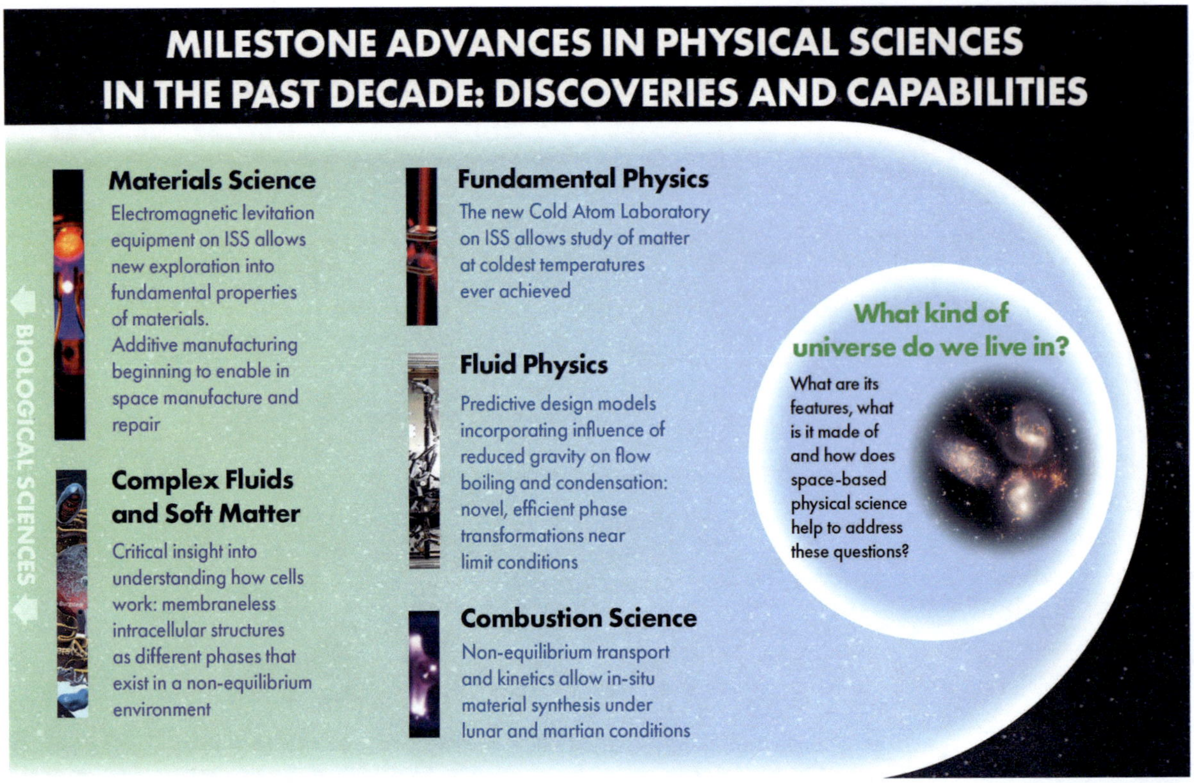

FIGURE 2-19 Milestone advances in physical sciences in the past decade. Remarkable progress in the physical sciences enabled by and enabling space exploration spans many disciplines, including those that can interface directly with biological sciences research and those that extend toward fundamental concepts of states of matter comprising the universe. SOURCES: Composed by Tim Warchocki. Image credits: *Electromagnetic levitator*: German Aerospace Center, DLR. *Eukaryotic cell*: Christoph Burgstedt/Shutterstock. *Magneto-optical trap and atom chip*: NASA/JPL-Caltech. *FBCE flight hardware* and *Burning droplet*: NASA. *Stephan's Quintet*: NASA/ESA/CSA/STScI.

time to develop new materials. (See Box 2-5.) Recent progress and communities can be organized by material classes distinguished by interatomic bonding (metals, ceramics, polymers, etc.), just as biological sciences are often organized around kingdoms or species.

Materials Fundamentals, Including for Metals and Semi-Metals or Semiconductors

Microgravity environments, whether in orbit or on sounding rockets, parabolic flights, or drop tubes, remove the confounding factors of sedimentation and thermal convection, allowing many fundamental questions about materials behavior to be probed. The two new levitators operating on the ISS have already provided an array of new capabilities for determining thermophysical properties, as well as enabled studies of nucleation, crystal growth velocity and phase selection over a wider range of conditions than is achievable on the ground. In addition, the highly controlled conditions of directional solidification have long been a particularly valuable tool for studying solid/liquid interface behavior and microstructure formation during freezing or solidification of metals and metallic alloys, with space providing the necessary long duration, purely diffusive environment. Various recent efforts have considered eutectic solidification, the columnar-to-equiaxed transition in dendritic solidification, and immiscible alloy solidification. Crystal growth of semiconductor materials also continues to be an area of active research considering dopant diffusion and growth interface behavior, with potential for materials of low defect density and superior electronic conduction. An emerging application of microgravity directional solidification is

BOX 2-5
Materials Science in Space

Materials science includes both fundamental and processing-focused research, with sustained focus on materials in extreme environments and ISRU limited in the past decade. The three highest-priority recommendations in this area from the prior decadal survey, *Recapturing a Future for Space Exploration* (NRC 2011), were related to materials synthesis and processing, advanced materials for extreme environments, and ISRU. The decadal survey also noted that research efforts on the second and third of these priorities was in its infancy. While there was a more extensive history of research addressing the first point, very little research had been performed in the 8 years prior owing to shifting administrative priorities. Moreover, while biology and physics research and the associated broadening of funded researchers grew in the past decade, materials science research using the space environment has comparably withered. The 2018 midterm assessment noted that NASA funding for materials science research remained limited, with approximately 27 investigators supported in 2017 versus around 70 in 2004. The funded research included projects on directional solidification and coarsening to understand formation and evolution of dendrites and eutectics, crystal growth of semiconductors, and related modeling and simulation efforts. Electromagnetic levitation (i.e., the ISS-EML) was also mentioned as a new capability that was being funded as a part of the U.S. participation in partner-led investigations using facilities from international partners. Such research in space environments can contribute to knowledge used in processing science and manufacturing on Earth and is also key to enabling space-based repair of built environments and manufacturing in the locations targeted for space exploration by the United States in the coming decade.

freeze casting, which uses ice to produce materials with elongated, aligned pores, and has potential applications for ISRU.

Ceramics

Ceramics are critical for high-temperature applications, with ultra-high-temperature ceramics (UHTCs) maintaining structural integrity above 2000°C. UHTCs have been investigated for use on space vehicles, including for thermal protection systems and as the capsule nose-tip. A capsule nose-tip failed during an atmospheric reentry flight test owing to microstructural defects, highlighting the importance of advances in sintering (Golla et al. 2020). An area of recent focus for UHTCs has been in improving their sinterability and reducing sintering temperatures through use of sinter additives and new sintering techniques.

Polymers and Polymer Composites

Polymers and polymer composites offer high strength relative to weight, high mechanical compliance (i.e., low stiffness) relative to metals and ceramics, low-temperature processing, and a wide range of tailorable properties. Shape memory polymers can switch from a temporary deformed shape to an original shape upon application of appropriate stimuli. For spaceflight, shape memory polymers have been demonstrated as hinges, gravity gradient booms, actuators, and solar array substrates (Li et al. 2019). More broadly, NASA Glenn has run four Polymer and Composite Experiments (PCE-1 through PCE-4) on the MISSE-FF to learn how polymers and polymer composites respond to exposure to the LEO space environment, with a focus on radiation and atomic oxygen (De Groh and Banks 2021).

Over the past decade, polymers are increasingly fabricated through additive manufacturing (AM), both for rapid prototyping (e.g., 3D printing to extrude hydrocarbon-based thermoplastics such as polylactic acid and nylon) and for 3D printing without cells in the initially printed construct (e.g., UV-irradiation printing of engineered polymers with human neuron-like geometry and stiffness) (Espinosa-Hoyos et al. 2018) or 3D bioprinting now taken to mean extrusion printing of hydrogels mixed with biological cells in the bio-ink. Methods to 3D print biocompatible polymers also proceed by non-extrusive means, including UV-polymerization of prepolymers with physical or digital masks (Espinosa-Hoyos et al. 2018). Indeed, the German space agency (DLR) recently supported such bioprinting of bandages for astronauts (Everett 2022), requiring deep knowledge and technological progress in processing of such complex fluid-like materials. Earth-based advances in printing technologies and polymeric "inks" as printable materials has been rapid, fostering recent focus on printing that leverages the microgravity advantages of increased structural buoyancy, fluid dynamics, and reduced gas bubble formation that can facilitate and inform additive manufacturing of such relatively compliant materials.

Composites of polymers and inorganic materials exist with various levels of structural hierarchy, ranging from cross-plied glass fiber-reinforced polymer sheets to coaxially drawn fibers used for communication, or for functional textiles (also called e-textiles, smart textiles, and flexible electronics) that could power next-generation extravehicular mobility units (EMUs) or astronaut spacesuits (Buckner et al. 2020; Shi et al. 2019; Wicaksono et al. 2022). Composites also include cement, and specifically the cementitious calcium silicate-based materials that bind the aggregate in structural concretes. Experimental tests of cement solidification in reduced gravity were also carried out on the ISS and could leverage ISRU and facilitate in-space manufacturing (Collins et al. 2021).

In-Space Manufacturing and Repair

Manufacturing research and innovations such as AM, commonly known as 3D printing, received no mention in the previous decadal survey (NRC 2011), but AM has burgeoned into a research enabler and multi-billion-dollar industry today. As described earlier in this chapter (in the context of engineering and technology advancing biological and physical science research), AM offers the ability to fabricate structures with greater complexity and with less waste than traditional subtractive and formative manufacturing methods. For example, NASA is partnering with Aerojet Rocketdyne to create new additively manufactured hardware for liquid rocket engines with fewer individual components and lower mass (Guerges 2021). AM in space is fundamentally different from terrestrial AM because of the different levels of gravity and its effects on momentum and heat transfer (Reitz et al. 2021), and on the settling, positioning, and recoating of powder feedstock. A thermally driven material extrusion additive manufacturing system was installed on the ISS in 2014. Since then, additional 3D printers have been installed, including the ISS Additive Manufacturing Facility (AMF), which is a commercially available facility from Made in Space, Inc. (now a wholly owned subsidiary of Redwire Space). The majority of AM performed in space or space-simulated conditions (e.g., microgravity) has focused on polymer and ceramic feedstocks, although limited work has been performed in metal, biological, and regolith additive manufacturing for microgravity conditions (Sacco and Moon 2019).

Modest advances in more traditional manufacturing areas have also been realized, with recent microgravity experiments focused on brazing (Yu et al. 2021) and soldering in reduced gravity conditions. Ultrasonic AM maintains the metals in the solid state to effectively weld them in microgravity conditions.

In Situ Resource Utilization

The extraction of oxygen from regolith is one of the most pressing ISRU issues to enable long-term exploration and responsible utilization of the Earth's moon. This process involves three steps: extraction, beneficiation for preparation of feedstock, and reduction. The current state of research regarding excavation (Just et al. 2020) and beneficiation (Rasera et al. 2020) have both been summarized in recent review papers. To facilitate coordination of efforts toward ISRU by the many disparate fields involved, a standard flowsheet and terminology (e.g., yield, recovery, conversion) were recently proposed (Hadler et al. 2020). For example, lunar regolith and its simulants may be used as a possible soil replacement for plant growth; as a silicon source for microelectronic device components and solar cells described in 2005 by researchers at NASA, academia, and a private company (Freundlich et al. 2005) and announced as reduced to practice in a private sector blogpost in 2023 (Blue Origin 2023); or as components of structural materials. In fact, a recent, rapidly growing area of research is in the use of lunar regolith as building material for habitats and other structures, particularly via additive manufacturing.

Techniques investigated to date include directed energy deposition and powder bed fusion using concentrated sunlight, microwaves, and lasers as energy sources (Just et al. 2020). AM with lunar regolith does not necessarily require substantial preprocessing, which greatly simplifies fabrication of structures. Powder bed fusion and directed energy deposition use lunar regolith as the only feedstock, while other AM approaches (e.g., material extrusion, binder jetting, vat photopolymerization) contain an additional, often organic, binder phase. AM with a binder is less energy intensive. However, binders generally cannot be sourced from the lunar environment, although plastic waste can be converted to binders in some cases (Isachenkov et al. 2021).

Coatings

Coatings are not a material class, but rather either single adherent layers of one material composition or multiple layers of multiple material compositions or classes; when the coating is of application-specific low thickness, it is also referred to as a thin film. These can be deposited through AM or can be fabricated through sundry deposition methods including those developed by the semiconductor fabrication and physics research communities. This physical form of material is important to space exploration at nanometer- to centimeter-scale thicknesses, because they protect the bulk of a structure from the extreme environments of space (e.g., radiation, high temperature, atomic oxygen, low-energy ions). Coatings can be made from any material class, in principle; the coatings need to be compatible with and adherent to the substrate that they coat. Coatings to aid in thermal control typically reflect light in the solar spectrum, so oxides are often used as pigments. Given the high-temperature gradients that thermal control coatings experience, low coefficients of thermal expansion (CTEs) are critical in addition to thermal stability, high emissivity in the infrared spectrum, and strong adhesion to the substrate. Optical coatings, depending on use case, may be highly reflective or anti-reflective (Hołyńska et al. 2018). Given the importance of coatings for protection of space structures as well as evolving regulations for the production and use of chemicals common to protective coatings and their processing (e.g., REACH in the EU, new TSCA in the United States), coatings development remains an active area of research.

Complex Fluids and Soft Matter

Over the past decade, there have been significant advances in soft matter science and our understanding of self-assembly processes. In particular, newly developed forms of active matter that can give rise to unique bulk material properties and perform useful tasks are envisioned to transform manufacturing, medicine, and robotics. Given the relatively weak forces that can both form and destroy the mesoscopic structure of these materials, space—reduced gravity environments in particular—presents a unique workspace to explore the interactions, physical properties, and transport phenomena of colloidal suspensions, gels, foams, granular materials, and liquid crystals.

Intracellular Condensation and Coacervation

A key area of complex fluids research in the microgravity environment has been the behavior of colloids and proteins near phase transitions. A related area where interest and understanding have grown greatly over the past decade is in condensation processes or phase transformations that occur in the out-of-equilibrium environment of biological cells (Hyman and Simons 2012). These processes, which create membraneless compartments with various functions, are liquid–liquid phase separations involving biopolymers (proteins, RNA, and DNA) and are often complex coacervation processes (Shin and Brangwynne 2017). This microcompartmentalization as well as recycling and self-repair (Théry and Blanchoin 2021) seen in cellular processes are critical concepts; because of intricate balances of weak forces and slight mass density differences, benefits to their study are found in microgravity.

Active Matter

Active matter is not a material composition class per se, and it can comprise one or more classes of materials (e.g., described by metallic, ionic, or hydrogen bonding)—often in the form of composites thereof. Regardless of composition, the shared characteristic is the *active* capacity of the matter to move and exert force, typically in response to applied field (e.g., magnetic field) or other changes in the physical environment (e.g., pH of the surrounding fluid). Active matter is visually fascinating and potentially useful because it seemingly moves under its own power and will, batteries not included. Some active matter can be described as composites comprising particles immersed in fluids, and thus as a subset of complex fluids; at sufficiently high viscosity, the phases can also be described as soft matter. To repeatedly or continuously reconfigure or change phase, these materials consume energy and are described as non-equilibrium states.

Beyond the anthropomorphism that motivates prediction of such indirect interaction processes through non-equilibrium statistical physics, recent studies of active matter advanced understanding of microscopic biological cell motion (from swimming bacteria to contractile muscle cells), mini-robot exploration of surfaces, mesoscopic filtration of porous media comprising Janus particles (Modica et al. 2022) that could later apply to ISRU, and macroscopic phenomena far from Earth such as dusty plasma physics (Bourgoin et al. 2020). Research intensity has increased over the past decade (Marchetti et al. 2013) to predict dynamic assembly or reconfiguration of the material volume, including the capacity for "active agents" or particles within the material to interact and assemble. Recent studies identified new forms of "phase separation" (Solon et al. 2015), including those that can be stimulated on demand by changes in applied field or physical environment, and others for which the attractive forces are not yet clear (Palacci et al. 2013).

In the search to understand and make material volumes that can move and assemble on demand, the space environment provides a *key control condition* to understand and predict how active matter works (i.e., reduced gravity conferring reduced sedimentation and convection that otherwise affects sufficiently dense and large particle interaction forces on Earth). This motivation is most justified for soft active matter systems governed by weak (thermal energy-level) interaction forces. The space environment also presents a *key demand* for materials that can move and change shape autonomously, at least in the absence of electricity that will be required of other functions on the way to and on surfaces such as the Earth's Moon and Mars.

Non-Equilibrium Statistical Physics

The ultimate goal of determining guiding principles has progressed for certain classes of materials—for example, in topological materials (Lubensky et al. 2015). Both intracellular condensation in the ATP-fueled cellular environment and materials processing owing to high pressure or confinement are non-equilibrium processes. To predict behaviors of biological systems at non-equilibrium, classical thermodynamics is useful but could be augmented. However, in the processing of many physical systems including granular and slurry materials, the concept of equilibrium appears irrelevant. Some important advancements in nonequilibrium statistical mechanics of densely loaded systems have been made, and determining the similarities and differences of jammed and glassy materials is a recognized challenge (Baule et al. 2018). The merging of granular and suspension flow at the jammed state has provided valuable insight (Boyer et al. 2011). Access to the space environment has proven key to advancing understanding of such complex fluids and soft matter in non-equilibrium in the past decade, particularly because reduced gravity conditions for sufficiently long times is required to understand the role that gravitational forces play in phase transitions. Accurate prediction of phase transitions among metastable material states in space environments when subjected to variable gravity *and* extra-terrestrial radiation remains unresolved.

ISRU- and AM-Related Topics

Slurries and dispersions are critical to ISRU and AM, and these will be supported by major advances in dense suspension rheology that show the importance of the full range of surface forces, including contact friction in shear thickening and jamming (Morris 2020). So-called soft matter may in fact be very resistant to permanent deformation in the jammed state but may also be mechanically compliant (deforming to large strains under low applied stress) and fragile. This reconfigurability is of interest across a range of network-forming materials, including granular material in robotic grippers (Brown et al. 2010) and polymeric materials (Webber and Tibbitt 2022). The capability to control complex fluid behavior is enhanced by a remarkable expansion of the types of particles and their interactions: there are now ellipsoids, cubes, Janus and patchy particles, and even shape-changing particles (Youssef et al. 2016). Rheological control through surface tension and contact line forces has seen a large growth in capillary suspensions (Koos and Willenbacher 2011), and the classic Pickering emulsion has seen a resurgence, opening the potential for use of particle-stabilized drop systems (Chevalier and Bolzinger 2013).

Recent work shows commonalities between granular materials and highly concentrated suspensions, and a growing emphasis is on the force networks that control properties (Papadopoulos et al. 2018). Such global understanding coupled with the influence of particle-level interactions provides guidance to additives that may alter surface interactions to facilitate gathering and processing in ISRU and use in forming parts by AM. Major theoretical advances in these amorphous materials include those in glasses (Berthier and Biroli 2011).

Enabling Methods

Advancements in fundamental understanding of complex fluids and soft matter in the past decade have been added by rapid advances in instrumentation and data analytics, for both space-based and Earth-based research. For example, developments in miniaturization, imaging and data analysis, and droplet-based microfluidic platforms open windows to the manufacture of pharmaceuticals and early-phase understanding of various materials (Marre and Jensen 2010). There is now the ability to perform large numbers of experiments in a self-contained apparatus with high-resolution images of the contents of 100-μm-scale domains (drops, for example) (Agresti et al. 2010). The combination of reactions in droplets and subsequent application by active matter methods opens pathways to material design (Shang et al. 2017). Microrheology instrumentation and data analysis methods to determine properties of sub-mm^3 scale sample volumes of soft matter and complex fluid systems has advanced greatly in recent years (Squires and Mason 2010). Laser tweezers can now probe properties and internal force generation by living biological cells (a type of active matter) (Berns 2020), while synthesis of probe particles and the ability to tune external fields now allow tracking of particles and cells at unprecedented levels and even in living subjects (Wu et al. 2020). The understanding of biological tissue mechanics and pathologies of relevance to space requires handling of vast data sets—for example, the handling of thousands of images in super-resolution microscopy (Schermelleh et al. 2019).

Data science has been key to recent advancement of experimental methods and findings in soft matter. Similar influence naturally has also been provided by advances in the complexity and speed of computer simulation, where understanding of various processes in soft matter as well as biological and other complex fluids (Spagnolie 2015) has advanced through data analysis but even more through new tools and algorithms in microrheology (Zia 2018) and glass dynamics (Ninarello et al. 2017). The large data sets generated by these calculations, as in molecular dynamics, have driven machine-learning approaches (Jackson et al. 2019). Both simulation and experiments in active matter systems also generate vast data sets, and robotic swarms have been shown to give rise to a myriad of dynamic structures (Rubenstein et al. 2014), such that the use of machine-learning approaches to extract guiding principles has taken hold (Cichos et al. 2020). The potential for moving to progressively more high-fidelity models and their linkage to other models for digital twinning of physical systems is of growing interest.

Fluid Physics

Two-Phase Thermal Management of Space Systems

Although a capable option for any thermal management challenge, systems capitalizing on phase change heat transfer are particularly attractive for utilization in space thermal-fluid systems where their orders-of-magnitude improvement in heat transfer coefficient allows for appreciable reductions in size and weight of hardware. Because of this potential, there is a push by space agencies worldwide to develop the technology further and allow implementation in both space vehicles and planetary bases. Current targets for adoption of phase change technologies include thermal control systems (TCSs), which control temperature and humidity of the operating environment, heat receiver and heat rejection systems for power generating units, and fission power systems (FPSs), which are projected to provide high power as well as low mass-to-power ratio (Chiaramonte and Joshi 2004; Lee et al. 2016).

Compared to single-phase schemes such as free and forced liquid convection, two-phase schemes offer very high heat transfer coefficients and the ability to dissipate large amounts of heat while maintaining safe system temperatures (Bryan and Yagoobi 1997; Castaneda et al. 2023; Patel et al. 2013, 2016). Managing fluctuations associated with phase change is a challenge for ensuring reliability and consistency of operations in pressure fluctuations, flow instabilities, and cross talk between components. Often, high-pressure drops in the flow architecture are utilized to manage these fluctuations. Liquid-to-vapor phase change (boiling) is possible using a variety of competing schemes such as pools, macro-channels, micro-channels, jets, and sprays. The vast majority of prior NASA microgravity studies were focused on pool boiling (Dhir et al. 2012; Raj et al. 2012; Xue et al. 2011). However, the absence of buoyancy in microgravity has shown that pool boiling is associated with formation of very large bubbles on the heat-dissipating surface, which severely degrades heat transfer performance (Mudawar 2017). For space applications, channel flow boiling is preferred over the other schemes owing to its simplicity, dependence on flow inertia to lessen vapor accumulation, and adaptability to cool multiple devices in series in a fully closed loop while requiring relatively low pumping power.

Investigators have relied on a variety of platforms to explore the influence of gravity on flow boiling, including those providing short-duration microgravity (drop towers, parabolic flights, sounding/ballistic rockets), as well as long-duration microgravity on the ISS. Most notable among ISS experiments are those by the Japanese Aerospace Exploration Agency (JAXA) (Inoue et al. 2021a,b; Ohta et al. 2016). Two key limitations of the JAXA experiments have been low power input and low flow rate, which limit testing to very small ranges of the operating parameters important to space applications.

Flow Boiling and Condensation

The flow boiling and condensation experiment (FBCE) was conceived in 2011, largely in response to recommendations included in the last decadal survey (NRC 2011), with intent of developing an integrated two-phase flow boiling/flow condensation facility for the ISS. The motivation for this research in the space environment was the recognition that going to and exploring the Earth's moon or Mars requires generation of much electrical power, and that will produce attendant heat that must be managed (Zudell 2021). As a two-phase process that uses the heat to boil a moving liquid until it changes it into a moving vapor, flow boiling is an effective heat transfer method. While it sounds simple, this phase transformation phenomenon is not well understood fundamentally or well predicted in the space environment. By comparing the microgravity data against those obtained in Earth's gravity, it will be possible to ascertain the influence of body force on two-phase transport phenomena in pursuit of empirical correlations as well as both theoretical and computational models. Preceding delivery of FBCE to the ISS, extensive flow boiling and flow condensation work was performed in both parabolic flight (Kharangate et al. 2016; Konishi et al. 2015a,b) and Earth gravity (Devahdhanush et al. 2022). These studies employed flow boiling and flow condensation modules very similar in design, construction, and instrumentation to FBCE's flow boiling module (FBM) and flow condensation module for heat transfer measurements (CM-HT).

Initial flow boiling experiments were conducted in Earth's gravity (Kharangate et al. 2016) by tilting the flow channel to different orientation, thereby changing the magnitude of body force components parallel to and perpendicular to the flow direction. At low flow rates, drastically different interfacial behavior was observed for the different orientations. Increasing flow rate was observed to reduce those differences, culminating in virtually identical interfacial behavior and equal critical heat flux (CHF; upper safety limit for boiling) values above a particular flow rate threshold.

During 2020 and 2021, computational fluid dynamics (CFD) models (Lee et al. 2019, 2020) were also developed to successfully predict interfacial behavior, flow patterns, and heat transfer parameters from prior FBM experiments. Mission sequence tests in 2021 were conducted in vertical up-flow in terrestrial gravity for representative operating conditions of the ISS data matrix. Initial parabolic flight experiments tests with CM-HT showed that flow behavior of the condensate film is sensitive mostly to flow rate (Lee et al. 2013a). A model was proposed to predict the condensation heat transfer that accounts for damping of turbulent fluctuations near the film interface.

The parabolic flight experiments were followed by a series of terrestrial gravity experiments conducted in the horizontal orientation (Lee et al. 2013b). Four dominant flow regimes have been identified—smooth-annular, wavy-annular, stratified-wavy, and stratified—whose boundaries were in fair agreement with published flow regime maps (Chen et al. 2006; Soliman 1986). The local condensation heat transfer coefficient was highest near the inlet, in the annular region, where the film is thinnest, and decreased monotonically in the axial direction in response to the film thickening.

Capillarity and Contact Line Phenomena

In microgravity, the influence of capillary forces, especially at three-phase contact lines, can have a dominating influence. These capillary forces play a role in a number of complex fluid phenomena as noted above, but here the focus is on the relation to fluid management in space. Of course, the underlying research findings here also extend to biophysics and biological science research. In microgravity, the dominance of these forces occurring in multi-phase systems can lead to significantly different behavior than seen in terrestrial gravity (Bryan et al. 1997; Castaneda et al. 2023; Patel et al. 2013, 2016). The difficulty of detachment of bubbles in boiling is an example. While complications induced in systems relative to terrestrial counterparts is one focus, potential benefits of capillary force dominance has also been a topic of interest for NASA (Conrath et al. 2013).

The past decade has seen significant advances in the ability to modify interfacial properties, with pronounced effects on heat transfer of special interest to boiling and cryogenic fluid management (Attinger et al. 2014). Once detached, the behavior of bubbles of vapor in contact with their partner liquid phase is not in all respects the same as that with a different gas—for example, steam bubbles in water may be different from air bubbles in water (Conrath et al. 2013). These topics are difficult to explore experimentally, requiring dedicated systems such as FBCE and other ISS installations for extended study. Numerical simulation of the topic has provided a crucial path forward, and the capabilities in this direction are expanding rapidly (Sui et al. 2014), but there remain substantial issues to resolve when specific interactions of materials and surface treatments play a role—for example, in contact line hysteresis and dynamics (Erbil 2014).

Fluid Physics and Thermal Management of Cryogenic Fluids

Cryogens constitute a unique class of fluids that are clearly distinguishable from water, refrigerants, and fluorochemical coolants by virtue of their very low saturation temperatures, and therefore high susceptibility to change phase by boiling. Cryogens also exhibit general thermophysical property trends, including low surface tension, low latent heat of vaporization, and low liquid viscosity, the effects of which are reflected in the physics of both fluid flow and heat transfer (Ganesan et al. 2021).

Early research focused on enabling technologies for controlling liquid fuels to accommodate positioning the fuel to the tank outlet and the coupling of liquid sloshing with the overall spacecraft (Salzman et al. 1973; Sumner 1978). Handling cryogenic fluids, particularly near their critical points, is needed for further understanding. The cryogenic fluid management (CFM) roadmaps (Johnson and Stephens 2018) outline the 25 relevant technologies for a range of space missions. Thus far, only reduced gravity cryogenic transient (quenching) data have been procured, which are only applicable to "line chill-down" CFM technology. Accurate sizing of any cryogenic feed system where two-phase cryogenic flow occurs depends on reliable correlations at the fundamental level. Routinely, thermal/fluid design codes such as GFSSP and SINDA/FLUINT are used to design cryogenic propellant transfer systems; these are lumped node codes that use correlations to model both single-phase and two-phase flow, heat transfer, and pressure drop. However, it has been shown recently that the existing correlations used in these two models do not agree well with available cryogenic flow boiling data in the quenching configuration (Hartwig et al. 2016), overpredicting LH_2 heat transfer by as much as 200 times when compared to data. Overall, the penalty for poor models translates into higher margins, higher safety factors, and ultimately higher cost, in design. In extreme cases, lack of accurate cryogenic transfer predictive models may result in vehicle failure (Steward et al. 1995).

Development of accurate predictive tools for both two-phase pressure drop and heat transfer that are applicable to different cryogens, different flow geometries, and multiple thermodynamics conditions has been exceptionally difficult because of (1) sparsity of reliable cryogen data, even for terrestrial gravity, and (2) extreme difficulty in performing two-phase cryogen experiments in reduced gravity. A set of BPS-related experiments called zero boil-off tank (ZBOT) experiments were also carried out on the ISS, in part to provide validating data for CFD models important for cryogeneic fluid management relevant to propellants under conditions specific to the space environment. This includes potential effects of zero contact angle on phase change or evaporation of cryogenic fluids, which may differ in LEO propellant depots compared with further Earth distances on the way to Mars. This current limitation reflects lack of understanding of multi-phase systems in the space environment, despite the important uses of such systems.

Combustion Science

Combustion research is an active area driven by fundamental questions and its relevance to fire safety and propulsion. Combustion and fire safety research in microgravity/reduced gravity can have major contributions in three key areas: (1) improved understanding of microgravity fire safety, leading to improved material qualification, fire detection, and fire suppression in spacecraft and outposts; (2) novel use of combustion technology in space exploration, such as new propellants for spacecraft propulsion, extreme-pressure and low-temperature combustion, plasma-assisted ignition, and high-pressure (i.e., supercritical-pressure) water oxidation for waste disposal and water reclamation; and (3) fundamental understanding of combustion physics leading to more-efficient and less-polluting terrestrial combustion devices (e.g., engines and burners).

Fire Safety

Spacecraft are not exempt from fire risk, and fire accidents have been reported (Friedman 1993, 1996). NASA has developed a fire safety strategy in which the main focus relies on the reduction of the fire risk, either by designing the ambient composition or selecting materials based on flammability behavior (Rojas-Alva and Jomaas 2022). The last decadal survey, *Recapturing a Future for Space Exploration* (NRC 2011), identified the development of fire safety protocols for spacecraft and space habitats—an important area that is integral to astronaut safety. It was also mentioned that the understanding of material flammability in reduced gravity is incomplete, and concerns existed about the effectiveness of fire detection and fire suppression systems designed for reduced gravity. Reviews have emphasized the advantages of material screening methods based on a flammability index such as maximum oxygen concentration (MOC) or limiting oxygen index (LOI) in comparison to the pass/fail test based on NASA STD 6001B (NASA STD 6001B 2011) (Fujita 2015). Material flammability in microgravity (not just in 1 *g* but in partial gravity), however, can be higher than that in normal Earth's gravity. Therefore, the quantification of the difference as a function of partial gravity is important to universally use flammability index as the material screening method. The minimum limiting oxygen concentration (MLOC) in microgravity can allow comparison between MLOC and MOC/LOI to determine the quantitative differences. MLOC can be affected by flow velocity, flow direction, material thickness, presence of conductor, external heating, ambient pressure, and so on, and therefore requires a thorough assessment. NASA's Burning and Suppression of Solids (BASS) program investigated the burning and extinction characteristics of a wide variety of fuel samples in microgravity and covered some of the parameter space. (See Figures 2-20 and 2-21.)

Cool Flames

The recent observations of prolonged "cool flame" burning of isolated large-diameter droplets under the flame extinguishment (FLEX) experiments program on microgravity combustion and advanced combustion microgravity experiments (ACME) (Nayagam et al. 2012) have simulated substantial interest (Ju 2021). (See Box 2-6.) The FLEX experiments for the first time showed two-stage (hotter and cooler) combustion behavior in which the

FIGURE 2-20 NASA astronaut and Expedition 66 Flight Engineer Mark Vande Hei configures the Combustion Integrated Rack in the U.S. Destiny laboratory module to support a pair of fire safety experiments known as SoFIE, or Solid Fuel Ignition and Extinction. SOURCE: Courtesy of NASA, https://www.flickr.com/photos/nasa2explore/51884456920, CC BY-NC-ND 2.0.

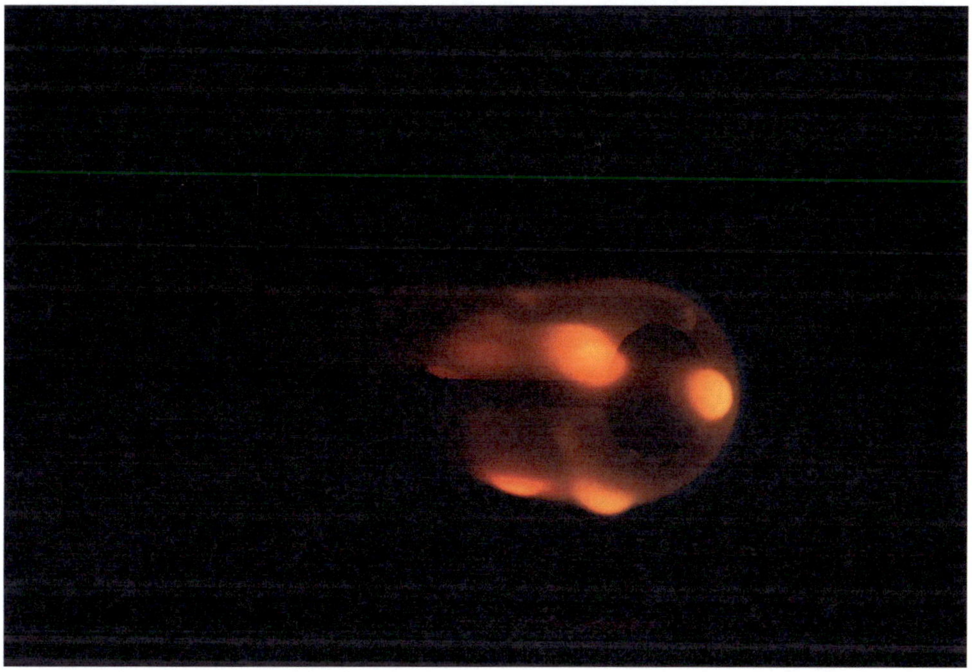

FIGURE 2-21 A close-up image photographed during a run of the Burning and Suppression of Solids (BASS) experiment onboard the International Space Station. Following a series of preparations, NASA astronaut Chris Cassidy (out of frame) conducted runs of the experiment, which examines the burning and extinction characteristics of a wide variety of fuel samples in microgravity. SOURCE: Courtesy of NASA, https://www.flickr.com/photos/nasa2explore/8639043958, CC BY-NC 2.0.

BOX 2-6
Combustion Science Required to Address a Flaming Surprise on the International Space Station

Note that "cool flame" combustion was not identified as a topic of study in *Recapturing a Future for Space Exploration*. However, its accidental discovery onboard the International Space Station (ISS) in the past decade resulted in a surge in research that was not only driven by the goal of developing fundamental understanding of the phenomena, but also by its implication in the development of fire safety protocols, because low-temperature cool-flame combustion cannot be visibly observed. Onboard the ISS, the Multi-User Droplet Combustion Apparatus (MDCA) Combustion Integrated Rack (CIR) was utilized to conduct the isolated droplet work for a range of initial droplet diameters, with different fuels (single component: alcohols, n-alkanes, primary reference fuel (nheptane + iso-octane)) over a very limited pressure range (0.5–3 atm). These validated sub-models could easily be extended toward multi-physics-based spray combustion modeling, which generally does not consider the entire low- to high-temperature combustion regime (Sirignano 2014). Such observations changed the way fuel combustion is understood and predicted (Nayagam et al. 2012).

low-temperature "cool flame" is self-sustaining for long durations (e.g., up to 40 seconds, depending on the choice of fuel and droplet diameter) (Farouk et al. 2017). Mathematical modeling of FLEX experiments indicates that the observed second-stage burning occurs at temperatures between the turnover temperature and the hot ignition temperature, within the negative temperature coefficient regime, and is driven by the excessive heat loss during the first-stage high-temperature burn.

Reducing emissions and improving fuel efficiency that can be achieved at high pressure by employing low-temperature combustion strategies has served as a primary motivator for cool flame research because homogeneous charge compression ignition (HCCI) and reactivity-controlled compression ignition (RCCI) had demonstrated the application of low-temperature combustion in engines (Dryer 2015; Reitz and Duraisamy 2015). Combined experimental and numerical studies have provided insight on the unique chemical kinetics that allows low-temperature cool flame to form and sustain (Curran 2019). This progress has also allowed the development and refinement of transport and chemical kinetic models specific to low-temperature regimes, with future potential to develop alternatives to petroleum-derived fuels.

Supercritical Combustion

The high-pressure oxidation environment has the potential to serve as a wastewater reclamation process during long-duration spaceflights. While the concept of supercritical combustion/oxidation has been known for decades (Augustine and Tester 2009), it has recently received increased attention because of its importance in high-efficiency power generation (e.g., the Allam cycle, in which supercritical carbon dioxide is used as the working fluid (Allam et al. 2017)) and also as a potential process for recovering energy and reclaiming water from wet waste streams (e.g., hydrothermal flames, in which combustion flames are produced in aqueous environments at conditions above the critical point of water, pressure $P > 221$ bar, and temperature $T > 374°C$) (Augustine and Tester 2009; Cui et al. 2020). Studies on supercritical fluids onboard the ISS under a joint collaboration between NASA and CNES (French Space Agency) have further intensified the interest in advancing supercritical combustion—specifically, hydrothermal flame activities. Detailed mechanistic behaviors of supercritical combustion phenomena are not yet fully understood, partly because of the technical challenges associated in ground-based and/or flight experiments, diagnostic limitations for characterizing supercritical combustion processes, and limited insight on the kinetics, transport, and thermodynamics parameters required to conduct mathematical simulations. The lack of fundamental understanding of the dynamics and the associated chemical kinetics continues to severely limit implementation of supercritical oxidation. The scientific understanding of supercritical water oxidation and the associated kinetic processes is, therefore, critical to advance solid waste treatment and/or wastewater recovery and management for long-duration spaceflights and advanced space exploration systems.

Fundamental Physics

As noted in Chapter 1, fundamental physics research aims to discover and explore the physical laws governing matter, space, and time. Quantum technologies and devices operate according to the principles of quantum mechanics (Safronova and Budker 2021). An early example is the atomic clock, which currently has an accuracy equivalent to 1 second in the age of the universe. What atomic clocks have accomplished for the measurement of time, laser interferometers have accomplished for the measurement of length; atom interferometers have accomplished for the measurement of fundamental constants, rotations, and accelerations; and atomic and diamond-based magnetometers have accomplished for the measurement of magnetic fields.

These advances triggered interest in using quantum technologies to also study the fundamental laws of physics, the foundations of our understanding of the universe: the standard model of elementary particles, combined with general relativity (NASEM 2020). Specialists in atomic and molecular physics, particle physics, gravitational physics, nuclear physics, and astrophysics have developed ground-based quantum technologies while working toward space qualification (Alonso et al. 2022). In addition, there has been rapid commercial development of quantum technology. This section summarizes both quantum technologies and their fundamental physics applications, as well as recent space deployment of modern quantum technologies.

Fundamental Physics Research Since the 2011 Decadal Survey

The past 15 years have seen revolutionary developments in quantum technologies, particularly those related to the ability to precisely control quantum states of photons, atoms, molecules, and even solids. There have been many significant successes in space, in drop towers, and on the ground. For example, the precision of optical atomic clocks has improved by three orders of magnitude in the past 15 years, reaching the level at which the best clocks will not lose 1 second during 30 billion years. These advances will lead to the development of or drastic improvements in a plethora of quantum sensors (optical atomic clocks, atom and laser interferometers, atomic and diamond-based magnetometers, optomechanical systems, and many others), opening completely new and unexpected avenues for probing the fundamental nature of the universe. In 2018, NASA launched the Cold Atom Laboratory (CAL) to the ISS (Elliott et al. 2018). It uses lasers to cool atoms down to near absolute zero to study the Bose-Einstein condensate (BEC) state of matter. A BEC makes the quantum properties of atoms macroscopic, so scientists can easily observe them, and CAL produced the first BECs in orbit. Multiple groups are conducting experiments inside CAL, which is operated remotely from NASA's Jet Propulsion Laboratory.

Recent years also brought forth developments of smaller, portable, and autonomous high-precision quantum devices, as well as a rise in the commercial development of various quantum technologies. Together, these quantum technologies can provide a next generation of space-based ultra-precise sensors for transformative basic and applied science, ranging from long-wavelength gravitational wave measurements and searches for beyond-standard-model physics to precision navigation throughout the solar system, including on and near planets, moons, and other bodies.

Important successes in precision measurement in space were obtained in the prior decade by testing the Einstein equivalence principle. The Micro-satellite à traînée compensée pour l'observation du principe d'equivalence (MICROSCOPE) has tested the universality of free fall with a precision of $\sim 10^{-15}$, an improvement of about 100-fold relative to ground-based tests (Touboul et al. 2017). Further improvements are possible—for example, the STEP proposal aims at reaching 10^{-17}, six orders of magnitude better than has been achieved on the ground (Blaser et al. 1993). Quantum tests of the equivalence principle by measuring the free fall for two atoms in an atom interferometer (Albers et al. 2020) now have an accuracy of a few parts in 10^{12} on the ground (Asenbaum et al. 2020). Spaceborne tests have been proposed that might reach an accuracy of 10^{-15} and could also provide a measurement of absolute atom masses and the fine structure constant, which, in turn, is important as a test of the standard model and the search for possible extensions of it (Aguilera et al. 2014; Williams et al. 2016).

Finding 2-8: Quantum sensors in space will ultimately enable transformative searches for new physics not possible on Earth. Space offers higher stability and accuracy for quantum sensors, as atoms, ions, or molecules do not have to be suspended against Earth's gravity but can be held by gentler forces. This reduces unwanted influences and boosts coherence and signal-to-noise ratio. In addition, operation in space allows modulation of the velocity and position of the sensors more strongly than is possible on Earth, which is important in testing the theory of relativity and in detection of dark matter and dark energy.

Astronauts recently helped upgrade the ISS CAL facility with a new tool called an atom interferometer that uses atoms to precisely measure forces, including gravity. The team recently confirmed that the new instrument is working as expected. The German Aerospace Center's (DLR's) Matter-Wave Interferometry in Microgravity (MAIUS) sounding rocket experiment produced the first Bose condensate and atom interferometer in space, and the Cold Atom Clock Experiment in Space (CACES) aboard China's Tiangong-2 laboratory demonstrated the first cold atom clock in space. Entanglement in space was demonstrated by sending entangled pairs of photons to three ground stations across China, each separated by more than 1,200 kilometers; this phenomenon was first demonstrated in 2017 (Yin et al. 2017).

Research in Quantum Gases

Study of the properties of matter under extreme conditions of low temperature where quantum effects dominate is an extremely broad field: BECs and Fermi-degenerate gases (De Marco et al. 2019; Giorgini et al. 2008; Inguscio et al. 2008; Regal et al. 2003) are two examples for quantum matter in different states, which exhibit opposite behavior: in BECs, particles all condense in the lowest-energy state, whereas in Fermi-degenerate gases, the Pauli exclusion principle forces particles to occupy higher-energy states. Generation of these states enables studies of macroscopic quantum effects that will act as quantum simulations to further understanding of, for example, superfluids (Greiner et al. 2002), high-temperature superconductors, nuclear physics, and quantum chromodynamics (Bloch et al. 2012; Feynman 1982; Jordan et al. 2012; Zohar et al. 2012).

In microgravity, such research can operate in regimes not achievable on Earth—for example, the creation of low-Kelvin condensates in a drop tower (Deppner et al. 2021). Space enables the use of weak traps, reducing perturbations, allowing the fundamental properties to be observed and manipulated. The ISS has already been used successfully as a platform for space-based quantum-gas research, gelation and phase separation in colloidal suspensions (Kodger et al. 2017), and tests of critical phenomena in the Lambda Point Experiment (LPE) and the CheX (Larson et al. 2002) and LTMPF (Lee and Israelsson 2003). Free-flying spacecraft would avoid gravitational and electromagnetic field perturbations present on the ISS and will be important for precision measurements in fundamental physics. Lunar or martian bases would enable seismographic and other studies of these celestial bodies, their magnetic fields, and atmospheric phenomena. These bases would also provide platforms for large telescopes, and could furthermore become stable, long-term laboratories for reduced gravity experimentation.

Finding 2-9: Most milestone advances in physical sciences, both applied and fundamental, have relied strongly on research access to the ISS. Continued access to LEO environments for research, as well as focused attention on alternative platforms extending to lunar and martian surfaces, will be critical to maintaining this rapid and important pace of progress in physical science research in space.

OPPORTUNITIES AND CHALLENGES OF RESEARCH IN THE SPACE ENVIRONMENT

Discoveries in the past decade have set the stage for research directed toward enabling organisms to survive and thrive in space. This is a credit to the perseverance of the BPS research community including NASA because it is difficult to conduct space science: the environment is hostile and nonintuitive, and crew time is highly restricted.

Practical effects of reduced gravity become important—for example, while sequencing has been successfully performed on the ISS, it is difficult to mix and manipulate the fluids properly owing to changes in the relative magnitudes of the forces acting on the fluids. In plant experiments, water follows hydrophilic paths to places it was not intended to be, making it more vulnerable to either overwater or underwater plants. In addition, gas transport is limited by diffusion unless there is forced convection. Heat exchange is limited owing to the loss of convective heat transfer unless there exists a phase change. Power, volume and mass, and crew time are highly restricted in space, meaning that some of the best instruments cannot be flown, while there is still a need for new sensors, particularly for volatile organic compounds.

Over the coming decade, miniaturization and automation promise to help improve the capabilities of space science, bringing them in line with what is done on the ground today. As humans advance toward the Moon and Mars, continued attention and inspiration needs to come from sensor development, automation, AI, and enabling big data to be done in space. Integrated analyses across experiments can be enhanced by the use of AI and machine learning to identify common changes or key factors associated with the space environment. Note, however, that current telemetry will not support the ever-increasing data streams that will be produced. (See Figures 2-22 and 2-23.)

An exciting opportunity exists to use the space environment as both laboratory and motivation to expose principles of non-equilibrium statistical physics and complexity in matter that are both enabled by the space environment and will be critical to future space habitats and logistics. The former topic ranges from non-equilibrium

FIGURE 2-22 Computer systems in space become old and outdated just as they do on Earth. Here, Expedition 65 Commander Thomas Pesquet of the European Space Agency installs a Joint Station Local Area Network router and its associated components inside the International Space Station's U.S. Destiny laboratory module. SOURCE: Courtesy of NASA, https://www.flickr.com/photos/nasa2explore/51557563472, CC BY-NC-ND 2.0.

FIGURE 2-23 The International Space Station (ISS) is an incredibly complicated laboratory in space that has acquired much new equipment over its lifetime, like the components being replaced by astronaut Jessica Watkins. Before the end of the decade covered in this report, the ISS is expected to be retired. SOURCE: Courtesy of NASA, https://www.flickr.com/photos/nasa2explore/52213756408, CC BY-NC-ND 2.0.

processes induced by energy utilization (biological activity) in cells and tissues, to the extreme rate dependence and morphological changes that can take place in flows of slurries and dispersions exposed to processing flows, such as in mixing. Furthermore, the reduced gravity can unmask phenomena that are often hidden by gravitational convection, such as Marangoni convection. The limited proof of concept studies of ISRU enabled by advances in additive manufacturing will transform in-space fabrication and maintenance tasks. Increased complexity, whether in soft matter systems that can recapitulate the structures and/or functions of biological systems or in multi-material systems that can be fabricated using additive manufacturing, will enable new capabilities, such as design of new multicomponent systems or in situ monitoring and control. This deep science will support engineering and innovative approaches in a way that is expected to prove crucial in considering environments where experience is lacking and there may be very limited time and resources to develop empirical knowledge.

There are many opportunities for transformative science enabled by the space environment. For example, quantum mechanics—the physics of the very small—and general relativity—the physics of the very large—are two of the great scientific revolutions of the 20th century. They are essentially perfectly confirmed yet incompatible. Putting state-of-the-art atomic clocks onto multiple space platforms will provide the opportunity to make unprecedented tests of general relativity and other fundamental concepts, as well as providing the basis for space-enabling technologies like navigation. Similarly, the space environment presents opportunities to explore the extent to which fundamental biological processes are shaped by, or even depend upon, the physical environment of Earth. Thus, research in the coming decade can capitalize on the unique features of the space environment to understand fundamental biological processes such as gene-environment interaction-based adaptations over one lifetime and over generations, gravity sensing, time sensing, ecosystem formation, and the function of fluid dynamics in functional phase separation in cells.

Experiments in space are unquestionably demanding of technology, time, and other resources in all scientific disciplines. Yet for BPS disciplines, research in space offers unique opportunities to advance space exploration and make fundamental discoveries not possible in any other way. Surprising discoveries in the past decade demonstrate the importance of such fundamental research and bring into focus new questions to be answered in the next. The growth of interdisciplinary or multi-disciplinary studies have been particularly exciting. Many new opportunities for fundamental, paradigm-shifting discoveries that may cross the conventional boundaries of scientific disciplines await in the next decade. Importantly, the accomplishments outlined above were achieved with a relatively narrow slice of research talent in the United States; investing intentionally in the broadening of science perspectives and lived experiences that the next generation of BPS researchers will bring to space science will be as important as investing in the research activities if the United States is to remain collaborative and competitive in this research environment over the coming decade.

This overview thus sets the stage for the current unknowns and research priorities. Chapter 3 next summarizes the report framework, organized by three themes of critical and outstanding scientific priorities that motivate the prioritized KSQs. The KSQs are more fully elaborated in Chapters 4 and 5. The research campaigns and initiatives are described in Chapter 6, and the associated strategic challenges of the research ecosystem are discussed in Chapter 7. This framework organizes the BPS research priorities of the coming decade 2023–2032 required for the United States and others to thrive in and make the most of the space environment for decades to follow.

3

Framework for Thriving in Space by Prioritizing Themes and Scientific Questions

Reflecting on the impressively broad scope of biological and physical sciences relevant to space (Chapter 1) and the impressively deep insights for science and society over the prior decade (Chapter 2) motivates a major decision point for the coming decade. That decision pivots on how the United States frames the opportunity to focus finite resources of scientific discovery to make the most impactful advances of knowledge in the most complex laboratory known on Earth: the space between our planet, the Moon, and Mars. In the 2011 decadal survey, *Recapturing a Future for Space Exploration*, during a markedly different point for the U.S. biological and physical sciences (BPS) research community and for NASA, this framework could be described retrospectively as topical in that prioritized research topics were described within many disciplines from plant biology to condensed matter physics, each with the potential to make important advances within that discipline. This was an understandable approach that bridged survival of research disciplines with sufficient expertise, interest in, and access to the space environment by the U.S. research community and facilitated even the possibility of international collaborations by U.S.-based researchers. In the present decadal survey, the framework is thematic and integrative. This is an approach that builds on the BPS research gained over the prior decade, so that both the BPS research community expertise and wider society can thrive in the years of scientific cooperation and competition to come. The framework described below establishes the themes of recommendations to follow in this chapter and in Chapters 4–7. This approach, organized around three inspiring themes, focuses collective attention on the key scientific questions (KSQs) for which answers are required to both explore and benefit fully from global access to the space environment that extends from the surface of Earth to the surface of Mars.

KEY SCIENTIFIC QUESTIONS AS PRIORITIES FOR BIOLOGICAL AND PHYSICAL SCIENCE RESEARCH

Themes

Three themes bind together the scientific priorities of BPS in space environments for the coming decade. These three themes are not unique to this decade, yet the progress in human exploration of space, the scientific achievements and insights of the previous decade, and the rise of commercial development in space exploration and research all bring these themes into sharp clarity as descriptors of scientific importance.

Answering the most fundamental questions about how living organisms function and the universe itself works will continue to require access to the unique features of space. At the same time, the past decade has realized blurring of the boundaries between BPS research to great advantage. For example, answering key questions about vascularized plants or humans in space environments requires awareness of and even automation of the physical principles governing complex fluid flow in microgravity, and answering questions about new materials used to propel or protect humans on the Earth's Moon or on the way to Mars need to consider the impact of any waste solvents or particulates on future generations of life.

Focusing on the scientific priorities in the following three themes positions the United States to not only travel to but also to thrive in the space environment sustainably, all the while returning benefits to societies on Earth:

- *Adapting to Space*—Understanding what happens to a terrestrial organism as it leaves Earth. This theme examines the nature of the processes involved in physiological adaptation to space, especially the limitations of that acclimation process, limitations that might affect survival or offer opportunities for decreasing the risk of being in space.
- *Living and Traveling in Space*—Searching for those principles that guide long-term and safe habitation of space. This theme probes the interactions among life, living systems, and physical structures that are designed for life and life support in space habitation and travel. Included in this theme are the physical phenomena whose understanding is required to safely transit among destinations in space.
- *Probing Phenomena Hidden by Gravity or Terrestrial Limitations*—Quantifying underlying, fundamental physical and biological processes that are revealed when the experiment is freed from the limitations imposed by Earth. This theme seeks to learn in space what cannot be learned on Earth.

These themes are not exclusionary, as there are phenomena and principles that connect them—principles that ripple through the fundamental fabric of biological and physical structures and functions. These themes do, however, coalesce and categorize specific sets of KSQs—questions that could enable further space exploration and be uniquely enabled by being in space. (See Figure 3-1.)

Criteria Guiding the Selection of Key Scientific Questions

The following criteria served as decisional guidance in the development of the themes and KSQs—perspectives that arose from the 2011 decadal survey and were updated to reflect the current state of BPS science in 2022. These criteria were informed by expert presentations and community input to the steering committee and the panels. These criteria were therefore critical in the development and selection of the KSQs that underpin this report (Chapters 3–5) and the research campaigns that were chosen in part to support rigorous answers to those KSQs (Chapter 6). Such criteria also encouraged clearer articulation of the "so what?" or shared understanding of the importance and implications of answering such scientific questions—clearer to other scientists with varied expertise, and also clearer to decision makers and U.S. taxpayers who support this work through both taxes and career choices. These criteria were given no rank order or weight, yet all served as direct prompts for both identifying and then crafting the narratives for the KSQs, by considering the extent to which the described research developments in candidate scientific questions will, within the next decade:

- Answer grand science challenges and usher transformational change.
- Require access to the space environment to achieve pivotal discoveries.
- Reduce uncertainty about both benefits and risks of space exploration.
- Solve terrestrial problems or yield societal benefit.
- Reduce the costs of space exploration.
- Possibly lead to entirely new options for exploration missions, to bring revolutionary change to space exploration.

Note that the BPS research community recognizes high value in addressing questions that recognize biological processes as being driven by the physical world and describable by physical concepts and measurement, and

FIGURE 3-1 During the past decade, the ability to reach and resupply the International Space Station (ISS) has changed in significant ways. Here the *SpaceX Crew Dragon Endeavour* spacecraft is lifted onto the *SpaceX GO Navigator* recovery ship shortly after it landed with NASA astronauts Robert Behnken and Douglas Hurley on board in the Gulf of Mexico off the coast of Pensacola, Florida, on August 2, 2020. The Demo-2 test flight for NASA's Commercial Crew Program was the first to deliver astronauts to the ISS and return them safely to Earth onboard a commercially built and operated spacecraft. Behnken and Hurley returned after spending 64 days in space. SOURCE: Courtesy of NASA/Bill Ingalls, https://www.flickr.com/photos/nasakennedy/50185336517, CC BY-NC-ND 2.0.

that the physical sciences can be biology aware for both inspiration and application. Additionally, in the coming decade there will be some KSQs that seem "purely biological science" without need for new physical insights and capabilities, and there will be some that seem "purely physical science" without regard to biological inspiration, analogy, or application. However, there will also surely be some KSQs that by this coming decade seem to naturally draw from both disciplines and research communities, with biological science–motivated studies being much more physical science-aware, with physical science-focused research being much more cognizant of the implications of research outcomes on biological systems, and with biophysics-grounded research framed as more than an academic fringe topic or subset of BPS.

These three themes help to frame the KSQs that BPS research will address in the coming decade—a decade when space science research will contribute to the canon of knowledge that will help society thrive both on Earth and in the increasingly ambitious space exploration frontiers pursued by the United States and other nations.

Theme 1: Adapting to Space

From the earliest days of the programs dedicated to human exploration of space, there have been keen questions about whether terrestrial biology is capable of functioning in space. Many of those questions were directly associated with the impact of spaceflight on human beings. Most of the data collected until the space shuttle era were therefore medical in nature. However, once the initial survival of astronauts was assured, science began to expand the theme of space adaptation toward more fundamental questions regarding the entire range of biological processes that can be affected by going into space, including the plants and microbes that will be a part of habitation systems. Major science gains through the space shuttle and ISS decades have been strong increases in understanding what happens to organisms that go to space, at the physiological, cellular, and molecular levels. Three KSQs in this theme for the next decade remain or have arisen from those gains:

- How does the space environment influence biological mechanisms required for organisms to survive the transitions to and from space, and thrive while off Earth?
- How do genetic diversity and life history influence physiological adaptation to the space environment?
- How does the space environment alter interactions between organisms?

These questions begin with an organismal focus, working toward an understanding of what happens to an organism as it leaves Earth and enters the spaceflight environment. Life in space operates differently from life on Earth. Each organism goes through a process of physiological adaptation, a process that has been well discovered in the previous decade and begs to be well understood in the coming decade. Data from the space shuttle and ISS eras strongly suggest that genetic differences imply that not every person, every plant, or every microbe will experience the same adaptive process, or even the same ability to thrive in space. Thus, the direct assessment of the role of genotype in the process of an individual's adaptations to spaceflight becomes a key question. Moreover, data also suggest that being in space alters the epigenetic characteristics of spaceflight organisms—DNA repair—as well as the interactions between organisms. It is these interactions between organisms that govern complex issues, including pathogenicity. This theme concludes with a KSQ that deals with complex biological systems as they interact to adapt to establish a new stable ecosystem in space. (See Figure 3-2.)

FIGURE 3-2 The International Space Station has now been in orbit for 25 years. It is regularly serviced by resupply craft. Here Northrop Grumman's Cygnus cargo craft is pictured attached to the Unity module shortly after being captured with the Canadarm2 robotic arm. SOURCE: Courtesy of NASA, https://www.flickr.com/photos/nasa2explore/49558409347, CC BY-NC-ND 2.0.

Theme 2: Living and Traveling in Space

The advent of the ISS era was a major inflection point in the trajectory of space biological and physical science. It signaled the transition of science from short-term, sortie missions lasting a few days into the science of long durations of being in space. The ISS created the opportunity for science observations lasting weeks, months, and even years. Moreover, the ISS serves as a test-bed for science and technology for long-duration deep-space missions, such as to the Moon and Mars. This transition in science practice, together with the exploration goals that include missions to the lunar surface, creates a realm of science needs that reflect longer-term interactions between biology and hardware systems, as well as the fundamental physical processes needed to sustainably explore new space locations. Answers to four KSQs are needed to meet these requirements:

- What are the important multi-generational effects of the space environment on growth, development, and reproduction?
- What principles guide the integration of biological and abiotic systems to create sustainable and functional extraterrestrial habitats?
- What principles enable identification, extraction, processing, and use of materials found in extraterrestrial environments to enable long-term, sustained human and robotic space exploration?
- What are the relevant chemical and physical properties and phenomena that govern the behavior of fluids in space environments?

These key questions span a range of investigations from biology through physical science, as well as investigations that integrate across disciplines. These questions formally address the topics of biology and biological systems within the environments built for space explorations. They begin by recognizing that establishing long-term habitation in space requires an understanding of multi-generational effects that arise from reproductive cycles taking place in space. They also address the physical sciences that enable the design and engineering of spaceflight systems. It is understood that scientific knowledge gained by answering these questions is framed specifically to enable society to thrive when accessing the space environment to live for extended times or transit from LEO to Mars. It is also historically likely that this knowledge will be applied in potentially unrelated ways to improve daily life on Earth.

Theme 3: Probing Phenomena Hidden by Gravity or Terrestrial Limitations

Science from the space shuttle and ISS eras has demonstrated clearly that fundamental phenomena in physical and biological systems can be uniquely approached when gravity is removed from the equations. Processes and phenomena that are not observable on Earth can be readily seen in spaceflight. Moreover, spaceflight platforms also present approaches to physical measurements that are limited by the physical distances available on the surface of Earth. There is, therefore, recognition that major science gains are available to space-based laboratories, and those gains are prioritized by these KSQs:

- What are the mechanisms by which organisms sense and respond to physical properties of surroundings and to applied mechanical forces, including gravitational force?
- What are the fundamental principles that organize the structure and functionality of materials, including but not limited to soft and active matter?
- What are the fundamental laws that govern the behavior of systems that are far from equilibrium?
- What new physics, including particle physics, general relativity, and quantum mechanics, can be discovered with experiments that can only be carried out in space?

These questions include consideration of both natural and engineered systems, and may apply to both biological and physical phenomena. For example, cells and tissues can deform under the same mathematical descriptions that predict deformation of complex engineered fluids. These questions uniquely require aspects of the space environment to provide thorough and useful answers. For example, plant root responses to blue light are observable only under microgravity conditions, and spaceflight experiments have determined threshold values for gravity responses in plants only possible with such space environment access; also, useful observation times of complex fluids when far from thermodynamic equilibrium can require low disturbance levels unique to microgravity. These questions expand Earth-based knowledge about how materials form and change when temperature or other physical conditions change suddenly, how fluids flow differently in space conditions whether to sanitize water, conduct experiments in nanoscale channels, or store propellants, and even how to measure time so precisely that communication and locations can be shared with fellow citizens who participate in space exploration in future decades.

Taken together, these 11 KSQs organized across three themes also reflect divided attention to research impact that is (1) *enabling of safe and productive exploration and habitation of space* from LEO to Mars, and (2) *enabled by access to the space environment*—often requiring that access owing to inaccessibility of such environmental features on Earth or predictable by simulation. There is distinct yet equal value in pursuing both types of KSQs in the coming decade, rather than focusing attention only on space mission readiness and sustainable habitation, or only on fundamental research inquiries of potential but currently uncertain societal benefit. Implicitly, this approach also develops a diverse BPS research community and connects increasingly with NASA, other international space agencies, and other U.S. government agencies whose research and readiness missions, and industrial research and development efforts, benefit from answers to these questions.

Table 3-1 summarizes the KSQs for the next decade, discussed more fully in Chapter 4, focusing on space science research that enables safe and productive exploration of space from the ground through LEO and on to the Moon and Mars; and in Chapter 5, focusing on space science research that is uniquely enabled by the characteristics of the space environment.

Recommendation 3-1: NASA should direct its research resources toward the key scientific questions identified in this study (Table S-1 and Chapters 4 and 5).

Recommendation 3-2: NASA should work with other U.S. government agencies and other nations' space agencies to coordinate research resources toward the key scientific questions, as relevant to multiple-agency missions.

TABLE 3-1 Key Scientific Questions in Biological and Physical Sciences Space Research Over the Decade 2023–2032

Themes	Key Scientific Questions
Adapting to Space (Chapter 4)	• How does the space environment influence biological mechanisms required for organisms to survive the transitions to and from space, and thrive while off Earth? • How do genetic diversity and life history influence physiological adaptation to the space environment? • How does the space environment alter interactions between organisms?
Living and Traveling in Space (Chapter 4)	• What are the important multi-generational effects of the space environment on growth, development, and reproduction? • What principles guide the integration of biological and abiotic systems to create sustainable and functional extraterrestrial habitats? • What principles enable identification, extraction, processing, and use of materials found in extraterrestrial environments to enable long-term, sustained human and robotic space exploration? • What are the relevant chemical and physical properties and phenomena that govern the behavior of fluids in space environments?
Probing Phenomena Hidden by Gravity or Terrestrial Limitations (Chapter 4)	• What are the mechanisms by which organisms sense and respond to physical properties of surroundings and to applied mechanical forces, including gravitational force? • What are the fundamental principles that organize the structure and functionality of materials, including but not limited to soft and active matter? • What are the fundamental laws that govern the behavior of systems that are far from equilibrium? • What new physics, including particle physics, general relativity, and quantum mechanics, can be discovered with experiments that can only be carried out in space?

This finite and intentionally prioritized list of KSQs leaves many important lines of BPS research inquiry or conceptual demonstrations on the table; some of these lines of research are more naturally treated as components within integrated research campaigns discussed in Chapter 6. Such a limited list of KSQs is not resilient to the many scientific, economic, and geopolitical surprises that the United States is certain to encounter in the coming decade as in past decades; a recommended decision process to augment and reprioritize KSQs is offered in Chapter 7.

> Finding 3-1: The need for science data on exploration of space exceeds the pace of current experiments. Moreover, as exploration and occupation of space increases in the next decade, the rate and replication of science experiments could increase to meet that demand.

> **Recommendation 3-3: As activity in low Earth orbit increases, and lunar and martian missions are increasingly likely, NASA should increase resources dedicated to understanding the answers to these key scientific questions.**

These themes reflect a significant recasting of scientific thought for the BPS community, framing the science not as a collection of phenomena needing to be understood, but as codified pursuit of the underlying principles at play in the space environment. The resulting KSQs represent a deep prioritization of the science within each of the themes. Those KSQs are fully elaborated in Chapters 4 and 5, and are supported in part by the campaigns and initiatives presented in Chapter 6. (See Figure 3-3.)

FIGURE 3-3 The Expedition 62 mission patch floats inside the seven-window cupola, the International Space Station's "window to the world." The International Space Station has served as the basis for much biological and physical sciences research in the past decade, hosting numerous expeditions, astronauts, and experiments. The next decade in space offers new and different opportunities. SOURCE: Courtesy of NASA, https://www.flickr.com/photos/nasa2explore/49596884056, CC BY-NC-ND 2.0.

4

Science to Enable Space Exploration

Enabling human exploration of space, particularly beyond low Earth orbit (LEO), is a science challenge on the grandest of scales—and one that requires the very best integrated efforts of experts skilled in dozens of different disciplines. Engineers and physical scientists contributing to the development of rockets, propulsion systems, and spacecraft design need to be cognizant of biological limits regarding tolerable gravity forces and life support requirements that take in the wide variety of extraterrestrial conditions of space. Similarly, life scientists working to optimize long-term adaptations to the space environment need to be aware of the tight constraints on crew time, as well as mass and volume of the supplies that can be accommodated aboard spacecraft, and the altered physical processes driven by the space environments. *This chapter focuses on the key scientific questions (KSQs) that need to be answered in the coming decade to enable space exploration, including the rationale and possible research areas that support such investigations.*

The utilization of LEO in the coming decades also brings its own grand science challenges. The opportunities for large numbers of people visiting and living in LEO in commercial and government space stations are likely to increase within the coming decade, highlighting the need of a strong science knowledge base supporting spaceflight. Increased occupation of LEO with additional orbital platforms further opens LEO to unprecedented higher levels of research, development, and manufacturing in space. Therefore, the commercialization of LEO and beyond creates both the need for science support as well as the opportunity to accomplish that necessary additional science with more platforms, researcher work hours, research facilities, and increased flight capacities.

In the next decade, more humans will be traveling regularly to and from space, while also living longer in LEO for science, exploratory, and recreational purposes. They will be traveling to the Moon and living for at least short periods on the lunar surface and preparing for longer-duration exploration missions to Mars or other similar destinations targeted for the coming decades. The guiding principles to meet the critical science needs for these scales of human occupation of space and travel into deep space are summarized within two overarching themes that enable space exploration and present the priorities of science for the coming decade:

- *Adapting to Space:* Optimizing biological adaptations for surviving in, and adapting to, the space environment
- *Living and Traveling in Space:* Creating and maintaining safe, sustainable built environments and building a stable human presence

These two themes provide useful and pragmatic categorizations of the science needed to support the human exploration, habitation, and utilization of space. However, these themes are not independent of each other, nor do they exclusively draw on separate scientific principles. Rather, these themes interact with each other. These themes also draw from the third and final theme represented in Chapter 5, *Probing Phenomena Hidden by Gravity or Terrestrial Limitations*, to meld into a broad scientific endeavor that integrates the many areas of science needed to enable spaceflight successfully and responsibly while benefiting life on Earth.

Each of these themes are built from the specific KSQs that are required to address priority needs. The KSQs of Chapter 3 are described in detail below, to guide research that enables the future of space exploration and continue to drive interesting scientific discoveries that are still needed for exploration. Many questions posed in this chapter on enabling space exploration overlap with those of the following chapter, which addresses those questions enabled by the space environment. Yet the fundamental principle that brings these questions together in the themes enabling space exploration is that these are more applied questions, in both biological and physical sciences (BPS), that are essential to address to support continued success of space exploration in the coming decade.

Several of these BPS KSQs afford some intersection with the Human Research Program (HRP) at NASA, and directly impact the science of human health. (See Appendix D.) However, all BPS investigations are based on fundamental biological and physical sciences that are organized to identify the underlying principles of spaceflight phenomena and phenotypes. Such KSQs are therefore based on fundamental biology across a broad array of organisms, rather than being based on human studies. It is expected that the BPS and HRP programs will continue their history of interactive science in drawing the basic science results of BPS as quickly as possible into relevant human health and life support applications.

Finding 4-1: Interaction between the BPS Division and HRP is critical to ensuring the transfer of basic science knowledge into hardware, policies, and procedures that increase the health of astronauts.

Recommendation 4-1: NASA should continue to strengthen the science exchange between the Biological and Physical Sciences Program and the Human Research Program. Such effort may include establishing a coordinating body and shared research initiatives as well as the two-way exchange of technologies, data, mission science, specimen banking, and plans.

Ideally, in-depth science needs to precede development and increased occupation of space. Specifically, a range of KSQs by BPS could be considered co-requisites or even prerequisites to the scale of space occupation and travel that is envisioned for the next few decades with NASA's HRP and other spaceflight providers. To date, science has identified important biological and physical phenomena that affect human spaceflight, generally negatively. Many of those phenomena have yet to be mitigated or addressed to the point of eliminating risks posed by those phenomena. If the scale of increasing occupation of LEO continues at its current pace, then a concomitant pace of science progress would also increase to safely lead to space-based habitation. Moreover, the current funding pace for BPS cannot support the current science needs, let alone the anticipated scale of needs for the future decade. The science proposed in the themes and KSQs exceeds the current funding trajectory of BPS. These themes in Chapter 3 can be viewed as sequential in importance, even if occurring concurrently. Adapting to space may take prominence early in the coming decade, whereas living and traveling in space will likely receive increased scientific attention toward the midpoint of the decade. (See also Chapter 6.)

Table 4-1 summarizes the KSQs presented in detail in Chapter 4, addressing two of the three themes of this decadal survey. These two themes focus on research outcomes that will enable space exploration in the decades to come. Historically, such research will also confer benefits to life on Earth (including the BPS space research community of U.S. experts), but the driving motivation behind answering these questions is the imperative of national competitiveness for safe and ambitious space missions of increasing duration, complexity, and application. Ironically, that imperative exists with a currently low level of scientific understanding of risks, benefits, and strategies for humans and human-built ecosystems to not just survive, but also to thrive, in that transition. Likewise, researchers must grapple with the limited knowledge about risks and benefits to long-term exploration of potential adverse consequences of our research activities on the lunar and

TABLE 4-1 Key Scientific Questions for Which Answers Will Enable Space Exploration

Themes	Key Scientific Questions
Adapting to Space: Optimizing biological adaptations for surviving and thriving in the space environment	• How does the space environment influence biological mechanisms required for organisms to survive the transitions to and from space, and thrive while off Earth? • How do genetic diversity and life history influence physiological adaptation to the space environment? • How does the space environment alter interactions between organisms?
Living and Traveling in Space: Creating and maintaining safe, sustainable space habitation environments	• What are the important multi-generational effects of the space environment on growth, development, and reproduction? • What principles guide the integration of biological and abiotic systems to create sustainable and functional extraterrestrial habitats? • What principles enable identification, extraction, processing, and use of materials found in extraterrestrial environments to enable long-term, sustained human and robotic space exploration? • What are the relevant chemical and physical properties and phenomena that govern the behavior of fluids in space environments?

martian environments. Answers to questions posed in this chapter will be important to close that gap of data and knowledge over the coming decade.

BPS KEY SCIENTIFIC QUESTIONS THEME 1: ADAPTING TO SPACE—OPTIMIZING BIOLOGICAL ADAPTATIONS FOR SURVIVING AND THRIVING IN THE SPACE ENVIRONMENT

Many gains were made over the past 60 years of space exploration in understanding how organisms, including humans, physiologically adapt to relatively short stays in the space environment (Afshinnekoo et al. 2020, 2021). Indeed, nearly all data from prior decades involve experiments conducted on organisms that made the trip to from Earth to space, after they had spent a few days or weeks in space. Put another way, these data are derived from the transition from Earth to space. Understanding mechanisms underlying those adaptations is important to promoting positive adaptive responses and mitigating negative responses to living in space. Future work will also benefit from the inclusion of ground-based simulations of microgravity, combined with other relevant aspects of spacecraft and planetary conditions, such as the chronic low-dose-rate radiation of galactic cosmic rays, elevated ambient carbon dioxide levels, and altered light–dark cycles. The KSQs of this theme are focused on understanding the initial responses of terrestrial systems to these space environments, the responses that characterize the acclimation and physiological adaptation of an individual to a new environment. Multi-generational genetic adaptation is part of theme 2.[1]

The concepts embodied within this theme of *Adapting to Space* transition lead from the phenomenology of space adaptation into the second theme of sustainable *Living and Traveling in Space*. These themes complement the third theme of *Probing Phenomena Hidden by Gravity or Terrestrial Limitations*, which pushes beyond our current knowledge of, and ability to harness matter, energy, and the universe itself.

Developing a codified understanding of how terrestrial organisms acclimatize and physiologically adapt when they move into space requires answers to essential questions that have been exposed by the research productivity of the previous decades. Some questions are near to answers that will soon translate into mitigation strategies for exploration, such as consistently elevated cytokines (e.g., specific interleukins) that can be targeted by countermeasures. Other questions are only beginning to be asked (e.g., epigenome and epitranscriptome changes) (Saletore et al. 2012), and these questions will help to guide future generations of researchers on advanced mitigations.

[1] See definitions at https://www.merriam-webster.com/dictionary/acclimation and https://www.biologyonline.com/dictionary/physiological-adaptation.

In the following discussion, there is an emphasis on the need to study a wide variety of organisms, larger numbers of individuals, and the need for a better understanding of interacting systems within and between organisms. There remain many unknowns about how the space environment interacts with living organisms, and the following questions aim to increase understanding across multiple species such that knowledge and understanding are based on fundamental biological processes that can be robustly modeled.

Understanding the processes of *Adapting to Space* is guided by the following KSQs:

1. *How does the space environment influence biological mechanisms required for organisms to survive the transitions to and from space, and thrive while off Earth?* Survival in the extreme environments of space requires an organism to alter regulation of critical biological functions, including but not limited to growth and development, repair of wounds, modulation of structures and metabolic pathways, and generation of robust and appropriate defense and immune responses.
2. *How do genetic diversity and life history influence physiological adaptation to the space environment?* Individual differences in biological responses to the space environment have been consistently observed in many organisms and among different genotypes of an organism. These primary observations strongly suggest that genetic and epigenetic factors affect survival and may confer optimal functioning advantages in space.
3. *How does the space environment alter interactions between organisms?* Survival under any circumstances requires effective communications within and among cells and organisms. Much of the data from the past decades strongly suggest that such communications are altered during spaceflight; however, the fundamental principles mediating those alterations have yet to be developed. Understanding the way spaceflight changes how cells communicate both within an individual organism and between multiple organisms is therefore critical to managing disease and multi-species ecosystems in space systems.

The impact and rationale for each of these three KSQs are provided below. Within each KSQ, potential research areas outline the scientific sub-questions that need to form the basis of BPS research projects or programs. While prioritized by the committee, these are denoted as potential research areas in recognition of the expectation that other research areas responsive to these high-level questions may emerge and mature over the coming decade. (See Chapter 6.)

Recommendation 4-2: NASA should increase resources dedicated to producing and understanding the answers to the key scientific questions that address the transitions to and from space. The committee sees potential for significant advances in space exploration if a biological and physical sciences portfolio in the coming decade is aimed at understanding

- **Biological responses that occur during transitions between the Earth and space environments over extended duration and distance to fundamentally enable space exploration;**
- **Genetic diversity to understand positive and negative responses and long-term adaptations to spaceflight to accelerate the identification of risks, mechanisms of adaptation, and potential positive adaptations that could improve life in space; and**
- **How cells, systems, and organisms concurrently adapt to the spaceflight environment and develop mechanisms for encouraging positive and countering negative communicated responses.**

Question 1: How Does the Space Environment Influence Biological Mechanisms Required for Organisms to Survive the Transitions to and from Space, and Thrive While Off Earth?

Impact and Rationale

Space is an isolated, confined, and extreme environment that induces extreme changes and significant physiological adaptations in a variety of living organisms, including microbes, plants, and animals. Many of these adaptive changes resemble pathophysiological changes. Plants grown in space exhibit gene expression patterns similar to plants experiencing both biotic and abiotic stress on Earth (Manzano et al. 2022). Humans and other animals in space exhibit muscle atrophy and bone resorption that resembles accelerated aging of the musculoskeletal system (see also Chapter 2) (Juhl et al. 2021). Rather than years, these animal adaptations occur on a timescale of weeks to months. Some of these changes owing to weightlessness can be reduced with countermeasures. Others can be mitigated with adaptive technologies such as glasses to mitigate the symptoms of Spaceflight Associated Neuro-Ocular Syndrome (SANS) (Wojcik et al. 2020). On the other hand, optimal preventive measures to combat effects owing to radiation have yet to be defined. Some, but not all, of these biological adaptations appear to be reversible after returning to Earth, while others (such as subsets of gene expression changes) are not (Garrett-Bakelman et al. 2019). Understanding such adaptations in more depth is necessary for optimizing survival of organisms during space exploration, especially beyond LEO. Knowledge gained through research on these adaptive processes will provide mitigations for adapting to space, while also contributing to better treatments for related pathological states on Earth and to basic knowledge of normal variations in biological adaptation and physiological stress (Afshinnekoo et al. 2021).

The changes that are observed in the space environment are sometimes referred to as maladaptations, precisely because of the resemblance to pathologic states. However, a distinction can be made between changes that are appropriate physiological responses versus those that indicate the induction of pathology. For instance, browning skin in response to ultraviolet light exposure is an appropriate adaptive response caused by the oxidation of melanin (Brenner and Hearing 2008). Conversely, developing melanoma is a pathological response caused by cumulative UV damage to DNA (Teixido et al. 2021). Both are responses to UV light exposure. Plants exhibiting stress transcriptomic responses in space may show altered growth patterns, but it is unclear whether those responses and patterns of gene expression will negatively affect plant productivity. While decades of research have shown myriad changes in response to the space environment, less is known about whether these changes are appropriate adaptations to stimuli or reflect underlying damage to homeostatic or repair mechanisms.

The adaptations to space that are of high societal interest (see Figure 4-1) are those that are functional and familiar, and in many ways reminiscent of Earth-based challenges of isolation, disease, or aging (e.g., impaired memory or sight or bone loss). However, the understanding of why these functional adaptations occur—and thus the means to mitigate them so that humans and other living organisms can thrive when transitioning to new space environments—is rooted in fundamental biology at the level of tissues and cells and molecules. A good example of a homeostatic mechanism is the ability of cells to sense typical weight-bearing loads and to remodel structures accordingly (Robling and Turner 2009). Because the chronic lack of weight bearing in the space environment alters how cells sense mechanical loading, it is critical to find approaches to ensure structural integrity on long missions. Indeed, the physiological repair and maintenance of the integrity of tissues in both animals and plants is an essential consideration for all space travel beyond very short-term stays in space. Similarly, spaceflight changes in cellular signaling constitute a particular concern. Several lines of investigation have reported significant impacts on neuroplasticity, cognitive functions, neurovestibular system, short-term memory, cephalic fluid shift, control of motor function, and psychological disturbances, especially during long-term missions (Marfia et al. 2022; Mhatre et al. 2022). The perturbation of brain and neural function is perhaps one of the most important effects of the space environment because the nervous system controls many essential body functions (Roy-O'Reilly et al. 2021). In plants, cellular recognition of being in the space environment involves apparent signaling of stresses and environmental conditions that are seemingly at odds with the actual environmental data, suggesting that interpretation of cellular signaling is affected by space (Barker et al. 2020).

Biological Mechanism—Fluid Shifts

A shift in the distribution of body fluids is a feature of the space environment. On Earth, gravity pulls fluid downward, inducing hydrostatic pressure differentials within a given organism that are substantially reduced in microgravity. For example, in humans, the changes in hydrostatic pressures that occur when traveling from Earth to space result in physiological changes such as sustained headward redistribution of fluid (about 2 liters) and a decrease in arterial pressure (Baran et al. 2021; Nelson et al. 2014). Fluid shifts produced by changes in

FIGURE 4-1 Research on biological organism adaptations. Research that quantifies effects of space environment on biological organism adaptations spans from plants to microbes to worms to mammals, with attendant challenges in physical systems that enable such scientific experiments. SOURCES: Composed by Tim Warchocki; adapted from S. Bhattacharya, NASA, presentation to the committee, January 13, 2022. Image credits: *Fruit fly*: Erik Karits, Pixabay. *Baby mouse*: Pexels. *Flower, Microorganism, DNA, Greenery, and Pollen*: NASA. *Dandelion*: Daniel Dan, Pexels. *Adult hermaphrodite*: adapted from R. Sommer, Max Planck-Institute for Biology Tuebingen.

the gravitational environment can, therefore, alter the pressure, volume, and flow of fluids in living systems (Diedrich et al. 2007; Najrana and Sanchez-Esteban 2016; Villacampa et al. 2022). With a transition to a reduced gravity environment, living systems adapt to maintain homeostasis. Data collected from microbes, plants, and mammals show that microgravity imposes significant consequences on virulence, growth, physical strength, and stamina (Bacci and Bani 2022; Nickerson et al. 2004; Villacampa et al. 2022). Yet, it is nearly impossible to uncouple the effects of fluid shifts on cells, tissues, and whole systems. This is because the effects of fluid shifts arise from a complex network of events that occur over multiple timescales to govern the synthesis, transport, and conversion of molecules that may each be disproportionate to what is deemed normal under Earth's gravitational pull. Such pressure differences regulate molecular transport, while mechanical forces in turn regulate cellular activity and trigger tissue remodeling. The reduced hydrostatic pressures and mechanical loading couples the cell interactome, the local microenvironment, and global transport processes to produce the observed mechanistic and functional changes.

Other ways that fluid shifts disrupt mechanisms that arise in living systems are by controlling thermodynamics, reaction kinetics, and transport processes. For example, oxygen transport from the air to the bloodstream in the lung is dependent on hydrostatic pressures, differential tissue perfusion, plasma volume, and ion/protein concentrations (Dunn et al. 2016). Reduced oxygenation capacity, blood pressure, and blood volume results in loss of physical strength and stamina and may impair cognition and vision. In contrast, certain cells have evolved to have mechanosensory functions (e.g., hair cells of the inner ear and certain bone cells) and alter signaling to the brain or nearby cells, triggering vertigo, nausea, and atrophy of muscle and bone (Cohen et al. 2019; Pagnotti et al. 2019; Wang et al. 2022). Similarly, plants evolved mechanosensory statoliths that are responsible for the transport of the growth hormone auxin (Su et al. 2020). Indeed, this same disruption can occur in engineered fluidic and microfluidic systems in space, complicating the way scientists measure and model biological systems using instrumentation. The effects of fluid shifts in microgravity are complex; microgravity will alter hydrostatic pressures, flow, and mechanosensory pathways that drive changes in overall maintenance, growth, and repair processes.

Potential Research Areas—Fluid Shifts

How do shifts in fluid dynamics impact the movement and metabolism of molecules?

Astronaut performance is dependent on cognitive and physical function, which can be compromised by exposure to microgravity (Garrett-Bakelman et al. 2019). Earth's gravitational force pulls fluid down to the feet when the individual is upright; this effect is removed when they are lying in a supine position. These changes do not result in significant pathology on Earth. Sustained exposure to microgravity, however, will alter natural physiological gradients and can have debilitating consequences, including the movement of drugs or food additives within the body. In general, fluid shifts alter molecules from reaching and leaving their desired destinations, which may disrupt sensing, concentrations, and gene pathways. In addition, changes in mechanosensory pathways may alter the resistance of flow of molecules across biological boundaries. Thus, molecules that circulate or are entrained locally may exhibit new rates of synthesis or transport phenomena that may readily impact molecular absorption, degradation, metabolism, and excretion. Elucidating mechanisms that are influenced by pressure gradients and mechanics, coupled with changes in local concentrations, may determine opportunities to correct or ameliorate negative health effects or cognitive impairment.

How do fluid shifts in the lungs affect oxygenation, mechanics, and resident immune and microbial cells?

Pulmonary perfusion (Najrana and Sanchez-Esteban 2016) is idealized by two models: zone and slinky. The zone model establishes that regional perfusion depends on the balance between pulmonary arterial pressure, pulmonary venous pressure, and alveolar pressure (Gold and Koth 2016). Hydrostatic pressure caused by gravity is critical to determine pulmonary perfusion. In the lower lung, blood flow depends on the difference between arterial and venous pressure. At the top of the lung, a decrease in hydrostatic pressure owing to microgravity can cause pulmonary pressures to fall below alveolar pressure and compromise blood flow. Thus, there is a vertical gradient in blood flow in different regions of the lung. In the slinky model, the lung is represented as a spring, where gravity causes uneven ventilation in the lung through the deformation of lung tissue (Yamada et al. 2007). In either model, the pressures of perfused blood and mechanics of the lung will influence oxygenation, release of carbon dioxide, and potential changes in extracellular matrix, immune cell population, and microbial populations. Understanding how to increase oxygenation and maintain healthy lung mechanics in reduced gravity environments is critical and aided by basic science of fluid flow in such contexts. In addition, lunar regolith can also impact lung and immunology; these are important areas of study.

Note that other potential research areas have increasing relevance to astronaut health and safety, including effects of fluid shifts on vascular remodeling and on visual acuity—particularly with combined space effects including microgravity and radiation. These potential research areas are outlined in Appendix D, BPS Theme 1.

Biological Mechanism—Plant Growth and Survival

Many spaceflight studies have attributed the changes observed in plant growth in spaceflight solely to the reduced gravity environment. However, the reported effects are also likely to be attributable to some or many of the other variables encountered on flight. These include the different hardware utilized as well as altered convection and gas exchange, leading to changes in exposure to stress-inducing signals generated by the plant itself or arising from other elements of the flight cabin including radiation. Several plant transcriptomic studies reveal alterations in pathways involved in oxidative stress, defense responses, and DNA damage (Barker et al. 2020; Manian et al. 2021; Manzano et al. 2022).

Potential Research Areas—Plant Growth and Survival

How do the altered fluid shifts experienced during space exploration affect plant growth and survival?

Gravitropism causes plants to grow in a specific orientation by triggering a transduction cascade that affects the distribution of a growth hormone that plays an important role in plant growth and development (Leyser 2005). Mechanosensory statoliths, which evolved to respond to gravity, normally modify the transport of the growth hormone, auxin, by acting on specific auxin transporters (Konstantinova et al. 2021; Levernier et al. 2021). On Earth, statoliths utilize sedimentation to achieve concentration gradients. In microgravity, sedimentation does not occur, altering plant growth behavior (Boonsirichai et al. 2002; Kiss et al. 2000; Teale et al. 2006). Understanding how fluid shifts affect growth and plant structures will help identify gene pathways for enhancing plant growth. (See Figure 4-2.)

What are the outcomes of biological interactions with planetary surface materials?

Plants, especially crops, are normally grown in soil in terrestrial environments. While aeroponic and hydroponic systems can be deployed in spaceflight and on planetary surfaces, the use of lunar or martian surface materials drives development of support bases systems that further reduce transport costs by utilizing local resources. Several terrestrial inert support mediums like perlite, rockwool, or clay aggregates are commonly used to support the hardy plants in hydroponics-based systems. In these systems, an aqueous growth medium provides the cocktail of macro- and micro-nutrients required for plant development. The nutrients can be stored in their raw chemical/powder form such that they occupy a relatively small area, whereas the support mediums are more voluminous. Plant biology researchers have extremely limited access to the lunar regolith collected by the past lunar missions carried out by the United States (Jerde 2021) and others. This has placed constraints on research; greater access is required to carry out plant growth optimizations and scale up the studies on crop plants. While the recent lunar regolith-based plant study (Paul et al. 2022) on Arabidopsis showed no effects of the space environment on the seed germination, post-germination development was severely impacted compared to the lunar regolith simulant JSC-1A. The chemical composition of the lunar regolith (Lindsay 1976; Papike et al. 1982; Taylor 1992) suggests that it may be a great support medium for early seeding, but not necessarily a good source of bioavailability of nutrition and ions required for plant growth. Much of the plant mineral and water intake from the environment and its movement inside the plant body depends on the physical properties, including temperature, humidity, pressure, and gravity potentials inside the plant and in the growth environment. Therefore, to maximize the mineral utilization and extraction for plant development and/or find alternate

FIGURE 4-2 Studying plant growth and survival is a key area of biological and physical sciences research. Here, fresh chili peppers are pictured growing inside the International Space Station's Advanced Plant Habitat shortly before being harvested. SOURCE: Courtesy of NASA, https://www.flickr.com/photos/nasa2explore/51818701986, CC BY-NC-ND 2.0.

FIGURE 4-3 Fresh food, such as vegetables, provide essential vitamins and nutrients that will help enable sustainable deep-space expeditions. SOURCE: Courtesy of NASA, https://www.flickr.com/photos/nasamarshall/20191358350/in/album-72157667439026641, CC BY-NC 2.0.

strategies and avoid carrying and shipping the large voluminous media needs, research is required to assess the effectiveness of extraterrestrial (lunar and Mars) regolith as a replacement for soil/support medium and nutrition (Duri et al. 2022). (See Figure 4-3.)

Biological Mechanism—Metabolism

A comprehensive examination of data from human tissue, cultured cells, and animal models suggests a genetic basis for the effects of altered gravity (da Silveira et al. 2020). This study included an examination of the largest cohort of astronaut data to date. Analyses of findings indicated significant effects on metabolic genes, which would be expected to directly and indirectly impact all tissue types. In mouse models, lipid metabolism and liver function have also been shown to be impacted by the spaceflight environment (Beheshti et al. 2019; Blaber et al. 2017; Jonscher et al. 2016). Indeed, the effects of altered gravity on muscle has also been linked to changes in expression of metabolic gene and peripheral insulin resistance in mice (Gambara et al. 2017; Vitry et al. 2022) suggesting whole body effects of tissue-specific gene expression changes.

Fundamental understanding of mammalian cell metabolism includes consideration of mitochondria, organelles present in most eukaryotic cells that are the locus for the biochemical processes of cellular use of oxygen for energy production. Thus, they are essential to many cellular functions, including calcium signaling, cell cycle control, cell growth and differentiation, and cell death (Nguyen et al. 2021). Microgravity has been shown to increase glycolysis and affect mitochondrial metabolism (Nguyen et al. 2021). The tricarboxylic acid (TCA) cycle that occurs within mitochondria is a core metabolic pathway that produces the majority of the reduced coenzymes used to generate adenosine triphosphate (ATP) in the electron transfer chain (Cox 2013). The TCA cycle is also upregulated in microgravity, as are reactive oxygen species (ROS) levels and nicotinamide adenine dinucleotide phosphate (NADPH) oxidase activity (Nguyen et al. 2021), which in its reduced version is an essential electron donor in all organisms (Spaans et al. 2015). Conversely, microgravity results in the downregulation of the oxidative phosphorylation system by which ATP is produced, resulting in reduced ATP production and mitochondrial respiratory chain components (Nguyen et al. 2021). Given the pivotal role mitochondria play in cell functions, it is essential to understand how the metabolic production of ATP using oxygen (along with the by-product of ROS) in the space environment is altered and how this impacts on the organism's health.

Potential Research Areas—Metabolism

How are cellular oxygen uptake and utilization changed by the radiation experienced in the space environment and what are the downstream impacts?

The mitochondrial response to galactic radiation is less described, but recent evidence using simulated galactic radiation (dosing with heavy ions) has demonstrated that levels of key proteins involved in mitochondrial fatty acid metabolism are decreased, while levels of proteins involved in various cellular defense mechanisms, including antioxidant defenses, were increased (Jain et al. 2011). When mice were irradiated with simulated galactic radiation, they exhibited increases in steady-state levels of mitochondrial ROS and mitochondrial and non-mitochondrial respiration, without any changes in mitochondrial mass, suggestive of a metabolic rewiring of mitochondria (Kim et al. 2021).

Overexpression of the enzyme mitochondrial catalase (which protects against ROS) in mice significantly decreased cognitive dysfunction after proton irradiation, ameliorating adverse effects via a preservation of neuronal morphology. This demonstrates the neuroprotective effect of reducing mitochondrial reactive oxygen species through the targeted overexpression of catalase.

Can metabolism be deliberately slowed (e.g., shallow metabolic depression or even inducing torpor) to protect organisms from the negative impacts of the space environment (including radiation) and to reduce consumables needs on exploration-length missions?

Studies dating back 50 years demonstrated that hibernating ground squirrels are partially protected against gamma radiation effects (Musacchia and Barr 1968). More recent work has demonstrated radio-protective effects of a torpor-like state (called synthetic torpor) in non-hibernators like zebrafish and in rodents. If synthetic torpor could be safely achieved in humans, this may help protect against, for example, carcinogenic effects of ionizing radiation, and the upload mass of food and supplies might be reduced by a significant amount. In addition, the induction of synthetic torpor may protect (as it does in natural hibernators) against the disuse bone and muscle atrophy observed in long-duration flyers. Although these goals are more relevant to exploration-class missions of several years' duration, early proof-of-concept work in rodents and non-human primates has already started with grants from the Translational Institute for Space Health (TRISH).

Biological Mechanism—Structural Biology and Wound Responses

Wounds can occur during space exploration, including internal wounds often associated with bone fracture. Thus, it is necessary to know whether the healing of such fractures will proceed normally or will be delayed. Few fracture healing studies have been conducted in space, and those completed do not capture the entire duration of fracture healing. To date, it is unknown whether healing to prefracture/pre-spaceflight mechanical properties can be achieved in the space environment, and no studies have been performed that considered fractures that had been induced in space. Thus, it is critical to perform fracture healing studies in bone that has adapted to the spaceflight environment, and to consider both the effects of microgravity and radiation exposure.

Appropriate wound healing requires coordinated interactions between immune cells, soluble factors, and skin cells, all of which are individually impacted by microgravity (Jemison and Olabisi 2021). Microgravity has been shown to impair the functionality and wound healing response of so-called soft tissues (e.g., muscle, skin), but the mechanisms meditating such space environment–specific changes are not known. There is evidence that wound healing is further delayed by subclinical hypovolemia, a state in which astronauts arguably remain throughout spaceflight missions. Experiments with ex vivo human constructs are on the horizon, but these cannot incorporate the astronauts' hypovolemic state, characterized by a 10–15 percent decrease in blood volume, and such platforms do not capture the complex interplay between platelets, the soluble factors the platelets release, and immune cells and their factors.

Although plants and animals differ in their development and approach to repair damaged tissues, cells, and organs, they both undergo repair and regeneration throughout organismal life (Ibáñez et al. 2020; Vining and Mooney 2017). In general, plants will experience damage owing to pathogens and/or physical wounding like breakage, laceration, and localized cell/tissue death. Plants cannot move away from stressors in their local environments, nor can damaged or new cells be moved in long-distance transport within an organism owing to the presence of cellular structures like cell walls. However, the damaged cells or the cells within the same neighborhood can return to a stem cell state. This ability can facilitate cell repair and/or initiate the formation of new cell/tissue/organ through regeneration and differentiation phases (Perez-Garcia et al. 2018). The regaining of stem cell activity, regeneration, and differentiation are largely dependent on sensing stress; activating repair genes; and the synthesis, transport, and regulation of growth hormones and downstream factors (Hernández-Coronado et al. 2022; Ikeuchi et al. 2020; Mironova and Xu 2019; Reid and Ross 2011). While wound healing and repair are considered trauma states, it is likely that many clonally propagated (asexually produced) plants and plantlets use the same repair and regeneration mechanisms to survive, thrive, and continue to produce food and nutrition for sustaining humans in the space environment. Therefore, it is vital to study these processes in plants as well as animals.

Potential Research Areas—Structural Biology and Wound Responses

How are the mechanisms governing wound healing of mineralized and soft tissues altered in the space environment?

Actual wound response mechanisms in animal tissues could be studied to understand impacts from the space environment and to identify specific healing mechanisms at play in space environments that may differ from those on Earth. Because wound responses are seen in plants in spaceflight, even in the absence of actual wounding, increased knowledge of the plant defense responses may reveal the overall principles involved in recognition of tissue damage in space and the repair mechanism enacted as a response to that perceived damage. Indeed, analogy to other organisms including extremophiles has the potential to provide Earth-based knowledge that can be applied to enabling space exploration.

> Finding 4-2: Extensive space research to date clearly indicates that extreme environments of space require an organism to regulate critical biological functions, creating responses that are still not fully understood. However, the consequences of these responses will have a dramatic impact on the ability of life to survive and thrive in the spaceflight environment, including deep-space vehicles and in extraterrestrial surface structures. Understanding of these consequences at a basic science level is critical for developing countermeasures, particularly when considering the extended times or distances away from Earth envisioned for missions to the Moon and Mars.

Question 2: How Do Genetic Diversity and Life History Influence Physiological Adaptation to the Space Environment?

Impact and Rationale

Genetic diversity among individuals of any one species can influence both short- and long-term adaptation to the space environment; this includes the modulating impact of preexisting phenotypic adaptations or disease. Intrinsic individual variability in biological responses to the space environment, independent of factors like activity level or nutritional input, are consistently observed in many organisms and can even exceed any sex- or age-related differences. This poses both a conundrum and an opportunity. (See Figure 4-4.) In particular, if the individual differences promoting a positive (or the smallest negative) adaptation can be identified, this may lead to strategic selection of plants or even humans best suited to thrive in the space environment (Rutter et al. 2020). It might also

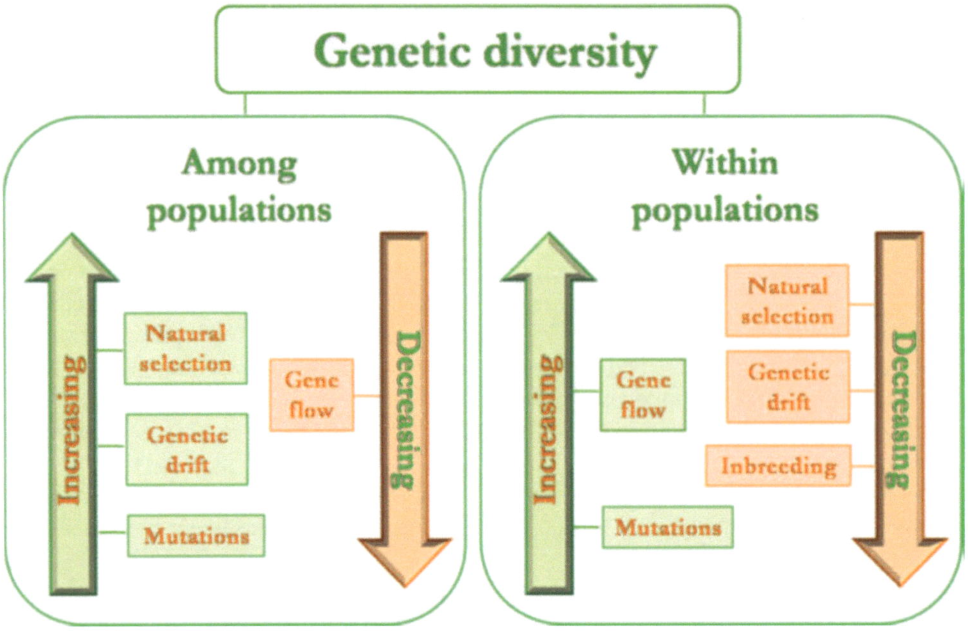

FIGURE 4-4 Individual and community life history, epigenetic changes, and environment drivers. SOURCE: Modified from M. Nonić and M. Šijačić-Nikolić, 2019, "Genetic Diversity: Sources, Threats, and Conservation," in *Life on Land*, Springer Nature.

suggest how individual genes could be manipulated to create organisms more resilient to the physiological stressor of microgravity, galactic cosmic radiation exposure, and other relevant aspects of the space environment. Similarly, given that many astronauts are older than 45 years of age, the extended duration (2–3 years) of exploration class missions and the opening of opportunities for more humans to travel in space owing to privatized spaceflight, it is increasingly critical to understand if and how preexisting chronic conditions that may go undetected before launch will impact on that individual's health and performance in the space environment.

Genetic Diversity in Spaceflight Response

Two decades ago, the only astronaut health issue for which a large enough data set exists to allow valid conclusions to be drawn about gender differences was orthostatic intolerance following shuttle missions, in which women have a significantly higher incidence of presyncope during stand tests than do men (Harm et al. 2001). Although more findings on gender responses have been reported, the sample size for long-duration female astronauts is still relatively modest, and it remains a challenge to make robust conclusions. In 2014, a series of reports were prepared and published as a group, including an executive summary of the series (Mark et al. 2014). The studies covered neurosensory systems, behavioral health, reproductive health, the immune system, musculoskeletal health, and cardiovascular alterations. Tentative conclusions included that spaceflight-induced visual impairment is more typically associated with male astronauts, while orthostatic intolerance manifests more frequently in women. However, in the study (conducted in 2013) women comprised only 57 of the 534 astronauts (11 percent). The sample size for women is still low, but in a recent (2022) metabolomic study on 51 astronauts, the percentage of female subjects doubled compared to the 2014 study. The metabolic profiling of the astronauts revealed that there were numerous statistically significant differences between male and female astronauts in energy metabolism, bone mineral and muscle regulation, immunity, as well as macromolecule maintenance and synthesis (Stroud et al. 2022). Sample size among other variations among humans also remains low, as the totality of crewed missions in the space environment do not reflect the demographic and associated genetic diversity among Earth-based populations. Mindful broadening of the representation of humans deployed in crewed missions by NASA and other organizations can confer this data generation over time, such that important fundamental scientific questions of genetic diversity to stressor acclimation in the space environment can be formulated and assessed sooner than 2 decades from now.

Evidence gathered from multiple organisms have shown that different genotypes respond differently to environmental stresses (Friedman et al. 2020; Krebs and Loeschcke 1994; Martin et al. 2017; Roger et al. 2012). For example, genetic variation in mice resulted in differences in the skeletal response to unloading from immobilization (Friedman et al. 2019, 2021). *Caenorhabditis elegans* strains that carry the same allele of *npr-1*, but with different genetic backgrounds, showed different pathogen avoidance behavior and immune responses toward *Pseudomonas aeruginosa* infections. Even in large ecosystems, genetic diversity may play a particularly important role in the face of multiple stressors. If different genotypes differ in their ability to deal with certain stressors, the capacity of a community to withstand a stressor depends on the presence of a resistant genotype. Genotypic richness increases the probability of the presence of such a genotype, and hence the capacity to uphold ecosystem functioning under stress conditions. Thus, knowledge of the effects of space-related stresses and of genetic variation for such stress resistance could be necessary for decision making with respect to choices of reserves designed to protect a particular species, choosing the individuals to introduce to a reserve, and estimating long-term performance of populations in the space environment that may change unpredictably.

Genetic variations occur naturally to enable plants to adapt to their ecological environment. Among the wild accessions (ecotypes) of the model plant *Arabidopsis thaliana*, there is a wide range of genetic diversity and trait variations (Shindo et al. 2007). Studies with Arabidopsis have contributed the bulk of the fundamental insights into the genes important to the physiological adaptation of plants to spaceflight, even to the level of organ- and cell-specific responses (Barker et al. 2020; Manian et al. 2021). But there is also a wealth of plant spaceflight data from other plant species and genotypes, which ranges from observational studies of apparent health and vigor to molecular analyses of transcriptomes and proteomes that categorize genomic responses. For example, the monocot plant model (*Brachypodium distachyon*) (Hasterok et al. 2022) comprised a new plant model for spaceflight experiments in 2021 (Masson, APEX-08[2]) and will contribute to evaluations of genetic diversity in a model grass species that will be relevant to future exploration agriculture.

In the spaceflight environment, genotypic variation in *Arabidopsis* impacts the plant's ability to adapt. For instance, between Arabidopsis ecotypes (Columbia [Col-0] and Wassilewskija [WS]), there are morphological differences in how these plants grow in microgravity. In the presence of directional light, the WS ecotype roots tend to skew strongly to the right, while the Col-0 roots do not (Califar et al. 2020; Paul et al. 2012). Moreover, various ecotypes of *Arabidopsis* show different transcriptomic responses to the spaceflight environment (Choi et al. 2019; Paul et al. 2017; Sng et al. 2018). Comparative analysis between Col-0 and WS transcriptome showed that there were substantial differences in the number of genes engaged to physiologically adapt to the environment of the ISS, and further, that manipulating the genetics by even a single gene could impact the metabolic cost of space adaptation. In addition, evidence indicates that between these ecotypes, there are differences in the way plants regulate their molecular processes (i.e., alternative splicing) to adapt to the spaceflight environment (Beisel et al. 2019). Thus, investigating different ecotypes of the same plant species may provide insightful discoveries on how plants adapt and thrive in space.

For any organism, experiments with known defective or intentionally mutated genotypes test whether the spaceflight response can be influenced by manipulating the genome, and can help elucidate the processes most important to plants in space with respect to viability, stress responses, and productivity. It is important to note that novel environments may also induce responses from signals that are inappropriately activated or misinterpreted. Thus, it is also important to identify which responses are counterproductive and impose an unnecessary load on the plants as they try to adjust, and then determine whether they can be eliminated to reduce the load on the adaptive process. A refined understanding of the genomic responses of plants to spaceflight can enable genetic manipulation to produce varieties that are better adapted to growth in spaceflight through the elimination of unnecessary

[2] See information on Advanced Plant Experiment-08 (APEX-08), https://science.nasa.gov/biological-physical/investigations/advanced-plant-experiment-08.

responses and through the introduction of new important traits. While there is no single gene that defines the physiological adaptation to spaceflight, there are metabolic pathways that appear important to almost every genotype that has been assayed. The genes associated with cell-wall remodeling, ROS signaling, and unique application of light-sensing pathways suggest that targeting these pathways for investigation may help in the development of "space-adapted" plant genotypes. Before genetic manipulation can take place, the genotype needs to be better correlated to phenotype. Future work will need to focus on which traits to enhance and which phenotypes to study, manipulate, or vary; this can be guided by the genetic variants of desirable traits and their associated loci, which can then be engineered with molecular genetic tools. Of note, there exists a wealth of genetic variation in natural populations of plants, and many of those traits are possibly relevant for deep-space exploration. For instance, even within a single species, there are significant differences in the degree of tolerance to DNA damaging UV radiation (Piofczyk et al. 2015) or heavy metal stress (Li et al. 2019).

Examples of how a single gene mutation can elicit a different response to environmental stresses or treatments are found in abundance in the field of molecular biology. For example, in *C. elegans*, mutation of a single age-1 gene causes not only 65 percent longer life expectancy but also increased intrinsic thermotolerance (Lithgow et al. 1995). In *Drosophila*, overexpression of a single gene, SOD1, in the motorneuron extends normal lifespan by up to 40 percent and rescues the lifespan of a short-lived Sod null mutant (Parkes et al. 1998). In plants, the overexpression of a single gene can confer resistance to a myriad of abiotic and biotic stresses (Parmar et al. 2017) as well as influence pollinator preference (Lüthi et al. 2022). However, in response to spaceflight, there are only a handful of studies that have investigated how a single gene mutant impacts the organism's response to the environment. Plants with a single gene mutation show different responses to the spaceflight environment in terms of physiological development or molecular changes (Angelos et al. 2021; Califar et al. 2020; Paul et al. 2021; Wang et al. 2021). In *C. elegans*, a Dystrophin-like dys-1 gene, which increases body wall muscles, showed that gene expression in spaceflight was less affected than in wild-type *C. elegans* (Xu et al. 2018). Such investigations will broaden our understanding of how organisms physiologically adapt to space and how a gene(s) can confer increased or decreased survival or fitness in the spaceflight environment, and these insights will be valuable in providing organisms with added advantages to thrive in space for extended periods of time.

Potential Research Areas

Given that many biologic adaptations are affected by multiple gene loci, how best to identify genetic variations that are best suited to adapting to the space environment?

One approach to understanding the impact of individual variability is to quantify responses to real or simulated space environmental factors across many genetic strains within a species. These studies generally illustrate a marked variability, depending on the complexity of the outcome measure, but cannot pinpoint the precise gene(s) responsible for those variable responses. Another approach, more useful in mammals/humans, is to contrast those sub-populations with the largest and smallest decrements in function and explore underlying genetic variations. The growing database housed by GeneLab provides a wealth of information that can be used in this fashion.

With all of the data now available from model biological systems in space, how best to use those data to enhance biological usefulness in space exploration?

In addition to the genetic diversity approach to examining the physiological adaptation to spaceflight, an approach particularly important for plans is understanding the pathways to implementing that genetic information to produce biological systems that directly support space exploration. How best to deploy the currently available genetic information to producing strains that better support exploration?

Does the space environment induce epigenetic changes in some individuals, and, if so, how much does this contribute to individual variability of responses?

Stress tolerance and vulnerability varies widely across biological organisms, species, and individuals within a species. There are many factors that influence tolerance to stressors within a lifetime and across generations.

There is emerging evidence that epigenetic changes in gene expression may contribute to tolerance to a wide range of stressors, including those associated with the space environment. For example, NASA's Twins Study (Garrett-Bakelman et al. 2019), which provided a unique opportunity to compare molecular profiles of identical twin astronauts, revealed how the structure and function of numerous physiological processes are altered by space exploration (e.g., changes in DNA methylation, telomere length, immune response, microbiome, biochemistry, and metabolomics), and these results were confirmed by chromatin profiles in SpaceX's Inspiration4 crew. Further research is required to understand how the physical variations that naturally occur between individuals impact the induction of such changes.

Can gene manipulation be harnessed to improve/optimize stress-resistance function genes?

Much is known about genes, proteins, and enzymes; genetic differences within and across species; and the nature of stressors that trigger cellular stress responses, including several common to the space environment (Kültz 2020a,b; Liu et al. 2021). For example, ionizing radiation triggers the DNA damage response; heat, cold, and chemical toxins evoke the unfolded protein response; and hypoxia, respiratory poisons, and nutrient deprivation activate the mitochondrial stress response. Cellular stress response pathways and stress-adaptive specific gene hubs have been identified in highly stress-resistant organisms, including the tardigrade (Kirke et al. 2020). Selective breeding and gene editing to increase expression of cellular stress response pathways do elevate tolerance to specific environmental stressors such as salinity, heat, cold, and nutrient deprivation in plants, animals, and simple organisms.[3] More work is needed to identify cellular stress responses activated to space-associated stressors (Liu et al. 2021). The results of this work can accelerate the evolution of space-tolerant phenotypes across the tree of life.

Finding 4-3: Genetic differences in biological responses to the space environment are consistently observed in many organisms, strongly indicating that genetics plays a role in survival and optimal functioning in space. In addition, epigenetics and life experience likely have both positive and negative impacts on survival in the spaceflight environment.

Question 3: How Does the Space Environment Alter Interactions Between Organisms?

Impact and Rationale

The function of "cross talk" and interactions between cells and tissues within a species, as well as between organisms and between species (e.g., host/microbe interactions), is to regulate homeostasis and to communicate a response to stressors. The space environment is well known to produce stress on organisms through the action of microgravity, space radiation, fluid shifts, and other stressors (Afshinnekoo et al. 2021), making the understanding of intercellular and extracellular communication in organisms in the space environment a priority for the future.

Intercellular communication is an essential hallmark of multi-cellular organisms (see Figure 4-5) and can be mediated in three main ways: direct cell-to-cell contact, transfer of extracellular molecules, and transfer of extracellular vesicles (EVs) like exosomes. For example, one form of cell-to-cell communication, gap junctions, has been implicated in unloading induced bone loss (Lloyd et al. 2012, 2013). Extracellular molecules and extracellular vesicles are transported to distant sites by fluid systems within the organism, and these include blood plasma, lymphatic, and other fluids such as cerebrospinal fluid (CSF) in animals and phloem in plants. The study of bioactive molecules and EVs in eukaryotes and prokaryotes has the potential to greatly increase our knowledge of the effect of the space environment on organisms. At a basic level, they are a key step in the response of organisms and are therefore important in understanding fundamental and critical questions about life in space. They can also be used as a diagnostic tool to detect imbalances within the organism and predict possible future pathologies. Furthermore, there is also a potential for their use as treatments for countermeasures in space and on Earth.

[3] See Branco et al. (2022), Du et al. (2022), Guimarães et al. (2021), La Spina et al. (2021), and Zhang et al. (2021).

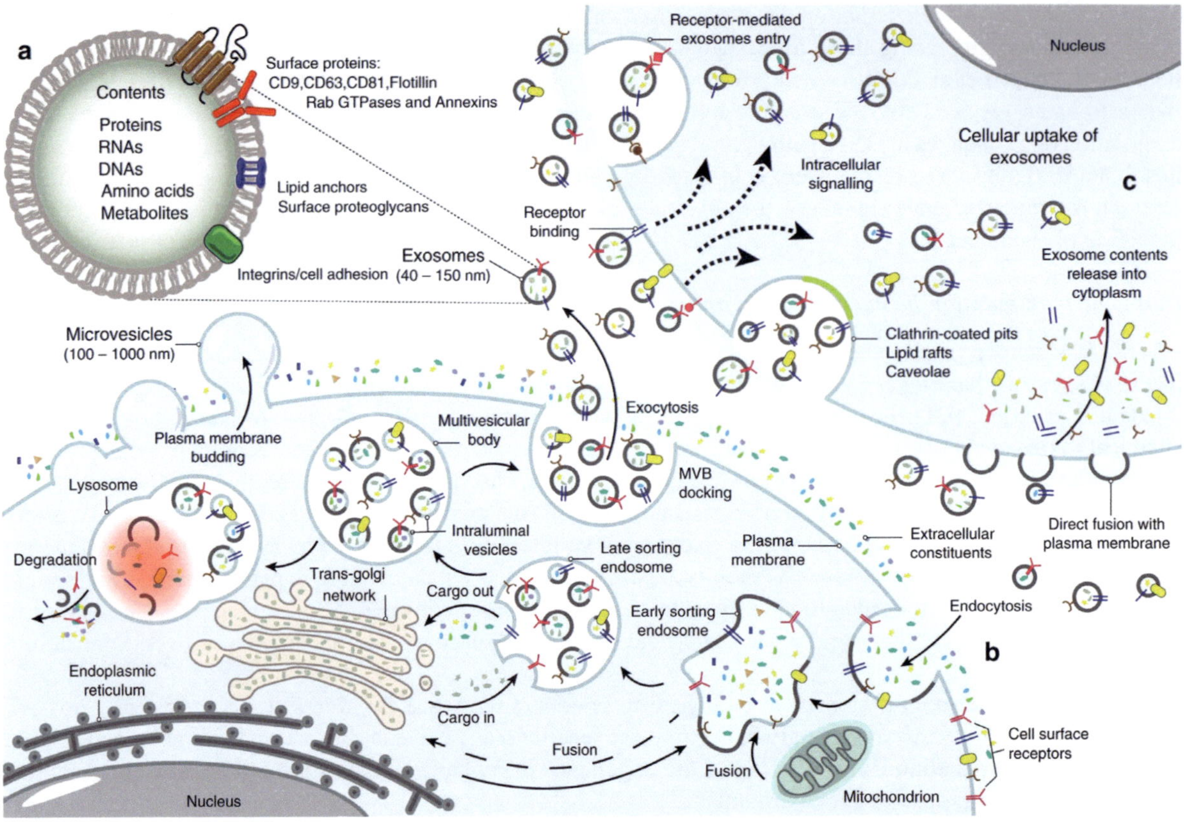

FIGURE 4-5 Intercellular communication. Biochemical cross talk among cells enables communication within and among tissues, organisms, and species—including via exosomes. SOURCE: From Ni et al. (2020), https://doi.org/10.1038/s41413-020-0100-9, CC BY 4.0.

Impact on Extracellular Molecules and Extracellular Vesicles

Extracellular vesicles are prime candidates for cross-talk vectors, and their study in organisms in space is critical for an understanding of the biological effects of the space environment. Extracellular molecules are the communicators in intra-organism cross talk and can be either small molecules, protein-based, or nucleic acid–based. Some proteins like insulin or inflammatory response molecules like cytokines have been known for some time, whereas others are still being discovered (e.g., histone variants). Cell-free nucleic acids include several types of DNAs and RNAs, and also have shown responsiveness to spaceflight (e.g., mitochondrial DNA, mtDNA) (Lo et al. 2021). First recognized as biomarkers in cancer patients (Schwarzenbach et al. 2011), these molecules are potentially biomarkers for many more pathologies and homeostasis changes in organisms in the space environment. Indeed, Malkani and colleagues have identified and validated a spaceflight-associated microRNA (miRNA) signature that is shared by rodents and humans in response to simulated short-duration and long-duration spaceflight (Malkani et al. 2020). Additionally, a subset of these miRNAs was found to regulate vascular damage caused by simulated deep-space radiation.

Extracellular vesicles are a heterogeneous group of membrane-limited vesicles loaded with various proteins, lipids, and nucleic acids. Release of extracellular vesicles from its cell of origin occurs either through the outward budding of the plasma membrane or through the inward budding of the endosomal membrane (Abels and Breakefield 2016). Initially, it was shown that the release of extracellular vesicles is part of a disposal mechanism to discard unwanted materials from cells. Subsequent research has further shown that the release of EVs is also an important mediator of intercellular and interorgan communication that is involved in normal physiological processes as well as in pathological progression. So far, extensive evidence on all these different types of vesicles indicates that extracellular vesicles are a key player in the intercellular communication between cells. Because they can fuse with a cell membrane of specific tissues, the contents can be efficiently delivered to a target cell, and thus EVs are also potentially biomarkers for homeostasis changes and pathologies in organisms in the space environment. Identification of extracellular molecules and extracellular vesicles secreted during spaceflight or planetary life is the first step in understanding key cellular cross-talk mechanisms.

Potential Research Areas—Extracellular Molecules and Extracellular Vesicles

What are the downstream phenotypic consequences of this crosstalk as altered by the space environment?

Much research has identified a role for extracellular vesicles in disease processes, either in contributing or responding to pathogenic mechanisms, or as a source of biomarkers for disease. For example, exosomes have been found to contribute to diverse biological processes, such as angiogenesis, inflammation, morphogen transportation, and programmed cell death (Colombo et al. 2014). In the central nervous system (CNS), with the ability of extracellular vesicles to transport cargo packaged by the originating cell, their role in the pathogenesis of neurological conditions, and particularly neurodegenerative diseases associated with misfolded proteins, has become a growing area of interest. Neurodegenerative diseases, such as Parkinson disease, Alzheimer disease, Creutzfeldt–Jakob disease, and amyotrophic lateral sclerosis, share a common mechanism in which distinct proteins become misfolded and deposited in specific regions of the body during the pathogenic process. Another common feature of these disorders is that these misfolded proteins "spread" to defined brain regions, suggesting that the disease process involves intercellular movement of these proteins (Braak et al. 2003). Extracellular vesicles can carry the many different proteins associated with neurodegenerative diseases.

Extracellular vesicles also help control cardiovascular homeostasis and remodeling by mediating communication between cells and directing alterations in the extracellular matrix to respond to changes in the environment (Hutcheson and Aikawa 2018). There is also growing evidence for cytokines acting in a paracrine fashion to impact nearby tissues. For instance, osteokines produced by bone and altered by unloading (e.g., RANKL), have demonstrated anabolic or catabolic effects on myogenesis the development of new muscle fibers. Conversely, many myokines produced by skeletal muscle are altered by unloading (e.g., myostatin), which may in turn have detrimental effects on bone by promoting osteoclast activity while inhibiting osteoblast activity. Thus, cross-talk factors are normally directly involved in homeostasis and potentially in pathologies. As such, they are an important factor underlying spaceflight adaptations.

Impact on Complications of Host–Microbiome Communication

There is also cross talk between organisms, such as between the gut microbiome and a host. It will be important to understand how the deep-space environment alters human, plant, and animal microbiomes over time. The host microbiota plays a crucial role in maintaining the balance between health and disease, influencing host immunity in both innate and adaptive immune function (Zheng et al. 2020). This includes the mucosal immune system, which is relatively understudied, but serves a critical role as the first line of defense against mucosal pathogens, as well as dietary antigens (Wright et al. 2019). During deep-space travel, biological organisms are exposed to the full range of solar and galactic cosmic radiation in addition to other stressors, such as microgravity. These environmental perturbations have the potential to further disrupt immune responses beyond those observed in LEO, where organisms are afforded some radiation protection by Earth's magnetosphere. For example, changes in immune function have been shown to impact the sensitivity of cells and tissues to radiation, and conversely, radiation can also alter immune function (Fernandez-Gonzalo et al. 2017; Laiakis et al. 2021; Lopatkin et al. 1992; Pecaut et al. 2003). A study on mice demonstrated that spaceflight-related changes in the host microbial community were correlated with shifts in metabolism-related gene expression in the host (Jiang et al. 2019). There is currently a large gap in our understanding of how deep-space stressors may alter microbiota-immunity interactions and the pathobiological consequences of those alterations, such as increased infectious disease risks and autoimmune responses. Connecting changes in microbiome composition to function will be particularly important when considering how to optimize positive adaptations that could improve host resistance to deep-space stressors. These tests could also evaluate the potential for pre- and probiotics or nutritional countermeasures to boost positive adaptations or mitigate negative adaptations.

Potential Research Areas—Host–Microbiome Communication

How can important microbial commensals be harnessed while also preventing exposure to pathogens?

Short-duration growth of biomedically important microbes in LEO and in spaceflight analog culture has been shown to induce global changes to a wide range of cellular processes, including growth and metabolism, gene expression, virulence, and pathogenesis-related stress resistance (e.g., acidic pH, oxidative stress, thermal stress, biofilm formation, antibiotic resistance) (Schiwon et al. 2013). Amplification of the disease-causing properties of microbes poses a potential risk to the success of deep-space missions, especially given the negative impacts of space travel on some aspects of immune function (Crucian et al. 2018; Guéguinou et al. 2009). An enhanced understanding of how the stressors associated with deep-space environments impact microbes, their disease-causing properties, and their susceptibility to treatment with antimicrobials will be important to enable deep-space exploration (Tierney et al. 2022). In parallel, research into how the deep-space environment alters host immunity and how immune defenses can be bolstered through the use of countermeasures, including those that help maintain and/or restore balance to the microbiome, will be critical.

In addition to studying alterations in microbiome composition, another key focus could be to investigate whether the functional properties of commensal microbiota are maintained in the deep-space environment. As an example, multi-drug-resistant *Acinetobacter pittii* was found adapting and evolving onboard the ISS, which indicates that such challenges may arise again for other strains (Tierney et al. 2022). ISS strains often contain functions and traits that enable them to survive in harsh environments, including the transcriptional regulator LexA, among others, and matching these changed genetic profiles will be key to long-term tracking of any biomedical risk.

NASA performs routine microbial monitoring on the ISS such as Microbial-Tracking projects 1–3, but culture-based methods currently in use do not provide a comprehensive assessment of microbial diversity. NASA is actively working to increase monitoring capabilities by testing different in situ protocols for nucleic acid extraction for downstream molecular analyses on the ISS, and such work can be continued and expanded. High-throughput sample processing onboard the ISS followed by next-generation sequencing (NGS) would enable astronauts to know (in near real time) what microbes are present, their properties, and possible risk. This was shown by the Biomolecular Sequencer mission (Castro-Wallace et al. 2017) and subsequent work by ISS crews using small commercial sequencers. Many terrestrial species sent from Earth to grow on the ISS have become more virulent and more antibiotic resistant, and have formed more biofilms, properties that can affect astronauts' health and the stability of the spacecraft. (See Figure 4-6.) The ability to quantify microbial load associated with crew, in real time, will provide an efficient means to monitor immune function and to allow for proper countermeasures to be implemented, such as medication strategies, prophylactic antibiotics, or stress-reduction therapies (Kim et al. 2013; Knox et al. 2016; Singh et al. 2018).

Moreover, additional integration and mapping of microbial metadata for crew, wastewater, air sampling, and environmental monitoring could link to a range of new tools that can leverage the data for further biological functions and molecular discovery. This can include tools to mine Clustered Regularly Interspaced Short Palindromic Repeats (CRISPR) arrays and new Cas enzymes, biosynthetic gene clusters (BGCs) prediction and annotation, anti-microbial resistance (AMR) genes and virulence markers, novel taxa and strain-typing of space-based variants, phased genetic variants from the genomes, access clinical versus environmental data from crews to look for associations, other nucleobase modification detection tools, annotation of putative mobile genetic elements, support for long reads and other NGS or spatial omics data types, and integration with various other microbial (EMP, MetaSUB, HMP) and model data sets.[4]

Of course, microbes can interact with a given host as either a symbiont, commensal, or pathogen. The analysis of the plant-microbe interaction has demonstrated and underscored the importance of microorganisms and their biochemistry in the healthy functioning of terrestrial plants (Saad et al. 2020). For example, nitrogen-fixing rhizobacteria, which attach to plant roots and form biofilms, are important players in a healthy plant–microbe relationship. To date, 12 Veggie space crop experiments have been conducted on the ISS, and of these, so far, the VEG-01A,

[4] See Danko et al. (2021), Foox et al. (2021), Liu et al. (2021), McIntyre et al. (2019), MetaSUB International Consortium (2016), and Parker et al. (2020).

FIGURE 4-6 With new equipment, it has now become possible to sequence DNA in space as shown here by astronaut Raja Chari. SOURCE: Courtesy of NASA, https://www.flickr.com/photos/nasa2explore/51805058680, CC BY-NC-ND 2.0.

VEG-01B, and VEG-03A "Outredgeous" Red Romaine lettuce samples were evaluated by microbiome analysis using next-generation sequencing on the Illumina platform (Khodadad et al. 2020). Here, robust microbial communities were observed along with no human pathogens. The microbiological counts on spaceflight-grown produce during VEG-01A, VEG-01B, and VEG-03A were determined to be no more abundant than store-bought produce. This baseline survey begins to build our understanding of the microbiome associated with space crops and will lend to further microbial tracking efforts. Additional studies evaluating the genomic and physiological effect on plants after controlled pathogen infections will show if any altered mechanisms of pathogenesis are unique to the spaceflight environment. Zinnia plants growing in the Veggie unit, which were under an excess water stress, were highly susceptible to infection by an opportunistic fungal pathogen *Fusarium oxysporum* (Schuerger et al. 2021). Decreased plant resistance and increased microbial virulence are potential problems likely to be encountered in the space environment.

Potential Research Areas—Plant–Microbial Interactions

What are the plant–microbial interactions that enable a thriving space agriculture, and what are the mitigation strategies needed to maintain it?

On Earth, plants are colonized by a variety of micro-organisms, including mutualistic or pathogenic bacteria, rhizobacteria, mycorrhizal fungi, and mycoparasitic fungi. These plant-associated microorganisms can have profound effects on seed germination, seedling vigor, plant growth and development, nutrition, diseases, and productivity. Plants can employ mutualistic microbial interactions and can regulate different signaling molecules, attractants, or stimulants for the purpose of establishing selective relationships with different microbes, such as the plant growth promoting (PGP) microbes, and for defense against pathogens (Baetz and Martinoia 2014; Schuerger et al. 2021; Shigenaga and Argueso 2016). These dynamic processes are also co-dependent on the environment in which the plants are grown (Mendes et al. 2013). Advancing our understanding of how deep-space environment associated stressors impact plant growth as well as their microbiota and their functions would be key in establishing space agriculture that will support safe food production. Research focused on molecular studies to characterize the plant and microbial changes alone or in combination will fill knowledge gaps on plant–microbial interactions relevant to the spaceflight environment. In parallel, research into how space-stressors affect disease development in crops will be critical. Apart from investigating plant-resistance mechanisms or pathogen-virulence, focus on enabling technology to diagnose, monitor, and counter these challenges will be needed to ensure plant health, human health, and overall ecosystem health to maintain a thriving space agriculture.

> Finding 4-4: Cells, tissues, systems, and organisms communicate through a variety of mechanisms, including biochemical and mechanical processes. Spaceflight data suggest that these communication processes are affected by spaceflight, potentially impacting host pathogen relationships and other community biological processes.

BPS KEY SCIENTIFIC QUESTIONS THEME 2: LIVING AND TRAVELING IN SPACE—CREATING AND MAINTAINING SAFE, SUSTAINABLE SPACE HABITATION ENVIRONMENTS

As humans travel farther from the planet for increasingly extended durations (approaching years), they will rely more heavily on in situ resource utilization (ISRU). This aspect of sustainable habitation on the lunar surface and other planetary surfaces will include the recycling of waste products into oxygen, water, food, and other critical materials as well as utilizing resources available in the surrounding location. At present, astronauts still rely heavily on the routine transport of materials and consumables to the ISS, which will cease operations in the coming decade. For example, food is regularly transported to the ISS, although there have been recent opportunities for astronauts to eat space-grown vegetables. Additionally, while interdependent recycling systems are used onboard ISS to produce breathable air and potable water for the crew, the cycle is not near closure and water is routinely launched to the station for replenishment. However, as deep-space outposts are being established and multi-year missions are considered to Mars, continuous resupply of critical resources will become increasingly more difficult and expensive as compared to LEO. (See Figure 4-7.)

The theme of *Living and Traveling in Space* fully integrates the biological and physical sciences in enabling space exploration, including a completely integrated system of biology encased within unique physical systems, which is seeking to eventually thrive without resupply. The concepts embodied within this theme pull from the phenomenology of the theme of *Adapting to Space* by extending the idea of adaptation toward the notions of a consistent and sustained system existing in space. The concepts are therefore very systems-based and deeply examine physical phenomena that support spaceflight and biological systems as impacted by the physical systems that provide space transit and habitation. These concepts also extend to those in Chapter 5, seeking answers to questions enabled uniquely by access to the space environment, in that the fundamental physical science knowledge allowing communication and navigation in space underpin technical advances to live in and explore space more deeply.

Understanding the processes of *Living and Traveling in Space* is guided by the following KSQs:

- *What are the important multi-generational effects of the space environment on growth, development, and reproduction?* Meeting the goals of long-term and effective habitation of space requires an understanding of the reproduction and development of micro-organisms, plants, and even animals as affected by the spaceflight environment across multiple generations. This knowledge formalizes and initiates the conceptual leap from the adaptation of individuals going to space to the adaptation of a species to living in space within new space-induced equilibria.
- *What principles guide the integration of biological and abiotic systems to create sustainable and functional extraterrestrial habitats?* The logistically isolated, environmentally closed, risk-intolerant, and functional

FIGURE 4-7 NASA astronaut and Expedition 66 Flight Engineer Kayla Barron works with the Veggie PONDS space agriculture experiment that explores how to grow fresh food during long-term spaceflight missions. SOURCE: Courtesy of NASA, https://www.flickr.com/photos/nasa2explore/51874369450, CC BY-NC-ND 2.0.

requirements for minimal environmental impact drive a need for efficient and flexible design in the provision of environmental services derived from local resources. Furthermore, these services need to operate as close to closed-loop as possible.

- *What principles enable identification, extraction, processing, and use of materials found in extraterrestrial environments to enable long-term, sustained human and robotic space exploration?* Resources from planetary materials, atmospheric, and mission wastes can all be captured and harnessed for production of mission-critical/high-value chemicals, materials, and biological feedstocks. Effective chemical, physical, and biological methods for locating, extracting, and processing these resources for use in downstream production requires development of new mechanisms and supportive infrastructure.
- *What are the relevant chemical and physical properties and phenomena that govern the behavior of fluids in space environments?* Fluids may behave very differently in microgravity or other spaceflight gravity levels. In addition, several factors—including elevated radiation, reduced pressure, extreme temperature, and so on—may significantly alter the behavior of fluids whether contained in biological systems or outside in free form. In some cases, the phase of the fluid may be hard to determine, or the phase-change phenomena can be very complicated. On the other hand, complex fluids may play an important role in additive manufacturing, space transport technology, and biological repairing processes.

The impact and rationale for each of these four KSQs is provided below. Within each KSQ, potential research areas outline the scientific sub-questions that are needed to form the basis of BPS research projects or programs.

While prioritized by this BPS decadal survey, these are denoted as potential research areas in recognition of the expectation that other research areas responsive to these high-level questions may emerge and mature over the coming decade. (See Chapter 6.)

Recommendation 4-3: To ensure the long-term survival of life in the spaceflight environment, NASA should ramp up investigations into space impacts on sustained human presence in space by investigating:

- **Reproduction, development, and evolution within all relevant biological systems;**
- **The relationships between biology and space hardware to ensure structural integrity, optimized recycling, and utilization of local resources;**
- **Effective chemical, physical, and biological methods for locating, extracting, and processing local resources, especially from the Moon, for use in local habitation and downstream production; and**
- **Fluid physics, combustion, and related sciences to enable sustainable space exploration and habitation.**

Question 4: What Are the Important Multi-Generational Effects of the Space Environment on Growth, Development, and Reproduction?

Impact and Rationale

Vital to long-term space exploration, but largely unexplored to date, are multi-generational impacts of living in the space environment. (See Figure 4-8.) For life to be sustainable in space over the long term, it is critical to understand how the combined stressors of spaceflight affect reproductive health, growth, and development of

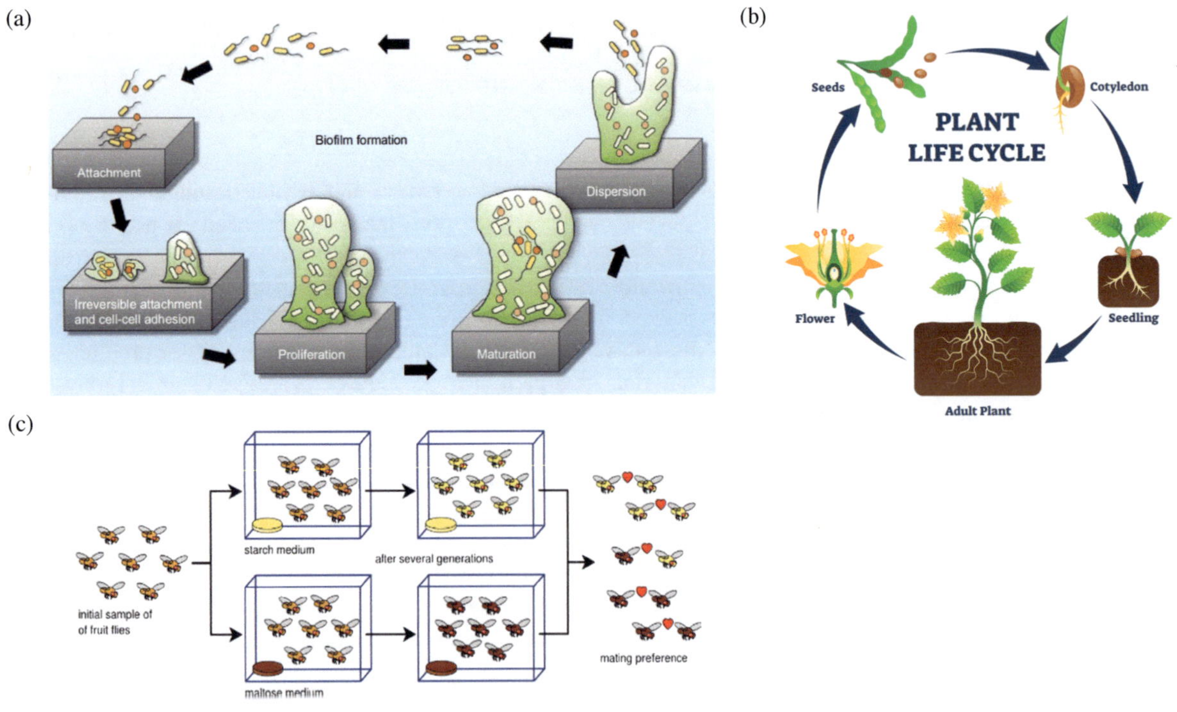

FIGURE 4-8 Organism life cycles. Multi-generational effects of the space environment can impact microbes, plants, and animals' growth and development. SOURCES: (a) Courtesy of Perfectus Biomed Limited. (b) © VectorMine, Dreamstime.com. (c) Adapted from © UC Museum of Paleontology Understanding Evolution, www.understandingevolution.org, CC BY-NC-SA 4.0.

subsequent generations. It is known that, with every cell division cycle, new rarely lethal mutations are integrated in the newly replicated genome. Thus, there is an accumulation of mutations that may or may not create additive/negative effects on functional elements. Environmental stressors have been associated with increased genome-wide mutation rates in eukaryotes and prokaryotes (Galhardo et al. 2007; Horneck et al. 2010; MacLean et al. 2013; Vijayendran et al. 2007). Given the unique combination of stressors present in the space environment, long-term growth in space may induce unique mutagenic events relative to ground controls. Terrestrial studies have shown that, in addition to primary DNA sequence changes, epigenetic modifications (e.g., DNA methylation, histone modifications, chromatin restructuring) also play a significant role in the phenotypes expressed in subsequent generations. Over the past decade, an increasing number of studies have discovered spaceflight-induced epigenetic alterations in a variety of biological organisms in LEO.[5] Given the added impacts of long-term radiation exposure in spaceflight during deep-space exploration, it is expected that there will be generational effects in long-duration space habitation. Owing to generation times and the opportunities and needs for reproduction, the generational effects of spaceflight are most likely to be seen in microbes, plants, and other organisms with short life cycles (e.g., worms and flies), although the concepts and consequences equally apply to any animal reproduction in space. Large numbers of multiple generations of microbes are already inhabiting the ISS, simply as part of the built environment of that spacecraft. For plants, seed production and germination in space is likely to be integral to crop production in support of human occupation of space and the Moon and Mars. Multiple generations of animals, such as worms and flies, may be needed for waste recycling and/or food production in space, as well as a catalog of microbial functions. Also, while the potential of multiple human generations in space is likely to occur only beyond this decade, the general principles that guide mutation, epigenetics, and development could be approached and understood.

Multi-Generational Effects—Microbes

Microbes cycle through generations at rapid rates, making multi-generational adaptation a very active part of microbial adaptations, including to the spaceflight environments. In contrast to plants and animals, multiple generations are a given in the durations of spaceflights, making any distinction between organismal and multi-generational adaptation meaningless. Across generations, microbes exhibit a diversity of mechanisms for adapting to their environment, through immediate control by means of phenotypic plasticity through transcription and translation, and evolutionary mechanisms, such as mutation, epigenetic modification, and horizontal gene transfer by way of conjugation, transformation, or phage-mediated transfer. Over the past decade, there have been a few long-duration microbial culture experiments in spaceflight or spaceflight analog culture that have observed changes in survival, growth, and metabolism, as well as in the genome and/or gene expression (Bai et al. 2022; Fernander et al. 2022; Horneck et al. 2012; Nicholson and Ricco 2020). Microbes grown for short duration (hours to days) during spaceflight have been extensively studied, and in ground-based microgravity analog culture systems exhibit unique changes in fundamental characteristics (e.g., morphology, growth, metabolism, gene expression, gene transfer susceptibility) and biomedically important phenotypes (e.g., antibiotic resistance, pathogenesis-related stress responses, in vitro infections, virulence). While the precise mechanism(s) responsible for these changes remain to be fully elucidated, key features associated with the spaceflight environment include alterations in gravity, fluid dynamics, decreased hydrostatic pressure, and increased radiation. Bacterial responses to the spaceflight environment are anticipated to operate as a function of both the internal and external environment of the cell, as previously observed for microbes in terrestrial environments (Benoit and Klaus 2007). Microbes rapidly sense and respond to physical, chemical, and biological cues in their environments. The quiescent, low fluid shear environment that exists as an indirect effect of microgravity is proposed to be an important environmental signal in liquid microbial cultures (Nickerson et al. 2004). Related external influences also include changes in the availability of nutrients and the ability to remove metabolic byproducts. It has also been proposed that reduced convection and the secondary substrate concentration gradients that form for nutrients such as phosphate and oxygen may be an important mechanism governing microbial alterations in microgravity (Kim et al. 2013). These changes are not

[5] See Chen et al. (2021), da Silveira et al. (2020), Higashitani et al. (2021), Hughes-Fulford et al. (2015), Waters et al. (2021), and Zhou et al. (2019).

only important when considering the responses of naturally occurring microbes, but also members of synthetic microbial communities, where the individual microbial members and their intended functionalities may be manipulated to achieve a desired endpoint (e.g., improvement of plant health or production of a needed biomolecule). In addition, microgravity could impact animal and plant physiology, gene expression, and overall health. Thus, it will be critical to gain a broader understanding of how different microbes adapt to the deep-space environment, which includes the combined effects of altered gravity, radiation, pressure, low fluid sheer stress, vibration, and so on. The effects of spaceflight on microbial genetic evolution also needs to be examined to avoid the spread of potentially hazardous biological entities beyond the environment in which they were intended.

Potential Research Areas—Microbes

How does the spaceflight environment shape microbial genotypes, phenotypes, and their relationships over time?

Prokaryotic genomes are arranged in operons and are relatively simple compared to larger eukaryotic genomes. However, genetic principles that apply to higher organisms can be tested in microbes. The regulatory processes of microbial gene expression such as epigenetic modification, chromosome compaction, and post-translational modifications, and the formation of inclusion bodies can be observed and assessed using molecular techniques in cultures of microbes in spaceflight and contrasted to ground controls to elucidate the effect of the altered stressors experienced in this unique environment. A short-term study can reveal the frequency of mutations, insertions, duplications, translocations, and deletions within genomes. In addition, a long-term multi-generational study of microbes, either as a monoculture or mixed consortia, can reveal how the spaceflight environment shapes the microbe or microbiome and can reveal the forces that shape the mechanisms of mutation, genetic drift, and natural selection. Prior work in the past decade using microbial consortia focused on the ISS built environment as a microbial observatory. Environmental monitoring was conducted on surface, air, and water samples from the ISS, and both genotype and phenotype assessments were made. It was found that, on average, the studied ISS isolates were no more virulent or pathogenic than terrestrial control strains as defined by their genomes and by phenotypic assessments (Blaustein et al. 2019; O'Rourke et al. 2020). However, because the founding populations for these microbes were not known, it was not possible to make assessments of mutation, genetic drift, and natural selection because, as this work has also shown, microbes hitchhike on crew and payloads (Lee et al. 2021; Morrison et al. 2021), providing new genetic influxes to the population. This knowledge gap can be addressed by deploying a controlled multi-generational assessment of individual microbes and microbial consortia during sequential continuous cultures or chemostats. These experiments can first be designed on the microfluidic scale and subsequently be scaled up as required. The resulting increased understanding the microbial responses of key species to ambient and engineered spaceflight environments will be imperative for supporting robust biological life support systems in future surface habitats on the Moon or Mars. These experiments can leverage new advances in multi-omics profiling and sensor technologies, as well as classical microbiological analyses, to obtain a more complete view of any changes that occur in space relative to ground control.

Future long-term culture experiment designs could consider the renowned long-term evolution experiment (LTEE) in *E. coli* from Lenski and colleagues (Lenski and Travisano 1994). This experiment began with six replicates of two founding strains of asexually reproducing *E. coli*. The experiment was started with the goal of observing and then recapitulating the evolutionary processes of mutation, genetic drift, and natural selection. Each day, 1 percent of the culture was diluted into fresh media, and then every 7 days the remaining 99 percent of the culture was preserved to archive the population at that timepoint's population so it could later be revived for further experimentation. Today, the LTEE has run for more than 60,000 generations (Good et al. 2017), and its results have demonstrated that rapid changes occur early on as the culture adapts to new conditions. In theory, this technically simple experiment could easily transition to flight. However, limitations associated with crew time, safety, and hardware have previously presented challenges for liquid cultures. Automated hardware is likely the best approach for long-duration experiments of this nature. As such, the hardware design needs to incorporate multi-functional capabilities that enable repeatable transfer of small volumes (microliters) of culture or conversely, the retention of small, defined culture volumes in chambers that can be flooded with larger volumes of media or preservatives. Biofilm formation is another key consideration that can also present a technical challenge to hardware operation, particularly for long-term experiments.

How do microgravity-induced changes in mechanical forces, including alterations in fluid dynamics, affect microbes?

Microbes rapidly and frequently sense and respond to physical cues in their environment, including fluid shear, quorum sensing, and surface contacts (including with other cells). In response to changes in fluid shear, microbes have been observed to alter key characteristics such as gene expression, stress resistance, biofilm formation, adherence, invasion, and virulence (Nauman et al. 2007; Nickerson et al. 2004; Persat et al. 2015). Microbial structures implicated in sensing of mechanical changes associated with environments in terrestrial settings include fimbrae, pili, flagella, mechanosensitive ion channels, and the cell envelope.[6] Not all microbial adaptations to mechanical forces are intuitive. Some studies have had surprising and important consequences, such as the discovery that increased fluid flow enables certain bacteria to adhere more tightly to a surface, as is the case for *E. coli*. Thus, the microgravity environment possesses fluid characteristics that regulate microbial responses in unexpected ways.[7] However, there are still major knowledge gaps in this area. Future research needs to focus on the identification of the molecular mechanism(s) used by microbes to sense and respond to the unique physical environments associated with space and how this may change over multiple generations. Because the mechanisms used by different microbes on Earth to sense and respond to physical forces have not been fully elucidated, this also presents an opportunity for spaceflight researchers to contribute new mechanistic discoveries to this area of research.

How best to harness favorable spaceflight-induced microbial phenotypes to make beneficial microbiomes for humans, plants, and the greater ecosystem?

Microbes are of particular interest and utility for mission science because their rapid generation times will allow experimental assessment of evolutionary effects of spaceflight in short time periods, and also because of their dominant beneficial effects as symbionts with plants, fungi, and animals in microbiomes, and their deleterious effects as pathogens. The past 10 years have underscored the importance of microbes as ecosystem unifiers and demonstrated that they are ubiquitous in the spaceflight environment. Outstanding basic questions include: What are the fundamental effects of the spaceflight environment on horizontal gene transfer mechanisms utilized by bacteria, and what is the effect of the altered physical environment on cellular machinery involved in DNA replication, RNA transcription, and translation, and DNA repair? How can spaceflight biofilm production be managed to reduce unintended effects and to optimize beneficial properties (e.g., wastewater purification systems)? Additional questions related to the utility of microorganisms for beneficial applications include a wide range of unknown biology. For example, what combinations of microorganisms that evolve and thrive in the space environment best benefit growth of specific plants in space? How are those beneficial phenotypes maintained over long durations in space? How is the natural skin and gut microbiome of animals and humans altered over generations in space, and how can beneficial microbiomes be maintained—for example, by probiotic or prebiotic ingestion? Can some of these microbes be used for ISRU applications or even for carbon sequestration in a habitat or spacesuit? Which microbes most efficiently enable fermentation as a nutrient preservation measure in space environments?

How does the spaceflight environment influence microbial community dynamics and the ability of members of microbial communities to coordinate their metabolism toward carrying out key metabolic processes, such as nutrient cycling?

Microbes on Earth primarily exist as members of interacting communities to carry out fundamental processes that are vital for ensuring life on our planet. These beneficial processes include support of plant growth, cycling of carbon and other nutrients, degradation of pollutants, decomposition of waste, and purification of water. These beneficial processes will also be needed for long-duration space journeys and eventual habitation on the Moon and Mars. It is not yet understood how specific communities of microbes that carry out needed processes in the space habitat are impacted by space conditions; how composition of microbial communities are impacted over generations in the space habitat; and how beneficial outcomes of microbial communities for plant growth, water purification, or other practical uses to humans may be optimized in the space environment.

[6] See Blount and Iscla (2020), Dufrêne and Persat (2020), Mathelié-Guinlet et al. (2021), Nickerson et al. (2004), and Persat et al. (2015).

[7] See Bijlani et al. (2021), Fajardo-Cavazos and Nicholson (2021), Nickerson et al. (2004, 2016), and Zea et al. (2016).

Multi-Generational Effects—Plants

Successful crop production for astronaut life support functions absolutely depends on having viable seeds for the next cycle for the sexually produced crop plants, as well as the continued maintenance of the fidelity of the genome, which could impact viability of the subsequent generations (including nutritional quality). Decades of plant research have demonstrated that plants can successfully develop and set seed in spaceflight environments.[8] However, there are also abundant data that illustrate that plants perceive the spaceflight environment as an abiotic stress, and engage diverse genomic responses[9] to physiologically adapt to stressors that include microgravity, radiation, and habitat constraints. With every meiotic (reproductive) and mitotic (developmental) cycle of a cell division, the full copy of the genome is replicated and provides the functional elements for transcribing expressed genes and translated proteins. Plants contributing to advanced life support (ALS) systems included in space and other exploration habitats are envisioned to thrive and multiply, and continue to provide food and recycling functions over multiple generations. However, little is known about how the accumulation of spaceflight-associated mutations may compromise downstream crop yield phenotypes over multiple generations, other than the confirmation that plants can indeed successfully reproduce in space (Mashinsky et al. 1994; Musgrave et al. 2000). What is yet to be studied in spaceflight environments is the long-term life cycle and multi-generational effects on sustaining plant health and development, and crop yield and nutritional quality, while monitoring any mutational load and epigenetic changes imposed on the genome. Therefore, it is important to study the multi-generational effects of the space environment on plants.

Potential Research Areas—Plants

How does the spaceflight environment influence pollination/fertilization and seed development/viability of various crops over multiple generations, and what strategies are needed to ensure continuous crop production?

Some of the earliest orbital plant experiments failed to set seed, not because of any gravity-sensing mediated pathway per se, but because the biophysical behavior of fluids and gases in the spaceflight environment can impose physiological and biochemical stress. In the microgravity environment, without convective mixing, localized ethylene accumulation and oxygen depletion can compromise plant functions, such as seed set (Musgrave et al. 1997; Salisbury 1997; Wheeler et al. 1996). For many crop plants requiring insect pollination, it has been shown that humans can effectively play this role in orbital habitats, and *Brassica* plants hand-pollinated by astronauts on space shuttle mission STS-87 successfully set seed at a rate comparable to the ground controls (Kuang et al. 2000). However, while gravity is not required for seed set and development (Musgrave et al. 2000), there can be delayed development and altered compositions of lipid and carbohydrate storage in seeds set in spaceflight habitats compared to the ground controls (Kuang et al. 2005; Popova et al. 2002). The quality of seed storage compounds is crucial to the next generation of the plants, and to humans who may be relying on the seeds as the primary crop, thus further investigations to thoroughly understand how the spaceflight environment impacts seed set and development is essential to the success of sustained space agriculture. Decades of research in plant science on the space shuttle, MIR, and the ISS reinforce that the absence of a gravity vector itself may not be the most crucial stressor or impediment to successful plant growth and development, but rather the attending features of a microgravity environment need to be reconciled (lack of convection, fluid physics of water delivery, modified tropisms, and gravimorphogenesis) with the needs of a plant habitat. Thus, it is essential that research in the next decade coordinates the engineering of spaceflight habitat development with plant genotype selection and development.

What are some of the space-relevant epigenetic changes that affect plants over multiple generations, and how can plants develop fitness and resilience to these changes?

Epigenetic changes are one of the many mechanisms that plants and animals employ to cope with environmental stresses (Crisp et al. 2016; Kakoulidou et al. 2021). Plants can trans-generationally inherit epimutations that

[8] See Ivanova et al. (2001), Link et al. (2003), Mashinsky et al. (1994), Musgrave (2002), and Musgrave et al. (2000).

[9] See Barker et al. (2020), Johnson et al. (2017), Kruse et al. (2017), Meyers and Wyatt (2022), Paul et al. (2013, 2017), and Sugimoto et al. (2014).

prime them to better deal with continuous stressors, although current data suggests that the dynamic epigenome in one generation plays a much larger role in plant adaptation. Work in plant communities challenged with recurrent stressful events suggests that the epigenetic variation of the plants in these ecosystems are part of the adaptive strategies (Weinhold 2018). Epigenetic responses have also been well documented as important to strategies for pathogen resistance, and in coping with abiotic stressors such as salt and drought (Hewezi et al. 2018; Kumar et al. 2017; Pandey et al. 2017; Wong et al. 2017), and at least some of these epigenomic alterations appear to be transmitted through seeds. Moreover, *C. elegans* has also shown some trans-generational inheritance of phenotypes like metabolism and viral resistance (Rechavi et al. 2011, 2014), indicating that these phenotypes are not just restricted to plants. The nature of those epigenome changes can inform genetic modification strategies to breed plants that are more resistant or adaptable to that stress, and the same strategy can be applied to any stressor that elicits an epigenomic response, including spaceflight.

What do these epigenetic landscapes look like in the spaceflight environment and how can crop resilience be enhanced to ensure productivity?

Spaceflight induces changes in the epigenome of plants, specifically in DNA methylation patterns and chromatin states, and these disruptions can affect plant growth in space (Paul et al. 2021; Zhou et al. 2019). What is not yet known is the extent of trans-generational stability of these epigenetic changes and whether establishing spaceflight methylation patterns within seeds can improve next-generation adaptability to spaceflight.

Understanding the role of epigenetic modifications in the spaceflight environment has practical application to exploration needs. In terrestrial environments, acquired DNA methylation patterns can transfer to the next generation, where they may contribute (Heard and Martienssen 2014) to the ability of the progeny to cope with the stress experienced by the parental generation. Should that occur in spaceflight, it will be possible to identify genetic elements that increase the adaptability of plants to spaceflight—the first steps in developing spaceflight cultivars bred to better serve the roles of plants in life support.

What are some of the characteristics of space-induced mutational load over multiple generations, and what techniques reduce these effects on subsequent populations to sustain crop yield?

Mutational loads can accumulate over multiple generations in the presence of a set of specific stresses (Bragg et al. 2015; Lu et al. 2021). Mutation load can also be lessened by any factor that causes more mutations to be removed per selective death, such as inbreeding, synergistic epistasis, population structure, or harsh environments (Agrawal and Whitlock 2012). Once a comprehensive understanding of how space stressors impact the mutation load of crops is established, then the spaceflight-practical strategies to detect and reduce them can be set in place to ensure deleterious mutations are not proliferated.

Multi-Generational Effects—Animals and Humans

One important set of questions in this area involves longer-term studies within the unique space environment to confirm the impact on the full cycle of conception, embryonic and neonatal development, adult fertility, and production of the next generation. Most human space travelers are expected to be of childbearing age, so it is of immediate relevance to determine the impact of the space environment on reproductive health of mammalian (and preferably human) adults. Limited data exist on the impact of LEO conditions on sperm production/viability; there is just one published study on ovarian function (Nguyen et al. 2021) as altered by time in LEO. Germline cells are among the most sensitive to radiation exposure, so this concern is exacerbated as missions planned in the coming decade include more time outside LEO, up to and including lunar surface missions. This critical gap in knowledge results in current crew members' assuming unknown risks to fertility with each space mission undertaken and thus can be considered the purview of NASA HRP; however, the fundamental understanding of space environment effects on biological processes in the BPS portfolio can provide key basic science data.

There have been no U.S.-led studies performed in-flight on LEO impacts on mammalian pregnancy/parturition since 1995. The three flight experiments published prior to this date exposed pregnant rodent dams during only a portion of their pregnancies and after fetal organ development was complete, so data in this area are quite limited (Ronca et al. 2021). Furthermore, animals need to be studied across two full generations to generate progeny developed entirely in the space environment. With the long view in mind, this research line is critical to long-term stays on planetary surfaces, including settlement and the survival of the species long-term outside Earth's 1g environment.

Potential Research Areas—Animals and Humans

How do spaceflight-associated epigenetic and epigenomic effects impact subsequent generations of animals?

Recent work from many taxa—from humans (Garrett-Bakelman et al. 2019) to mice (Ogneva et al. 2019) to nematodes (Higashitani et al. 2021)—suggest that the spaceflight environment can induce epigenomic changes in animals, and in the case of nematodes, those modifications appear to contribute to spaceflight adaptation. In terrestrial studies, environmental stressors and pressures can contribute to epigenetic effects that are transferred to the next generation in a variety of organisms (Freitas-Dias et al. 2022; Sangsuwan et al. 2022; Vogt 2022). Thus, it is important to consider the epigenetic impact of the sustained microgravity environment on not only humans but also the attending biology that share their habitat.

> Finding 4-5: Reproduction and development of microorganisms, plants, and animals in the spaceflight environment and across multiple generations need to be understood for life to be sustainable in the space environment, especially over long-term missions and within habitation outposts. Some adaptations to the spaceflight environment may manifest themselves only across multiple generations.

Question 5: What Principles Guide the Integration of Biological and Abiotic Systems to Create Sustainable and Functional Extraterrestrial Habitats?

Impact and Rationale

There are opportunities to strategically integrate biological organisms (e.g., plants, microbes, insects) and/or their molecular components into space habitats as resources for bioregenerative life support systems, health and medicine, and manufactured materials. These systems could be engineered to perform novel functions to fulfill the unique needs of living in space, such as withstanding harsh environmental stressors. Ultimately, the development of such systems can reduce costs, de-risk supply chain issues, and enable flexible and rapid responses to emergent needs, especially during longer missions (Drysdale et al. 1999; Ho et al. 2022). To achieve these goals, key research questions and technical hurdles could be addressed. Achieving precise and predictive control of living systems is no easy task, one that is further complicated by the harsh environment in which these biological systems need to safely and reliably function. In particular, it will be critical to understand how combinations of environmental stressors (e.g., gravitational, radiation, chemical, thermal) from different space environments can impact both engineered systems and native biological systems, as well as their interactions.

In biological systems, synthetic biology has been proposed to help overcome challenges associated with long-term deep-space exploration (Llorente et al. 2018; Menezes et al. 2015). Examples of where synthetic biology could be useful include ISRU, life support, manufacturing of supplies and drugs, human health measures, and the design of biological sensors/control systems. For example, microbes and plants could provide a boost to traditional regenerative life support systems in the production of oxygen and water, as well as in waste management (for solids, liquids, and gases). Microbes could further be utilized to bolster plant growth and crop production and protect against pathogen colonization. These organisms and others could also be beneficial in the production of "on demand" food, probiotics, nutritional supplements, and medicines.

Another opportunity is the incorporation of organisms or their molecular components alongside or embedded within traditional habitat materials to facilitate self-repair or function as biosensors that either signal the degradation of the material or its contamination with harmful chemicals or pathogens. The combination of real-time remote biosensing with deep learning and computational approaches could potentially help detect (and even predict) adverse events that could lead to ecosystem failure. In manufacturing applications, microbes and plants could aid in the extraction of valuable metals from regolith (planetary surface dust) in a process called biomining (Cockell et al. 2020; Santomartino et al. 2022). Biomining combined with additive manufacturing (3D printing) technologies could provide resources for habitat materials and electronics in the space environment, thereby reducing the need for transport from Earth. Conversely, if space biomining of planetary surfaces is successful, extracted materials could be returned to our planet to replenish dwindling resources.

As the research in this area progresses, it will be important to investigate how organisms, and their molecular component, can be stably and safely integrated into space habitats. For all of the aforementioned applications, gene editing technologies like CRISPR-Cas9 (clustered regularly interspaced short palindromic repeats) could be applied judiciously to modify living systems in such a way that optimizes their yields, functional efficiencies, and resilience to variable space stressors, such as radiation and partial gravities. The stability and safety of genetic modifications could be evaluated, to avoid unwanted gene transfer to other living systems and to minimize unwanted competition with other organisms for nutritional resources. Studies into whether the space environment alters the production of synthetically produced biomolecules and the mechanism(s) by which alterations occur could help shape strategies to maximize yield and purity, where relevant. Given the broad range of characteristics shown to be altered in biological organisms during spaceflight, it is anticipated that synthetically modified organisms will likely also follow suit. Understanding how short- and long-term exposure to altered gravitational forces, radiation, and chemical stimuli (e.g., chemicals found in regolith) could impact key biological characteristics of these systems will be critical. (See Figure 4-9.)

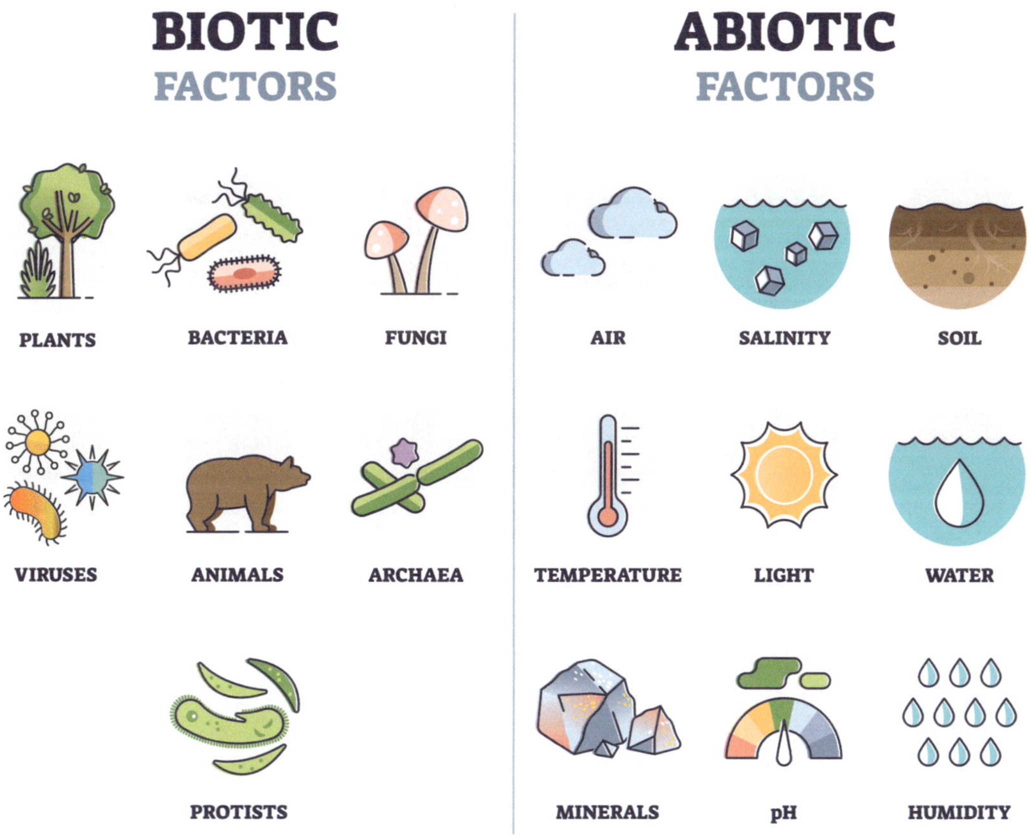

FIGURE 4-9 Examples of biological and abiotic system interactions that are potentially altered by space environments. Biological and abiotic systems likely interact in the space environment, although principles and opportunities of interaction are not yet well understood. SOURCE: Modified from iStock/VectorMine and iStock/lexashka.

Potential Research Areas

How can developments in synthetic biology be used to engineer reliable and functional biological systems to enhance life support systems, food and nutrient production, health, and manufacturing?

Significant advancements in the fields of cellular, molecular and systems biology led to the emergence of an area of research known as synthetic biology. Synthetic biology involves the systematic design or redesign of biological components (e.g., DNA, RNA, protein, whole cells) for solutions to challenging biomedical problems (El Karoui et al. 2019; Khalil and Collins 2010). Studies performed in this field have led to the creation of new products and biotechnologies and have also yielded important fundamental new insight into how nature works (Voigt 2020). Discovering new and unconventional ways to repurpose biological organisms and their molecular building blocks can help to overcome challenges associated with sustainability, health and medicine, food production, and manufacturing. Synthetic biology approaches can be applied at different scales, ranging from the engineering of single macromolecules (e.g., proteins), to the synthesis of entire microbial genomes, and even the complex integration of multi-species and cross-kingdom biological ecosystems.

How can biodiversity be created and maintained in beneficial biological communities in the space environment?

Specific research could explore the effects of long-term, deep-space exposure on phenotypes appropriate for the biological community under study. Longitudinal studies combined with multi-omics profiling (e.g., genomic, transcriptomic, proteomic, metabolomic) and deep learning approaches could reveal key changes in community features that could eventually be computationally predicted and used to improve control of the system. For microbial communities, examples of phenotypes that could be profiled could include viability, metabolism, DNA damage/repair, motility, population dynamics (interspecies communities), virulence, antimicrobial resistance, biosynthetic gene clusters (BGCs), volatile production, and biofilm formation. It will also be important to investigate how community diversity and biological balance can be maintained, including resilience to prolonged dormancy and other adverse events (e.g., plant and animal infections, phage infection of microbial cultures) and ability to recover from these events.

How best to identify and mitigate potential harmful effects or by-products associated with biological systems in space habitats?

It will be critical to identify, prevent, and mitigate any potential harmful effects or byproducts associated with biological organisms in space habitats (Nickerson et al. 2022). For example, biofilms have previously interfered with life support and operational systems (e.g., communication systems) in spaceflight (Yang et al. 2018; Zea et al. 2020). As this poses a major risk to the crew safety and habitat integrity, investigations need to be performed into how best to prevent the biofouling and corrosion by microbial biofilms in life support systems (water, surface air) and operational systems (e.g., communication) to maintain crew safety and habitat functionality under a range of conditions (e.g., variable gravity, radiation). Additionally, it is unknown how the deep-space environment can alter cell-to-cell communication and horizontal gene transfer among community members that could lead to changes in function and resistance to treatment with antimicrobials (disinfectants, antibiotics). For both pure and mixed cultures of microbes, it will also be important to study how the deep-space environment may shape their interactions with higher organisms (e.g., plants, animals, humans). These studies could be facilitated by cross-disciplinary collaborations across the space biology community between microbiologists, botanists, animal physiologists, and/or 3D tissue engineers to evaluate interactions using plants, animals, or models of human tissue (the latter derived from organotypic, organoid, or organ-on-a-chip models). In addition, given the findings from previous investigations that have identified increased disease-causing potential for select microbes during culture in LEO (Gilbert et al. 2020; Wilson et al. 2007, 2008), it will be important to screen candidate microbes (or microbial communities) for potentially negative impacts on animal, plant, and human tissues. In addition, there is evidence that terrestrial mixed species biofilms could act as a reservoir for potentially harmful microbes (Nath et al. 2010), but it is unknown to what extent this occurs in the space environment.

Can microbes and plants be exploited for biomining in space?

Exposure of living organisms to lunar or martian regolith is also an important environmental consideration that could impact their viability and function. Lunar dust was a major concern during Apollo missions, owing to the

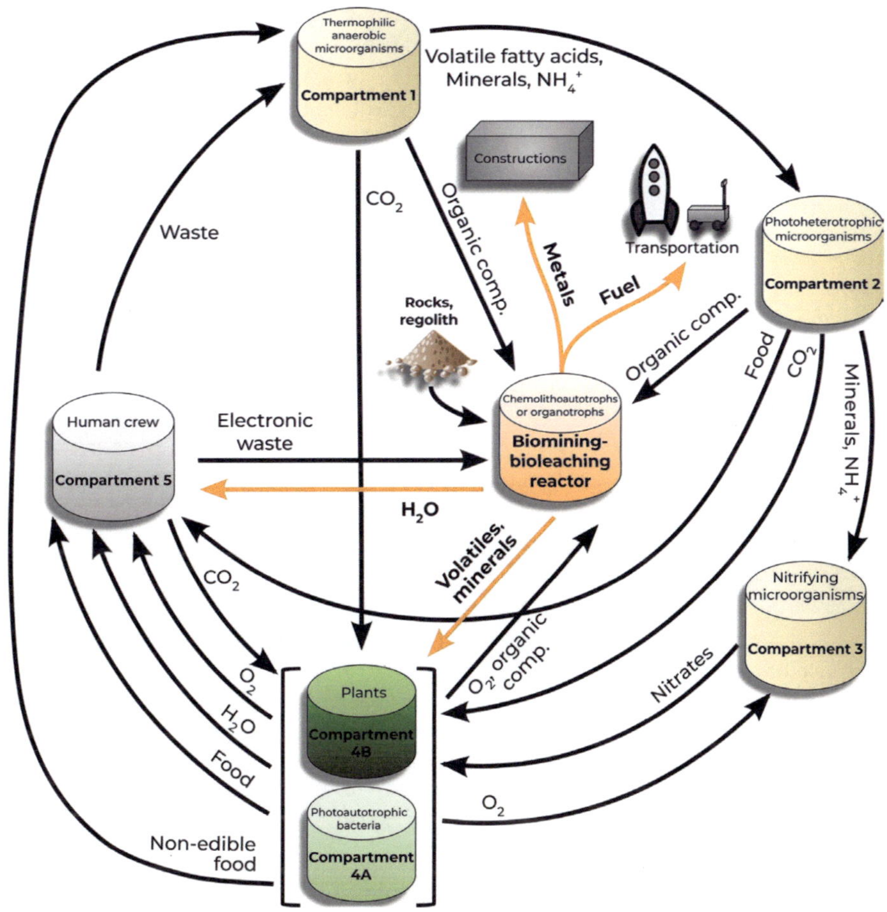

FIGURE 4-10 Schematic of a biomining/bioleaching compartment in the context of a life-support system. NOTE: Potential useful elements produced by this compartment are shown in orange. SOURCE: From Santomartino et al. (2022), https://doi.org/10.1007/s00792-021-01253-w, CC BY 4.0.

fact that the very fine, abrasive material adhered tightly to all surfaces (instruments, spacesuits, humans) (James and Kahn-Mayberry 2009; Lam et al. 2013). As it stands, biological organisms within a lunar habitat or vehicle will likely be exposed to regolith transported upon spacesuits or equipment. It will be important to characterize regolith-biology interactions within the context of the spaceflight environment for both human health as well as within the context of using plants and microbes to intentionally harvest minerals from regolith in biomining applications. For biomining, it will be important to investigate whether and how altered gravity/radiation environments alter leaching and degradation processes that present minerals to biology. (See Figure 4-10.)

How will the spaceflight environment impact the interaction between microbes and the surfaces of the built environment (i.e., a space, lunar, or Mars station)?

Microorganisms are present in every environment on Earth, ranging from the upper atmosphere to the deep-ocean subsurface. Through their metabolic activities, or even via interactions of cell surface structures (McLean and Beveridge 1990), microorganisms can alter the chemistry of the local environment and contribute to global chemical changes (Falkowski et al. 2008), as has also been shown experimentally in Biosphere 2 (Allen et al. 2003). Examples include enhancing soil fertility, generation, or depletion of various gases (e.g., carbon dioxide, methane, hydrogen sulfide), remediation of materials that are toxic to other life-forms, and solubility or immobilization of metal ions including helpful and potentially toxic materials. A number of microorganisms are extremophiles and so can live in regions that are inhospitable to other forms of life (Coker 2019). There have also been demonstrations of microbial survival under spaceflight conditions (Cockell et al. 2011). The assumption is that on exposure to regolith and other materials in extraterrestrial environments, microorganisms have the potential to survive and possibly thrive.

How and by what mechanisms do ecosystems change and adapt to novel and/or extreme environments, and what are the consequences for beneficial and deleterious interactions within and between species?

Healthy ecosystems are critical for the well-being of humans, animals, and plants on Earth and in space. The closed environments of spaceflight platforms, predominated by physio-chemical-based life support systems, have the potential to integrate bioregenerative life support systems to achieve closed-loop ecosystems such as those tested as a part of the Micro-Ecological Life Support System Alternative (MELiSSA) project[10] and the Chinese Lunar Palace (a lunar closed environment mock-up). To create healthy ecosystems, there is a need to monitor and understand the mechanisms by which ecosystems synergistically change and adapt to environmental perturbations (such as rapidly changing Earth environments and the extreme conditions of space environments). Experimentation in space environments thus enables answering the fundamental questions of the nature and mechanism of ecosystem change by providing artificial and controllable conditions for sustaining life. Monitoring of ecosystem components could be conducted at both organismal and molecular levels, and over time courses, both short and long. This knowledge is fundamental to understanding how ecosystem change impacts resources and is vital for developing evidence-based policy and management (Sparrow et al. 2020). Once appropriate monitoring approaches have been established, experimental intervention to create and test theories of ecosystem change derived from monitoring can be explored. Part of the evolution toward these systems will be testing smaller components or subsystems in space settings like the ISS, lunar and martian transit, and early lunar and martian surface missions (Johnson et al. 2021). Successful synthesis and monitoring of closed and controlled ecosystem environments can afford platforms for critically testing hypotheses of ecosystem change.

How does the spaceflight environment impact the microbes of the built environment (MoBE)?

Monitoring of—and research regarding—microbes has illustrated the contribution of individual community members to the robustness of polymicrobial biofilm formation. It is known that polymicrobial evolution occurs and that there are changes in microbial interactions between cohabiting species over time (Danko et al. 2021), as well as long-term adaptation and evolution within these communities.[11] Furthermore, microbial genomes from the ISS have been described and some isolates clearly show presence of drug-resistant genes (Damkiær et al. 2013; Singh et al. 2018) along with the presence of plasmids, suggesting an ability to carry drug resistance (Carattoli 2013). However, the mechanisms by which resistance and virulence factors are spread in single species and polymicrobial biofilms are not well understood in the spaceflight environment.

What are the spaceflight effects on biofilms?

Biofilms can be used for in situ resource utilization (ISRU) applications such as extraterrestrial biomining and growth of plant material. However, an understanding of bacterial growth dynamics in liquid culture in the spaceflight environment is better understood than bacterial cultures on substrates such as surfaces or agar. Thus, biofilms found in natural ecosystems could be studied more, even though they are primarily polymicrobial. Indeed, it is unknown to what extent colonies or biofilms associated with wet or dry surfaces are affected by the spaceflight environment and how they might adapt. In general, the ISS microbial community has been shown to consist of human-associated microbes, which can be transient or enduring, and consist of various types of bacteria and fungi.[12]

---

[10] The organizations partnering ESA in the MELiSSA Memorandum of Understanding are the SCK•CEN research center in Mol, Belgium; the VITO technology center also in Mol; Universitat Autonoma de Barcelona; the University of Guelph; University Blaise Pascal in Clermont Ferrand, France; SHERPA engineering in Paris; and IPStar in the Netherlands. SOURCE: https://www.esa.int/Enabling_Support/Space_Engineering_Technology/MELiSSA_s_future_in_space.

[11] See Botelho et al. (2019), Damkiær et al. (2013), Darch et al. (2017), Frydenlund Michelsen et al. (2016), and Limoli et al. (2017).

[12] See Bijlani et al. (2021), Castro et al. (2004), NASA (2013), Perrin et al. (2018), Rosenzweig et al. (2014), and Wang et al. (2021).

What is the effect of biology on inorganic substrates in the space exploration environment?

As biofilms thrive in continuously wet or moist conditions, dry surfaces tend to represent an unfavorable environment for biofilm formation. However, dry hard nonporous surfaces on the ISS can be intermittently damp, owing to fluctuating humidity levels, and experience high touch contact during normal usage.[13] It is critical to understand the means by which microbes become deposited on surfaces, which is likely through means of direct contact and air circulation because aerosol settling is not a factor in microgravity (Haines et al. 2019). Furthermore, it is critical to evaluate how the space environment may modulate microbial interactions with their substrate or surface and how these interactions may affect the subsequent development and morphology of the biofilm (Wang et al. 2021; Zea et al. 2018). This question of surface interaction further extends beyond corrosion to the role microbes can play in various ISRU procedures. The robustness of biomining—a blanket term for the processes by which a biological system (typically a bacterial biofilm) extracts and recovers desired metals from rock ores—in spaceflight environments could be further explored.

> Finding 4-6: The logistically isolated, environmentally closed, risk-intolerant nature of space habitats, together with requirements for minimal environmental impact, drives a need for efficient and flexible design in the provision of environmental services derived from recycling or use of local resources. Biological systems will closely interact and be affected by the built space environments, while the space vehicles and habitats will be affected by the enclosed biology.

Question 6: What Principles Enable Identification, Extraction, Processing, and Use of Materials Found in Extraterrestrial Environments to Enable Long-Term, Sustained Human and Robotic Space Exploration?

Impact and Rationale

Exploration and long-term sustained presence beyond Earth will be enabled using planetary surfaces, materials, and environment, particularly for ISRU. This is evident for activities that take place on those surfaces but will also enable long-term exploration for other planets and moons through the use of materials for fuel, chemicals, structures, and more. The cross-disciplinary nature of the problem requires integration of materials handling and beneficiation, robotics, chemical and materials engineering, and other space systems. This success also requires an understanding of the surfaces and available materials (including mineralogy, physical characteristics, heterogeneity on the surface), the environments in which materials originate and are used, the processes required to harvest and modify materials for use, and the long-term suitability for use. Some questions can be addressed by current ongoing exploration, but many require surface exploration with landed assets that can perform materials characterization in situ or return samples for more complex ground-based analysis and experimentation. The development and testing in relevant environments (including low-gravity, vacuum, extreme thermal variation, high-radiation, high solar energy flux) will be essential for full understanding of raw, processed, and manufactured materials and structures. (See Figure 4-11.)

Potential Research Areas

What are the important features of the composition and structure of lunar (or martian) regolith and the accessible subsurface that influence its use and processing?

Additive manufacturing with lunar regolith does not necessarily require substantial preprocessing, which greatly simplifies fabrication of structures. Regolith powder size, shape, and size distribution are important to control to achieve fully dense prints. The chemical composition of lunar regolith varies across the Moon's surface, which introduces an additional challenge because feedstock chemistry contributes to how a powder will interact with a given energy source (Isachenkov et al. 2021). Powder bed fusion and directed energy deposition methods of additive

[13] NASA Technical Standard, NASA-STD-3001, NASA Spaceflight Human-System Standard, Volume 2: Human Factors, Habitability, and Environmental Health.

FIGURE 4-11 Identification, extraction, processing, and use of materials, including biological. Materials that are extracted, processed, and recycled in space environments are key to sustainable and habitable ecosystems and require research to understand and manage space environment-specific challenges. Leveraging means to achieve this with biological or living systems can confer some unique processing capabilities and potential for lower impact of processing and reprocessing. SOURCE: From Heveran et al. (2020), https://doi.org/10.1016/j.matt.2019.11.016, CC BY-NC-ND 4.0.

manufacturing can fabricate structure using lunar regolith as the only feedstock, while other approaches (e.g., material extrusion, binder jetting, vat photopolymerization) contain an additional, often organic, binder phase that could be derived from a number of potential sources including biological. Additive manufacturing with lunar regolith when there is a binder phase would be less energy intensive because the lunar regolith would not be sintered. However, binder feedstocks generally cannot be sourced from the lunar environment, although plastic waste can be converted to binders for some of these additive manufacturing methods. In addition, the martian atmosphere is composed of carbon dioxide (95 percent CO_2), molecular nitrogen (2.8 percent), and argon (2 percent). With the high solar energy flux on the martian surface, non-equilibrium plasma assisted synthesis using CO_2 and nitrogen with solar energy will enable new opportunities for non-equilibrium synthesis of fuels, chemicals, and materials for space travel and living.

What are the long-term responses of expected materials of construction to the space environment?

Much is currently unknown regarding the secure lifetime of materials in the space environment, including how radiation, strong thermal gradients, and thermal cycling may degrade their structure and performance. There may be composites of materials that are adaptable to and resilient to these environments, and there may be in situ materials that would be more useful in construction and repair. Beyond structural materials, the effect of radiation and other space environment features on organic materials including engineered polymers is less understood, as are the key features required of radiation shielding, radiation-hardened and radiation-resistant materials required of electronic or photonic computing and communication. Complex multi-component materials of interest could also include functional fibers and textiles for human protection and augmented mobility. In LEO, radiation from radiation belts is an important consideration, but beyond an altitude of approximately 36,000 km, the radiation belts' protective influence declines. Galactic cosmic rays and associated radioactive particles (e.g., ^{16}O) are a significant concern for materials stability during long-term missions beyond LEO.[14] Furthermore, while the use of lunar regolith as a source of silicon and other elements within solar cells or protective cover glass (which can be used in powered habitats) has been predicted for more than 2 decades (Blue Origin 2023; Freundlich et al. 2005), whether the resulting materials are more or less resilient to space use environments remains unknown and presents a strong candidate for accelerated exposure testing.

What research is needed to improve mechanisms of construction and repair in space?

Robust construction of the built environments in space is critical to the success of any mission. Control of the interior environment relies on the integrity of the surrounding structure, and thorough understanding of the challenges that a material would confront is essential in assessing the materials and methods of construction for various space environments. Temperature variation and thermal cycling, as well as the impact of extended periods of radiation exposure, are examples of issues that could be considered to establish the viability of materials and methods of their use. Because damage will inevitably arise, repair methods need to be considered, and these need to be flexible with regard to materials and able to shape materials on demand because the ability to carry spare manufactured parts is limited. Thus, flexible methods for forming and/or placement of various materials is an essential engineering science research direction to enable space exploration. Some of these issues clearly intersect with basic engineering, for which abundant experience exists, but there are significant new challenges and also some new opportunities—for example, reduced gravity levels on the lunar and martian surfaces may open windows to use of materials that are not sufficient in terrestrial gravity.

How does the space environment impact the joining of materials—for example, by welding or the application of various coatings?

What material properties are critically needed to make a weld survive strong gradients; for example, is a composite structure with ductile elements relevant? Are there surface treatments that can make a generic joining material work for multiple materials? Understanding the limits and opportunities for joining dissimilar materials (metal and otherwise) safely in these space environments of low atmospheric pressure and potentially high space dust particulate concentrations will help determine the limits of habitats and infrastructure in the built environment on the Moon and Mars, with plausible benefit of new materials and processes also advancing Earth-based joining and construction approaches.

How can chemical propellants be manufactured onsite (the Moon or Mars) and used for return flights?

Propellants could be manufactured from the available resources on Mars (e.g., CO_2 and nitrogen) or the Moon, and they need to be able to meet the certification to be used in rocket engines. For this to occur in future decades, basic research regarding safe and benign chemical processing in these environments needs to advance in the coming decade.

> Finding 4-7: Resources from planetary materials, atmosphere, and mission wastes can all be captured and harnessed for production of mission-critical and high-value chemicals, materials, and biological feedstocks. Moreover, biological systems can also be used as sensors and tools for mining local resources.

Question 7: What Are the Relevant Chemical and Physical Properties and Phenomena That Govern the Behavior of Fluids in Space Environments?

Impact and Rationale

Fluids and combustion behave significantly differently in microgravity compared to conditions on Earth. With the reduction of gravity, the buoyancy force is diminished, and other forces, such as surface tension and electrostatic forces, surface properties, and geometries, play a larger role (Balasubramaniam 2021). With the reduction

[14] See Berger et al. (2020), Chancellor et al. (2018), Durante (2014), Onorato et al. (2020), and Wilson et al. (2001).

in buoyancy, molecular diffusion and thermal radiation play a greater role in combustion and dramatically modify the burning limits. This in turn affects such a critical issue as fire safety. While the fluid properties are generally well-understood, and their motions are governed by well-established equations and/or correlations, fluid behavior in reduced gravity is very often unpredictable without solving the governing nonlinear equations numerically, and surface forces associated with contact lines can become much more influential than in normal gravity. Simplifying assumptions that work well in Earth's gravity well may need to be reexamined in reduced-gravity environments.

Rapid two-phase flow calculations tailored to the relevant geometries could be a critical support for fluid management strategies. Reduced gravity changes the burning rate and instability of combustion and enables strong coupling between combustion with radiation and molecular diffusion. This is relevant for high-performing liquid propellants such as cryogenic methane or hydrogen and the oxygen with which they react. To enable spaceflight, these or other propellants need to be stored properly, both under the reduced gravity of Moon and Mars and in space under microgravity, for extended periods. Propellant transfer is a critical process that extends the range of space travel by reusing a rocket stage, providing a new range for every propellant transfer where it is possible in orbit, on the Moon or Mars. Under microgravity, propellant is not oriented inside a tank; it may attach to surfaces or float in fluid masses, drops of different sizes.

Propellant transfer, however, requires a specific orientation of fluid inside the tank, such that it can move into a pump inlet or can be pushed out (by pressurization) through an outlet. The current methods to orient a fluid are, for example, thrusting to collect the propellant at the bottom of the tank, or using propellant management devices such as internal structures or a bladder. Additional methods could be based on centrifugal forces or gravity gradient utilization, or other effects. Depending on the method of fluid orientation, there are additional considerations with respect to heat transfer into the cryogenic propellent. Heating cryogenic propellant is generally not desired, because a potential phase change increases pressure in the tank and may require venting, losing vaporized propellant. Insulation reduces heat transfer, but does not cancel it completely, and is in fact difficult to predict under reduced gravity in the space environment. For long flight durations, additional active cooling such as zero-boil-off or reduced-boil-off strategies, is required. There are several factors (Jayawardena et al. 1997) to be considered, such as the efficiency of cooling under microgravity and how to improve it, the impact of sloshing or external acceleration, and the potential presence of two-phase flow. The burning rate of propellants in space is also critical for space exploration and rocket propulsion. Gravity plays an important role in affecting the growth of the Rayleigh–Taylor instability, flame surface area growth, and the burning rate.

Storing and moving fluids is not limited to propellants. Space exploration requires water, refrigerants, heat pipes, pressurization gas, oxygen, waste, and other fluids. Fluids need to be stored without leaks and, for gases, efficiently (i.e., under high pressure), causing safety concerns for long-term applications such as composite overwrapped pressure vessels (COPVs). Water, in particular, could be reclaimed and purified for sustainability. All liquid transfer in space involves similar challenges as those described above for cryogenic propellants. Special precautions are necessary for potentially toxic fluids such as some storable propellants and to generally avoid contaminations. In fact, any research benefits long-term space exploration, but also sustainability on Earth.

Propellant production under partial gravity is of particular interest on Mars and the Moon and a key enabling method to increase the range of space travel. Water mining is crucial for propellant generation, and the water needs to be stored and treated under partial gravity, similar to propellants. Thermal management is required to keep both water and propellants in liquid phase. The additional process of converting water and CO_2 extracted from the Mars atmosphere is ultimately one of the highest priorities for Mars exploration, including human return missions. Predicting fluid physics properly, and ultimately utilizing it while mitigating detrimental effects such as the cryogenic boil-off mentioned above, is therefore critical to long-term space exploration missions and necessary for living and traveling in space.

Issues related to fluid transfer and convection also affect power cycles, but in terms of power cycle design, the effects of the extraterrestrial environment are not limited to partial gravity. Availability of a specific fluid/gas may vary between environments—on an ice moon there would be an abundance of water, somewhere in the outer solar system ammonia might be easily available, and Mars has both water ice and CO_2. The optimal implementation of the power cycle would also depend on available sources of energy (chemical reaction, a radioisotope generator, solar heat, etc.) and waste heat disposal mechanism (which would be limited to radiative outside of a planetary-like environment).

FIGURE 4-12 Fluid behavior in space environments. Fluids in space environments are critical to biological functions, engineered materials processing, and safe transportation, but can confound expectation and prediction based on Earth-based fluid dynamics. Artemis astronaut Christina Hammock Koch illustrating this on the International Space Station. SOURCE: Courtesy of NASA, https://www.nasa.gov/sites/default/files/thumbnails/image/iss060e001486_0.jpg.

Thus, fluids in space environments are critical to biological functions, engineered materials processing, and safe transportation, but can confound expectation and prediction based on Earth-based fluid dynamics. (See Figure 4-12.)

Potential Research Areas

What aspects of fluid behavior will improve fire safety onboard spacecraft?

The most conspicuous difference between fires on Earth and in space is caused by the presence/absence of buoyancy. This produces spherical flames (see Figure 4-13) whose study is essential to fire safety in space: this flame behavior is impossible to study on Earth. It can also simplify analysis by being amenable to development of reduced-order models and reveal information about soot formation and flame extinction useful beyond orbital environments. Ongoing experiments consider a coflow laminar diffusion (CLD) flame (Smooke et al. 1999), where the burning flame is surrounded by laminar flow of the oxidizing gas (e.g., air). In a gravity field, the coflow would be directed upward. In orbit, the velocity of the coflow becomes an important control parameter affecting the flame shape, stability, and burning product generation.

Previous testing at microgravity and lunar gravity indicated that some materials burned to lower oxygen concentrations in low gravity than in normal gravity. Similarly, the data obtained under martian gravity conditions also suggested that there is a partial gravity level at which materials burn more readily than on Earth. Flames, especially in microgravity, can also be strongly affected by electromagnetic fields, as preliminary studies show (Chien et al. 2022).

FIGURE 4-13 Near-spherical flame in orbital environment. SOURCE: Courtesy of NASA, https://www.nasa.gov/sites/default/files/thumbnails/image/microgravity_flame.jpg.

Another important issue is material flammability. The present flammability standard accepted by NASA[15] establishes requirements for evaluation, testing, and selection of materials that are intended for use in space vehicles and associated ground support equipment (GSE). It is a pass/fail ground test, where an upward flame is spread over a material sample. The known drawbacks of such tests are as follows: First, cabin-air conditions onboard spacecraft (pressure, oxygen content) can change. Second, there are at least two factors that are known to change material flammability in microgravity. Natural convection does not occur, and the flow surrounding the flame may be owing to the HVAC system or movement by the astronauts and have a very low and/or irregular velocity. Second, combustion products remain around the area where combustion took place.

While some studies of material flammability in microgravity exist—refer, for example, to Fujita (2015) for a summary along with a description of the shortcomings of existing flammability tests—more systematic investigations are essential to improve fire safety relevant to space missions and to develop predictive models and new fire-resistant materials.

How best to store fluids onboard spacecraft for long-duration travel, managing temperature and pressure in the presence of different thermal environments?

Answers to this question address storage conditions, insulation mechanisms and materials, and how much propellant is lost over time if not cooled actively. This can include consideration of actively cooling cryogenic propellants, which has its own scientific and technical challenges owing to acceleration of spacecraft and in the presence of sloshing.

[15] See NASA STD 6001B, https://standards.nasa.gov/standard/NASA/NASA-STD-6001.

How best to transfer fluids, either from water or propellant generation for use (thrust generation) or to other spacecraft?

Refilling a spacecraft in orbit is incredibly efficient, but it is also difficult and costly; reusable vehicles are a significant advantage by lowering overall launch costs, allowing launch from orbit with a fully loaded vehicle. Challenges are vehicle docking, propellant orientation, propellant transfer at a reasonable rate avoiding or minimizing heat transfer, and repeating the process until the target vehicle has been filled up sufficiently. Furthermore, when transferring cryogenic liquids, an initial part of the liquid is used to bring the tank structure to cryogenic temperature. It boils off and is vented; ideally, the structure could be prechilled, and the use of propellant minimized. Predictive capabilities in fluid transfer would speed progress over current technology design and development iteration cycles.

How does partial gravity affect the combustion instability and limits?

This needs to be understood to predict the combustor heat release and instability in rocket propulsion. With reduced buoyancy, the combustion limits will be narrowed, owing to the increasing impact of thermal radiation and molecular diffusion, and thus need to be quantified. Compared with the terrestrial reference frame, several instabilities that affect combustion are considerably different on board a moving spacecraft. Stratification-driven Rayleigh-Taylor Instability (RTI) will not manifest on board a coasting spacecraft, while it will be quite prominent in the presence of acceleration/deceleration. The same would be true for the multi-phase analog of RTI (Vorobieff et al. 2011). Moreover, extension of studies of multi-phase effects in combustion instability to altered-gravity environment can lead not just to improvements in space propulsion but also to better understanding of combustion efficiency on Earth.

It is also possible that in microgravity, different instability mechanisms may become prominent, such as fingering instability because of imbalance between heat release owing to local fuel combustion and radiation heat loss (Ijioma et al. 2015). Combustion limits and extinction boundaries are also different in microgravity (Okuno et al. 2016), as well as flame stability, requiring further microgravity studies.

Furthermore, replacement of petroleum-derived fuels with liquid alternative fuels would necessarily involve a combination of carbon sequestration and renewable fuel production. However, research including that in the space environment will be required to consider how alternative fuels (e.g., bio-derived fuels) influence the combustion behavior and supercritical oxidation. In fact, the lack of fundamental understanding of the dynamics and the associated chemical kinetics of supercritical oxidation limit its implementation, as critical to advance solid waste treatment and or wastewater recovery and management for long-duration spaceflights and advanced space exploration systems. Research on such science underpinning combustion in space environments will be important to achieving the projected high-efficiency and low-emission signatures.

How can power cycles be adapted to extraterrestrial environments?

Modern deep-space spacecraft usually reject heat directly into space via radiation, without a refrigeration system. Such cooling can be quite effective, with passive cooling of the James Webb Space Telescope (JWST) sufficient to reach temperatures suitable for operation of mercury-cadmium-telluride (HgCdTe) detectors at temperatures around 37 K. On the same telescope, however, some detectors operating at 7 K require a closed-cycle cryocooler relying on a combination of three cooling effects: pulse tube cooling, the Joule-Thomson effect, and a traditional refrigeration cycle. The ISS uses water and ammonia refrigeration cycles and radiates heat into space.

However, human space exploration, such as missions to the Moon and Mars, will put much tougher demands on refrigeration systems for spacecraft and power-cycle systems, both for in-space and for planetary environment use. Cooling/power cycle systems may be inferior in performance to alternatives such as the vapor compression cycle (VCC) (Brendel et al. 2021) as well as magnetic, thermionic, thermoacoustic, thermoelectric, thermotunnel, and similar technologies (Brown et al. 2012). Lunar regolith can be utilized for extracting noble gases (e.g., He, Ar, Kr, etc.), which can then be utilized for supercritical power thermodynamic cycles. The large range of temperature variations in lunar environment can be utilized for thermal energy storage schemes that can then provide supplementary heating or cooling schemes (e.g., for thermal management of habitats, supplement power cycles, etc.).

Last, for exploration of planetary-like environments, ISRU options for coolants or heat transfer media may exist. Such options may include water in which process contaminants are dissolved, ammonia, noble gases extracted from luna regolith, and carbon dioxide. Associated refrigeration and power cycles are likely to require study and optimization or modification for use on these extraterrestrial surfaces.

Finding 4-8: Microgravity, radiation, reduced atmospheric pressure, and extreme temperature all significantly alter the behavior of fluids and flames, whether contained in biological systems or outside in free form. Understanding these altered behaviors is critical to providing safe space environments and to effective processing and manufacture in space.

COMPLETING THE SCIENCE IMPERATIVES

It is increasingly recognized that the key scientific questions of the BPS do not stratify easily into the categories of those enabling space exploration and those enabled by having access to space. The KSQs were developed based primarily on the themes guiding the overall science priorities for the coming decade; however, these KSQs were also articulated by a research community holding a constant institutional memory that there is value in recognizing the distinction between exploration science and fundamental science. Chapter 4 concludes having addressed those KSQs that most directly support exploration. Chapter 5 then transitions to that science that can be accomplished only with access to space, with the further recognition that fundamental science supports both Earth-based knowledge and applications as well as exploration.

5

Science Enabled by the Space Environment

Some of the most amazing things known about life on Earth and in this universe cannot be learned by experiments conducted solely on Earth. Space-enabled fundamental science represents the opportunity to understand fundamental biological and physical mechanisms that cannot be studied on Earth, and that can only be understood by access to the space environment beyond this planet. From recent history, it is clear that such knowledge helps society not only better understand our place in the universe with a sense of awe but also take major technological leaps forward to make daily life on Earth safer and easier.

As discussed in Chapter 1, the reduction of gravitational forces and high levels of ionizing radiation are two major factors distinguishing the space environment from terrestrial conditions. Gravity causes buoyancy, which profoundly alters flow motion, material transport, and segregation in both biological and physical systems. In the familiar gravity on Earth, particles settle to a plane; microgravity releases them to circulate in a three-dimensional space. In addition to reduced gravity, the radiation and chemical environment of space is unique, and variable, depending on the distance from Earth and the nature of the environment. Other space environment features related to electromagnetic field strength, presence of dust particles, and cycles of temperatures and pressures far beyond the "standard temperature and pressure" that serve as Earth's baseline for experiments also play a role (see Chapter 1). The systematic difference in transport phenomena as a function of variable gravity enables solid materials and fluids to react differently with each other, invoking different distance and time scales, and affecting kinetics and dynamics. These features unique to the space environment and inaccessible on Earth enable the search for novel synthetic pathways and new materials and chemicals made in space and can also surprise scientists with non-intuitive behavior governing transport phenomena. Furthermore, microgravity in low Earth orbit (LEO) and beyond enables careful studies of situations that are not at thermodynamic equilibrium, and thus not well predicted by existing theories and equations—for example, studies of colloidal systems such as active particles (Khadka et al. 2018) designed to serve as microscopic models for interactions among living objects such as birds, fish, bacteria, and people.

Chapter 5 is thus organized around a single theme that presents the priorities of science for the coming decade:

Probing Phenomena Hidden by Gravity or Terrestrial Limitations: Revealing underlying biological and physical processes that cannot be quantified on Earth.

This theme describes the space-enabled fundamental science that will form the underpinning of both space-enabling and Earth-based applications for future generations. Measuring, modeling, and understanding these phenomena are sometimes referred to as fundamental science; the primary motivation is deeper understanding of how the world and universe around us works. While the related research may seem abstract or academic, such fundamental science is the "seed" for crops of end-use applications and practical benefit to society that historically follow over future decades. This utility may take a human lifetime to manifest, but who can doubt the value of investing in that anticipation (Simmons 2014) with examples of that promise in today's energy storage materials (including now ubiquitous electric vehicle batteries dreamed up more than 20 years ago), medicines (including proteins that bind with precision based on knowledge of nucleic acid sequences), and navigation technologies (including data from the space-based radio-navigation system connected to every mobile smart phone)? Indeed, this fundamental knowledge—the understanding that will become the stuff of textbooks and cutting-edge products for the next Earth-based generations—also advances space exploration when harnessed in an engineering context. That is, the research that is enabled by access to space eventually—and sometimes quite predictably and quickly—also enables new approaches to travel between, communicate across, and live in space environments.

> Finding 5-1: Fundamental research in biological and physical sciences serves the basis of many indispensable societal capabilities and conveniences on Earth, in addition to allowing society to understand more about how the surrounding world works.

The space-enabled scientific questions within this theme are complementary to and often intersect the space-enabling scientific questions described in Chapter 4, which are in turn organized thematically around *Adapting to Space* on slipping the bonds of Earth, and then *Living and Traveling in Space* for longer durations and over greater distances from Earth. This chapter focuses on the key scientific questions (KSQs) that are enabled by access to space environments including and beyond LEO, as well as the rationale and possible research approaches for such investigations.

Chapter 5, organized thematically around *Probing Phenomena Hidden by Gravity or Terrestrial Limitations*, includes KSQs related to biological phenomena that can be understood completely only when some features of the space environment are considered. However, this chapter also places emphasis on KSQs regarding physical phenomena, the answers to which impact human life but may be entirely about abiotic or engineered materials. Table 5-1 summarizes these four KSQs enabled uniquely by access to space environments.

These four KSQs admit some overlap with one another, regarding experimental systems of interest or method of inquiry. This can be viewed as a strength in identifying broadly relevant scientific needs. These questions also inherently cross disciplines, requiring knowledge from physics and chemistry to materials science and engineering. The questions apply to many classes of materials and phenomena, including biophysical phenomena, and are also relevant to the development of space-based technologies of broad utility, from autonomous spacecraft navigation to improved life-support systems for humans in space. Quests for biological science principles need to be aware of and draw from physical principles; and impactful physical science-motivated inquiries need to be aware of and leverage knowledge gained from analogy to biological systems. This is not a novel concept: the electrochemistry of today's portable batteries was inspired in form and function by the "animal chemistry" stored in twitching

TABLE 5-1 Key Scientific Questions Enabled by Access to Space Over the Decade 2023–2032

Theme	Key Scientific Questions
Probing Phenomena Hidden by Gravity or Terrestrial Limitations: Revealing underlying biological and physical processes that cannot be quantified on Earth	• What are the mechanisms by which organisms sense and respond to physical properties of surroundings and to applied mechanical forces, including gravitational force? • What are the fundamental principles that organize the structure and functionality of materials, including but not limited to soft and active matter? • What are the fundamental laws that govern the behavior of systems that are far from equilibrium? • What new physics, including particle physics, general relativity, and quantum mechanics, can be discovered with experiments that can only be carried out in space?

animal muscles (Tretkoff 2006). Living organisms, intrinsically out of equilibrium (as an organism at equilibrium is dead), have harnessed and refined physical and chemical processes for biological function to sustain homeostasis, so there is strong overlap between biological and physical sciences in any discussion of fundamental phenomena, especially involving transport, soft matter, interfaces, heterogeneity, and phase separation. Fundamentally, there is strong overlap between physical and biological phenomena, especially in the area of interfaces, phase separation, self-assembly and colloids; materials synthesis and processing are intertwined with heat and mass transport; and organic and inorganic materials span interatomic interactions ranging from weak to strong, with many properties in common. Each KSQ takes the form of a fundamental question, followed by a discussion of scientific impacts and rationale, as well as potential research approaches in space. Advances in answering these KSQs rely strongly on access to computational tools and databases.

BPS KEY SCIENTIFIC QUESTIONS THEME 3: PROBING PHENOMENA HIDDEN BY GRAVITY OR TERRESTRIAL LIMITATIONS—REVEALING UNDERLYING BIOLOGICAL AND PHYSICAL PROCESSES THAT CANNOT BE QUANTIFIED ON EARTH

This theme can be referred to in shorthand as *phenomena hidden by Earth's environment*. The path to identifying the scientific principles otherwise obscured by 1 *g* gravity, relatively low radiation levels, and other features that define "standard conditions" of Earth-bound research is guided by the following four KSQs:

- *What are the mechanisms by which organisms sense and respond to physical properties of surroundings and to applied mechanical forces, including gravitational force?* To address this fundamental question thoroughly, including the capacity to cope with or leverage differences in those properties and applied forces in space environments, sustained access to the space environments of interest is required. As just one example, the response to gravitational forces by all organisms ranging from microbes to plants to mammals occurs along a spectrum of gravity (not just 1 *g* and 1/6 *g*), so mechanistic answers require access to that spectrum.
- *What are the fundamental principles that organize the structure and functionality of materials, including but not limited to soft and active matter?* The interatomic and electronic forces that govern the structure and many properties of materials on Earth are well understood at standard temperature and pressure for many material classes including elemental metals. However, in the altered space environment and in the context of complex alloyed, inorganic, and organic materials (so-called soft matter or active matter, whether natural or synthetically engineered), as well as their solid composites or fluid-solid suspensions, these fundamental principles have been elusive. The space environment (including microgravity) allows for control or relaxation of the physical forces that obscure those principles.
- *What are the fundamental laws that govern the behavior of systems that are far from equilibrium?* Classical thermodynamics predicts the stability of phases based on laws established to describe the energy or work of a system, as well as the reversibility of that system shifting from one equilibrium state to another. When systems or materials are from that equilibrium state, as can occur under high rates of change in pressure or temperature, assumptions of how work and energy are exchanged or how heat is dissipated can fail miserably. To harness such states, new understanding is needed.
- *What new physics, including particle physics, general relativity, and quantum mechanics, can be discovered with experiments that can only be carried out in space?* The space environment facilitates experiments impossible on Earth, including those required to understand dark matter and dark energy. The current "standard models" of physics and relativity fail to predict and describe these dominant features of the universe. Moreover, the currently incomplete knowledge of atomic, molecular, and optical physics and phenomena such as quantum entanglement also gate progress in technologies needed for space communication and navigation, motivating practical implications of such fundamental understanding beyond the coming decade.

These KSQs are further described below, in terms of the impact and rationale motivating this use of the precious space environment resources for fundamental research. Possible research areas addressing sub-questions and specific systems of study are also described, with full understanding that these are a subset of the possible research areas that will best address these KSQs in the coming decade of access to space for BPS research.

Recommendation 5-1: NASA should substantially increase resources dedicated to producing and understanding the answers to the key scientific questions detailed in this report. This investment recognizes the potential for significant societal impacts utilizing the space environment for the biological and physical sciences portfolio in the coming decade, aimed at

- **Identifying the mechanisms by which organisms sense and respond to the surrounding environment, including gravitational force;**
- **Advancing knowledge of material structure, self-assembly, and stability of materials, including but not limited to soft/active matter, in space environments, cognizant of but distinct from the applications of that knowledge to space exploration and habitation (e.g., manufacturing in space);**
- **Supporting ground-based and microgravity research on understanding the fundamental laws of systems far from equilibrium, especially those that underlie the existence of life; and**
- **Identifying new principles of physics that can only be discovered through experiments in space, including those governing particle physics, general relativity, and quantum mechanics.**

Question 8: What Are the Mechanisms by Which Organisms Sense and Respond to Physical Properties of Surroundings and to Applied Mechanical Forces, Including Gravitational Force?

Impact and Rationale

Plants, microbes, and animals directly sense and respond to gravity as a so-called body force, and also respond to applied mechanical forces such as stretching (tensile forces or stresses) and compressing (compressive forces or stresses) at the cell and molecular levels. (See Figure 5-1.) Many cell types within these organisms can also exert mechanical force on their surroundings, with different molecular components acting as both force generators or force sensors; these mechanical effects are unrelated directly to gravity, particularly at biological length and force scales. In the space environment, it is possible to differentiate between direct and indirect gravity effects. On Earth, these effects are coupled, and thus it remains poorly understood how different environmental components interact with gravity sensing. Access to the space environment in LEO and beyond can reveal the mechanisms of force sensing that operate independently of or differentially in response to changes in gravitational force, in different organisms and at the level of cells and molecules. This mechanistic knowledge informs our understanding of phenomena taken for granted in a 1 *g* society, helps to clarify whether and when those phenomena will be replicated in different space environments, and may even elucidate why some biological processes are more rapid or energy efficient in space environments.

Animals and plants sense gravity directly by use of specialized sensors; all organisms sense mechanical stresses, due either to indirect effects of gravity or to other internal and external forces, in all of their cells and organs. Mechanosensing is fundamental to growth, development, gene expression, cellular growth, cellular differentiation, cell migration, bone modeling, muscle and vascular function, and endless other biological effects, and it is poorly understood. In some cases, the proteins that sense cell adhesion forces and the genes they control are known, but how cells detect and measure levels of stress (normalized force) and strain (normalized displacement) and respond to these forces is almost entirely unknown. The ability to experiment in space environments gives a unique opportunity to separate direct sensing of gravity from other mechanical effects. Little is known about how general environmental stresses—radiation (Boerma et al. 2005; Nagayama and Fukuei 2020), DNA damage, changes in atmospheric composition—alter the responses of cells to mechanical forces, and if the effects of these mechanical stresses are therefore mediated in part by otherphysical stresses.

Experiments on plant growth and development in space have been carried out for decades in pursuit of questions focused on the feasibility of using plants for food and regenerative life support functions. More recent studies have taken advantage of the spaceflight environment to pursue questions related to the fundamental processes of

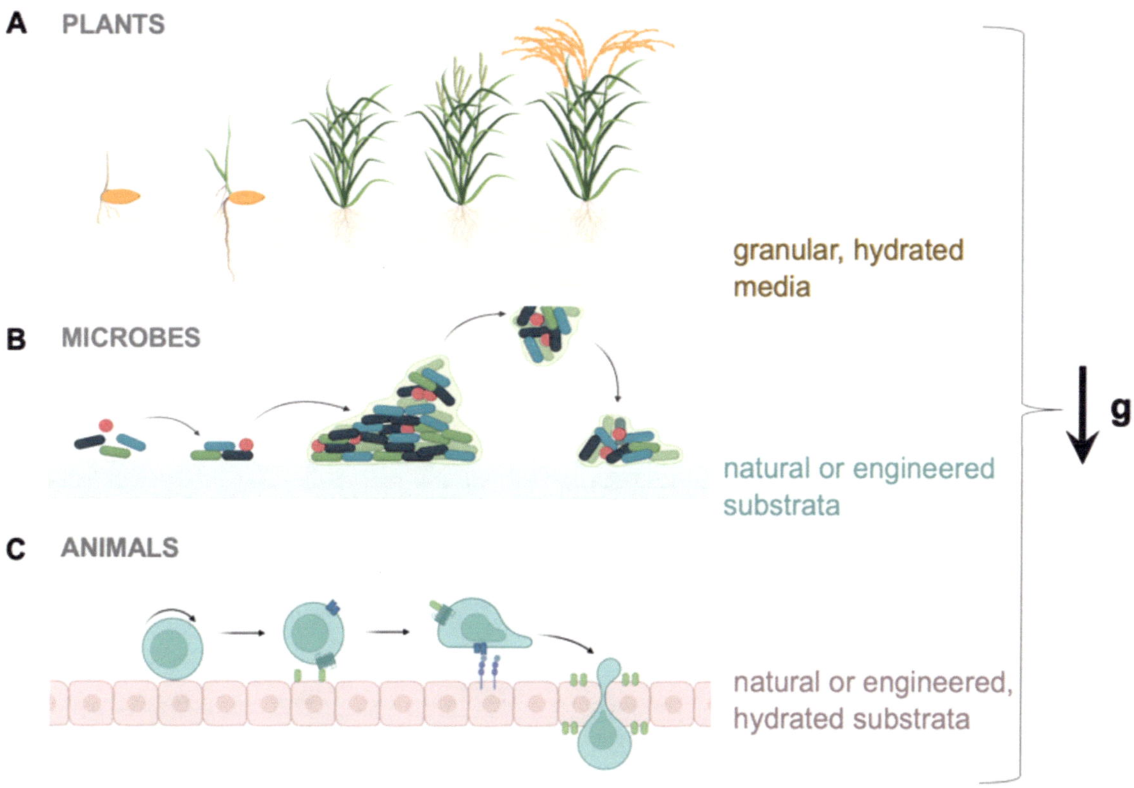

FIGURE 5-1 Complex molecular responses to variable gravitational, applied, and biophysically generated mechanical forces. Gravitational and applied forces are cues that elicit complex molecular responses of (A) plants during developmental stages, (B) microbial cells and systems during adhesion and biofilm formation, and (C) animal cells and systems, including cell-generated forces against materials to which cells adhere. Current knowledge about these systems' responses to external cues has been mostly limited to experiments in Earth's gravity (1 *g*). Access to the microgravity environments of space is vital to elucidate fully the extent of these effects on Earth and in space exploration environments. These include space environment effects on the systems with which plants, microbes, and animal cells, tissues, and systems interact. SOURCE: Created with BioRender.com.

gravity and light perception and response pathways. Regardless of the motivation, access to the spaceflight environment has enabled questions regarding plant growth and development not possible on Earth. Experiments in low Earth orbit (LEO) have established that plants exhibit significant transcriptional reprogramming in response to the space environment, including differential gene expression (Barker et al. 2021; Chandler et al. 2020; Paul et al. 2017), as well as alterations in epigenetic states that control gene expression (Paul et al. 2021; Xu et al. 2021) and in alternative splicing of transcripts (Beisel et al. 2019). Because these studies compare space to ground controls, the responses may only be attributable in part to changes in gravity. The continued advancement of these questions will require additional hardware and more sophisticated analytical tools to begin to unravel the multitude of variables encountered by plants in the spaceflight environment.

While it is known that plants and animals can sense gravity directly—for example, by otoliths in the mammalian ear or by sedimentation of starch-filled plastids in specific plant cells—important aspects of the mechanisms remain unknown. Studies in the near-absence of gravity will enable the separation of gravitational forces from other mechanical forces necessary to understand the fundamental biological mechanisms by which gravity sensing occurs and the degree to which gravity sensors influence other mechanical responses. Because mechanosensing has a major role in guiding structural growth in biology, this understanding will have potential Earth translational relevance to agriculture through plant morphogenesis under various environmental conditions and to human health through amelioration of the effects of aging, bone density decreases, and fibrosis.

Answering these questions requires experimentation in space as well as on Earth. Different gravity levels from micro- to hyper-gravity, including moon and martian levels, will enable distinction of graded responses and threshold effects of gravity and mechanical stresses whose sensation is affected by gravity, and will permit mechanical effects to be quantitated in the absence of gravitational load and of gravity sensing. This will allow a distinction to be made between the causal mechanisms of different effects, and their separate study. The known effects of the radiation environment on mechanosensing can only be partitioned between direct gravity effects and other mechanosensory effects in space, and the thresholds of response for any effect of gravity or gravitational loading can only be determined in orbit with appropriate centrifuges, and in the lunar and martian environments.

The proposed experimental approaches will answer this KSQ concerning the mechanisms and thresholds of gravity sensing and mechanical response in different organisms: plants, as critical for sustainable and climate-robust agriculture; and mammals including humans, as critical for understanding in vivo tissue de/regeneration and in vitro growth of cells as chemical production systems or as medicine. While not motivated primarily by space exploration risk mitigation, the fundamental research herein will also reduce the risks of space exploration by facilitating plant growth in space (and therefore providing critical nutrients such as vitamins for long-term space missions), or by pointing to methods to ameliorate muscle wasting, bone resorption, and therapeutic cell production and regenerative medicine for NASA-trained and civilian astronauts (Levine 2010).

Potential Research Areas

The experiments described in this subsection require access to the space environment for microgravity and radiation exposure, including access to centrifuge facilities in space for application of <1 *g*, as well as a means to capture response kinetics/behavior and preserve tissue for molecular analyses on Earth. The ability to resolve reporter gene expression and/or other physiological changes in real time during exposure to altered gravity treatments will allow understanding the dynamics of the responses, including -omics and epigenetics analyses. As a complement to the space experiments, more Earth-based hardware is required for preliminary experiments and ground controls, including centrifuges for hypergravity, clinostats and random positioning machines to mimic weightlessness (and to test by comparison to space responses), and equipment to replicate the vibrational effects encountered in the space environment as a control. Plant growth in space will require sealed growth chambers with controlled light, watering, atmosphere and ethylene scrubbing, and spectral analysis of plant health. Animal growth in space will require growth chambers suitable to the animal type. Last, a data depository and software approaches that allow a uniform analysis and presentation of data from space experiments and their ground controls is necessary for community utilization and analysis of results. This implies not only the maintenance and growth of GeneLab (see Finding 4-4), the current data repository, but also an increase in its analytical toolset, and coordination with other public databases of experimental results from genomic and other types of experiments. These potential research areas draw from different communities and require different technical capabilities in LEO and beyond and are thus organized by organism type without implication of ranking priority.

Plants

Mechanical and Gravitropic Mechanisms

Plants respond to gravity and to other mechanical forces. The gravitropic responses are thought to result from signals caused by sedimenting amyloplasts, starch-filled plastids that are found in gravity-sensing tissues and that move within cells when the gravity vector changes. The other responses, such as to stem bending during vibration or touch and meristem cell polarity that responds to local cell expansion or applied pressure (Hamant et al. 2008; Heisler et al. 2010), do not depend on gravity sensing through amyloplast sedimentation. They may (or may not), however, depend indirectly on gravity (as, for example, stem bending from wind, which results in shorter and thicker stems (Braam 2005), and may or may not require the additional force and stress distribution owing to gravity acting on a stem that has been moved from the vertical). Some responses may depend independently on gravity and other mechanical stimuli, for example reaction wood (Groover 2016). Reaction wood forms asymmetrically in branches to reinforce them. It forms parallel to the gravity vector, when the branches are stressed by being bent. Does gravity determine the direction, and does bending stress determine the magnitude of the reaction in woody tissue, or do each contribute to both? The ability to remove or change the magnitude of the gravitational vector in the space environment will allow tests of these hypotheses regarding this long-term adaptation to altered gravity and non-gravitational mechanical forces.

The mechanism of gravity sensing in plants is poorly understood, and the transduction of gravity signals from the cells with amyloplasts is also only partially understood. The mechanisms by which bending is sensed, by which meristem cells polarize, and by which microtubules align to stress are largely unknown, despite active study. Microgravity and partial gravity, as assessed using centrifuges in LEO, and in the lunar environment, make possible a quantitative understanding of amyloplast sedimentation (e.g., by measuring sedimentation dynamics in partial gravity of different strengths). They enable the differential and combined effects of gravity and other mechanical forces to be recognized in all of the many mechanical responses of plants.

Decoupling of Gravity and Other Tropisms

The microgravity and partial gravity environments also allow for the decoupling of tropisms such as phototropism and gravitropism in plants. Plants respond to red light in a directional manner in microgravity but not in nominal 1 g on Earth (Kiss et al. 2012; Millar et al. 2010). Red-light experimentation was examined in combination with gravity-sensing thresholds (Kiss et al. 2000) made possible by the retired European Modular Cultivation System (EMCS), which had onboard centrifuges allowing for gravitational accelerations ranging from microgravity to reduced gravity to a 1 g control. These experiments demonstrated a robust positive phototropic curvature in response to unidirectional red illumination in the stems (i.e., hypocotyls of young seedlings) of plants that developed in microgravity. Time course studies showed a positive phototropic curvature relative to red light in microgravity and 0.1 g conditions, and the mean curvature was not significantly different between these treatments. In contrast, the response at 0.3 g was not significantly different from the response of the control (1 g), and red-light phototropism was attenuated in both. This range of reduced gravity (relative to Earth) is in the approximate range of gravity levels on the Moon (0.17 g) and Mars (0.38 g). These initial results suggest that lunar gravity may pose issues for plant growth, while martian gravity may be adequate; they also raise the question of a sensing mechanism that creates the threshold effect. The testing of such hypotheses will be enabled by the in situ research in the spaceflight environments of the ISS, Gateway, and lunar surface, and eventually martian surface.

Root Growth and Response to Resistant Substrates

Similar to the decoupling of the phototropic response from gravitropism, a comparison of thigmotropic responses under 1 g, 0.17 g, 0.38 g, and microgravity can show plant response to resistant substrates. For example, if regolith is used as a growth substrate versus a hydroponics or aeroponics approach, is there a substrate density that is not permissible for plant growth? Will a plant under partial gravity behave like a plant in microgravity when it encounters a surface it cannot penetrate? Will the root grow without direction, and will lighting need to be used to promote a negative phototropic response to encourage root growth into substrate (Scherer and Pietrzyk 2014)? Can plant tropisms and plant development in the spaceflight environment be explained in the context of mechanosensitive channels? These channels have been implicated in the perception of important mechanical stimuli such as osmotic pressure, touch, gravity, and pathogenic invasion (Basu and Haswell 2017).

Transcriptional Responses to Spaceflight and Partial Gravity

Plant transcription and epigenetic marks are altered in the space environment. Some of these changes are very likely owing to gravity alterations in space and to indirect effects of gravitational load on mechanosensing. Others may be owing to gravity effects on the plants' external environment (such as alterations in the movement of water at the roots, or the effects of low gravity on convection and therefore gas exchange at leaf surfaces). A few studies utilizing an onboard centrifuge have focused on gravitational alterations, including interrogation of a dose-dependent response to fractional gravity as well comparisons of transcriptional profiles in response to microgravity, lunar and martian gravity. The latter experiments have highlighted the potential modulation of the transcriptional response to partial gravity by red and blue light and indicate that more work is needed to further delineate this interaction. The expansion of available space environments to the lunar surface and work in weightless environments such as ISS and on the way to Mars would be greatly enabled by provision of centrifuges with sufficient capacity for growing whole plants at partial gravity, as a control at 1 *g*, and in hypergravity. It is essential that such gene expression studies under partial gravity are expanded to include additional species of plants (beyond *Arabidopsis*), and that future work assess the combined effects of gravity, light, space radiation, and the atmosphere in space vehicles such as planned for lunar and martian exploration.

Animals and Unicellular Organisms

Fluid Regulation in Vascular and Extravascular Spaces

The regulation of body fluid is profoundly different in microgravity compared to Earth. On Earth, the fluids in the body are pulled down toward the feet by gravity, and the fluids are returned back to the heart by the activity of muscles. In the microgravity environment of space, fluid accumulates around the head and the chest. In addition, rapid change in total blood volume, plasma volume, and red blood cell mass occur because the area to contain blood is decreased. The plasma volume and total blood volume decreases during the first hours in space and remain at a decreased level throughout spaceflight (Hargens and Richardson 2009; Hargens and Watenpaugh 1996).

Mechanical Effects on Cell Differentiation and Development

Animal cells respond strongly to mechanical stresses in their organization and development. For example, the internal organization of muscle cells, and their self-organization as they form muscle fibers, is a response to tension (Mao et al. 2022). Tension-driven multi-scale self-organization is also observed in muscle fibers grown in vitro from human-induced pluripotent stem cells (hiPSCs). Even in cells that are not considered highly contractile (such as muscle cells) or migratory (such as fibroblasts during wound healing), it is now well established that many tissue cells and stem/progenitor cells exert force to maintain adhesion with surrounding cells, extracellular matrix (ECM), or other materials encountered in experiments in vitro or implants in vivo.

Gravitational forces also control animal cell behavior and function. Cell growth and differentiation respond to altered gravity through reduced response to epidermal growth factor (EGF) (de Groot et al. 1991). In addition, intercellular EGF and PKC signaling is altered in microgravity (Boonstra 1999). This emphasizes the need for studies of whole organisms and multi-organ chips, as well as single organoids and isolated cells in culture.

Cells respond to microgravity through cytoskeletal reorganization and tension on ECM components (Hughes-Fulford 2003; Wu et al. 2022). Cell–cell interactions are likely impacted by the cytoskeletal reorganization. Changes in direct interactions between the cytoskeleton (CSK) and nuclear membrane involving the LINC (linker of nucleoskeleton and cytoskeleton) complex have been reported (Neelam et al. 2020). CSK reorganization may differ depending on the duration of exposure to both short- and long-term altered gravity.

Inflammatory factors also likely play signaling roles. Mechanosensitive endothelial cells show decreased secretion of cytokines in response to microgravity (Li et al. 2018), although many studies in other cell types show no response or increases (Ludtka et al. 2021). Given the roles of cytokines in wound healing and immunological health, more studies in both simulated and actual microgravity are warranted.

Nuclear Mechanotransduction

Nuclear mechanotransduction, or signal transduction transmitted within the cell and over the nuclear membrane to effect genetic expression, is a rapidly growing field of study and would provide an explanation for how individual cells might detect gravitational changes (Discher et al. 2017; Maurer and Lammerding 2019). Mechanotransduction at the cellular level has been implicated in a wide array of cellular processes such as proliferation and apoptosis, and in diseases such as cardiomyopathy and muscular dystrophy (Jaalouk and Lammerding 2009) as well as bone growth and remodeling (Moosavi et al. 2021). Understanding the changes in gene expression in response to altered gravities would suggest mechanisms that could be used to develop space-specific countermeasures as well as Earth-based disease therapies.

Bacterial Mechanosensing

Bacteria are prokaryotic cellular organisms that share some structural and mechanistic features with cells in plant and animal systems.The roles and mechanisms of bacterial mechanosensing are understudied (Gordon and Wang 2019), leaving considerable scope for studies of the mechanisms of how mechanosensing works for these prokaryotic microbes in space environments, where fluid flow and consequently shear forces experienced by bacteria are expected to be altered by gravitational effects. Mechanosensing in bacteria is known to exist—in particular, in the formation of biofilms. The shear force experienced by bacteria after they transition from free-swimming to substrate-attached, and shear forces generated by bacterial motility on stiff substrates, can substantially change bacterial behavior and metabolite levels (Araújo et al. 2016). Flow velocity affects the characteristics of *Pseudomonas fluorescens* biofilms, and shear stress affects growth, adhesion, and biofilm formation of *Pseudomonas aeruginosa* with antibiotic-induced morphological changes (Fonseca and Sousa 2007). Long-term bacterial adaptation to gravity changes will point to the underlying cellular mechanisms and signaling pathways that bacteria use to sense their physical environments and can provide insights into how cells in multi-cellular organisms also respond to gravity changes at the cellular level.

> Finding 5-2: Access to microgravity and other aspects of the space environment is required to truly understand how microbes, plants, and animals respond to body forces and applied forces at the molecular level.

Question 9: What Are the Fundamental Principles That Organize the Structure and Functionality of Materials, Including But Not Limited to Soft and Active Matter?

Impact and Rationale

On Earth, transport is mediated by gravity (in addition to other applied forces). Gravity is the driving force for fluid flow—even on small scales such as when considering capillarity, surface tension, and Coulombic forces that are balanced by gravity. On Earth, the weight of a buoyant material is balanced by the weight of the displaced fluid. However, in microgravity, surface tension forces dominate in both cases. With respect to heat transfer, natural convection is driven by fluid motion owing to the differences in the relative densities of the fluids. Warmer fluids of lower density will rise, while cooler fluids with higher density will fall owing to gravity. In microgravity, differences in fluid density will not cause what is considered on Earth to be natural convection; this lack of natural convection affects temperature, constituent concentrations, and distributions of both temperature and concentration.

In the terrestrial environment, gravity obscures important forces and phenomena because of buoyant effects that govern most processes that involve the transport of mass, energy, and species. Vertical motion occurs owing to density differences that promote mixing within phases that disrupt the dynamics of reaction fronts or limit the effective range of other forces (i.e., both surface and body/volumetric forces), such as surface tension, electromotive forces, magnetic induction, and so on. There are two common types of buoyancy-driven motion: one is driven by a density difference between phases, such as a vapor bubble rising through a liquid, and the other is driven by the density difference of a reaction front like a flame or plasma discharge within the same phase and has been typified by the contrast in candle flames in normal and microgravity. As such, new discoveries have been found like the experimental evidence that first confirmed the existence of "cool flames" (Nayagam et al. 2012) and "warm flames." Flotation, seen as the bubble rising through the liquid, or conversely, sedimentation, limits the observation time in a static situation or equilibrium condition in a terrestrial environment. In addition, the object ceases motion when it encounters an obstruction such as a container boundary, and the structure of agglomeration is primarily affected by the physical boundary, as opposed to other forces and neighboring objects.

Natural convection is the motion owing to the density variation within a single phase and is typically attributed to temperature or concentration gradients. This motion promotes mixing and overwhelms diffusion of chemical reactants and products and disrupts the range and impact of the dynamics of reaction fronts and the charge effects.

This mixing alters chemical reaction rates, heat transfer, and mass transfer across surface and interfaces. In normal gravity, the mass of fluid impacts both itself and its contents. The fluid "weight" distorts deformable volumes such as bubbles, droplets, and cells. The three-dimensional symmetry of such spherical objects provides experimental data to validate simplified symmetrically based models for heat and mass transfer and the limiting case of the minimal surface area to volume ratio.

Access to the space environment is essential here, because in a wide range of applications, theories and models of complex fluid and chemical behaviors ignore, for the sake of simplicity, the gravity terms in the governing equations. Experiments in terrestrial laboratories, however, are (with few exceptions) all conducted in the presence of Earth's gravitational field. This gravitational field can obscure important phenomena and dominate the problem of interest rendering the direct comparison of experiment and theory/model difficult or even impossible. Experimentation in the low-gravity environment then achieves the idealized condition of a negligible gravitational force term and buoyancy-induced flow, thus providing an experimental capability that is nearly impossible in terrestrial laboratories (except for the reduced gravity facilities on Earth such as drop towers and aircraft flying parabolic trajectories), and allows investigation of the influence of other factors on determining the fluid flow, chemical reactions, vibrationally and electronically excited states, and phase change processes.

The transport of mass and energy affects both the structure and composition of all phases of matter (as well as phase change phenomena). While it is possible to impose larger local accelerations using centrifuges to assess the effect of gravity, it is not possible to ascertain these other factors (such as buoyancy driven flows, which will not only reposition material elements but also drive the flow around these elements, thus exposing them to varying conditions). The momentum, heat, and mass transfer are coupled within each phase, while the discontinuities at the interface between phases and across the reaction wavefronts can make the coupled and non-linear analyses of such scenarios more complicated.

In general, experiments conducted in space environments to date have been with relatively benign materials under a narrow band of conditions owing to safety concerns to both the crew and spacecraft. While there is a tendency to lump many chemical compounds as either "liquids," "solids," or "gases," and focus solely on traditional transport and thermodynamic properties (such as density, viscosity, specific heat capacity, thermal conductivity, surface tension, contact angle, latent heat, etc.), there are other factors that can significantly alter their behavior. For example, liquid hydrogen, which has an extremely low surface tension and density compared to water, has the largest flammability. Most aqueous solutions tend to have properties that are similar to water, but minute amounts of sodium lauryl sulfate or other surfactants dramatically change the surface tension and wetting characteristics. Some perfluorofluids are highly nonpolar and will not dissolve many organic and ionic compounds but absorb large quantities of gases. Experiments could be conducted considering different test fluids, varying concentrations of additives, and different configurations (of flow conditions, sample containers, and geometries) to improve the statistics. The varying concentrations of additives can serve as precursors for forming nanoparticles and colloidal solutions from in situ chemical reactions where these in situ chemical reactions can be triggered by external stimuli (such as heating, light, radiation, ultrasonication, sol-gel reactions, electro-magnetic induction, subjecting to microwave radiation, etc.).

While multi-phase flows have been viewed within the NASA biological and physical sciences (BPS) and engineering communities principally as gas–liquid or vapor–liquid flows, the definition needs to be extended to include the solid phase, immiscible liquids, and emulsion. The lack of gravity limits the settling rate of solids, and this prolonged suspension permits other forces and reactions to occur. Single and double "entity" or multiphase systems (such as bubbles, droplets, particles, etc.) need to be investigated in order to assess the impact of imposed fields on the ability to induce motion (enhance transport of mass, momentum, and energy), change phase, merge (enhanced mixing), and react with its host medium and the system boundaries. Ultimately, the experiment measurements and modeling need to be extended into systems that contain not just a huge quantity of particles but also a wider dispersion of sizes as well as heterogeneous interfaces in order to gather population behaviors and to identify the stability envelope of such systems.

For example, phase change materials are considered to be attractive for thermal energy storage by leveraging their high values of latent heat of fusion. However, salt-hydrates suffer from multiple issues that compromise their reliability, especially for initiating nucleation and during the solidification processes. This can include diffusion limited transport, chemical stratification, and so on. These effects can become acute under varying levels of gravity (that are nearly impossible to implement and study in Earth-normal gravity environments) as well as for different sample-size conditions and different configurations of heat exchangers (involving different flow conditions of the heat transfer fluids, geometries and containment architectures). Containment architectures within the heat exchangers can involve fins, foams (metal or carbon foams), acoustic/ultrasonic agitation (for enhanced mixing), nucleation promoters/additives, incorporation of gelling agents, and so on. The power ratings if such heat exchanges are used for thermal energy storage can vary significantly, owing to absence of natural convection, both during melting and solidification cycles. This cycling is akin to charging and discharging cycling, whether for thermal management applications for electronics chip cooling or for sustainability applications for reduced water usage in power production.

Last, materials and mixtures governed by weaker internal interactions can exhibit both liquid-like and solid-like characteristics for which the organizing principles of structure and phase change stability remain incompletely understood. Such materials that can flow readily under gravitational force (i.e., are easily deformable under body forces or applied forces, whether reversibly or not) are referred to as soft matter. Such materials that can self-assemble in response to fields or stimuli, or can generate force against surrounding environments, are referred to as active matter. (See Figure 5-2.) For such material types, access to the microgravity feature of the space environment is critical to gain such knowledge about how the functional units or "particles" interact to provide and change structure in response to applied stimuli. Because the interparticle forces are rather weak forces, gravitational stresses limit the size of the crystallites to tens of unit cells per side; in space environments, larger assemblies and masses suitable for fundamental experiments are possible. Moreover, many soft matter systems are microscopic, colloidal, and suspensions of particles in fluids. They are difficult to density match and, in a terrestrial environment, they sediment, which restricts experiments to two dimensions. Opening the three-dimensional world to these materials with microgravity allows a much greater variety of phenomena, types of organization, and different materials and processing. Furthermore, in terrestrial gravity the driving temperature, concentration, and magnetic, electric,

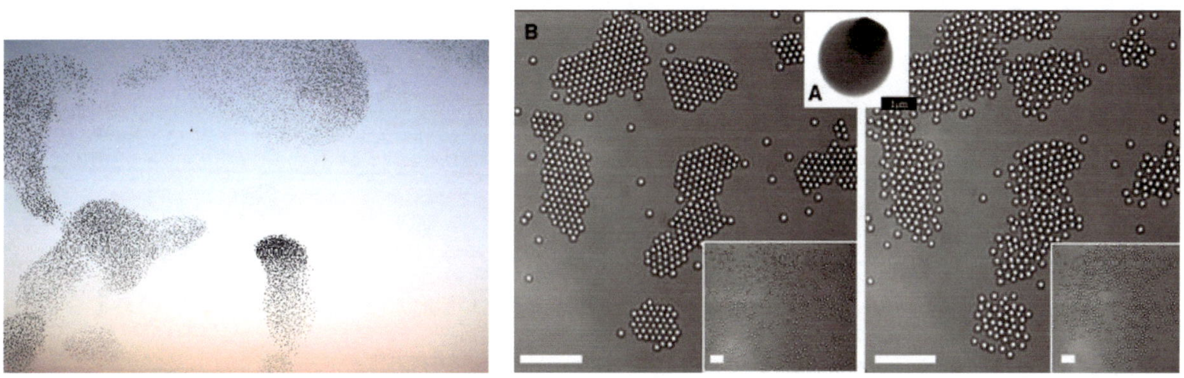

FIGURE 5-2 Active matter. Active matter includes the capacity to extract energy, exert work, and internally reconfigure owing to interactions among particles of all kinds. *Left:* Starling bird murmurations are a flocking phenomena common to many animal species. They arise not from attractive forces but rather from the animals slowing down as they enter a more populated region. *Right:* A non-living example: composite material particles acting as colloidal swimmers, also slowing down and flocking as more particles are encountered. This motility-induced phase separation is visible in (B). (A) Scanning electron microscope image of one particle, a polymer sphere with a protruding iron oxide–containing cube. (B) Colloidal particles flock under blue light, a phenomenon called motility-induced phase separation, from initially homogeneous particle distribution (inset). (C) Flocks disaggregate by thermal diffusion when the blue light is turned off (inset, after 100 s). Scale bars indicate 10 μm. Ground-based experiments are limited to two-dimensions owing to gravity-driven sedimentation. SOURCES: *Left*: From M. Ballerini, N. Cabibbo, R. Candelier, et al., 2008, "Interaction Ruling Animal Collective Behavior Depends on Topological Rather than Metric Distance: Evidence from a Field Study," *Proceedings of the National Academy of Sciences* 105(4):1232–1237. Copyright (2008) National Academy of Sciences, U.S.A. *Right* (A–C): From J. Palacci, S. Sacanna, A.P. Steinberg, D.J. Pine, and P.M. Chaikin, 2013, "Living Crystals of Light-Activated Colloidal Surfers," *Science* 339(6122):936–940, https://www.science.org/doi/10.1126/science.1230020. Reprinted with permission from AAAS.

and light fields create density gradients and produce flows that interfere with the basic phenomena of interest. Microgravity allows for isolation of the fundamental interactions and dissection of the phenomena.

There remain many important basic questions about soft matter in equilibrium. Such materials are constructed from building blocks that have internal degrees of freedom and internal sources of power and are different from conventional materials where the building blocks (atoms or simple molecules) and their interactions are well described. These engineered materials can exhibit a much wider set of properties and functionalities, and approach the functionality of natural, living materials. Rather than simple polymers, emulsions, colloids, and granular matter that interact through conventional forces, researchers have produced solid and liquid particles that interact through a host of forces. These now include DNA hybridization that can activate the motion of functionalized colloids and emulsions, biomolecules, proteins, and enzymes; entropically driven shape-dependent properties (lock and key colloids) that enable design of arbitrary shapes with artificial molecular motors that power the motions (translation and rotation) of particles; and nanocomposite tectons that self-assemble to create structurally distinct phases of matter (Zhang et al. 2016). Using the new forces and particles, great headway was made in Earth-based experiments over the past decade, including development of colloidal soft matter as photonic crystals and in the modeling and engineering of active soft matter. (See Figure 5-2.) These new tools will give rise to novel materials with exciting properties such as sensing and repair, shape-shifting, self-healing, and self-replication. But there is much we do not know about how these passive and active soft materials interact with the space environment; these unknowns are part of stated Grand Challenges in Soft Matter Science (Mezzenga 2021).

Aside from the processing and materials aspects that are established under steady-state or equilibrium conditions, the new building blocks allow production of dynamical systems for testing some of the most basic aspects of systems during transients and far-from-equilibrium (see Question 10). Using the design capabilities of soft and active matter, model systems can and have been made to test the role of entropy and entropy production and how it relates to where work can be extracted. Such techniques developed in soft/active matter have further been applied to biological systems. The quantitative analysis of these experiments is only possible owing to advances in machine learning, artificial intelligence, big data, and simulations. These developments affect not only the way data are currently analyzed but also how experiments are being conceived and executed.

Finding 5-3: Preceding decades of microgravity research have shown the promise of using the reduced-gravity environment to provide transformational change by identifying new states of matter, new physical phenomena, and emergent material properties.

Potential Research Areas

Chemical Kinetics and Combustion Processes

Experiments in microgravity have provided unprecedented insights into chemical kinetic phenomena, highlighting the importance of radiative loss in determining flammability limits, new steady-state regimes of low-temperature and intermediate-temperature fuel burning phenomena (cool and warm flames) and insights into sooting and the formation of large soot agglomerates that has enormous impacts on understanding the effects of wildfires on the environment and applications to material synthesis. Microgravity experiments also demonstrated that electric field and charges (ions and electrons) can also affect the energy transfer, reactions, and flow motion. However, large buoyant flows present in terrestrial laboratories prevent the attainment of the low strain environment necessary to study systems where flame structures or reaction wave fronts are determined by chemical kinetics and motion of charged molecules and particles. Fundamental experiments are necessary for developing predictive tools for chemical kinetics over a range of environments and fuels including biofuels, e-fuels, and low-carbon fuels.

Microgravity studies have shown that large soot agglomerates (super agglomerates) are preferentially formed in microgravity flames where, in the absence of buoyancy, the long residence time necessary for the formation of these structures is easily attained. Reduced-gravity studies of soot formation, including the formation of super-agglomerates, could further elucidate this phenomenon. The low stretch/strain environment is readily attained in reduced gravity studies in the absence of the buoyant flow present in all terrestrial studies. This environment provides the strictest test of chemical kinetic mechanisms, as the flame structure limits are kinetically controlled and not determined by the flow environment, providing insights into the fundamental physics of the processes involved.

Multi-Component Multi-Phase Behavior

Spaceflight systems utilize both single-component fluids such as propellants and multi-component fluids such as providing water and fertilizers to plants. However, many space experiments have selected single-component fluids for investigation, thus avoiding identifying and quantifying the effect of these components on the overall behavior.

Compared to a terrestrial laboratory, in a space environment, trace amounts of other components may affect the behavior in terms of contact line motion, coalescence, agglomeration and solidification, and interfacial heat and mass transfer. Experiments could quantify the scale of behaviors relative to the concentration of solutes, to understand the limits to which pure fluid analogy is sufficient with regard to "infinite dilution" assumptions. Another effect that can be clearly addressed in space environments is how the imposition of external fields—that is, electrical, magnetic, and so on—affects the phase change process, the orientation, motion, and separation of components and phases.

Effects of the Space Environment on Transport and Materials Synthesis

These microgravity-driven differences in transport phenomena can enable synthesis and fabrication of novel multi-component materials and structures. Additionally, microgravity enables containerless synthesis (Amselem 2019; Broutin et al. 1997) and processing. With respect to material synthesis and processing, the following narrower research areas could be considered:

- *Effects of gravity on nucleation, growth, phase transformation, and the evolution of microstructure.* While these questions have been probed for some single-component systems, further investigation is warranted of microstructure and its evolution for complex, multi-component materials fabricated in space. Furthermore, even in single-component systems, novel insights are emerging, owing to unique methods of processing and measuring these materials that are enabled by the space environment (Bracker et al. 2020).
- *New materials synthesis routes enabled by the space environment.* The space environment affords different processing methods and conditions than terrestrial environments. An area of future investigation may be focused on synthesis routes that are impossible/impractical on Earth but can be achieved in a space environment (e.g., nanoparticles formed in situ from additives when specimens are subjected to extraterrestrial radiation).
- *Effects of the space environment on complex crystal structures.* Possible areas of inquiry could include polymorphism and order–disorder transitions. Crystal polymorphism refers to the same molecules crystallizing in different ways, which can affect properties and performance. Microgravity affects dominating forces and transport phenomena, which may enable synthesis of new crystalline polymorphs. Most prior work on crystal polymorphism in microgravity has looked at a single protein (Amselem 2019), which could provide a foundation for more complex crystalline structures. Order–disorder transitions can be probed under microgravity conditions, leading to greater knowledge and measurement of the atomic interactions that lead to order–disorder in a system.
- *Special opportunities in space for the manufacture of drugs and expensive chemicals.* The unique environment afforded by space may enable synthesis and processing of molecules/macromolecules that cannot be practically manufactured on Earth. Possible advantages enabled by space include containerless manufacturing, manufacturing in water-free environments (for compounds with low water stability), and crystal polymorphism (Weber 2010).
- *Glasses and other amorphous materials made in space.* Amorphous materials could be cooled sufficiently fast to avoid nucleation and crystallization. Containerless processing avoids nucleation owing to contamination of the container surface. However, microgravity reduces heat transfer via natural convection, which may lower cooling rates if radiation is not significant. Fabrication of more complex, multi-component amorphous systems requires understanding how these materials differ from those made on Earth.

- *Material behavior during processing in reduced gravity.* Processing within this context may include extraction, refinement, manufacturing, and reuse/recycling. When gravity does not dominate, other forces may become more important, as discussed above. Identification of the governing physics and chemistry in reduced gravity will also enable measurement of these forces with resolution that cannot be achieved on Earth.
- *Critical heat flux conditions.* Conditions at fixed, transitory, and oscillatory pressure conditions could be examined to assess dampening and acoustic-like effects, with focus on changes owing to space environment including microgravity. For example, as the pressure is increased in a boiling system, the saturation temperature and heat of evaporation increase at a modest rate, but the vapor density varies linearly. Thus, the transition to film boiling is delayed because the bubbles accommodate higher rates of vapor generation, resulting in a peak in critical heat flux values with increase in operating pressure. However, at higher pressures, this behavior reverses, owing to the drastic reduction in surface tension near the critical point.
- *Fluid compressibility.* Given the lack of natural convection in microgravity, compressibility effects, especially at higher pressures, could be examined because many investigations have occurred at pressures near atmospheric. Solubilities, chemical reaction rates, and heat and mass transfer rates change at higher pressures (e.g., near critical pressures). At pressures approaching and above the critical point of the fluid, reductions in diffusive timescales governing thermal, mass, and momentum transport occur as well as changes in solubilities that impact chemical reaction rates and rates of phase transitions (e.g., phase change heat transfer).
- *Phase change processes.* Phase change processes could include studies that can take advantage of a quiescent environment that promotes both mass and thermal diffusion that may lead to different morphologies of bubble nucleation distribution as well as growth and departure. This includes solid–liquid and solid–vapor transitions as well as solid-state transitions. Materials that are elastic and compressible could be tested in microgravity, as the sag introduced by gravity does not complicate the analysis of the behavior. Where possible, the characteristics of extraterrestrial media such as lunar regolith could be measured and assessed for its processability.

Effects of the Space Environment on Material Properties

Key structural, physical, and functional properties of materials can change, owing to environmental exposure. On Earth, one observes this in everyday objects like rusting metal infrastructure or contact lenses that dry out when not stored in saline solution. Space offers environments unlike terrestrial ones, which can be hostile to Earth-familiar material stability owing to different (or no) gaseous environments, temperatures and temperature ranges, extraterrestrial radiation environments, electromagnetic fields, and varying levels of gravitation. While the role of different levels of gravity in material synthesis and processing and the science they enable are discussed in detail above, these other differences may also fundamentally change material properties and associated utility of materials in space. (See Figure 5-3A.) With respect to material properties in space, the following narrower research areas could be considered:

- *Effects of the ambient environment on properties of materials in space.* This includes considerations such as temperature, vacuum versus gas pressure, extraterrestrial radiation, and oxidizing versus reducing conditions.
- *Combined measurements including properties.* For experiments on materials in the space environment, interpretation of the results depends critically on the properties of the sample. For many classes of these experiments, the properties of the sample can be measured in situ with the same apparatus. This approach gives not only accurate properties of the material, but also accurate properties of that sample at the time of the experiment.

Colloidal Phase Behavior

Dispersed networks of stimuli-responsive colloids, polymers, and proteins can be actively driven to undergo phase transformations or maintain metastable states that produce metamaterials with unusual physical properties

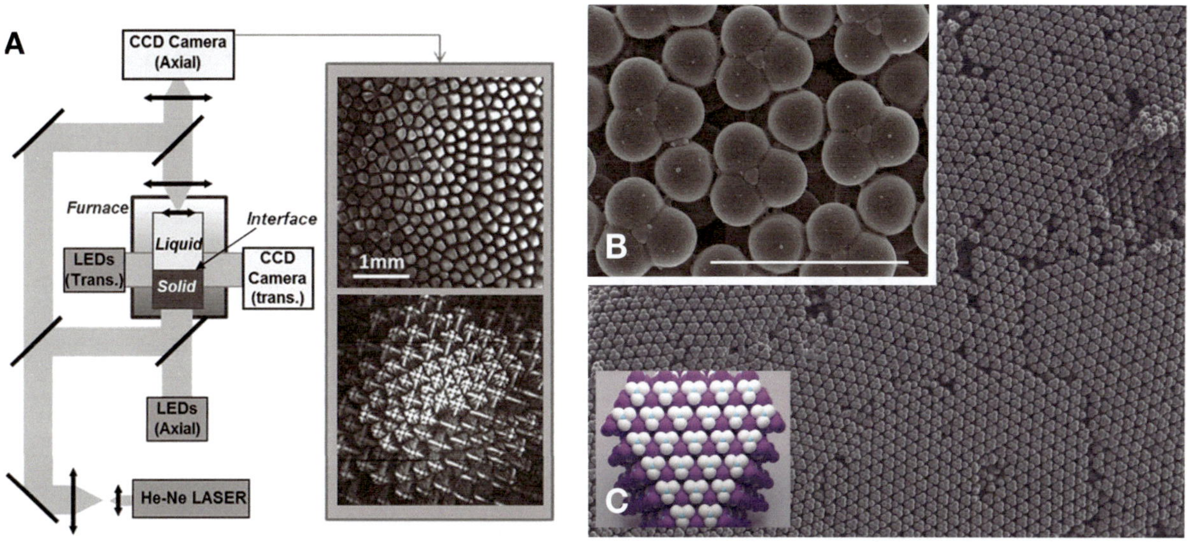

FIGURE 5-3 Examples of crystallization for which space environment-based studies confer specific scientific advantage. (A) Cellular and dendritic patterns obtained under microgravity are much more regular and ordered than those on Earth, as visualized through specialized differential interference contrast optical imaging apparatus. This difference allows more accurate measurements, improving our understanding of phenomena important to manufacturing and crystal growth on Earth. (B) Colloidal assemblies can exhibit emergent functional properties, here as photonic band gap materials. Colloidal diamond lattice formed from triangular, deoxyribonucleic acid (DNA) functionalized particles and (C) crystallization of cubic diamond colloidal crystals. Previous experience demonstrates that colloidal crystals are larger and less defected when grown in microgravity. SOURCES: (A) From F.L. Mota, Y. Song, J. Pereda, et al., 2017, "Convection Effects During Bulk Transparent Alloy Solidification in DECLIC-DSI and Phase-Field Simulations in Diffusive Conditions," *Journal of the Minerals, Metals and Materials Society* 69:1280-1288, https://doi.org/10.1007/S11837-017-2395-6, Springer Nature. (B, C) Modified from M. He, J.P. Gales, É. Ducrot, Z. Gong, G. Yi, S. Sacanna, and D.J. Pine, 2020, "Colloidal Diamond," *Nature* 585:524–529, https://www.nature.com/articles/s41586-020-2718-6, Springer Nature.

(Aubret et al. 2018; Soni et al. 2019; Spellings et al. 2015). The space environment enables exploration of model colloidal systems with diminished gravity and pressure that limits buoyancy-driven convection, sedimentation, hydrostatic pressure gradients, and liquid film drainage, and can stratify density gradients. The effects of increased radiation and temperature fluctuations can also play significant roles in colloidal behavior. In this environment, a wide range of phase behavior of fragile systems near and far from equilibrium can be effectively probed, including syneresis, coacervation, jamming, and bicontinuous particle-stabilized emulsion (bijel) formation.

Effects of Anisotropy and Self-Motility in Phase Separation

The time-reversal symmetry of low Reynolds number environments can be subverted by the spatiotemporal asymmetry of active colloids (Chen and Santangelo 2018; Chen et al. 2021), which can induce liquid–liquid and liquid–gel phase separation. New orderings of soft matter have recently been reported that have intriguing properties, such as the ability to output power rather than dissipating it. The reduced gravity of the space environment may enable the discovery of even more exotic forms of weakly bound network materials, phase transformations, and new phenomena.

Plug-and-Play Assembly Methods for Soft Materials

The ability to destroy and reassemble soft materials into structures with tunable properties has been proposed; however, design rules and methods to assemble these materials and structures in the space environment have not yet been developed.

Active Matter

For both passive and active soft matter systems, shape, internal structure, and distribution of electrical charges, chemical activity, flow, and other external stimuli affect the overall structure and properties of intermediate and final products formed through self-assembly processes. In addition, for both homogeneous and heterogeneous active systems, dynamics of individual building blocks can vary within a population and alter self-association, adding significant complexity to dynamics that give rise to bulk soft materials, which in some cases can possess properties not observed in natural materials (Mallory et al. 2018).

Actuation of Colloidal Suspensions to Form Scalable Metamaterials with Tunable Physiochemical Properties

Relatively weak external fields, including magnetic, acoustic, and phoretic, have shown great potential to drive colloidal suspensions into reconfigurable states with desired properties. (See, e.g., Figure 5-3B,C.) However, it is yet to be determined how these transmitted signals can be used to form, hold, and fix weak colloidal structures that would be otherwise destroyed by gravitational forces on Earth.

Such studies, along with ground-based counterpart baseline experiments at 1 *g*, could enable the first well-controlled measurements of the forces driving self-assembly from the independent effects of gravity, buoyancy driven flows, and hydrostatic pressure. One could construct phase diagrams for concentrated suspensions of anisotropic active matter with and without the presence of external fields (hydrodynamic, electromagnetic, pneumatic, and acoustic). That knowledge could in turn enable design principles for materials enabling multi-scale soft robotics, and micro- and nano-scale colloidal machines that are mobile, reconfigurable, sensing, and capable of manipulating their space environment.

Material Synthesis and Processing

The reduced gravity environment of space and the lunar surface leads to greater importance of other forces (e.g., capillary, electromagnetic, van der Waals, thermophoretic, and Coulombic) and also fundamentally changes dominating transport phenomena. Understanding the new balance of forces and how they affect energy, heat, mass, and momentum transfer in reduced gravity is critical to controlling synthesis and processing across levels of gravitational forces and can enable creation of new multi-component materials (e.g., alloys, organic molecules/macromolecules, including those with complex secondary structures, and composites), and structures ranging from nanomaterials to meter-scale systems, and chemical synthesis. The experiments necessary to generate this new knowledge require the space environment.

Question 10: What Are the Fundamental Laws That Govern the Behavior of Systems That Are Far from Equilibrium?

Impact and Rationale

All systems undergoing change, whether mass and heat transport or chemical reaction and phase formation, are out of thermodynamic equilibrium; systems tend to respond to bring them closer to equilibrium. When the deviation from equilibrium is small, conventional kinetics and statistical mechanics can provide theoretical underpinning. However, when systems are very far from equilibrium, such formalisms become tenuous and there is need for new theory, chemistry, and physics. The opportunity for well-designed space-based experiments is critical to identify and validate the laws that govern material behavior when systems are far from equilibrium. (See Figure 5-4.) Equilibrium thermodynamics is governed by one basic principle from which all else derives: in equilibrium, entropy (a measure of the number of equivalent equilibrium configurations in a system) is maximized. Non-equilibrium phenomena have no such unifying principle, challenging our ability to understand, predict, and control the behavior of materials that are far-from-equilibrium.

Two of the fundamental laws of nature are that matter can neither be created nor destroyed, and that energy can neither be created nor destroyed. Both laws assert a balance in both the total amount of matter and energy, even though the amounts of specific molecular arrangements or type of energy can transition (kinetic versus potential, electrical versus thermal, etc.). Thermodynamics has used this balance or equilibrium successfully for predicting potential results and behaviors where the deviations from equilibrium are small. However, some conditions may be a case of "quasi-equilibrium," whereby small perturbations lead to significant changes such as a docile mixture of highly reactive chemicals that require an activation energy to initiate the event. In addition, technological advances have occurred in a range of applications that use systems/processes where the deviation from equilibrium is large enough that methods to predict results based on equilibrium fail because other mechanisms supersede the slow and gradual changes that form the basis for an equilibrium approach. These mechanisms rely on mass and energy that are transported into a region, upsetting the local equilibrium state, and can extend into a macroscopic scale. Even energy storage materials such as those within batteries are subject to such far-from-equilibrium (FFE) scenarios during rapid, cyclic charging and discharging across material interfaces.

The microgravity environment provides nearly uniform hydrostatic pressure and a significant reduction in natural convection, which together enable research in high-pressure and non-equilibrium phenomena. These variations can affect thermophysical properties, reaction kinetics, and material solubility, as these are affected by the dominance of strong intermolecular forces and non-elastic multi-body collision processes. Descriptions of fluid behavior in the vicinity of the critical point critically rely on an assumption of equilibrium throughout the fluid domain, or that deviations from equilibrium are minimal (from a system-wide predictive viewpoint).

Indeed, access to the space environment, including sustained microgravity conditions, is required to advance this KSQ. Technologies are increasingly reliant on computational models to shorten development time; optimize designs; and reduce costs relative to more traditional strategies involving building prototypes, testing, and redesigning philosophies. This expectation requires a predictive capability from models that need significant refinement and sophistication. In some cases, predictive models are unavailable because of a lack of fundamental data on thermophysical properties and chemical kinetic processes and in other cases where conventional theories (based on equilibrium thermodynamics) do not apply. For example, rapid solidification of metals often leads to FFE conditions with consequences to the resulting microstructure and properties. In stainless steels, a metastable crystal structure forms with a composition different than predicted at equilibrium because the solidification velocity overwhelms the diffusion necessary to reach local equilibrium. Furthermore, the rapid solidification causes a large concentration of defects in the metastable crystal, observable as broader detection peaks in Figure 5-4. Access to reduced gravity for experiments in the space environment and many ground facilities over the past decade provided wider access to stirring and undercooling parameters. This in turn enabled theoretical modeling to support discovery of an additional energy term that can now predict this unanticipated effect on a well-known material. Space environment is required of such discoveries, but the practical impacts are realized first on Earth in terms of quality and properties of metallic alloys cast for many structural infrastructure applications.

The aforementioned studies have identified new states of matter and new physical phenomena that have had profound impacts on their respective disciplines (e.g., see Chapters 2 and citations in Figure 5-4); these discoveries can be fueled by a laboratory environment where the gravitational body force is essentially eliminated. Addressing this KSQ extends our fundamental knowledge of how all classes of materials change form and phase; such knowledge will enable predictive modeling and insights. History indicates that such fundamental understanding of material physics beyond that predicted from current thermodynamic laws will confer human impact in ways that are not yet imagined; our ages are often named by the materials that can be processed for human needs. Anticipated practical applications include processing and manufacturing of structures and machines and potentially creating materials that cannot be fabricated similarly on Earth.

Potential Research Areas

Chemically Reacting Systems (Gases) at High Pressure

For fluids at high pressure and rates of change in pressure, intermolecular forces are enhanced as the pressure increases and the molecule mean free paths are reduced. As a result, the likelihood of multi-body collisions is increased. Additionally, strong intermolecular forces contribute to increasing the complexity and dynamics of the chemical kinetics. Moreover, radiative transport becomes increasingly complex as gray-gas (optically thin) assumptions may fail.

This consideration extends to boiling: not only does the distribution of the surface tension forces at the liquid–vapor interface change, but also the solubility of noncondensable gases in the liquid phase changes (which affects the dynamics of bubble nucleation, growth, and departure). The mass transfer in both phases is impacted also

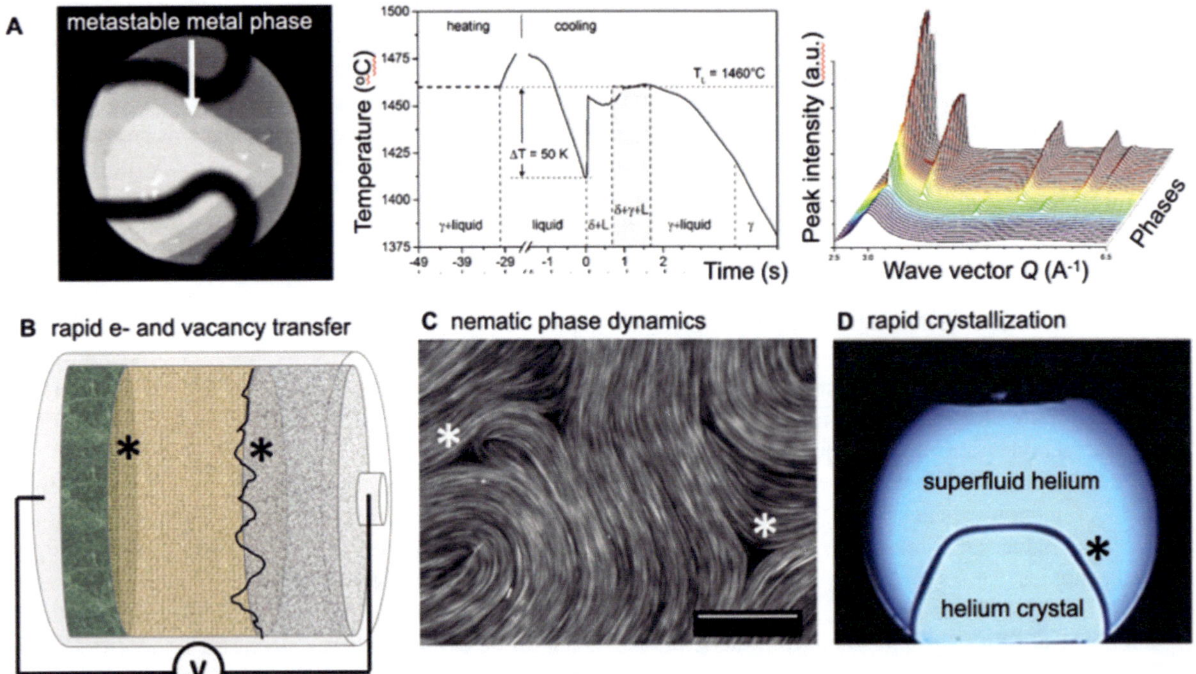

FIGURE 5-4 Materials far-from-equilibrium include complex fluids and solids with sundry applications in energy storage, information technology, and the built environment—but predictive descriptions of transformations and far-from-equilibrium states are lacking without access to the space environment under durations exceeding transformation timescales. (A) Rapid solidification in an undercooled metal such as stainless steel is far from equilibrium. A stable phase (bright area) grows into a metastable phase (medium gray), which in turn grows into the undercooled liquid (dark). Black lines are the sample holder. Non-equilibrium effects include crystalline defects that broaden the diffraction peaks. The defects retain free energy above the equilibrium state, accelerating the transformation among crystal phases. This additional energy is also affected by stirring the liquid just before solidification. (B) In rapidly discharged or miniaturized energy storage materials such as batteries, rapid or sharp changes in electron and vacancy transport across interfaces (*) between metals and metal oxides can occur that limit performance. (C) In materials with weaker interatomic interactions such as soft matter and complex granular fluids, responses far-from-equilibrium can occur with little perturbation and cause unexpected dynamics near topological defects (*), such as those shown in this frame from a movie illustrating dynamic reconfiguration of an active nematic phase. (D) Far-from-equilibrium, crystals and interfaces form even in Bose-Einstein condensates such as helium superfluids. SOURCES: (A) Modified from D.M. Matson, X. Xiao, J.E. Rodriguez, et al., 2017, "Use of Thermophysical Properties to Select and Control Convection During Rapid Solidification of Steel Alloys Using Electromagnetic Levitation on the Space Station," *Journal of the Minerals, Metals and Materials Society* 69:1311–1318, https://doi.org/10.1007/s11837-017-2396-5, Springer Nature. (B) Created by Krystyn Van Vliet. (C) Modified from P. Guillamat et al. (2018). (D) Modified from T. Takahashi, R. Nomura, and Y. Okuda, 2016, "^{4}He Crystals in Reduced Gravity Obtained by Parabolic Flights of a Jet Plane," *Journal of Low Temperature Physics* 185:295–304. https://doi.org/10.1007/s10909-016-1592-y, Springer Nature.

because of the increased probability of multi-component molecular interactions that can retard reactions because of the increased interaction between different chemical species.

Non-Equilibria Phenomena in Complex Fluids

Increasingly, new and emerging technologies involve phases of matter not easily classified into solid, liquid, and gas categories: "squishy" soft matter (with some examples also called complex fluids), as well as transcritical phenomena involving fluids near and exceeding the critical point and plasmas. These systems are in states characterized by weak interparticle interactions (as discussed in Question 9, at equilibrium on Earth), and the space environment can thus push such matter toward FFE behavior that cannot be predicted. Outcomes of such research are expected to be most relevant in engines moving to high pressure; advanced engine concepts—kinetically controlled engines; and non-equilibrium, plasma-assisted combustion and chemical manufacturing.

High-Pressure Transcritical Research

Experiments are necessary to provide fundamental properties and develop predictive models for single- and multi-component mixtures at high and transcritical pressures. An important research goal is to develop a predictive capability for reacting systems at high-pressure conditions, given the importance of three-body or multi-body collisional reactions and the increased importance of intermolecular force and radiative transport. It is important to consider (and in some cases, reassess) the fundamental laws that govern the phase equilibrium, phase interfaces, and heat and mass transport at high-pressure conditions extending beyond the critical points of the constituent fluids.

Scaling Predictions

This term has many connotations distinct and relevant to reduced gravity, ranging from length scales, population size or concentration, and phenomenological to system behavior. One such connotation involves the "jump" conditions of the temperature and concentration profiles at the interface in a phase change or across a chemical reaction wavefront process. Of practical interest to many applications are the controlling mechanisms in non-equilibrium microchannel flows where traditional continuum fluid mechanics do not adequately define the jump or slip boundary conditions behavior. Another scaling-related issue arises as the effects of forces on single- or two-body events are defined. The impact of multi-body interactions remains unclear—collisions and interference can either diminish or punctuate the effect of those forces.

For certain processes such as interfacial bonding among materials, it is important to determine if their response to acceleration or gravity is linear. The scale range here extends from zero gravity potentially to supergravity (giant-planet environment) and includes partial gravity environment (lunar and martian surfaces). Bonding is a useful example of a process affected by the residual gravity level, and studies of bonding among dissimilar materials in distinct space environments will be informative.

Dryout and rewetting on localized heat transfer surfaces impacts the stable operation of a two-phase system, and again, the role of acceleration, gravity, and partial gravity needs elucidation, in combination with the effects of surface geometry, pinning edges and flow paths upon the two-phase system stability. Other mechanisms may exist for the preferential accumulation and release of phases within a multi-phase system, impacting the stable operation of that system.

Flow and Structure

Even though matter may flow, when subjected to activating events—matter will react to these events and reorganize into new structures; these range in length scale of order from the microscopic to long-range macroscopic architectures. Determining the relationship of the motion of suspended bodies when subjected to external fields as well as localized charges and shapes may be deterministic in terms of the final structure. These fundamental questions can be studied in several different types of matter, leveraging the unique space environment characteristics, including metallic alloys, organic soft matter, and composite materials governed by colloidal interactions.

In a microgravity or partial gravity environment, it is possible to conduct clean experiments to determine how much and what type of impetus could be applied to a system at or near equilibrium to initiate a rearrangement. These experiments will help to identify the factors that affect the behavior rate of the morphing, translating, and/or rotating, as well as determine if there are multiple "end-states," and what determines which end-state is achieved. These end-states may be influenced by hysteresis and/or be "path dependent." Experiments that can quantify this in materials described as undercooled liquids or glasses can also advance predictive understanding and modeling of metastability and phase separation, as well as kinetic limitations in phase transitions.

Such understanding will contribute to a unifying goal of "model" systems for which generalized laws for non-equilibrium phenomena can be developed. One example of such a generalized law could be a framework uniting classical hydrodynamic instabilities (e.g., Richtmyer-Meshkov instability) and their multi-phase analogs (in the case of Richtmyer-Meshkov instability, the analog would be shock-driven multi-phase instability).

Statistical Mechanics for Active, Particulate Materials

There has been significant research into colloid interactions in simulated microgravity and on the ISS; however, since the 2011 decadal survey there have been numerous reports of novel active colloids. These particulate systems, whether powered through fuel or fuel-free energy sources, can assemble into network materials with unusual bulk properties (Okutani 2002; Weber 2010) and behavior, including self-healing, jamming, odd viscosity, auxetic behavior, and allostery. Currently, it is difficult on Earth to experimentally identify, decouple, and precisely measure the specific forces interactions that lead to the observed bulk behavior. The space environment provides an ideal test-bed to explore active colloidal interactions beyond and among equilibrium states. The reduced gravity states of the space environment will enable the first answers to important yet seemingly basic questions about these materials, including whether steady-state flow can exist between two different non-equilibrium states. This space-specific advantage is because sedimentation owing to Earth's gravity will be suppressed.

Question 11: What New Physics, Including Particle Physics, General Relativity, and Quantum Mechanics, Can Be Discovered with Experiments That Can Only Be Carried Out in Space?

Impact and Rationale

An intellectual crisis in fundamental physics provides an opportunity for space-based research. The 21st century brought forth this awareness of our collective limit in understanding fundamental physics—that is, our understanding of the physical laws governing matter, space, and time. On the one hand, the standard model (SM) of elementary particles and fields (Particle Data Group et al. 2022) has been extremely successful in predicting and explaining a plethora of physical phenomena. Also, with the discovery of the Higgs boson, all particles of the SM have been observed. At the same time, plentiful observational data on galactic and larger scales cannot be explained based on the observed matter content in the framework of the SM and "normal gravity" as described by the weak limit of general relativity. The prevailing view is that invisible "dark matter" (Figure 5-5) dominates the matter content of the universe (Bertone and Hooper 2018). Attempts to explain dark matter by simply "invisible" normal matter (i.e., faint stars, black holes, large planets, etc.) have failed—this possibility has been ruled out via detailed astronomical studies (Tisserand et al. 2007). In addition, observations indicate that the universe is expanding at an accelerating rate, an effect commonly attributed to "dark energy" (Frieman et al. 2008). A summary of all available data gives that only 4.93 percent of the matter-energy composition of the universe is "normal matter" (i.e., the SM particles) (Particle Data Group et al. 2022). Cold dark matter constitutes 26.5 percent, and the remaining 68.5 percent is dark energy, with both dark matter and dark energy of unknown origin (Particle Data Group et al. 2022).

Most dark matter searches to date look for a specific candidate, which has been seen as very promising from the perspective of theoretically favored extensions to the SM—that is, weakly interacting massive particles (WIMPs). However, despite much effort and advances in detectors, more than 2 decades of such direct WIMP dark matter searches have not yielded a discovery (Arcadi et al. 2018). As a result, the past decade has seen unprecedented effort in broadening the candidates for dark matter, extensive model building at all mass scales, and the design and implementation of numerous new types of ground-based detectors (Antypas et al. 2022; Bataglieri et al. 2017).

In addition to the dark matter and dark energy problem, the standard model fails to provide explanations for several other phenomena and issues. Examples include the hierarchy problem (why are the masses of known particles so much lighter than the fundamental energy scales of the unification of fundamental forces?); the matter–antimatter asymmetry problem (why is so much more matter than antimatter observed in the universe?); and the existence and origin of neutrino masses. Moreover, the class of beyond standard model (BSM) ideas involving supersymmetry predicted new particles within the observable range, which the Large Hadron Collider (LHC) so far has not detected (Canepa 2019). This presents a further crisis, now for the leading BSM theories.

Answering this KSQ, even partially, over the coming decade of space-enabled BPS research will help address and inform underlying subquestions, including: How are gravity and quantum mechanics connected? Does general relativity hold in extreme regimes, such as near or inside black holes? Are fundamental constants actually constant?

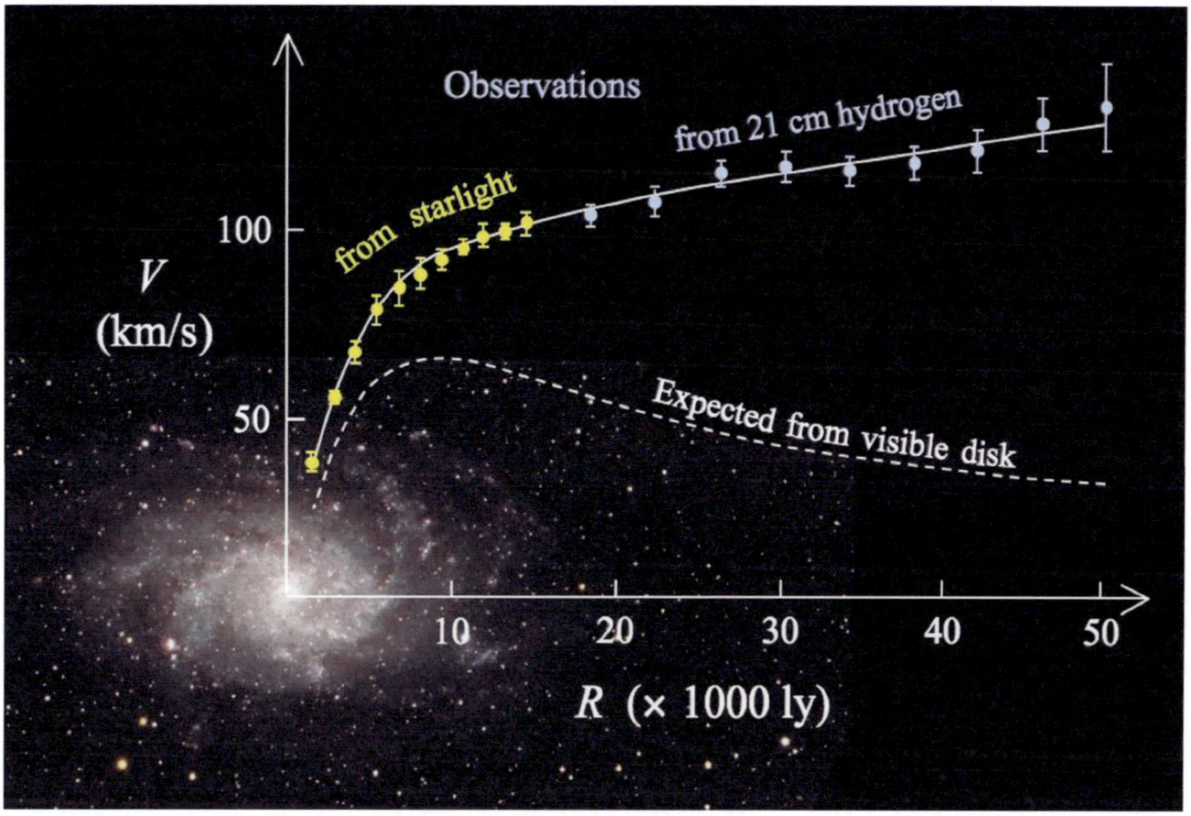

FIGURE 5-5 Rotation curves of M33 galaxy. The bottom curve is the prediction based on the mass of visible matter; the top curve is from actual rotational velocity observations. SOURCE: From P. Salucci, 2019, "The Distribution of Dark Matter in Galaxies," *Astronomy and Astrophysics Review* 27:2, https://doi.org/10.1007/s00159-018-0113-1, Springer Nature.

Are there violations of fundamental symmetries? Addressing any of these puzzles will be transformative even though the applications of such knowledge cannot be articulated today. This knowledge will change our current understanding of how the universe works and provide pathways to new frontiers. For example, solving the dark matter problem will not only reveal the composition of the universe but can also help us harness fundamental forces for technologies including quantum computing.

Indeed, while all of these questions seem to be purely fundamental, with minimal connection to practical applications and technology, it would be shortsighted to treat them as such. At the beginning of the 20th century, spectra of atoms, the photoelectric effect, and other puzzles of matter and radiation also seemed abstract and removed from human relevance. However, understanding these phenomena brought forth quantum mechanics, which became a cornerstone of modern technology. Atomic clocks enabling GPS navigation, electronics, lasers, nuclear magnetic resonance (NMR) spectroscopy, magnetic resonance imaging, superconducting magnets, electron microscopy, and much of modern chemistry are just a few examples. It is reasonable to expect that solving present fundamental physics problems—including via space-based experiments—will underpin future transformative technologies. Moreover, fundamental physics research in space extends beyond a quest for "new physics" and includes anticipated contributions to fundamental many-body quantum physics, predictions of phase transformations also embraced by the materials science and complex fluids communities, and methodological advances that may also advance adjacent fields such as astronomy and astrophysics.

As an example of the former, transformative advances in quantum technologies have led to a plethora of new high-precision quantum devices joining the search for dark matter, dark energy, and new forces; tests of gravity; searches for violation of fundamental symmetries; sensitive measurements of gravitational waves; and many other investigations (Antypas et al. 2022; Safronova and Budker 2022; Safronova et al. 2018). In addition, the development of quantum devices has allowed direct tests of the underlying principles and implications of quantum mechanics and the study of the quantum versus gravity interface. Importantly, the accelerated development of high-precision quantum sensors largely happened after the 2011 decadal survey, and thereby reshaped the landscape of new physics searches over the past several years. The availability of these new techniques also gave rise to much theoretical work in particle physics, atomic, molecular, and optical (AMO) physics, gravitational physics, and other fields exploring new ideas. Together, these developments led to broad recognition of the discovery potential for space-based fundamental physics experiments. The enthusiastic response by the BPS research community to the request for input papers related to this decadal survey clearly demonstrates this strong interest, with a diverse array of opportunities identified for transformative fundamental physics investigations exploiting the unique space environment, away from Earth's gravity. For example, some hypothesized dark energy and dark matter fields can be screened at the surface of Earth, reducing their measurable effects by many orders of magnitude. Furthermore, many fundamental physics experiments require a network of instruments placed at distances greater than Earth's diameter—for example, long-wavelength/low-frequency detectors of gravitational waves. Other solar system objects (including the Sun, the Moon, and asteroids) could also provide access to unique fundamental physics explorations, including through direct studies of their behavior or as locations for precision measurement instruments.

A shared tenet of several open, fundamental physics sub-questions is that deploying quantum sensors and related precision measurement technology in space will enable transformative searches for new physics and/or major advances in our understanding of fundamental physics in ways not possible on Earth. However, such investigations are broad and open-ended by their nature. The rapid ongoing progress in the development of quantum sensors and precision measurement technology leads to the need for extensive effort to answer the first key physics question formulated above.

Several well-defined ideas and corresponding mission concepts have already been proposed (Alonso et al. 2022; Derevianko et al. 2022; Schkolnik et al. 2022). However, these are likely to barely scratch the surface of what space-based quantum and precision measurement experiments can offer to address the many outstanding questions of fundamental physics. For example, the idea of using atomic clocks to search for dark matter was first suggested only in 2014 (Derevianko and Pospelov 2014), with several other ideas generated since then, and research ongoing (Antypas et al. 2022). In addition to specific mission designs where the goals are already clear, a well-funded, multi-disciplinary, ground-based research program—including both experimental and theoretical investigations—is needed to form a healthy and vital interdisciplinary community of fundamental and applied physicists and experts in other fields to seek further ideas, hone techniques, and develop optimal mission plans and goals.

Note that companion infrastructure and technology investment will be important to the efficient success of addressing this KSQ. For example, deployment of reflector mirrors on Earth's moon is an extremely low-risk/high-reward mission for improved tests of gravity and lunar conditions (Viswanathan et al. 2021).

Potential Research Areas

Search for New Physics with Clocks

As discussed in Chapter 2, the precision of space-based (optical atomic) clocks has improved by three orders of magnitude in the past 15 years, with expected additional orders of magnitude increased precision expected in the next decade (Brewer et al. 2019; Ludlow et al. 2015). Several such clocks, based on different types of atoms, have been demonstrated by multiple laboratories, with comparisons at the 10^{-18} level demonstrated over distance (Beloy et al. 2021).

Changes in gravity affect clock frequency, so comparing clocks in different gravitational potentials allows to test predictions of general relativity with exquisite precision as well as to map Earth's gravitational potential (relativistic geodesy). Figure 5-6 summarizes the benefits of such research conducted in microgravity. In general relativity, gravity will affect all clocks in the same predictable way, but BSM (i.e., new) physics can change that, inducing clock shifts that depend on the type of the transition on which the clock is based, time, and space

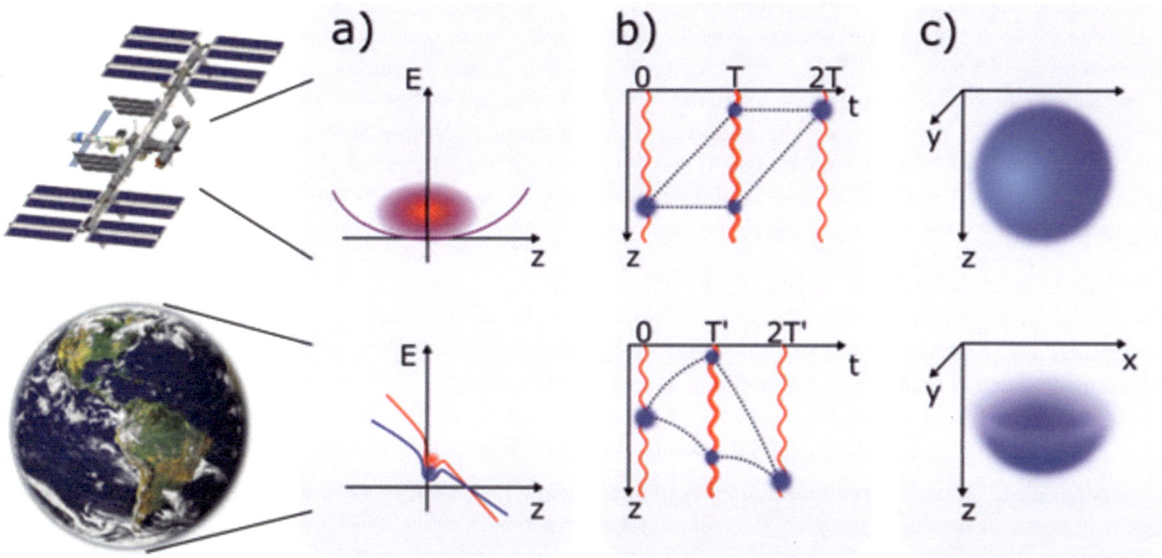

FIGURE 5-6 Benefits of microgravity for atomic physics. *Top row:* Experiments in microgravity. *Bottom row:* Experiments in Earth-bound conditions. In all figures, the z-axis points in the direction of gravitational acceleration g. (a) Atomic species of different masses m (red and blue) are confined in a potential. In the absence of any gravitational sag, the trapping potentials perfectly overlap, while in a gravitational field the two species experience a differential sag and the atomic clouds are (partially) separated. In addition, the traps could be steeper than in microgravity to prevent the atoms from falling out of the confinement. (b) Both graphs show a Mach-Zehnder atom interferometer. Laser pulses coherently split, reflect, and recombine the atomic cloud. In microgravity, the atomic trajectory is only determined by the interaction with the laser pulses, and long pulse-separation times T are accessible in a small setup. On Earth, gravity alters the trajectory of the atoms, and the free-fall distance of the atoms limits the pulse-separation time T. (c) In the absence of gravitational forces, new and complex trapping geometries can be realized, including shell-like 3D potentials. In an Earth-bound laboratory, gravity distorts such a shell trap, leading to an only partially filled shell. SOURCE: Frye et al. (2021), https://doi.org/10.1140/epjqt/s40507-020-00090-8, CC BY 4.0.

location. Then, measurements of ratios of frequencies of identical clocks in different space (and gravitational potential) locations, measurement of ratios of frequencies for two different types of clocks (based on different transitions), and looking for either oscillatory or transient changes in such frequency ratios can discover a broad range of "new" physics (Antypas et al. 2022; Ludlow et al. 2015). The sensitivities of presently operating and future optical atomic clocks to new physics that affects the fundamental constants have been calculated from first principles to high accuracy. The effect of various types of dark matter and other new physics on clocks and specific search protocols have been established and tested on Earth where possible. Moreover, high-precision clocks have excellent long-term stability with negligible drifts, enabling autonomous navigation of spacecraft, including deep-space autonomous operation. As a result, atomic optical clocks in space have been proposed for a wide variety of fundamental and practical applications. Fundamental physics clock applications include tests of gravity, searches for dark matter and variation of fundamental constants, tests of Lorentz symmetry and local position invariance, measurements of gravitational waves (GWs) in unexplored parameter space (long-wavelength/low-frequency), and others.[1] Practical applications include Earth-wide optical time and frequency metrology, ranging measurements, autonomous spacecraft navigation, relativistic geodesy, very long-baseline interferometry (VLBI), and others (Ely et al. 2021; Ludlow et al. 2015; Mehlstäubler et al. 2018).

Optical atomic clock operation outside the laboratory has been recently demonstrated (Ohmae et al. 2021), with several groups pursuing further work in this area. Automated operation, reducing clock size, weight, and power

[1] See Alonso et al. (2022), Derevianko et al. (2022), Fedderke et al. (2022), Ludlow et al. (2015), Kolkowitz et al. (2016), and Schkolnik et al. (2022).

(SwaP) is ongoing as a part of the development of portable applications (Takamoto et al. 2022) as well as owing to sharing of related technologies and components with quantum information science and other applications. This concerted recent effort makes optical atomic clocks the most mature and versatile of the precision quantum sensors for space applications. Nonetheless, major technological development will be needed to transition and make ready optical atomic clocks for space use. An important point to stress is that the ability to compare different clocks and clocks in different locations is required to make use of the fantastic precision and stability of optical atomic clocks—that is, significant technological effort will also be required to implement optical time transfer between space- and ground-based optical atomic clocks.

Finding 5-4: Beyond optical atomic clocks, there is a diverse array of quantum and precision measurement technologies that will be critical to addressing space-based fundamental physics questions and associated applications, as outlined above. This topic has great potential to be scientifically transformative and is both enabling of space exploration and enabled by space.

Gravitational Effects on Quantum Optical Systems and Detectors

It is not clear whether gravity influences quantum entanglement. This potential influence is of fundamental interest in terms of quantum mechanical states, but with practical implications. Research that assesses this will require access to the space environment in LEO and lunar surfaces or pathways to Mars, with implications for Earth-based and longer-term quantum computing.

Many of the most promising space-based quantum and precision measurement technologies involve ultracold (laser-cooled) atoms and molecules (Alonso et al. 2022). Therefore, in addition to space-enabling experiments searching for new physics, it is important to study the behaviors of ultracold atoms and molecules in microgravity. The space environment will also enable the use of new trapping potentials to produce colder and denser atomic and molecular samples than possible on Earth, which will be enabling for many of the fundamental physics investigations described above. In addition, both atomic clocks and atom interferometry in space have been proposed for new types of long-wavelength/low-frequency GW detectors—key messengers of fundamental physical processes in the universe that are in the GW bands not accessible on Earth (because of seismic noise and Earth's finite size) and outside the range of proposed space-based optical interferometer GW detectors such as LISA (Fedderke et al. 2022; Kolkowitz et al. 2016; Sedda et al. 2021). Therefore, extensive experiments to explore and understand cold atom and molecule behavior in microgravity are needed.

An intriguing possibility is the use of space-based quantum sensors as a part of multi-messenger astronomy—that is, looking for new physics signals correlated in time with GW detection (Dailey et al. 2021). In addition, atom-based GW space detectors can also serve as dark matter detectors (Abe et al. 2021; Antypas et al. 2022), so such instruments fall both within the "observation" and "fundamental physics" categories of space-based science.

Finding 5-5: In the search for new physics, fundamental physics and quantum science in space remain largely unexplored. In addition, quantum technologies are developing very rapidly, and the space program could adapt to benefit from new developments as quickly as possible. However, impactful progress and discovery requires synergistic and coordinated advances in ground-based and space mission infrastructure over more than one decade. That time horizon favors team-based science.

Recommendation 5-2: For fundamental physics in space, NASA should facilitate durable formation of collaborations and efficient knowledge transfer between researchers working in multi-disciplinary teams. This scope to address these multi-decadal key scientific questions should include ground-based infrastructure, theoretical and experimental physics, precision measurement and technology development with private sector participants, and should be coordinated with missions in which biological and physical sciences research is one among several whole-of-government objectives.

Impactful progress on the above four KSQs is enabled by predictable access to the space environment. In the coming decade as in past decades, access to space for BPS research will be a precious and limited resource. These KSQs represent the recommended focus of BPS resources and scientists based on current knowledge, not because the answers will definitely advance specific space missions but rather because the space environment is key to advancing useful and transformative knowledge about how our world works. The current pace of change in knowledge on Earth—on these four KSQs spanning biophysics, physics, and properties of solids and complex fluids, and fundamental physics that requires and also offers measurement precision—has been impressive over the past decade. Thus, prudent planning and use of the space-based science and associated ground-based analogs can also anticipate surprise discoveries and innovations, prompting entirely new key questions and directions of research that are enabled specifically by access to the sub-orbital, orbital, and lunar surface environments in the coming decade. History indicates that emergent areas of space-enabled research are unlikely to be part of any existing roadmap for space exploration; the less we can predict what these future space-enabled areas are today, the more transformative they are likely to be for society in future decades. Likewise, multiple studies of which the space science and technology industry is aware (Hofstra et al. 2020; McSweeney 2021; Phillips 2014) indicate that teams with greater diversity of perspective and experience are favored to generate creative solutions to technical challenges.

Recommendation 5-3: In all of the space-enabled research areas, NASA should allocate funding with an anticipation that new directions of research may arise.

Chapter 6 outlines research campaigns as comprehensive, multi-year efforts that address audacious goals supporting space exploration and knowledge. These research campaigns are anticipated to provide partial informational or infrastructural support needed to address the KSQs outlined in Chapters 3–5, as well as those identified by the BPS community in the coming decade. (See Figure 5-7.)

FIGURE 5-7 Biological and physical sciences in space research started with Skylab in the 1970s, continued during the space shuttle era through 2011, and is currently centered on the International Space Station. The future of BPS research is beyond all of these platforms and can even include the Moon. SOURCE: Courtesy of NASA, https://www.nasa.gov/sites/default/files/thumbnails/image/iss013e47643.jpg.

6

Research Campaigns

In keeping with the move of the NASA Division of Biological and Physical Sciences (BPS Division) to the Science Mission Directorate (SMD), where large space missions operating over a decade or longer are the norm, and in keeping with the rapidly expanding presence of human activity in low Earth orbit (LEO), the statement of task for this decadal survey[1] included development of notional, proof-of-concept research campaigns. As tasked, these research campaigns were intended to address prioritized research activities as part of complex or multidisciplinary missions and to operate in the space environment as well as in appropriate ground-based research. Furthermore, these large-scale activities were to be considered in broad cost categories to "assist NASA's understanding of the top-level scientific performance and resource options." These research campaign activities emanated from over 100 input papers collected from the research community and discussion among the steering committee and the panels of the study. From those input papers and discussions, four distinct elements were developed and advanced for further technical risk and cost evaluation (TRACE). The TRACE process is described in Appendix E.

These four elements—comprising two research campaigns, one multi-agency opportunity, and one research infrastructure concept—are to be considered notional, to be pursued only when specific campaign-level investments above the current operating budget are available for the BPS Division.

Two research campaigns were identified that are coupled tightly to the key scientific questions (KSQs) and aligned with NASA needs and objectives, which scale to currently envisioned space platforms. These research campaigns directly address many of the recommended KSQs and directly target mission needs for future space exploration and habitation. They are designed to be accomplishable by NASA largely within the next decade.

Bioregenerative Life Support Systems (BLiSS) is a recommended research campaign to firmly demonstrate fundamental knowledge of the impact of the space environment on individual organisms and the dependencies on and synergies between organisms and the diverse populations that could serve as biological life-support systems in space—systems that could be self-sustaining. Manufacturing Materials and Processes for Sustainability in Space (MATRICES) is a recommended research campaign to develop a nonliving, self-reliant, sustainable, circular economy of materials and processes through better understanding of the attributes and fundamental characteristics of solid materials and complex fluids in space. There are no full-scale, closed-loop cycles for materials on Earth, yet they are essential for operating independent of Earth for long periods of time. Development of the

[1] The statement of task is reprinted in Appendix A.

material systems, manufacturing techniques, and processing and reprocessing capabilities for intentional feedstocks and anthropogenic and natural source materials will not only enable living and working in space but would offer dramatic benefits to terrestrial sustainability.

Recommendation 6-1: NASA should pursue dedicated research campaigns that, through the coming decade, will drive resolution to specific groups of key scientific questions. Coordination beyond NASA, including other federal agencies and the private sector as well as public-private partnerships, should be considered for the dedicated new funding and materials to support the following two research campaigns:

- **BLiSS (Bioregenerative Life Support Systems) to build and understand the systems that would provide high-quality food, refresh air and water, process wastes and enable the creation of space environments sustainable for long periods of time independent of Earth.**
- **MATRICES (Manufacturing Materials and Processes for Sustainability in Space) to understand and harness the physical processes by which materials and complex fluids can be repeatably used in space, to enable sustainable exploration and circular production life cycles for the built environment on Earth and in space.**

Probing the Fabric of Space-Time (PFaST) is envisioned as a campaign-style, multi-agency opportunity that is only accomplishable through collaborative development between NASA and other U.S. government organizations and international partners. PFaST would use recent advances in atomic and optical clocks and spaceflight's ability to span large distances and large variations in gravitational fields to seek both validation of purely theoretical models as well as previously unobserved features of spacetime. It is an extremely large-scale research and technology effort that scales well beyond the sole domain of NASA. The committee therefore determined that this is a multi-agency effort that would provide important data for multiple agencies' technology needs that are derived from the biological and physical sciences (BPS) KSQs involving fundamental physics described in Chapter 5. This opportunity needs to be actively pursued, but only as a multi-agency effort where a substantial majority of the funding is provided by non-NASA sources.

Recommendation 6-2: NASA should pursue development of the Probing the Fabric of Space Time initiative in this decade only if it can obtain substantial (greater than 75 percent) funding from external (i.e., other than NASA) sources.

Polar Radiation of Model Organisms (PRoMO) is a notional concept of future research infrastructure that describes an opportunity for using a space vehicle not currently available to science—one that would allow a unique spaceflight experience that combines the effects of radiation and microgravity on mammals, plants, and cellular systems, thus underpinning the health risk–based decision process that is inherent in exploration beyond LEO. The PRoMO concept could enable crewless research investigations of physical systems, as well as on organisms including mammals, for extended exposure durations of interest to several KSQs described in Chapters 3, 4, and 5. This concept is presented for further study because of the long development time required to enable important BPS research, and the considerable unknowns in cost, especially as commercial spaceflight providers change the future cost landscape.

KEY TRACE ASSUMPTIONS

For the first time in the BPS in space research community planning, including the 2011 decadal survey, the committee utilized the TRACE process[2] to assess potential research campaigns. The Aerospace Corporation conducted the TRACE process under contract to the National Academies of Sciences, Engineering, and Medicine.

[2] The TRACE process is described in Appendix E.

The committee provided input to Aerospace and worked with that contractor to develop TRACEs for several research campaigns, which the committee then used as input to its deliberative processes. In assessing the research campaigns, the committee used the following assumptions:

- NASA retains its key role in providing infrastructure and support to researchers. The International Space Station (ISS) Program Office support, with a segue to commercial LEO destinations (CLD) Program Office support to continue at least at present levels.
- NASA will continue to provide upmass, downmass, crew time, integration and support in preparation for launch, and so on, at levels similar to the present.
- Costs associated specifically with mission integration and operations (MI&O) and launch of BPS research are assumed to continue to be borne outside NASA or others' funding of the BPS research activity itself. (See Chapter 7, Box 7-2.)
- Hosting and operations costs could be estimated from current NASA data, scaled by the number of experiments and projected crew hours. Some development reserves were included for phases B/C/D, but reserves for phases E/F were excluded under the premise that a research program is better managed by utilizing flexibility in the end date than by allocating dollar reserves to preserve the schedule.
- The current level of the core BPS Division research program is continued at least at the FY 2023 level, and so there is no break in research capability when the ISS is decommissioned. It is further assumed that the key facilities currently in use will be transitioned into the CLDs over time. If the current facilities are not transitioned, additional costs would be incurred for building, launching, and commissioning replacements.
- Crew time required to execute the core BPS Division research program and research campaigns can be estimated reasonably using NASA historical data. It is further assumed that, under the CLD model, crew time is an extensible resource (i.e., in principle, more crew time could be purchased if needed) and that arrangements will be made for private astronauts or civilian crews to use the NASA core facilities. These will enhance research productivity and may over time reduce NASA's cost, but that could not be predicted with sufficient certainty for this effort.
- Technical risk (regarding timeline) and cost estimates can be based reasonably on both NASA's historical costs (e.g., of crew time) and the best available published information from potential CLD providers.

A major new cost that has been considered is referred to as "Base CLD Facility Refresh" and represents the costs of moving some of the current facilities into a 51.6-degree orbit CLD and building fresh copies into a second CLD in another orbit. The costs of the proposed new facilities were estimated and scaled from the development of the legacy set, with an allowance for technology improvements. Note that the two research campaigns hold differing assumptions as to the transition time and implications of the transition from ISS to CLD capabilities. For example, the BLiSS campaign assumes experiments in parallel in the ISS and on the CLDs as they become available; the costs for continued ISS research and "copies" for research on the CLDs are included in the TRACE estimates for both the BLiSS and MATRICES campaigns.

Other considerations included strategic acquisition of multi-use infrastructure such as research racks, supporting space and resources for transport, environmental support, crew time, communication, and support for integration of experimenter-provided equipment and samples into launch transport facilities.

The research campaigns are described in the context of today's rapidly changing space economy, and based on the best estimates of when and how capabilities will emerge and the business models of the various prospective providers. For example, Box 6-1 describes what happens when SpaceX's *Starship* flies successfully.

MANAGING RESEARCH CAMPAIGNS TO ACHIEVE A STRATEGIC GOAL

The committee assumed that the core BPS research program will be managed similarly to recent practice. However, achieving the strategic goals of the two research campaigns will require a different way of managing the research efforts that goes beyond the standard Research Opportunities in Earth and Space (ROSES) open

BOX 6-1
What Happens When *Starship* Flies Successfully?

NASA has committed to the SpaceX *Starship* as part of the Artemis Human Landing System. *Starship* is architected to be fully reusable, including a reusable second stage with an objective of minimum refurbishment, with a vehicle lifetime of at least 100 launches, a large-scale cryogenic propellant/oxidizer depot, and in-space transfer. If these goals are met, *Starship* is likely to lower the launch cost per kilogram significantly (by a factor of 10 or more), coupled with a very large payload volume—that is, change the fundamental assumptions of experimentation in space. The timeframe will depend on success in making large investments and overcoming the technical challenges—large-scale cryogenic depots and transfers to enable lunar and interplanetary missions and a robust thermal protection system capable of supporting entry from Earth orbital and planetary return trajectories—as well as the competitive environment and how long it takes to amortize development costs.

The maiden launch was planned for 2022; however, development issues delayed the first launch into 2023. The potential for biological and physical sciences (BPS) research access specifically to Starship was not incorporated in the technical risk and cost evaluations of the BLiSS (Bioregenerative Life Support Systems) and MATRICES (Manufacturing Materials and Processes for Sustainability in Space) research campaigns, due in part to the associated uncertainties in development, deployment, and access by BPS researchers. This is potentially a revolution in the making, but with tremendous uncertainties of when and how the capability will become available and if the cost reductions will be achieved.

announcements for BPS Division grants. Research campaigns will require development and evolution of research and technology roadmaps—building on and extending the body of knowledge and processes, selection criteria, and an alignment of proposed efforts to the roadmaps. Also, Research campaigns will require decision criteria such as when enough is known about a topic to focus resources elsewhere, continuity in research programs to allow commitment to longer-term sequences of experiments with a high cadence, and a fast decision cycle to quickly follow up to lessons learned.

NASA has learned much in exploring alternative models of pursuing strategic goals such as the SMD Heliophysics DRIVE Centers, the NASA SMD/HEO Solar System Exploration Research Virtual Institute (SSERVI), as well as the Space Technology Mission Directorate's (STMD's) Space Technology Research Institutes and The Lunar Surface Innovation Initiative. Lessons learned could be used to develop the best approach for strategic management to achieve the goals and balance considerations above. Chapter 7 includes recommendations in the context of the full decadal survey, with associated decision rules, and the lessons learned from the above institutes and initiatives may inform this management strategy specifically for research campaigns.

IMPACT OF THE RESEARCH CAMPAIGNS BY 2033

NASA has adopted a set of Moon-to-Mars objectives (NASA 2022c). The KSQs and the research campaigns directly support and are traceable to these objectives. Each campaign is intended to make major and transformative scientific contributions with recognizable impact to that mission space and to society. For example, the establishment of a robust bioregenerative life support system is a key contribution to Transportation and Habitation Goals TH-2 and TH-3. In addition to meeting Moon-to-Mars goals, the body of knowledge built through these campaigns will be critical to establishing a vibrant in-space economy. While it is always difficult to predict the applications of research, the increased body of knowledge of microbe, plant, and animal systems, and of biological systems-of-systems responses, to the stressing space environment will surely benefit terrestrial applications—including climate adaptations. Similarly, advances in efficient manufacturing and work to enable truly closed-loop production processes and habitats will inform advances in terrestrial systems, improving the management of waste stocks and reducing impact associated with mining and refining planetary resources.

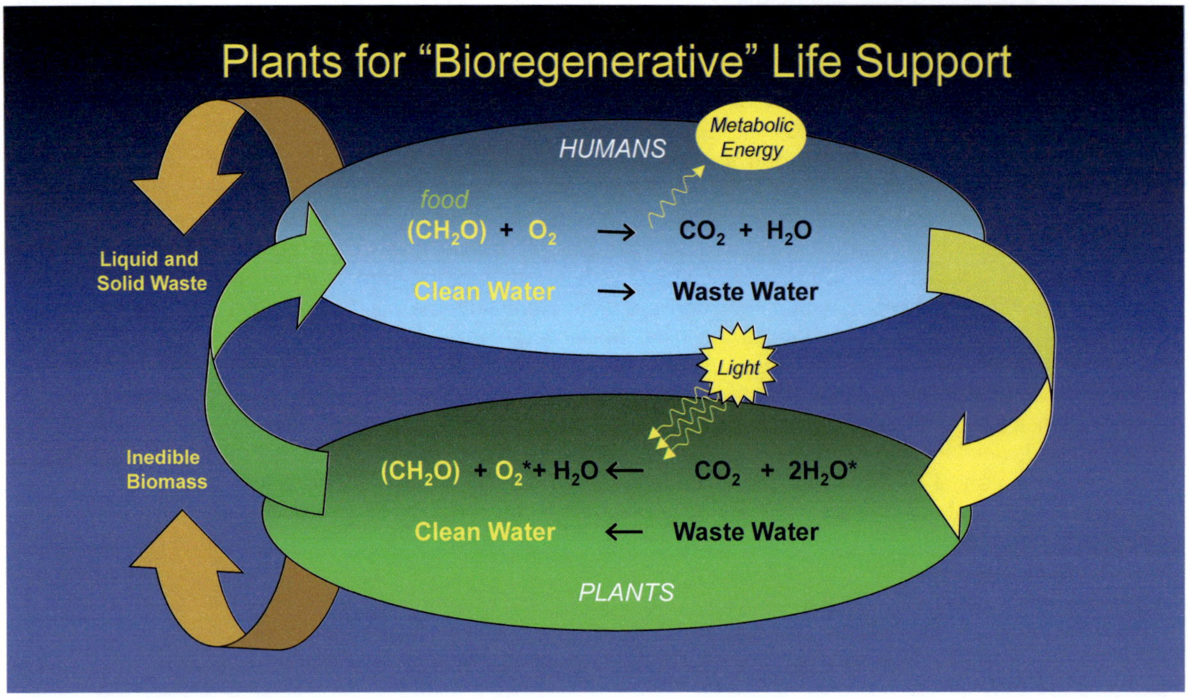

FIGURE 6-1 Fundamental concept behind bioregenerative life support. SOURCE: Wheeler et al., 2001, Proceedings of the Institution of Environmental Sciences (IES) Closed Ecology Experiment Facilities (CEEF).

RESEARCH CAMPAIGN: BIOREGENERATIVE LIFE SUPPORT SYSTEMS

The BLiSS research campaign is driven toward the following four high-level goals:

- Develop self-sustainable biological life-support systems that produce food, clean water, renew air, process waste, and create critical materials to meet the challenges of long-duration space missions.
- Harness beneficial properties of plants and microbes that will enable humans to live in space, independent of resupply from Earth.
- Create a highly functioning, robust, and resilient ecosystem and space environment that is self-sustainable under extraterrestrial radiation and gravity conditions.
- Enable long-duration (>3 years) exploration of deep space by providing a fully or partially closed–loop biological life-support system.

Future long-duration space missions of multiple years will require the ability to be self-sustainable without requiring resupply from Earth. (See Figure 6-1.) By harnessing the power of biological systems, the goal is to develop BLiSS for space exploration. Even while accepting that a full closed-system capability is unlikely to be achieved by the end of the decade, the BLiSS research campaign, as recommended, can provide appreciable offsets to resupply and quality-of-nutrient and quality-of-life benefits for the crews in that timeframe. The immediate benefits provided by BLiSS are production of vitamins and food, water purification, air revitalization, recycling of waste streams, and intangible mental benefits for crews living and working in space (Paradiso et al. 2014; Srinidhi and Turner 2021; Wheeler 2017). In addition, BLiSS can be a source of familiar biomaterials (e.g., plant fiber and wood) or more specialized materials such as drug precursors, specialty chemicals including epoxies, or bioplastics for three-dimensional (3D) printing.[3]

[3] See Aviles-Gaxiola et al. (2020), Buyel et al. (2021), Cestellos-Blanco et al. (2021), Haveman et al. (2023), Lu et al. (2020), and Shevtsov et al. (2023).

The NASA technology roadmap states that self-sufficient life support systems are crucial for sustaining life on long-duration missions (Kliss 2016). Several NASA needs assessments identify closed regenerative life support as an enabling technology for long-term, sustained human exploration, including the Lunar Human Exploration Strategic Knowledge Gap (SKG III-J-3), the Decadal Survey on Biological and Physical Sciences in Space Studies (DSBPS TSES6 and P3), the NASA 2020 Technology Taxonomy (TX06.3.5), and the Global Exploration Roadmap.

Earth is a worldwide BLiSS research campaign that is almost entirely self-sufficient and does everything from harvest energy from the Sun, renew the atmosphere, purify water, produce billions of different biomolecules, break down rock and inorganic and organic molecules, and produce the raw materials to build structures and feed future life cycles (Douglas et al. 2021; Drysdale et al. 2000). This is a system and capability needed to support life in deep space, at a vehicle scale, but that is currently not at a technology readiness level (TRL) that can be relied on in deep space. A BLiSS concept can be broken down in many different ways, but one approach involves modules or unit functions for primary production (e.g., plants), nutrient recovery and waste reprocessing, and secondary production to produce protein for consumption, or key biomaterials like drugs, bioplastics, biopolymers, or chemical feedstocks for other processes. Initially, some of these modules may be developed to augment or simplify functions that are currently done by the ISS environmental control and life support system (ECLSS), like water purification. As an example, plants can be watered with gray water, where a combination of plant microbiomes and plant roots recycle the nutrients and the water is transpired into the air. Recondensing the transpiration stream produces a water stream that could be fed into the final water polishing unit to make it potable rather than routing it through the entire ECLSS wastewater system. A square meter of plants produces about 5 liters of water a day, somewhat more than the allowance of an ISS astronaut. While it is not expected that a full, closed-system capability will be achieved by the end of the decade horizon described in this study (by 2032), the BLiSS campaign as recommended can provide appreciable offsets to resupply and quality-of-life benefits to exploration crews in that timeframe.

Research Thrusts

Overview

The major factors that impact crop systems and BLiSS for deep space are reduced atmospheric pressures with altered gas compositions, gravity, and radiation. Each factor needs to be considered in the context of a long-duration, multi-year mission. The first factor deals with the need to operate surface habitats at reduced pressures to reduce required prebreathing times for an extravehicular activity (EVA) in a space suit. These proposed atmospheres (e.g., the "exploration atmosphere" of 56.3 kPa atmospheric pressure and 34 percent oxygen) are not found anywhere on Earth. High carbon dioxide concentrations in spacecraft atmospheres can also be a challenge for plant performance (Burgner et al. 2020). The second factor relates first to gravity's effect on water, helping it to move in familiar ways, rather than following any hydrophilic path. This has a big impact on how soils wet and drain, hydroponics, aeroponics, and root zone hypoxia (Heinse et al. 2009; Liao et al. 2004; Monje et al. 2003). Gravity also impacts gas exchange with the shoot zone necessitating higher air flow rates at reduced gravities (Poulet et al. 2018). Partial gravity or microgravity also impacts living systems in more indirect ways, such as causing increased virulence of some bacteria and increased disease susceptibility in some plants and animals (Taylor 2015). A related factor is extreme isolation on deep space missions. For microbial communities, this can lead to permanent loss of diversity over time, possibly endangering a mission by losing or altering critical species of a plant or soil microbiome. There is a similar risk for loss of crop genetic diversity. On the other hand, extreme isolation presents a mental stressor for crew function and performance that gardening may help to alleviate. Radiation, the fourth factor, is elevated outside of Earth's protective magnetic field, with increased galactic cosmic ray (GCR) and solar proton events (SPEs) that degrade biological systems and materials over long periods of time. This can contribute to species loss, genetic damage, increased disease risk, reduction or elimination of seed germination, and destruction of the materials and hardware making up the BLiSS. Radiation exposure also increases the need for fresh foods, rich in antioxidants and vitamins, in the crew diet. The fifth factor is crop selection and optimization. Factors affecting crop choice include size of the hardware, ease of pollination, ability to produce relevant nutrition, crop hardiness in the relevant environment, and input needs of the crop. There may be other factors related to human responses over extended mission durations, particularly in helping with mental stress by establishing a sense of well-being with a familiar and tasty food, having colorful flowers, or producing food.

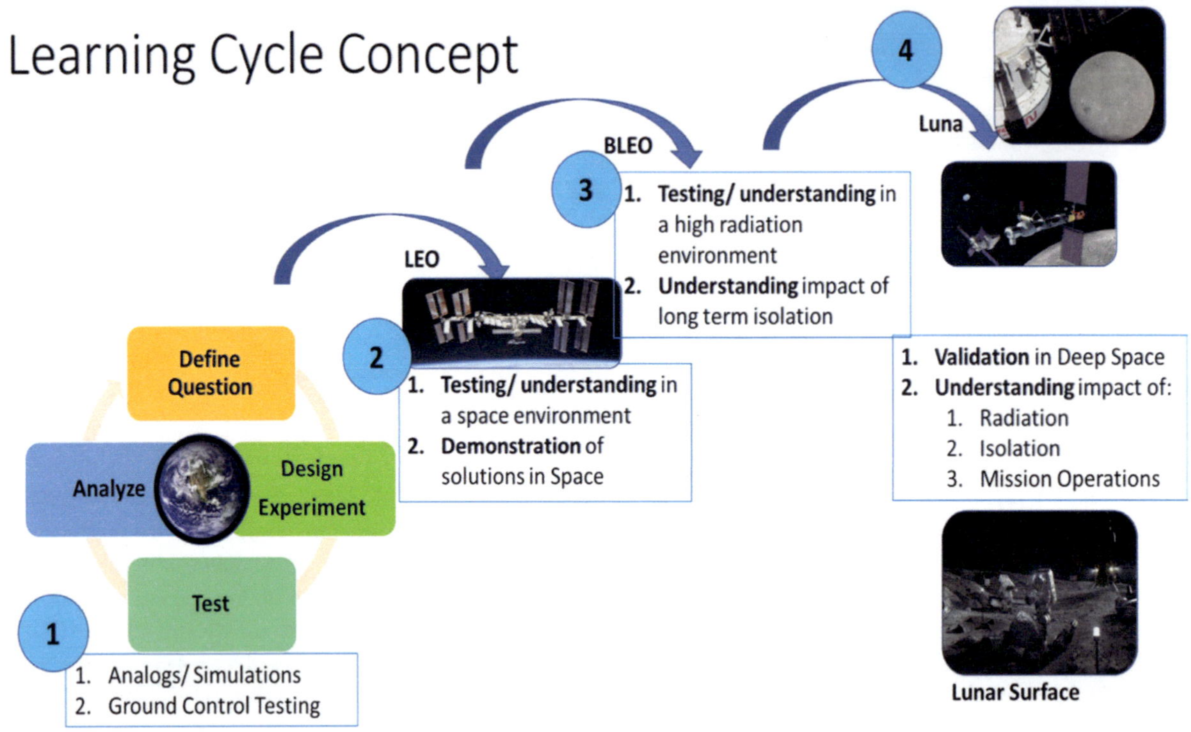

FIGURE 6-2 Learning cycle concept. NOTES: (1) Science questions are defined and refined to allow the design of valid experiments. Test results and data are analyzed to further refine the questions improving understanding by maximizing the use of Earth-based analogs. Many of these analogs can also be used for ground controls. (2) Validated learnings from ground tests are tested in low Earth orbit (LEO) to gain true understanding of the space environment or demonstrate that solutions work in microgravity. (3) The beyond LEO (BLEO) environment adds the variables of radiation and extreme isolation for further understanding how these variables impact the science. (4) Lunar opportunities will be rare and will be used to validate the learnings from the previous environments under mission relevant radiation, isolation, and mission operation conditions. SOURCES: Committee created with images courtesy of NASA; *Earth*: NASA/NOAA/GOES Project.

The BLiSS research campaign is designed to contribute answers to KSQs that are highlighted in Chapter 4, primarily under the second theme of living and traveling in space. That theme includes creating and maintaining safe, sustainable built environments and building a stable human presence. (See Table 7-1 for mapping to several KSQs.) Each KSQ could be explored using a learning cycle concept shown in Figure 6-2. Experiments need to be designed to test the concepts in meaningful and statistically relevant ways. This work can start with analogs or high-fidelity simulations on the ground more quickly and cost effectively than in space. Validated concepts can then be flown in LEO to gain an understanding of how space impacts the system, or to demonstrate that a solution works. Beyond LEO experiments are expected to be less available than LEO opportunities, so these could be used to add in the additional variables of a higher radiation environment and extreme isolation. Last, lunar environment (either lunar orbit or surface) opportunities will be least available but offer the opportunity to validate previous learnings under mission-relevant conditions (radiation, isolation, mission operations).

Objectives

BLiSS is designed to address KSQs and associated science and technology developments that lead to reliable bioregenerative life-support components. It is not designed to fully close the life-support loop in space. Therefore, key elements of BLiSS are considered to be primary production systems (e.g., plant or algae), nutrient recovery systems (bioreactors using algae, bacteria, fungi or mixed systems; insect or hybrid systems, and potentially mammalian cell systems); and physical principle systems (e.g., plasma, combustion, heating, syngas production, and benefaction). Additional modules may include secondary production modules that produce protein, drugs, or specialized biomaterials (e.g., silk, biomolecules, vitamins, acetates, chiral chemistries) or feedstocks for manufacturing. Work has already been done on many of these.[4] These systems could be fermenters, cell or tissue culturing systems, fungal production systems, or even plant-based, where the products could be varied through transformation or gene editing.

Bioreactors will be a critical part of any BLiSS because of the ability they provide to recycle nutrients. They offer the capability to help close the carbon, nitrogen, and water cycles while liberating critical locked-up minerals, including calcium, magnesium, potassium, and phosphorus. Additionally, bioreactors can be used to produce critical biomaterials or feed stock chemistries. Examples of types of products that can be produced include plastics made from poly-hydroxy alkanoic acid (PHA), poly-hydroxy butyrate (PHB), or polyethylene terephthalate (PET), used for high-strength plastics for spinning directly into fibers or solidified for later use. High-value chemistries or proteins include caffeine, aspirin, morphine, digitalis, insulin, taxol, or antibodies with therapeutic applications. Food waste and cellulosic materials may be converted to biofuels or used to produce high-value products. The type of bioreactor needed will be dependent on the nature of the starting material as well as the nature of the desired product and the gravity environment. In microgravity, bubbles do not float "up," making it difficult to separate gases from the liquid eliminating most classic bioreactor. Bioreactors need to be designed for space. Initial experiments could be ground-based, focusing on obtaining the desired production and then looking at bioreactor designs that achieve the same or superior results in microgravity. Innovation is needed to develop bioreactors that work, can be cleaned, and that operate effectively and reliably in space.

Space Crop Production

The most immediate need for BLiSS is primary production—the ability to provide supplemental nutrition and a fresh element to astronaut diets (Douglas et al. 2021). A key goal for the BLiSS research campaign is to develop small-scale production systems that can supplement both of these for crews on long-duration missions while simultaneously laying the groundwork for larger BLiSS elements. The goal is to demonstrate how to do this, and to ensure that the food is safe to eat and nutritious (Haveman et al. 2021; Khodadad et al. 2020). Systems are currently being designed to fit into hardware analogous to the Veggie unit on ISS (Figure 6-3; Massa et al. 2017), although they could be made substantially smaller, potentially allowing demonstrations on CLPS missions (NASA 2020a) or testing in the Orion or human landing system space vehicles. Incremental technology steps include the following: development and testing of low-volume systems for microgreen (or other primary) production; understanding, modeling and optimizing how to produce safe nutritious foods from system inputs and outputs; demonstration of crop safety (plant and human pathogens) and vitamin production; and automation of microgreen production.

Infrastructure Enabling Research on Earth and/or Space Asset(s) Over the Decade

Moderately Sized Crop Production Systems

Moderately sized crop production systems (e.g., Veggie [Massa et al. 2017; NASA 2020b] and Advanced Plant Habitat [APH; Monje et al. 2020; NASA 2017] analogous hardware) are useful for developing cropping and microbiome models for a BLiSS. Crops of interest have already been produced in these types of systems, but current watering and growth systems are not suitable for 3-year missions. There are also knowledge gaps about preventing microbial biofilm production on spaceship surfaces, establishing and maintaining healthy plant microbiomes, integrating plant and human pathogen management plans (Khodadad et al. 2020; Liao et al. 2004;

[4] See Clauwaert et al. (2017), Gòdia et al. (2002), Hao et al. (2018), Massa et al. (2017), Miles and Lunn (2013), Poughon et al. (2020), Salisbury (1992), Schwartzkopf (1997), Taulavuori et al. (2017), Walker and Granjou (2017), and Yuan et al. (2019).

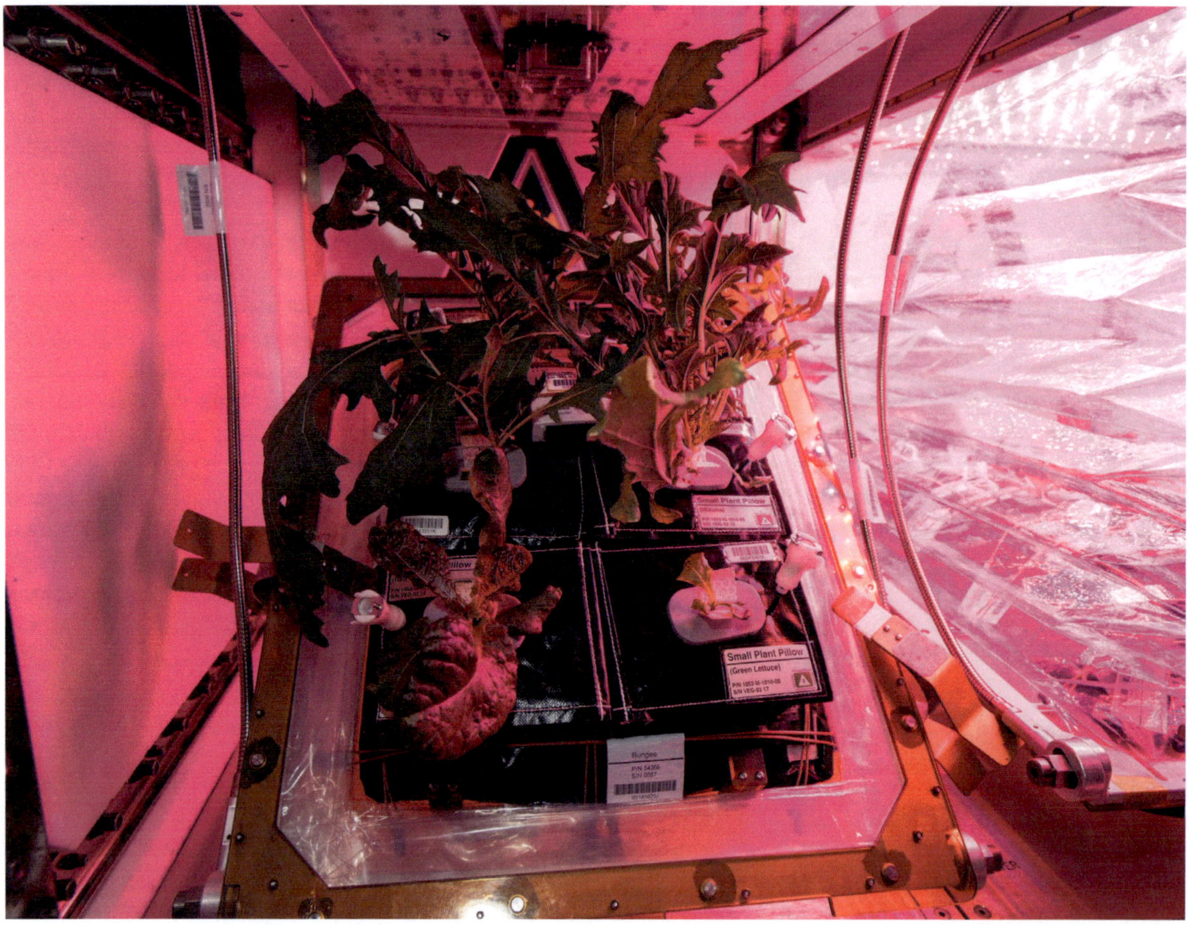

FIGURE 6-3 The Veggie hardware aboard the International Space Station. SOURCE: Courtesy of NASA, https://www.flickr.com/photos/nasamarshall/40077816484/in/album-72157667439026641, CC BY-NC 2.0.

Monje et al. 2003; Schuerger et al. 2021), and understanding the interrelationships between the crop and the microbiome. In addition, there are questions about the effects of altered atmospheres and radiation environments in deep space on crops and the associated microbiomes, as well as the effects of years in space on microbiomes on seeds. Additional goals for crop systems involve developing and testing models for predicting, monitoring, and responding to plant health (Shevtsov et al. 2023). Experiments need to be conducted in ways that will generate data sets that will be useful to future investigators and that can be used to create models and train artificial intelligence (AI) for optimizing plant growth and system response to off-nominal events (Escobar et al. 2023). Veggie and APH sized systems may also be useful for developing and testing automation of plant growth systems as well as for improving ease of use, reliability, robustness, and ease of repair.

Integration and Development

The BLiSS research campaign represents a major effort that will benefit from focusing on developing and modeling individual modules or functions as well as from large, coordinated multi-national campaigns (Fu et al. 2016; Lang and Bamsay 2023; Salisbury 1992). Developing a functioning BLiSS requires developing primary production modules and integrating nutrient recovery modules and secondary producer modules and bioproducts such as medicines, plastics, epoxies, and foundation chemistries. Models based on the inputs, outputs, and operations of these modules are needed so that they can be integrated into larger BLiSS designs. Additional testing and modeling of simple systems (e.g., a producing module and a nutrient recovery module) will uncover system dynamics and determine efficiencies. This will enable virtual testing of various designs to optimize systems before building and testing a larger BLiSS. Waste streams[5] and nutrient recovery methods could be explored in parallel, including ones based on physical principles that include combustion, biochar, syngas, plasma, separations, and benefaction (Bubenheim and Wignarajah 1997) using bioreactors and physical chemical systems. Additional concepts include insects for breaking down bulk waste materials while also providing a source of protein (Fu et al. 2016).

Ground Testing and Development

Large ground versions of full BLiSS systems have been built, including BIOS-3, Biosphere 2, and Yuegong-1 (Lunar Palace 1) (Salisbury 1999). Lunar Palace 1 highlighted water and solid waste recycling (Liu et al. 2020; Zhao et al. 2022), impact on microbiomes (Hao et al. 2018; Yang et al. 2022), and element and energy cycling (Dong et al. 2017). The ESA has the MELiSSA project (Hendrickx et al. 2006; Paradiso et al. 2014; Verbeelen et al. 2021), which is focusing on various elements of a BLiSS, while DLR in Germany has developed EDEN ISS (Poulet et al. 2021; Zeidler et al. 2021) for understanding the requirements for operating a greenhouse in space. The University of Guelph in Canada offers large, sophisticated pressure chambers capable of simulating the reduced atmospheric pressures (75 kPa with elevated oxygen percentages) being planned for the lunar habitat (University of Guelph 2022). The Japan Aerospace Exploration Agency (JAXA) has the Ecology Experiment Facilities (EEF) for studying material cycles, particularly in plant modules. Ultimately, it would be beneficial to have one or more cooperative facilities of a scale capable of supporting two to six crew members where designs could be tested on Earth.

Biological Flight Hardware

There are three substantial pieces of hardware for plant biology on the ISS: Veggie, APH, and Spectrum. Veggie is a very basic, two-MLE-size growth system providing a light, a fan, a platform for growing, and bellows that can be pulled from the light fixture to the base to provide some isolation of the plants from the ISS environment. APH is four MLE-size payloads (1/2 Express Rack). It is a sophisticated growth chamber that provides a completely controlled plant growth environment (humidity, temperature, light, CO_2, and automated watering) requiring little oversight from the crew but is considerably more complex. Spectrum is a multifunctional piece of hardware for biological experiments that uses different lighting to measure properties of the biology. It is specifically designed to stimulate and record the most used florescent proteins. Its capacity is limited to four 10 cm × 10 cm petri dishes that are mounted to a carousel that can be rotated and monitored automatically or remotely. It can be used with small plants (*Arabidopsis*), microbes, and worms (*C. elegans*). These are valuable types of hardware to have available in the deep space environment. Veggie is the easiest to get to the lunar environment, owing to its comparatively lower mass, but this comes with the cost of crew time. Human factors will play a large role in identifying appropriate mission options for this hardware.

It is important to emphasize that biological research has been done in modular formats (mid-deck locker equivalents, or MLEs) that fit into the same mounting hardware onboard the space vehicle (the space shuttle or the ISS). This allows details of scientific experiments to be changed without major alterations to the spacecraft. It is critical that some level of portability be incorporated early into the designs of CLDs and across all vehicles so that hardware does not become unique to specific space vehicles and obsolete or unusable on others; this will require considerations of interoperability standards with participation of the public and private sectors. For example, Figure 6-4 provides an overview of possible locations for experiment infrastructure to support the BLiSS campaign.

[5] See Ashida (1994), Atkinson (1997), Garland et al. (1997), Guntur et al. (1999), Mackowiak et al. (1996, 1997), Pisharody et al. (1996), Saulmon et al. (1996), Strayer and Atkinson (1997), and Stutte (1996).

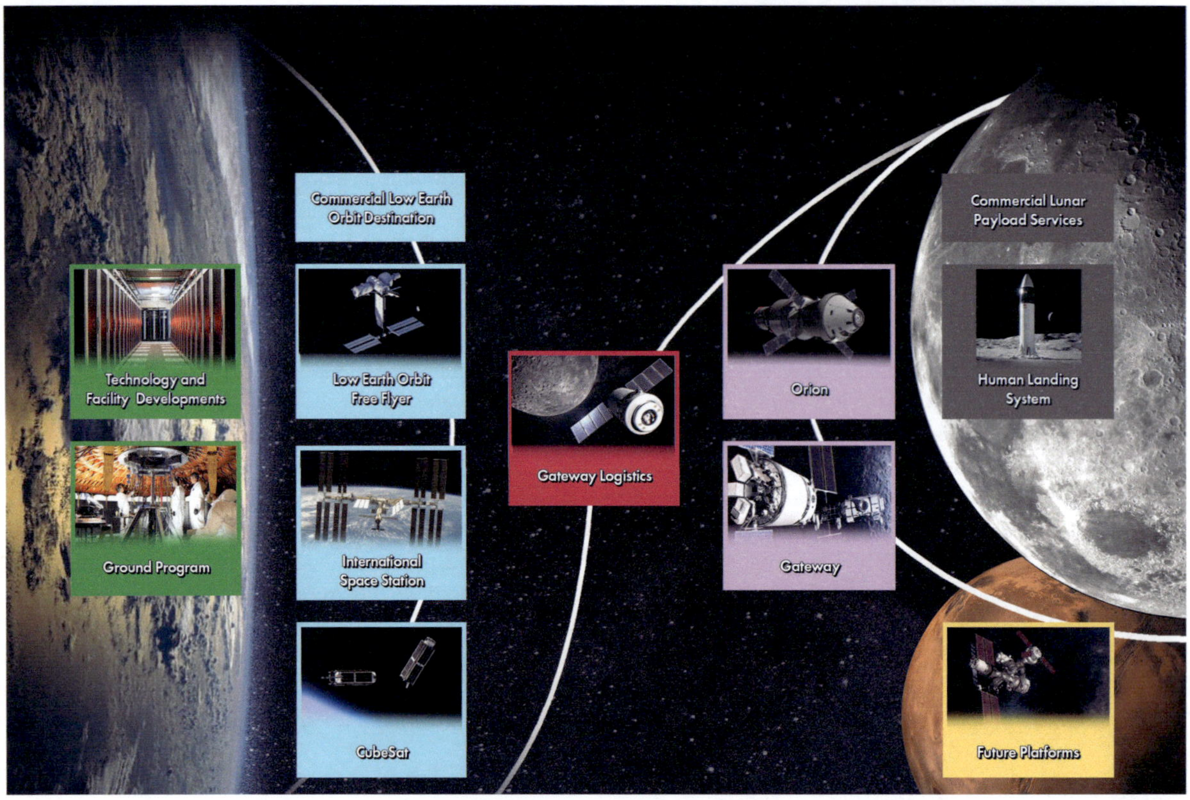

FIGURE 6-4 Overview of locations for experiment infrastructure to support the BLiSS (Bioregenerative Life Support Systems) research campaign. NOTE: NASA will continue to leverage all opportunities to expand the body of knowledge; however, the BLiSS campaign is primarily focused on ground-based, International Space Station (ISS), and Commercial LEO Destination (CLD) opportunities. SOURCES: Composed by Tim Warchocki. Image credits: *Technology and Facility Developments*: NASA Neil A. Armstrong Flight Research Center; NASA Photo/Tom Tschida. *Ground Twins Program*: NASA/Langley Research Center. *Low Earth Orbit Free Flyer*: Orbital Reef/Blue Origin. *International Space Station*: NASA/ISS. *CubeSat*: NASA Goddard Space Flight Center. *Gateway Logistics*: Courtesy of NASA, https://www.flickr.com/photos/nasa2explore/51046733772/in/album-72157716027881092, CC BY-NC-ND 2.0. *Orion* and *Future Platforms*: NASA. *Gateway*: NASA/Alberto Bertolin. *Human Landing System*: SpaceX/NASA. Background images—*Earth* and *Mars*: NASA. *Waxing Gibbous Moon 160321*: R. Pettengill, http://astronomy.robpettengill.org, CC BY-NC-ND 2.0.

Sample Handling and Storage Infrastructure

Handling of biological samples is often similar regardless of the source of the sample (plant, animal, microbial). This is because biological responses are typically transient and degrade over time. Mission architecture plays an enormous role in the design of biological experiments primarily because the most sensitive and informative measurements are challenging or not currently possible to do in space. Free flying, no-return missions require measurements to be done on board the spacecraft, limiting the type of experiment and the data that can be recovered.

To understand biology, it is increasingly necessary to understand what is happening at the -omics level. Traditionally this includes sequencing capabilities (genome, transcriptome, metagenomics/transcriptomics, epigenetics) as well as the abilities to look at proteins (proteomics), enzyme activities and metabolites (metabolomics), and phenotypes (phenomics). Omics capabilities are a rapidly evolving technology space (Biteen et al. 2016). Unfortunately, the capabilities to do all of this in a fully automated manner in the deep space environment do not yet exist. If that can be done, this would reduce the need to return samples to Earth for analysis.

Missions offering the possibility of returning samples may require the samples to be preserved by chemical fixation, flash frozen, dry down (e.g., of seed, spores, plant materials) or other methods (e.g., lyophilization or freeze-drying). These techniques could be useful on commercial lunar payload services (CLPS) missions to the Moon where the samples might be preserved until they could be returned by a crew. Considerations need to be made for how the samples will survive long lunar day/night cycles in a passive or dead spacecraft until a crew arrives.

Shuttle and ISS experience shows that for crewed or fully automated missions, refrigerators and freezers (MELFI, GLACIER, POLAR, MERLIN, Glovebox Freezer) are valuable for preserving samples and increasing the science value. These can be used in combination with fixatives that preserve the samples suitability for omics studies, or directly if samples can be rapidly frozen. Samples need to stay frozen once they are frozen, while fixed samples are often more tolerant of temperature variations.

There are examples of flight hardware with built-in refrigerators or freezers such as the Space Automated Bioproduct Laboratory (SABL; Bioserve), TangoLab (Space Tango), BioCulture System (NASA Ames Research Center), Advanced Space Experiment Processor (ADSEP; Techshot), Multi-Use Variable-g Platform (MVP; Techshot), Science Taxi (Yuri), as well as others. These serve as models for most likely, smaller, more compact units that could be used in deep space. Ultimately, the real challenge is to have room for enough experiments to do a statistically valid experiment, as well as the ability to repeat it to establish the flight-to-flight variability required for impactful science.

Recommended Capabilities

Nucleic Acid Analysis

There is a high need for nucleic acid (DNA/RNA) analysis in space to reduce or eliminate the need for sample return, and to allow timely access to the data. This can be essential when scanning for or identifying human or plant pathogens and for monitoring microbiomes in the built space environment. Knowing early that there is a problem allows for corrective actions that can prevent serious consequences. In addition, the ability to monitor the environment for long periods of time informs understanding of the establishment and evolution of microbial communities in the space environment. This will be essential to long-term survival in space, because most organisms will be impacted by the high-radiation environment that increases genetic damage and the extreme isolation that can lead to loss of biological community and genetic diversity. Fields of related expertise would include fluid physics, nano-fluidics, and capillary electrophoresis for automated analysis. Better ways to capture DNA from dirty environments and clean it—that are compatible with automation—are also needed.

Automated Chemistry

Understanding what is happening in biology requires an understanding of the physical chemistry and biochemistry present. The field of "omics" is based on understanding the proteins and metabolites being produced in addition to the gene pathways leading to their production. On Earth, scientists have a wide array of tools available, including gas chromatography (GC), liquid chromatography (LC)-mass spectroscopy (MS), Fourier transformed infrared spectroscopy (FTIR), Rama spectroscopy, nuclear magnetic resonance (NMR), and so on. There are specialties and subspecialties of all of these techniques that help scientists uncover the complex chemistry of biological systems. While there is a strong desire to have all of these instruments in space, the constraints of spaceflight prevent most of them from flying. A research campaign is needed that calls for the development of automated, miniaturized versions of instruments that can be used in space.

Currently there are versions of chemical analysis instruments being developed for deep space studies such as microchip electrophoresis (ME)-LASER-induced fluorescence (LIF) or "chemical laptop" technology[6] and capillary electrospray (CE)-MS and capillary electrospray ionization (CESI)-MS technologies being developed by the Jet Propulsion Laboratory[7] and used in the European Molecular Indicators of Life Investigation (EMILI)

[6] See NASA JPL-CalTech Microdevices Laboratory, 2022, "Chemical Laptop: A New Twist on Life," https://microdevices.jpl.nasa.gov/capabilities/in-situ-instruments-chemical-analysis/chemical-laptop.

[7] See NASA JPL-CalTech Microdevices Laboratory, 2022, "CE-MS: Combining Two Powerful Approaches," https://microdevices.jpl.nasa.gov/capabilities/in-situ-instruments-chemical-analysis/ce-ms.

(Brinckerhoff et al. 2022). While these devices are pushing back the boundaries on chemical detection in remote places, a recent review indicates that they are only able to detect some species and not others (Zhang and Ramautar 2021). Also, the table in the review indicates that many samples were extracted using reagents that are not usually allowed on the ISS owing to toxicity issues. Nonetheless, continuing to advance these types of analytic systems will advance researcher's abilities to do chemical analysis in deep space and on Earth in remote places. There is a high need to detect and classify volatile organic compounds (VOCs) because some of these compounds (e.g., ethylene) are bioactive in parts per billion quantities, while others (e.g., siloxanes) cause problems for ECLSS systems currently in use on the ISS.

Data Requirements

A common element to many of the instrumentation systems above, but especially to genomics, transcriptomics, metagenomics, and to chemical analysis, is the need for large data libraries; these can easily be petabytes for genome analysis. Data sets from single crop experiments on Earth often exceed 5 TB. Typically, on Earth, genomics work is done on multi-core processor machines (hundreds of CPUs) or using "cloud computing" where the new data set is uploaded and then analyzed using petabyte-sized data libraries housed in data centers. The cloud can also provide thousands of processors and very large memory assets to support the analysis—reducing the analysis time to reasonable time periods (less than a day for most applications). In space, this is a challenge owing to limited computational and memory assets, telemetry, bandwidth, and latency or delay. Typically, the farther you get from Earth, the longer it takes to transmit a data set, increasing the strain on telemetry. One way to reduce these challenges is to have the computational, memory, and data resources available on the space vehicle. Basically, future space vehicles are going to need to take the cloud, or at least pieces of it, with them. Having this type of infrastructure available will also enable AI development in space, because machine learning and other forms of AI also benefit from cloud computing architecture.

Impact by 2033 If the Campaign Is Successful

As far as scientists understand currently, the land portion of Earth remained largely uninhabitable for billions of years until early forms of microorganisms produced oxygen and early plants evolved and began to colonize it, providing both an energy source and habitat for other creatures. Earth's solar system is filled with distant worlds that are little changed and hostile to life for billions of years. The big idea of the BLiSS research campaign is to enable building fully functional biological support systems to support exploration of deep space. It is unlikely that this will be mastered or have the resources to build a functional habitat on Mars or the Moon by the end of the decade. Nonetheless, by 2033, the BLiSS campaign could deliver the science underpinning the modules of a bioregenerative life-support system that demonstrate augmentation, enhancement, or parallel replacement of portions of non-biological ECLSS systems. Obvious areas are in partial purification of water so that it only requires final polishing to become potable, and in production of critical vitamins that are needed in high levels in space, such as vitamins and antioxidants that are very difficult to synthesize especially without raw materials. Small-scale BLiSS modules will provide data for mathematical models that enable building larger, more functional systems and substantially de-risk the substantial food and nutrient gaps in a multiyear, crewed mission to Mars. These systems will help recycle some food, pull carbon dioxide gas out of the atmosphere, and contribute to water purification, particularly of gray water. Ultimately, this will reduce the resupply requirements for any bases. More importantly, these systems will provide the knowledge for designing and building far more capable systems that are less vulnerable to the space environment. They further promise to help solve food issues posed by climate change or extreme environments on Earth by providing optimized, stable environments for production that are much less dependent on weather conditions.

Broad Costs of the Research Campaign, Including the Associated Facilities and Platforms

As illustrated in Figures 6-5 and 6-6, the total estimated cost of a BLiSS research campaign thus scoped would be $1.7 billion over 10 years, with the expenditure peaking in ~2029. The cost estimate for this campaign was driven

	WBS Element			BLiSS Campaign Level Elements		BLiSS Basis of Estimate
Campaign Level	PM/SE/MA			$ 60		Scaled from NASA ARC/KSC yearly plant budget
	Science			$ 38		Scaled from historical KSC Plant Lab I budget
	Plant Campaign Technology			$ 37		Technology development of key hardware identified by NAS
	GeneLab			$ 3		Scaled from historical GeneLab grant data
	Training and Mentoring			$ 3		Average cost of two subject-matter experts
	Ground Facility			$ 32		Cost to equip facilities
	Ground Controls for Experiments			$ 100		Scaled from total number of experiments
	NASA Research Announcement Grants			$ 200		NAS estimate
	Total Campaign Level			**$ 473**		
	WBS Element	**ISS PO + CLD PO Contribution**	**Refresh/ Host Bio-facilities**	**BLiSS LEO**	**BLiSS Artemis**	**BLiSS Basis of Estimate**
Hosting Platforms	**Phase B/C/D**					
	Payload					
	Veggie HW/Facility		$6			Estimated from data provided by NAS/NASA
	APH HW/Facility		$36			Estimated from data provided by NAS/NASA
	Ohalo III HW/Facility		$17			Estimated from data provided by NAS/NASA
	Veggie Experiment			$7		Estimated from data provided by NAS/NASA
	APH Experiment			$17		Estimated from data provided by NAS/NASA
	Ohalo III Experiment (Exp)			$16		Estimated from data provided by NAS/NASA
	BIOSERVE/Bioculture HW/Facility/Exp			$32		Scaled from NSPIRES grant data
	Swabs/Air Samplers HW/Facility/Exp			$10	$6	NAS Estimate
	Processing Cost/Integration and Test			$54	$1	Historical ISS processing costs of payloads
	Phase B/C/D Reserves		$30	$30	$1	TRACE cost risk analysis
	Phase E/F					
	Hosting Accommodations		$816			Estimated from potential provider data
	Phase E/F Operations	$20		$28		Scaled from KSC Mission Integration, Operations, Sustainment
	Crew Time	$52		$900		Experiment cadence, duration, and operations intensity
	Launch Vehicle / Services ROM	$350		$43		NASA upmass/downmass costs
	Sample Return ROM	$50		$63		NASA upmass/downmass costs
	Total Hosting Platforms	**$ 472**	**$ 904**	**$ 1,200**	**$ 8**	
	Total BLiSS Unique Elements			**$ 1,681**		

FIGURE 6-5 Bioregenerative Life Support Systems Campaign (BLiSS) cost table with work breakdown structure (WBS) elements, infrastructure, and hosting platforms. NOTES: Cost estimates expressed as fiscal year (FY) 2023 in millions of dollars. Color elements in this table correspond with Figure 6-6. SOURCE: Provided by The Aerospace Corporation under contract with the National Academies of Sciences, Engineering, and Medicine.

from a set of specific goals to meet a set of capabilities needed to enable a bioregenerative life support system. For example, one such experiment series termed "Supplemental Crop Selection/Verification" was laid out as a series of 60–70 experiments in Veggie and expected to take ~4.5 years if there were two Veggie facilities available.

- BLiSS Campaign Level (red)—includes the costs of all crosscutting and ground-based activities to support the campaign including the ground-based experiments serving as controls for the flight experiments and science support to prepare for and build on the results of the space-based experiments. (See Figures 6-5 and 6-6 item in red.)
- BLiSS LEO (blue)—the total cost of executing the sets of experiments in space required to meet the campaign goals including support costs (crewtime, upmass, etc.) beyond those in the base support. (See Figures 6-5 and 6-6 item in blue.)
- Total BPS@FY 2023 level (yellow)—the total cost of the current BPS portfolio. To keep the campaigns separable, items 1 and 2 were not suballocated to each campaign. (See Figure 6-6 item in yellow.)

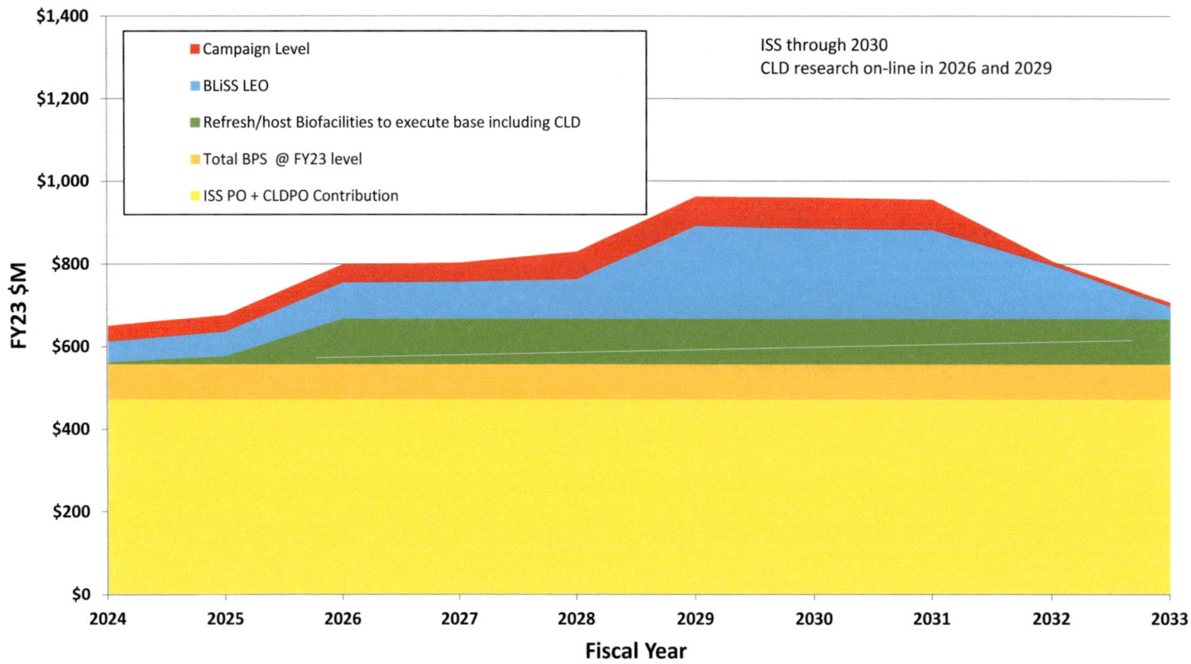

FIGURE 6-6 Bioregenerative Life Support Systems (BLiSS) research campaign, expressed as a cost profile incorporating the technical risk and cost evaluation. SOURCE: Provided by The Aerospace Corporation under contract with the National Academies of Sciences, Engineering, and Medicine.

- Refresh/host Biofacilities (green)—the total of the transition cost current capabilities to the CLD era. It is assumed that the CLD versions are new copies and may operate in parallel with those in the ISS. (See Figures 6-5 and 6-6 item in green.)
- ISS PO + CLD PO CLD Contribution (yellow)—a ROM estimate of the allocation of launch and return vehicle services, crew time, and integration and operations services provided for the total BPS program. (See Figures 6-5 and 6-6, items in yellow.)

For each of the top-level goals, the series of experiments was defined to include the location (e.g., ISS/CLD, Artemis or ground), the facilities to be used (e.g., Veggie, APH, OHALO, etc.), and the number of experiment repetitions or iterations. The calendar time was estimated, and crew time and upmass/downmass were then scaled using relevant NASA-provided cost estimates for LEO research based on the NASA Commercial Use Pricing Policy,[8] including midsize (1–4 MLE, EXPRESS rack) research payloads; EXPRESS rack payloads typically range from 25–150 kg, requiring power, data, and video links/storage and thermal management. Crew time requirements range from 1 hour for simple initiation/shut down operations to 200 hours for experiments requiring rodent maintenance and on-orbit dissections. A range was determined for each type of experiment. Launch and recovery costs were estimated at ~$20,000/kg for unpowered upmass, ~$45,000/kg for conditioned/late load upmass, $60,000/kg for powered upmass, $40,000/kg for downmass (unpowered, unconditioned), $45,000/kg for conditioned downmass, and $60,000/kg for powered downmass. Crew time is estimated at ~$130,000/hour. Estimated cost of integration, mission planning and operations, ground facilities and maintenance, and refresh of commercial LEO destination facility outfitting of repurposed/transitioned ISS

[8] See NASA Space Station, 2021, "Commercial and Marketing Pricing Policy," https://www.nasa.gov/leo-economy/commercial-use/pricing-policy.

equipment or equivalent capabilities (CLD refresh) were estimated in broad categories. No specific estimate was made for data transfer costs.

RESEARCH CAMPAIGN: MANUFACTURING MATERIALS AND PROCESSES FOR SUSTAINABILITY IN SPACE

The MATRICES research campaign is driven by an overarching aim with four goals, together enabling circular life cycles for materials in space and on Earth. The overarching goal is to develop a sustainable eco-system that allows materials and materials-enabled devices to have circular life cycles, thus reducing resource use and waste and harnessing materials in space via better understanding of the attributes and fundamental characteristics of the constituents and processes in the space environment. The component goals achieving this overarching aim are to

- Learn how materials and energy interact in the non-terrestrial environment and use that knowledge to design the infrastructure for space exploration.
- Enable a sustainable in-space economy by learning to make and build much of what is needed in space.
- Develop the nonliving aspects of sustainable circular systems of materials and processes via better understanding of the attributes and fundamental characteristics of its constituents and their formation.
- Understand and exploit the synergies of living and non-living systems for production of needed materials.

Throughout the era of human and robotic operations in space, the materials life cycle has been composed of carefully manufacturing finished items on Earth while ensuring they can withstand the harsh launch environment. Materials are used in space for a given lifetime and afterward are declared waste. The waste is either lost as debris or destroyed via burning in Earth's atmosphere upon reentry. Future long-term human activity in space and future sustainable behaviors on Earth could redefine our relationship with materials. MATRICES seeks a paradigm shift in the way space missions approach the use of materials by asking whether waste can be eliminated, as inspired by work by organizations such as the National Institute of Standards and Technology (NIST) on circular economies. Can all materials be selected for a multi-use life cycle that allows inputs to be formed, used to provide value, and then converted into another form for reuse? In addition, long-term human and robotic activity in space will be enabled by effective and well-thought-out use of resources available in locations beyond Earth, such as the Moon, asteroids, and Mars. To use celestial resources well implies having the science and engineering to adopt them while also understanding the limitations of the supplies in the locations where humans and robotic systems will operate. The research that is required to enable that future state of a circular economy of materials in space affords the opportunity to learn from the centuries of Earth-based failures and successes in extracting, processing, using, reprocessing, and wasting terrestrial resources (Figure 6-7) and to create transformative processes and materials that also benefit Earth.

To create a future with circular materials life cycles, effective and well-rationed use of in situ resources and a gradual elimination of waste in space and on Earth, specific research is needed that characterizes the performance of materials, manufacturing, recycling, and processing in space. This campaign shows how work done in the period of this decadal survey can bring value by demonstrating initial capabilities for partial manufacturing of space systems, additive manufacturing, recycling, and in situ resource management and processing.

The high-level vision of a circular economy outlined by NIST in Figure 6-7 is made even more concrete in Figure 6-8. To eliminate waste and develop methods to reuse materials on Earth and in space, methods need to be invented for space operations that allow life extension of products, distribution of products to the location and user that needs them, and reuse of products that reach the end of one life stage and need to be converted into another. Creating such a life cycle for products in space requires planning from the design phase of materials selection, logistics, and energy access. The remaining discussion highlights examples of research that could enable progress in achieving this vision. The space environment brings unique challenges to implement the circular economy while considering the effects of microgravity; the radiation environment; and limited access to power, volume, and water that may support manufacturing.

FIGURE 6-7 Natural resource extraction. NOTES: The National Institute of Standards and Technology proposes this conceptual model for the concept of a circular economy and studies how to prepare for Earth-based reduction of waste and reuse of materials. These approaches reduce continual and volatile dependence on primary minerals and materials, the Earth-based mining of which can impact environments and societies adversely. SOURCE: *Left*: Courtesy of B. Hayes/NIST; *Right*: Shutterstock/Insight-Photography.

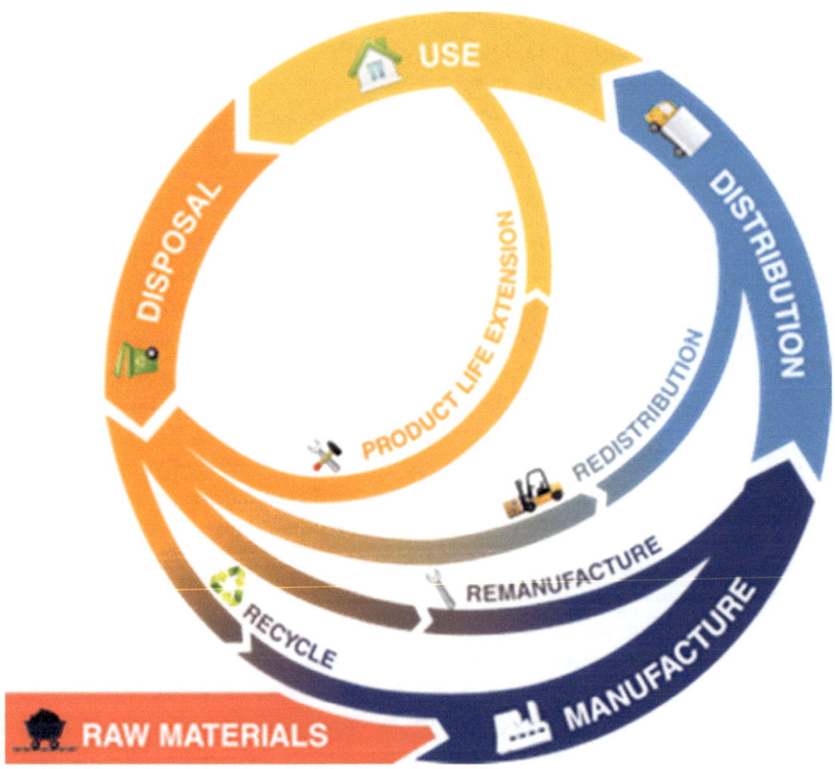

FIGURE 6-8 Alternative approaches to manage waste are needed to enable a circular economy. SOURCE: Courtesy of UK Research and Innovation.

Research Thrusts

Future human activity in space will be shifted by the capability to both manufacture raw materials into useful products and to reuse material via recycling and reprocessing. Future exploration can be greatly enhanced by identifying ways to use all or nearly all material as inputs to value creation and avoiding the creation or designation of materials as waste products. The pursuit of manufacturing will gradually enable humans to overcome traditional limitations of exploration that assume all materials are brought from Earth. In the process of making this feasible, fundamental discoveries about material properties, dynamics of manufacturing processes, and characteristics of in situ materials will be achieved.

The space environment is also an important context in which to phase out waste production and work toward a circular material life cycle; these capabilities will be impactful to change operations both on Earth and in space. This section discusses several facets of this experimental research that can be pursued within the coming decade to advance the capabilities for circular material life cycles and in situ resource processing. The work includes both non-living and biologically based material processes.

The MATRICES research campaign is designed to contribute to answering KSQs that are highlighted in Chapters 4 and 5, leveraging the opportunities of space-based experimentation to observe phenomena hidden by gravity and with an emphasis on creating and maintaining safe, sustainable built environments. (See Table 7-1 for mapping to several KSQs.) The work can begin by building on experience with additive manufacturing, circular material life cycles, material reclamation, and recycling on Earth. While additive manufacturing is not the only type of manufacturing that is relevant to space, it has a specific connection to recycling and reduced material waste. In both additive manufacturing and recycling, a material is processed into a feedstock that is often a limited set of pure ingredients that can be converted using various combinations of heat, binding, and mechanical force. There are several primary approaches to performing additive manufacturing on Earth. Some of these capabilities have been demonstrated in space and others are theorized to be feasible. Some of the key processes that enable additive manufacturing include heating and sintering or binder-based approaches. The research questions that inform progress in additive manufacturing depend in part on which process is under question.

Likewise, recycling is pursued on Earth with various approaches based on the materials. The methods used for glass, polymers, metals, paper, and other core ingredients each differ and may inspire distinct streams of research. For this discussion, consider the example of plastics recycling, which seeks to reuse engineered polymers. In this practice, the level of purity of the materials stream plays a role to influence the effort required to achieve recycling. When ingredients such as dyes, plasticizers, and antioxidants are used within a plastic product, it increases the challenge of separating materials to aid in recycling. Some sectors on Earth have developed standards for producing materials streams with limited ingredients; for example, this is done in the medical device industry, and the result is that products are easier to recycle. One consideration when designing a recycling process is whether the resulting output of the process seeks to maintain the same properties of the original feedstock or allows degradation. When degradation is allowed, the term "down cycling" is sometimes used to note that the mechanical properties of the next-generation product may have changed. Research to determine how to increase the number of times a product or feedstock can be reused in space would be valuable to enable circular value chains. During the plastic recycling process, one design decision is to consider whether the bulk material is broken down into polymers or monomers. The choice influences the energy requirements for the process and the options for products that can be created in the next generation. Molecular weight of the constituent materials is known to influence features such as stiffness, viscosity, and degradation temperature. Each of these factors needs to be considered when designing the process, and work is needed to determine whether aspects of the space environment may influence this.

Another category of recycling focuses on fluid management. This builds on past design approaches that reuse wastewater sources to create potable water for astronauts. In order to expand to larger-scale wastewater management, capabilities such as supercritical combustion to oxidize hydrocarbons in wastewater at low temperature need to be explored in the space environment.

The gravity environment, vacuum environment, and thermodynamics may each play a role to determine the performance of additive manufacturing. For example, in the area of thermodynamics, radiation is a key form of heat transfer and natural convection does not occur. The role of thermal radiative transfer is highly important for heat transfer in metals and plays less of a role in polymers, although the material influences the thermal options.

Owing to the lack of convection, heat will be retained within systems for longer periods relative to the case on Earth. The gravity environment influences the processing and handling of the media used in additive manufacturing. For example, if materials are in the form of powders or other loose small ingredients, these can be challenging to handle in the orbital gravity environment. Thus, uncertainties of these processes in the space environment include the following: How do thermodynamics in the space environment create obstacles or opportunities for additive manufacturing and recycling? What approaches to additive manufacturing and recycling are effective in a vacuum or near vacuum state, especially to reduce combustibility of feed stock that typically oxidize? What approaches leverage the microgravity environment to manage and process media for additive manufacturing?

The planetary surface environment brings new opportunities and questions for additive manufacturing. Further research is needed to characterize potential input materials from the surface of moons, asteroids, and rocky planets such as Mars. Work is needed to ask how the raw materials can be processed into new formats for manufacturing. Additionally, work is needed to design and engineer the processes that can be built, maintained, and fueled on a planetary surface to perform additive manufacturing and, eventually, recycling of planetary materials.

Limited technology demonstrations have pursued capabilities in this area. Made in Space (now owned by Redwire [Redwire 2020]) sent a 3D printer to the ISS in 2014 and demonstrated that a fused filament fabrication process could be performed with no significant defects on the products compared to Earth (Gaskill 2019). Later work included using ABS (acrylonitrile butadiene styrene), Green PE (polyethylene), and polyetherimide/polycarbonate as inputs for the Additive Manufacturing Facility operated by Made in Space. A 3D bioprinting facility was added to the ISS in 2020 (Sertoglu 2020), with anticipated future capabilities to print biologically generated materials (bioinks) as well as living cells within natural or synthetic bioinks.

Building on the current experiments, additional work could be done on the ISS to attempt new forms of manufacturing. As the ISS is retired, additional experiments on the scale of a benchtop module could be applied in commercial LEO systems that are either crewed or uncrewed. In later evolutions of the microgravity ecosystems, experiments could be included on research facilities in cis-lunar orbit or on planetary surfaces. The gravitational environment does impact the behavior of ingredients, and both microgravity and planetary gravity are of interest. Any facility needs to be able to record temperature, perform structural measurements, and enable video documentation. Current experience on the ISS allows the experiments to be designed to be autonomous; thus, crew are not required for manufacturing processes. There may be a need for crew to support the material characterization, such as with atomic force microscopy (AFM) and X-ray diffraction (XRD). In this case, launch cadence will be partly driven by crew availability. It will be valuable to gradually print larger objects, noting that large-scale, powder bed printers on Earth are on the scale of approximately 3 m × 1 m × 1 m, although some manufacturing technologies allow printed objects to extend far beyond the manufacturing platform. Expansion of this research thrust also includes the opportunity to compare different manufacturing approaches and materials. Future research will extend to robotic systems that can build additional manufacturing capability and expand or provide maintenance with some level of autonomy. The experimental system can be shared with other research topics such as cryogenic fluids.

Part of the research agenda to enable a circular material life cycle and the design of materials processing in space is to progress in knowledge about the behavior of materials in a setting outside the Earth environment. The fundamental principles for organizing the structure and functionality of materials are obscured by gravity. Traditionally, thermodynamics has used an approach based on equilibrium to predict potential results from chemical reactions and molecular rearrangements. However, many processes have significant gradients and/or rapid changes that deviate far from equilibrium conditions. The ability to "suspend" particles or fluid elements, such as bubbles and droplets, in their continuous medium allows for the investigation of other forces or field effects to manipulate the motion and orientation of these particles and fluid elements. This capability can produce structures and even chemical compounds by "building" the one element or molecule at a time. Conversely, the ability for targeted removal of portions of a structure or chemical compound can be used to either harvest desired products, degrade toxic compounds, or recycle waste.

The MATRICES research campaign will cover a wide field of fluid topics, including but not limited to colloids, gels, granular media, polymeric fluids, melts, and supercritical fluids. The operational concept for a research facility to support experiments in this area could be like that found in an academic or industrial research laboratory and conduct the iterative process of research that includes the ability to either synthesize or adjust sample composition on-orbit based on real time diagnostic measurements. As such, capabilities could include the ability to generate or alter the composition of fluid samples, to alter the experiment and analysis simultaneously, have real-time or near real-time interaction with the investigator, either remotely or in person. The instrumentation could include microscopes, high-speed cameras, hyperspectral imagers, extensional and rotational rheometers, gas-chromatographs, mass spectrometers, and devices with the ability to measure reactant and product properties.

Several examples from existing experimental facilities provide models for future capabilities to advance knowledge of the fundamental characteristics of materials beyond Earth to support manufacturing and recycling. A work volume similar to the Microgravity Science Glovebox (MSG) (Spivey et al. 2008) would provide the functionality of a chemical fume hood, fluid containment, and test cell cleansing to provide sample preparation and extraction/separation after testing. A second volume similar to an ISS ExPRESS Rack (Pelfrey and Jordan 2008) would house the instrumentation listed above. The Shear History Extensional Rheology Experiments (Hall et al. 2006; Jaishankar et al. 2012; Soulages et al. 2010) (SHERE[9] and SHERE-II[10]) were conducted in the MSG and examined flows and forces as a Boger fluid was stretched across an open volume. Similarly, the Observation and Analysis of Smectic Islands in Space (OASIS)[11] (Clark et al. 2011; Klopp et al. 2019) was conducted in the MSG and examined the behavior of defects on liquid crystal films to various perturbations. The Light Microscopy Module (LMM) in the Fluids Integrated Rack has been utilized to conduct the Advanced Colloids Experiments (ACE) Series,[12] the follow-up to the preceding Binary Colloidal Aggregation Test (BCAT) Series.[13]

Progress in this series of experiments will benefit from a semiautomated facility that can manipulate many samples, preferably simultaneously, in the desired environment. Specialized researchers operating both on Earth and in space will benefit from the ability to interpret results in near real time and adjust sample composition and manipulation parameters as necessary. Depending on phenomenological residence time, ground-based experiments may be used to refine techniques and test parameters prior to in-space testing.

Enabling long-duration space exploration independent of resupply from Earth will require not only establishing the biological knowledge and capabilities described in other sections, but also that of all of the physical systems interacting with and supporting the biological systems. Closed systems will require using physical/chemical systems to create and control the environments for the biological systems. Developing the understanding and methods to create the most efficient and effective end-to-end closed systems will require the ability to choose physical, chemical, and biological methods to execute steps in the subprocesses and create the right outputs and outcomes for successive steps. Doing so will require fundamental knowledge of pure materials and mixed systems, with a wide range of concentrations and parameters. For example, what are the changes in the physical parameters and behavior of water in microgravity when containing various kinds and concentrations of dissolved materials? How does the absence of gravity-driven convection and sedimentation affect the behaviors of waste in flow systems and reaction vessels? In microgravity, buoyancy forces are reduced, resulting in a reduction of sedimentation and natural convection. Consequently, diffusion and electrostatic and capillary forces increase in importance and allow larger 3D structures that are not possible on Earth. Are there as yet unidentified opportunities not available on Earth to process waste streams? A better understanding of fluid behavior to maximize fluid uptake in plants for developing the next generation of crop production and cell respiration is critical (Poulet et al. 2022). Similarly, if there is a need to recycle materials used for vessels, structures, and packaging, how is this best done? Going further, how can by-products of one reaction be utilized to feed into other critical processes? Some will fit into

---

[9] See NASA Glenn Research Center, 2023, "Shear History Extensional Rheology Experiment (SHERE)," https://www1.grc.nasa.gov/space/iss-research/msg/shere.

[10] See NASA Glenn Research Center, 2023, "Shear History Extensional Rheology Experiment-II (SHERE-II), https://www1.grc.nasa.gov/space/iss-research/msg/shere-2.

[11] See NASA Glenn Research Center, 2021, "Observation and Analysis of Smectic Islands in Space (OASIS)," https://www1.grc.nasa.gov/space/iss-research/msg/oasis.

[12] See NASA Glenn Research Center, 2023, "Advanced Colloids Experiments (ACE)," https://www1.grc.nasa.gov/space/iss-research/iss-fcf/fir/lmm/ace.

[13] For example, NASA Glenn Research Center, 2020, "Binary Colloidal Alloy Test-3 (BCAT-3)," https://www1.grc.nasa.gov/space/iss-research/mwa/bcat-3.

well-established bio-pathways supported by physical systems. Others may require classical chemical engineering approaches or hybrid processes with both. To remove otherwise intractable materials (such as per-and polyfluoroalkyl substances) via processes such as supercritical combustion need more study before becoming part of the "standard" repertoire.

A logical progression of research at the intersection of biological and physical sciences could be responsive to the questions derived from the KSQs in Chapters 4 and 5. One KSQ asks what physical properties and behaviors of biologically important materials affect how they participate in and affect biological processes. This work also asks how these behaviors may differ in the space environment. This work is partly motivated by the fact that biofilms have been discovered within the ISS Water Processing Assembly (Prakash et al. 2003), raising questions about the methods of generation, agglomeration, and attachment to solid surfaces in an aqueous environment. Further work is needed to understand the mechanisms of growth and promulgation.

Another KSQ asks about the fundamental characteristics of the non-equilibrium behavior of complex fluids, such as suspensions and slurries. This is of particular interest for those with water as the base fluid. This work explores whether there may be key physical states and conditions not exploitable on Earth that are important in the space environment. The work could investigate the surface physics and interactions with materials of interest under microgravity or low gravity conditions. In the space environment, it may be possible to differentiate between direct and indirect gravity effects, and to distinguish these from mechanical effects unrelated to gravity. Furthermore, studies could examine the chemical and molecular mechanisms that underlie the formation of functional subdomains in cells. This topic also considers the distribution of water flow among the evapotranspiration, plant growth, and fruit production. In the area of plant growth, other topics include the strategies to adequately hydrate and aerate plant roots while providing sufficient nutrition. This work may give insight into how the energy and mass balance can be optimized for fruit production and plant vitality. Another example for investigation in this area considers that in microgravity, buoyancy forces are reduced, resulting in a reduction of sedimentation and natural convection. Consequently, diffusion and electrostatic and capillary forces increase in importance and allow larger 3D structures that are not otherwise possible.

A third KSQ prompts studies of the detailed properties of in situ-acquired space materials and asks how the space environment affects their ability to be used directly by biological systems or processed into usable forms. For example, how does their microstructure affect water retention in plants and cells, root formation, microbial activity, cell respiration, and internal or skin effects on more complex organisms? A related question asks what physical, chemical, or biological processes create other useful materials.

Some of the research progress needed to advance this field includes exploring how to manufacture in space the supportive equipment such as pipes, tanks, and vessels to contain biological experiments. There may be opportunities for biologically based systems to enable the maintenance and repair of the non-living components of sustainable space environments to support long-duration missions. Biologically based systems may also contribute to recycling by participating in the conversion of a product into the base stock materials that can be manufactured. Both chemical and biological processes have been shown on Earth to assist with converting waste into reusable stock material. There is also potential to manufacture finished or intermediary products from bio-supplied base materials.

The types of experiments that explore the relationship between biological and mechanical materials processing will need a variety of equipment. For example, characterization and evaluation may be supported by Fourier-transform infrared spectroscopy (FT-IR spectroscopy), with libraries and software that helps the user to take a spectrum and capture the output. This tool could include multi-component search (MCS) algorithms. Other tools include systems to perform ultraviolet spectroscopy, shear rheology, nuclear magnetic resonance, and gas chromatography. Additional tools include a scanning electron microscope and mass spectrometers. For each of these systems, the capability could include automated sample acquisition and handling.

Infrastructure Enabling Research on Earth and/or Space Asset(s) Over the Decade

Two CLDs are assumed to provide platforms for continuing BPS research in space. A baseline assumption is that one of these CLDs will be in an ISS-like orbit (51.6 degree) and the other at a different inclination. For the purposes of cost profiles, it is assumed that research capabilities begin to become available as follows: one in 2026

TABLE 6-1 Overview of Legacy Facilities That Support the BPS Research Program in Materials-Related Studies

Facility	Reference Source
The Fluids and Combustion Facility	Weiland et al. (2005)
EXPRESS Rack for Microgravity Science with Glovebox	Thompson et al. (2014)
Materials Science Research Rack	Reagan (2017)
Light Microscopy Module	Sicker et al. (2016)

and the other in 2029. TRACE of the BLiSS research campaign assumed experiments in parallel in the ISS and on the CLDs as the latter access becomes available; cost estimates included continued ISS-based research capabilities and adapted "copies" for CLD-based research. TRACE of the MATRICES research campaign includes similar assumptions, specifically the assumption in these cost estimates that the high-mass, high-cost physical sciences facilities (Fluids and Combustion, EXPRESS Rack for Microgravity Science with Glovebox, Materials Science Research Rack, the light microscopy module, and the Cold Atom Laboratory) are transferred into the 51.6-degree orbit CLD from the ISS. These facilities were and will be expensive to build and are large and thus expensive to launch. Rebuilding and launching these to some other CLD would incur >~$1.5 billion in additional costs.

The legacy facilities that underpin the BPS physical sciences basic research program are well known and described in detail at the related sources. Table 6-1 provides reference material to further describe each legacy facility.

The facilities above are key and underpin the work of the MATRICES research campaign as well, but there is also a need for new capabilities to enable a more exploratory, highly interactive environment for rapid learning and repeat experiments. It is envisioned that these would be used by a combination of privately funded and government funded astronauts with the goal of maximum utilization for discovery and iterative development. The committee has identified a need for three new capabilities. All three of these are only described in concept and will require formulation studies and inputs from the scientific and commercial communities to define the base capability and evolutionary path. The first facility is designed to enable research on manufacturing in space. This capability builds on the concept of operations used in the EXPRESS Rack for Microgravity Science with Glovebox, but it is focused on enabled manufacturing processes. It is envisioned that a fully evolved facility could handle a wide variety of feedstock (ceramics, metals, plastics, wood, bioprinting, etc.). The second facility is a highly interactive wet laboratory and characterization facility to allow for experiments in a wide range of soft matter, fluids, suspensions, and slurries. In addition, the instrumentation could allow characterization onsite of a variety of samples. The third facility is designed to allow combined biological, chemical, and physical processing of multiple inputs. The inputs may include wastes, in situ resource utilization (ISRU) or other feedstocks to produce materials suitable for reprocessing into needed forms. Ideally, such a facility could handle waste or ISRU material and convert them into forms suitable for the manufacturing facility. Experiments in this facility would likely be synergistic with the wet laboratory to understand systems at different scales.

Impact by 2033 If the Campaign Is Successful

The combined vision for use of materials found in space and circular life cycle for materials in space proposes a capability that is vital to enable Earth-independent, long-duration exploration, especially beyond the Earth–Moon system. During the coming decade, the progress is expected to be uneven in disparate areas. For example, as observed on Earth, recycling is less complex when materials streams have a small set of pure ingredients. This implies that early demonstrations of recycling in space may occur on specific sets of objects that can be composed of a well-defined materials list. The campaign will also show the distinctions in complexity, cost, and effectiveness of additive manufacturing with different material classes, such as metals and polymers. The work will encompass both basic research and demonstrations of pilot capabilities, with low and medium TRLs (2–6). As this series of exploratory studies and demonstrations concludes, it will enable subsets of the materials management tasks for future human missions to be designed with additive manufacturing and recycling as part of the typical life cycle. This lays a foundation for a long-term path toward closed-loop material life cycles.

The biological aspects of the research agenda outlined in this campaign are expected to contribute to the design of life support systems for long-duration human exploration capabilities, such as those needed to travel to Mars or uses in situ materials from the Moon. The progress that is feasible by implementing the proposed campaign is targeted to demonstrate a variety of disparate capabilities that illustrate the potential for future designs. By 2033, further research will be needed to prepare fully designed systems; however, substantial uncertainty will be reduced by showing what may be feasible when combining mechanical, chemical, and biological properties of materials to create new life cycles. While fully closed-loop sustainable space habitats are unlikely within the decade, the research in biophysical systems could provide major advances in waste recycling to reduce replenishment from Earth and in-space disposal needs. Advances in understanding and technology for converting complex waste in a combined biological-chemical engineering environment into manufacturing feedstocks would be highly beneficial to terrestrial recycling and reducing impact on the environment.

Broad Costs of the Research Campaign, Including the Associated Facilities and Platforms

As illustrated in Figures 6-9 and 6-10, the total estimated cost of a MATRICES research campaign thus scoped would be $3.7 billion over 10 years, with the expenditure peaking in ~2029. The cost estimate for the MATRICES campaign was based on the overall goals to advance the science and technology to enable circular, sustainable space systems. Because these goals are more general than those for the BLiSS campaign, the experimental goals and cadence were defined using analogies to the existing program but at levels projected to deliver significant results over the decade.

- MATRICES Campaign Level (red)—includes the costs of all crosscutting and ground-based activities to support the campaign including the ground-based experimetns serving as controls for the flight experiments and science support to prepare for and build on the results of the space-based experiments. (See Figures 6-9 and 6-10 item in red.)
- MATRICES LEO (blue)—the total cost of executing the sets of experiments in space required to meet the campaign goals including the new facilities plus support costs (crewtime, upmass, etc.) beyond those in the base support. (See Figures 6-9 and 6-10 item in blue.)
- Refresh/host Physics facilities in CLD (green)—the total of the transition cost current capabilities to the CLD era. (See Figures 6-9 and 6-10 item in green.)
- Total BPS@FY 2023 level (gold)—the total cost of the current BPS portfolio. To keep the campaigns separable, items 1 and 2 were not suballocated to each campaign. (See Figure 6-10 item in gold.)
- ISS PO + CLD PO CLD Contribution (yellow)—a ROM estimate of the allocation of launch and return vehicle services, crew time and integration and operations services provided for the total BPS program. (See Figures 6-9 and 6-10 item in yellow.)

The cost drivers for the MATRICES campaign have similarities to and key differences from those for the BLiSS campaign. In comparison to the biological sciences facilities, the core physical sciences space facilities underpinning both the BPS base physical sciences program as well as this campaign were very expensive to develop originally. They are also large, driving up launch costs. It was assumed that NASA will sponsor activities in two CLDs and that one of these will be in an ISS-like orbit (51.6 degree) and the other at some other inclination. To avoid approximately ~$1.5 billion in the cost of building and launching new versions, it is assumed that the Fluids and Combustion, EXPRESS Rack for Microgravity Science with Glovebox, Materials Science Research Rack, the add-on light microscopy module, and the Cold Atom Laboratory are transferred into the new 51.6 inclination CLD (see Figure 6-11). For the purposes of cost profiles, it is assumed that research capabilities begin to become available as follows: one in 2026 and the other in 2029. This campaign is exploratory in nature and is expected to benefit greatly from the opportunities for more interactive experimental settings. Again, by analogy to the previous interactive facilities, concepts are described and costed for the three new facilities described above. The crew time and upmass/downmass were then scaled using from relevant NASA-provided cost estimates

	WBS Element		MATRICES Campaign Level Elements		MATRICES Basis of Estimate
Campaign Level	PM/SE/MA			$ 235	Wrap based on new facility cost
	Science			$ 30	Wrap based on new facility cost
	Training and Mentoring			$ 3	NAS estimate
	Ground Controls for Experiments			$ 75	Wrap based on total facility cost
	NRA/Grants			$ 150	NAS estimate
	Total Campaign Level			$ 493	
	WBS Element	**ISS PO + CLD PO Contri-bution**	**Refresh/ Host Physics Facilities in CLD**	**MATRICES LEO**	**MATRICES Basis of Estimate**
Hosting Platforms			**Phase B/C/D**		
	Payload				
	Fluids and Combustion Facility		Transfer		NASA data
	Express Rack for Microgravity Science w/glovebox (MSG)		Transfer		NASA data
	Materials Science Research Rack (MSRR)		Transfer		NASA data
	Light Microscopy Module		Transfer		NASA data
	Cold Atom Lab		Transfer		NASA data
	Wet Lab and Precision Measurement Bench			$464	Scaled from NASA data and technology investment
	Materials and Manufacturing Workbench			$265	Scaled from NASA data and technology investment
	Bioreactor - Materials Processing			$265	Scaled from NASA data and technology investment
	Processing Cost, Integration, and Test			$103	NASA data scaled for facilities and experiments
	Phase B/C/D Reserves			$99	Wrap
			Phase E/F		
	Hosting Accommodations		$1,210	$806	Scaled from provider information and NASA data
	Phase E/F Operations	$20	$195	$340	NASA data scaled by NAS crew time estimates
	Crew Time	$52		$561	NASA data scaled by NAS crew time estimates
	Launch Vehicle / Services ROM	$350		$300	NASA data scaled by NAS crew time estimate
	Sample Return ROM	$50		$45	NASA data scaled by NAS crew time estimate
	Phase E/F Reserves				Schedule extension
	Total Hosting Platforms	$ 472	$ 1,405	$ 3,248	
	Total MATRICES Unique Elements			$ 3,741	

FIGURE 6-9 MATRICES (Manufacturing Materials and Processes for Sustainability in Space) research campaign cost table. NOTES: Table includes work breakdown structure (WBS) elements, infrastructure, and hosting platforms. Cost estimates expressed as $FY 2023 in millions of dollars. Color elements in this table correspond with the graph presented in Figure 6-10. SOURCE: Provided by The Aerospace Corporation under contract with the National Academies of Sciences, Engineering, and Medicine.

for LEO research based on the NASA Commercial Use Pricing Policy,[14] including midsize (1–4 Middeck Locker Equivalent, EXPRESS rack) research payloads; EXPRESS rack payloads typically ranging from 25–150 kg; requiring power, data, and video links/storage; and thermal management. Crew time requirements range from 1 hour for simple initiation/shut down operations to 200 hours for experiments requiring rodent maintenance and on-orbit dissections. A range was determined for each type of experiment. Launch and recovery costs are estimated at ~$20,000/kg for unpowered upmass at ~$45,000/kg for conditioned/late load upmass, $60,000/kg for powered upmass, $40,000/kg for downmass (unpowered, unconditioned), $45,000/kg for conditioned downmass, and $60,000/kg for powered downmass. Crew time is estimated at ~$130,000/hour. Estimated cost of integration,

[14] See NASA, 2021, "Space Station: Commercial and Marketing Pricing Policy," https://www.nasa.gov/leo-economy/commercial-use/pricing-policy.

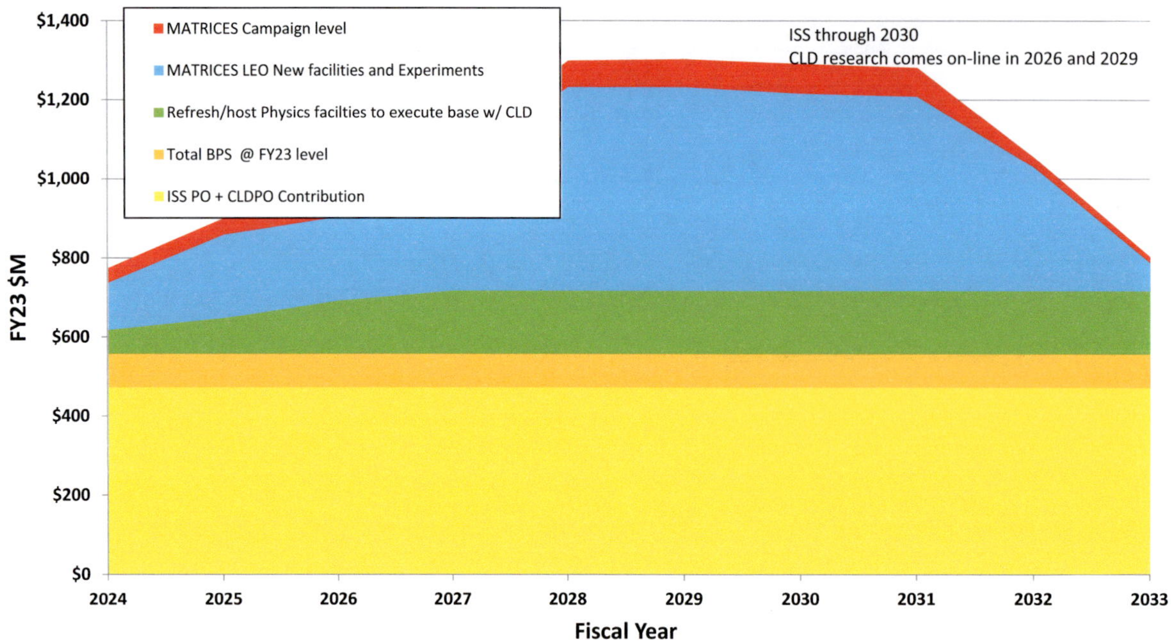

FIGURE 6-10 MATRICES (Manufacturing Materials and Processes for Sustainability in Space) research campaign, expressed as a cost profile incorporating the technical risk evaluation. SOURCE: Provided by The Aerospace Corporation under contract with the National Academies of Sciences, Engineering, and Medicine.

FIGURE 6-11 The Materials Science Research Rack, currently on the International Space Station, could be transferred to a new commercial station. SOURCE: Courtesy of ESA/NASA, https://www.flickr.com/photos/nasamarshall/9085541909/in/album-72157667439026641, CC BY-NC 2.0.

mission planning and operations, ground facilities, and maintenance were estimated in broad categories, and no specific estimate was made for data transfer costs.

The opportunities to pursue the research described above are enabled by an expected ecosystem in LEO that includes several additional years of ISS operations, CLDs, and planetary space experimental platforms. The vision assumes that both public and private sector resources will combine to enable new science and engineering outcomes. For example, government funds may serve as an anchor tenant for commercially operated orbital science laboratories; both government-selected and private-sector astronauts may contribute to scientific work; and research facilities may be multi-use, perhaps combined with facilities used for tourism and commercial manufacturing. Many types of space experiments can be carried out in multiple space regimes, regardless of orbital altitude or distance from Earth. Some experiments are uniquely suited for a specific planetary context, such as the surface of the Moon. As technology matures, new options for research may be available, owing to the availability of increasingly automated and reusable laboratory facilities. More research will be feasible if the need for crew time is reduced by increasing the autonomy of experimental systems and allow remote operation and monitoring. The research needed to enable future circular material life cycles will involve a variety of dissimilar experiments. Within the areas of in situ resource use, in-space manufacturing, recycling, and bio-assisted materials, there is a wide variety of research needs. For this reason, establishing and using modular laboratories in this research campaign that can be adapted to support other research in fluids, materials, and biological sciences are efficient investments.

MULTI-AGENCY OPPORTUNITY: PROBING THE FABRIC OF SPACETIME

In addition to the two research campaigns recommended above, the committee also considered a multi-agency opportunity that would ultimately include research campaign–related efforts important to NASA's BPS contributions to fundamental physics, and resulting in data and capabilities of interest to several other parts of the U.S. government and to society.

The goal of PFaST is to exploit the distance and gravitational characteristics unique to spaceflight to advance knowledge of fundamental physics leading to novel approaches for energy, communications, and computation challenges. Initial R&D thrusts are to develop space-rated, quantum technology optical lattice clocks (OLCs) to enable ultralong baseline measurements and develop optical time transfer (OTT) techniques and technology to enable space-space and space-ground synchronization of ultra-high-performance clocks.

Achieving the overall goal, including deployment of OLC and OTT into space missions, and creating an intercomparative network will require not only completing the technology development but also fully space qualifying units of low enough size, weight, and power (SWAP) to allow their incorporation into any medium-sized to large science mission. That work will require more than this decade, and this opportunity could be pursued only through a U.S. interagency, and potentially international, consortium to develop the spacefaring units.

While basic scientific research has led the way to great advances in technology, vastly improving the human condition, many fundamental questions remain to be explored. For example, the formation of galaxies and the accelerated expansion of the universe cannot be understood by currently known physics. Explaining them requires the existence of dark matter and dark energy, respectively, neither of which have ever been detected in the solar system. Some argue that the fundamental theory of gravity and the universe, Einstein's general relativity, may have to be modified. Besides being a paramount scientific question in and of itself, fundamental physics can also become instrumental in studying the universe in new ways. This is certainly true of detection of gravitational waves, which were long known to be a prediction of general relativity, but which had not been observed directly until 2015. Since then, they have opened up new windows into the universe. For example, next-generation gravitational wave observatories will be able to detect gravitational waves emitted at greater "redshift" (and thus much earlier in time, and hence closer to the Big Bang) than the electromagnetic sources that can be detected by using the James Webb Space Telescope (JWST). Gravitational waves are thus opening new windows into the cosmos, and are already challenging our fundamental understanding of stars, black holes, as well as galaxies. We live at the beginning of an era of quantum technology changing the way we live and think. The development and deployment of space mission–rated ultra-high-performance clocks with commensurate OTT techniques can provide new windows into the fundamental nature of the universe.

Scientific Landscape: Quantum Sensing and Fundamental Physics

Historically, the return on investment from exploring what is not fundamentally understood has been huge; ultimately, harnessing the effects of quantum mechanics (within transistors, communication technologies, and a myriad of other technologies) has led to the digital age, and engineering quantum effects are leading to a quantum age.

Effects like quantum tunneling, where particles can traverse barriers that they could never pass classically, were initially controversial but are now underlying everyday technologies like flash memory (which is used everywhere from USB sticks and smartphones to solid-state drives in computers) to Josephson junctions. Josephson junctions, which permit quantum sensing at levels that are often more than 10 million times more precise than ever before and have revolutionized military sensors, geologic exploration, and medicine, and are now one of the prime contenders for developing large-scale quantum computing. Josephson junctions are the quantum bit, or "qubit" in major quantum computers today (e.g., those developed privately by IBM, Google, and other companies).

Quantum technologies quietly made their way into the mainstream. Atomic clocks, and thus the Global Positioning System (GPS), are early successes of quantum physics that are now taken for granted. These clocks were first-generation quantum devices, based on the quantum physics of individual particles and atoms. Meanwhile, pioneering "experiments with entangled photons, establishing the violation of Bell inequalities and pioneering quantum information science" (the 2022 Nobel Prize in Physics) opened the door to "quantum 2.0," where intricate properties of multi-particle systems lead to groundbreaking applications. Examples include quantum cryptography (like the Chinese QUESS project), LIGO (which is now using a non-classical "squeezed" quantum state of light to detect gravitational waves arising from sources that are further away than what would otherwise be possible), and quantum computing (which is being rapidly advanced by researchers in academia and industry). Quantum computing is both a huge opportunity and a substantial threat, because it may be used in the future to crack advanced encryption, which is essential.

While unpredictable, advances in fundamental science are seldom accidental; "chance favors only the prepared mind," as Louis Pasteur said.[15] Preparing minds requires steadfast development and deployment of advanced technology, pushing to reach new technological heights, so that scientific discovery is favored. The noted science fiction writer Arthur C. Clarke's second law is, "The only way of discovering the limits of the possible is to venture a little way past them into the impossible" (Clarke 1962).

The Justification for Space-Based PFaST

This profound journey of discovery continues. Quantum physics is now enabling investigations that leverage the space environment in multiple ways.

The proposed research requires solar system distance scales to accumulate measurable signals from weak gravitational waves, dark matter and dark energy, or as-yet-unknown physics. Operations in deep space provide distance scales of literally 100 million miles (comparable to the Earth-Sun distance) over which tiny effects from gravitational waves, dark matter and dark energy, or as-yet-unknown physics can stack up to generate a measurable signal. For example, a pair of ultraprecise clocks that use "quantum 2.0" effects to gain unprecedented precision, one in LEO and one in a solar orbit trailing Earth (at the L5 Lagrange point), may be used to search for low-frequency gravitational waves that are currently undetectable, and that may provide clues about the origins of the universe. These clocks can also detect when stars, neutron stars, and/or black holes are on a collision course, long before an actual collision occurs. Earth-based detectors are often not capable of this task.

Weightlessness in space enables quantum effects to be scaled up in duration and spatial extent, beyond terrestrial boundaries. Fundamentally, in any quantum technology, a quantum process would be accomplished before unwanted interactions with the environment destroy the quantum states. The weight of particles in Earth-based laboratories produce background forces, making it hard to isolate the quantum process from Earth's environment. Operating quantum systems in space avoids the need to hold a system against Earth's gravity. This will enable

[15] During an 1854 lecture at the University of Lille, French microbiologist and chemist Louis Pasteur said, "In the fields of observation chance favors only the prepared mind."

minutes-long or even hours-long quantum processes, well beyond what can be achieved on Earth, as well as superposition states that span large spatial distances (Xu et al. 2019). One application may be finding out whether the gravitational interaction is itself a quantum effect (Carney et al. 2021).

Access to large variations in gravitational potential or space-time curvature allows the use of gravitational potential hills, present in the solar system, for the curved spacetime (gravity wells) that exists around massive objects, such as planets, and the Sun. In this environment, relativistic effects of curved spacetime can be probed in ways that are simply unavailable on Earth. This can be used (in combination with 1 and 2) to probe for dark matter or dark energy.

As the accuracy and resolution of time and distance measurements improve, seismic activity on Earth masks the fundamental physics details that scientists want to observe. Deployment of an optical clock on a free-flier orbital platform, or even possibly on the lunar surface, would greatly reduce these adverse effects of vibration.

Research Thrusts

Overview

The ultraprecise quantum sensing network proposed here will provide the collective "prepared mind" of the advanced physics communities to be able to discover new science that will undoubtedly become the basis for the development of the most advanced engineering systems in decades to come. PFaST is about harnessing the power of quantum technology to search for dark matter and dark energy in the solar system to probe the origins of the universe by listening to gravitational waves of currently unmeasurable frequencies, and to explore if the gravitational field—described by space-time curvature in Einstein's theory of general relativity—is fundamentally a quantum entity unto itself. More generally, it is also about developing U.S. leadership in quantum technologies, which is essential to many breakthrough technologies in quantum computing and in quantum sensing.

The approach is centered on deploying an advanced quantum sensing network whose performance will be many orders of magnitude better than previously thought possible, enabled by scientific discoveries that occurred only recently and that will be continuously refined throughout the decades to come. Establishing space-based OLCs for ultraprecise quantum sensing will enable breakthrough science for probing the fabric of spacetime in the following ways:

- *Precisely measuring the curvature of spacetime, and thus testing the understanding of fundamental physics at levels of precision that are 100,000 times what is possible in terrestrial experiments.* This includes testing relativistic gravity by measuring the curvature of spacetime near planets and other massive objects, searching for violations of fundamental symmetries, searching for additional fundamental forces, and searching for time-variations of the fundamental constants.
- *Probing the hidden nature of the universe by searching for dark matter and dark energy and other exotic low-mass fields.* In the same way that relativistic physics revolutionized our perspective of the world (and is now instrumental in the engineering of GPS and in many other advances in technology), a future understanding of dark matter and dark energy may produce models that similarly impact our worldview, and those refined and verified theories of nature may also have unexpected engineering benefits.

  A quantum network of OLCs may permit an early detection of gravitational waves produced by massive inspirals (such as neutron-star and black-hole collisions). The currently unmeasurably mid-band (mHz to 10 Hz) gravitational waves may provide an early warning of major cosmic events (e.g., stellar-mass binary in-spirals, intermediate-mass black hole binary mergers, and intermediate mass ratio in-spirals) long before terrestrial detectors can pick them up; combined with the solar system baseline, mid-band sensing allows precise localization of the events (Kolkowitz et al. 2016). This network will be sensitive to gravitational waves that have been emitted much sooner after the Big Bang than the associated electromagnetic signals detected by the JWST, which allows this network to probe the dynamics of the universe in the epoch after matter condensation.

- *Testing whether the gravitational field has quantum aspects*. Perhaps the most fundamental open question of fundamental physics is whether the gravitational field itself is a quantum entity. While this question has long been regarded as being beyond the reach of current experiments, the rapidly developing field of quantum information has opened doors for answering the question by testing whether the gravitational field can be in a superposition state and mediate quantum entanglement (Carney et al. 2021; Kumar and Plenio 2020). The basic time-scale needed for such experiments is given by the gravitational constant G and the density of available materials ρ as $1/(G\,\rho)^{1/2}$, which is about 15 minutes. Maintaining the required quantum state over such times is impossible on Earth, but is possible in weightlessness (Panda et al. 2022).

Numerous societal benefits are already linked to technology developed for space exploration, ranging from solar panels to cancer therapy. PFaST would be at the forefront of developing quantum technologies with a huge indirect Earth benefit, notably quantum sensing, computing, and quantum information processing. In addition, the quantum network of OLCs will allow a "solar-system GPS" for autonomous spacecraft tracking and navigation, including true one-way ranging and doppler; space-based synchronization/comparison links for high-precision optical clocks on Earth; and master clocks in gravitationally unperturbed reference frame for a redefinition of the SI unit of time (the second). Developing mastery in optical and microwave time-transfer, which is important in geodesy, navigation, and time-keeping, and the development of quantum sensors in deep space will enhance understanding of the behavior of quantum gases.

Key Steps

Atomic clocks and atom interferometers are two basic building blocks of quantum sensing. The quantum sensing network proposed here involves precise inter-comparisons of advanced, optical clocks that will eventually be deployed throughout the solar system. Clocks based on atoms held by radiation forces within laser beams (OLCs) have reached differential instabilities of better than 10^{-18} in 1 second of averaging time and better than 10^{-20} with averaging over longer time-scale (Bothwell et al. 2022; Zheng et al. 2022). The first goal of this multi-agency opportunity is to make such technology available for space-based operation by ground-based research. At the same time, optical lattices have also proved instrumental in generating the longest coherence times in atom interferometers, and observed scaling shows that these coherence times can be strongly extended by operation under weightlessness in space (Panda et al. 2022; Xu et al. 2019). Another goal of this multi-agency opportunity is to develop optical-lattice technologies for use in space-borne atom interferometers.

While these developments are under way on the ground, space-qualified clocks based on a different technology are already available. These are ion clocks with a 10^{-14} stability in 1 second, such as the Deep Space Atomic Clock (DSAC; a miniaturized, ultraprecise mercury-ion atomic clock). Flying such a clock—perhaps on a slightly improved version—in a highly elliptical Earth orbit will be an initial demonstration of highly accurate OTT as well as exploring the ultimate performance of microwave timing links. Such a mission will probe the fabric of spacetime by testing relativistic gravity at accuracies up to 30,000 times that of current work, enhance searches for dark matter and drifts in fundamental constants, and establish a high-accuracy international time scale and geodesic reference. The mission would also demonstrate space-based operation of optical frequency combs, as needed for OTT to the ground. Recent results from the Defense Advanced Research Projects Agency's (DARPA's) Micro-Mercury Trapped-Ion Clock (M2TIC) suggest that this technology may continue to improve dramatically over the next decade, achieving very low size, weight, and power.

By suspending atoms in an optical lattice made of the mode of an optical cavity under conditions of weightlessness, atom interferometry can reach coherence over much larger stretches of space and time. Ground-based experiments have shown that reducing the depth of the optical lattice as well as the sample temperature can extend coherence. However, doing so on the ground is impossible, as it causes the atoms to fall out of the lattice. In space, reaching coherences over time scales of 15 minutes to hours is expected, enough to show whether the gravitational field is quantum mechanical and consists of gravitons. Even with a lower-performing instrument, a mission could be searching for dark energy in the solar system.

Ultimately, a pair of ultraprecise clocks that use "quantum 2.0" effects would be used to gain unprecedented precision. Such clocks will be able to achieve the grandest goals of this multi-agency opportunity, with detection of mid-band gravitational waves and a search for dark matter and dark energy at unprecedented sensitivity. Several deep-space trajectories are interesting for such a mission. Going to the L4 or L5 Lagrange points offers long baselines with minimal variation of the relative position relative to Earth, which is beneficial for time transfer. This is favorable for gravitational-wave detection as well as many types of dark matter and dark energy searches.

Infrastructure-Enabling Research on Earth and/or Space Asset(s) Over the Decade

Reaching PFaST objectives requires two key capabilities: space-qualified stable and precise high-resolution clocks—operating at atomic microwave or optical frequencies—and a time transfer system capable of synchronizing (or comparing) these accurate clocks from spacecraft to Earth, or from spacecraft to spacecraft, without loss of stability, accuracy, or resolution. The specific campaign elements selected attempt to balance risk and usefulness of the achievements.

Space-Capable Optical Clocks

There have been significant improvements in the implementation of microwave, space-capable, trapped-ion atomic clocks (Burt et al. 2021). Their performance level, however, is below that of demonstrated optical ground-based clocks, which are currently quoted as having a fractional frequency instability (FFI) of $1 \times 10^{-16}\ \tau^{-1/2}$ and fractional inaccuracy of 1×10^{-18} (Abdel-Hafiz et al. 2019; Ludlow et al. 2015; Sanner et al. 2019).

Space-qualifying optical clocks is not easy; challenges include not only the size, weight, and power, but, among others, radiation resistance of the electronics, robotic vacuum maintenance, and reloading of ion sources (when used) and packaging to survive and resist launch environments and loads.

Opinions on the required performance of a space-based optical clock range (looking only at FFI as an example) from $1 \times 10^{-14}\ \tau^{-1/2}$ to that of the ground-based clocks mentioned above (Derevianko et al. 2022). A possible compromise is to use the intermediate value $1 \times 10^{-15}\ \tau^{-1/2}$ as an achievable goal.

Furthermore, there is an irreducible sensitivity to vibrations owing to the mechanical nature of several components, including optical paths. This complicates (but not totally eliminates) deployment in a crewed spacecraft, such as the ISS, where a close-proximity free-flyer or magnetically coupled "remora" platform may be required.

But perhaps the greatest challenge to a space-qualified optical clock is the ability to operate for its desired lifetime without the periodic maintenance that current Earth-based optical clocks require. For Earth-based clocks, maintenance limits the length of experiments, but maintenance is possible; for a spacecraft-based unit (with a possible exception of the ISS, which has other problems such as vibration and crew-induced seismic disturbances), human-tended maintenance is impossible, and "self-maintenance" could be a design feature.

For developing such a clock, this multi-agency opportunity could pursue one or two alternative approaches: (1) deployment to the ISS (with "analytical" corrections for the ISS environment) or (2) a small dedicated 1-year lifetime LEO platform with an OTT as described below.

If option 1 is pursued, then it will be possible to systematically develop long-term functionality in the lattice optical clock by developing new control technology that seldom requires crew intervention in the weightless laboratory environment of space. The crew will be available to intervene when necessary to service the clock, but the time between such crew interventions will become a measure of the effectiveness of newly designed systems to stabilize the clocks for long-term unsupported missions in space. New methods will need to be developed to correlate and compensate for the effects of vibrations, especially below 10 Hz, where an effective ability to passively and actively vibration-isolate the clock from the ISS vibrational environment is not readily available. Vibration in this low frequency regime may be measured by an accelerator, such as the Space Acceleration Measurement System (SAMS) on the ISS and correlated with deviations in the OLC stability. The effects of the ISS vibrational environment have been studied at length on other human spaceflight missions that supported ultra-sensitive experiments on the quantum fluids. (See the Lambda Point Experiment [LPE] and Confined Helium Experiment [CheX] experiments on the space shuttle.) Also, see the Science Requirements Document for the Critical Dynamics in Microgravity (DYNAMX) experiment that was prepared for the ISS (Science Requirements Document, JPL Document Number D-18698). Systematic variation of the clock's performance with steady acceleration may be determined experimentally. These data may be useful to estimate the systematics of uniform acceleration in future lattice optical clock designs. Such an OLC facility on the ISS may be operated later within this program and compared against a free-flyer optical clock to further characterize the effects of vibrations, and other aspects of the crewed space environment on optical clock stability and systematics.

Deep-Space Atomic Clocks

It is essential to develop ruggedized OLCs in space during this decade, in order to start the process that will achieve the advantages listed above. This is a top priority. But the development of space-qualified OLCs will likely take until 2030, and possibly beyond. Currently, "Quantum 1.0" clocks, also known as microwave atomic clocks (Burt et al. 2021), have already been miniaturized, ruggedized, and deployed in deep space, and less-accurate atomic clocks are an integral part of the GPS constellation of satellites that are deployed in 12-hour, medium Earth orbits. This DSAC is currently being prepared for a second deep-space deployment (DSAC-2), and while the performance of these already space-deployed DSAC devices is about 100 times less precise and accurate than the anticipated performance of OLCs, the DSACs are immediately ready for launch, and hence may be deployed on upcoming missions to the outer planets, such as the mission to Uranus (which was the top priority from the recently published planetary science decadal survey). This will permit the less-capable DSACs to be deployed on a very long baseline within this decade. The committee proposes this as a less-expensive, more immediate opportunity, but again, it considers this to be a lower priority than the development of the space-based OLCs over this decade.

One scientific advantage of a long-baseline deployment of less-accurate DSACs will be the ability to test time transfer back to terrestrial or Earth-orbital clocks over very long baselines, and this will permit the identification of technical challenges that will be encountered when space-based OLCs become available to be deployed over much larger baselines in the future. While the fundamental physics scientific return from DSACs is not expected to be high, the engineering advantage of such clock deployments on long-baseline planetary missions—for example, as a "rider" that requires low-size, weight, and power—will likely provide future engineering and deployment advantages. Furthermore, the measurement accuracy and size, weight, and power of DSAC technology continues to improve at a rapid rate, but this rate of improvement will definitely pale in comparison to the capabilities that future space-based OLCs are anticipated to provide. DARPA's M2TIC, for example, has provided Alan Deviations of about 10^{-14} in a volume of less than 10 L, and with a power requirement of less than 20 W. This miniature space-based clock is scaling with volume in a manner that is similar to OLCs, which suggests that miniature mercury trapped ion clocks may show dramatic improvements over the coming decade.

Some scientific return may be realized in a DSAC-like clock deployment over long baselines. The size of the dark matter signature, if it exists at all, is unknown, so the better OLCs performance will provide a better chance for major scientific discovery. This is why the committee places top priority on the systematic development of OLCs in space, as discussed below. But the variation of the density of dark matter in the solar system may depend strongly on the distance from the Sun, and on the distance from large massive planets, because dark matter is thought to interact only gravitationally. If so, then any detected signature from less-precise DSACs that are distributed over a large baseline of a few astronomical units would be a major discovery of its own, and it would strongly motivate the future deployment of OLCs over these long baselines in the future. But a null scientific result over these long baselines would also motivate similar dark matter searches using much more accurate OLC deployments over large baselines in the future.

Optical Time Transfer

The next element is the time transfer capability (Ashby 2003). There have been recent demonstrations of OTT in the atmosphere at distances of about 30 km (round-trip). This provides confidence that at least a surface to LEO OTT (LOTT) can be developed and implemented within the decade. Such a LOTT can enable synchronization of two distant Earth-based optical clocks in co-visibility with the platform hosting the LOTT. This co-visibility has an inherent low duration, thus limiting the time available for synchronization, averaging, and comparison.

The next step would be to extend the range capability to geostationary orbit (GEO), therefore enabling GEO OTT (GOTT) capability. This not only increases the maximum co-visibility distance between the Earth clocks, but also allows long synchronization and averaging times.

The two next steps in the development of time transfer capabilities are deep-space OTT (DOTT) and space-to-space OTT (SOTT). DOTT increases the maximum distance between the spaceborne optical clock and an Earth-based one, enabling very long baseline measurements. The main difficulty of achieving this capability is the combination of the large ($1/r^2$) signal loss and the atmospheric losses. It is questionable if this capability could be developed in the next decade unless major breakthroughs occur in the development of LOTT and/or GOTT. Similarly, SOTT has "only" the $1/r^2$ problem to handle and many be feasible in the next 10 years. Of course, one could conceive of a relay system whereby a spacecraft-to-spacecraft link is relayed from a GEO platform to Earth.

Practices to Manage Technical Priority and Risk Over the Decade

As mentioned, there have been significant recent advances in Earth-based optical clocks and trapped-ion space-capable clocks by non-NASA organizations, such as NIST. In addition, the Department of Defense (DoD) has considerable interest in precision space clocks for communications and cryptographic applications.

The first step in managing the development risk is to set reasonable expectations; as mentioned above, an FFI of $1 \times 10^{-15}\ \tau^{-1/2}$ seems a reasonable compromise between improving the state of the art and an unrealistic 10-year goal. Along these lines, deployment at the ISS or its commercial replacements seems to reduce the risk by simplifying launch and hosting costs, but at the expense of having to cope with the relatively high-vibration environment inevitable in a crewed spacecraft. Two possible avenues of managing this risk are using the ISS deployment as a development and validation step, correcting the results for the effects of vibration, or developing a suitable vibration-free deployment, such as a proximity co-free-flyer.

Another risk mitigation approach is to host in the ISS the LOTT system, using it as a short-duration relay between co-visible ground optical clocks. This would both demonstrate the performance of the LOTT (and OTTs in general) and enable Earth-based experiments.

Last, NASA's development could coordinate closely with both the Earth-based clock developers and DoD efforts in this area to avoid duplication and take mutual advantage of the other partner's advances.

Impact by 2033 If PFaST Is Successful

Of the various benefits, the capability of GPS-like autonomous navigation throughout the solar system and space-based synchronization of Earth-based optical clocks are very likely to be achievable by 2033. In particular, an ISS-based "relay" OTT would be useful and relatively straightforward to deploy on the ISS, because it would not be affected by the ISS environment the way an optical clock would.

Detection and measurement of dark matter or dark energy within the solar system does depend on achieving effective OTT beyond LEO or developing the clocks to the suggested performance goal. The risk of achieving this objective is higher but is nevertheless within reach of a 2033 target date. Another possible 10-year result enabled by a GEO OTT or deep space OTT would be to demonstrate deep-space quantum links. The speed of achieving either approach with U.S. leadership and anchorship participation depends strongly on collaborative interagency progress on the base clocks. But more importantly, a millihertz to 10 Hz gravitational wave observatory within the next 20 years can only be possible if this opportunity is started now.

Potential Use of PFaST-Critical Infrastructure to Others

PFaST infrastructure could benefit national security-related communications and cryptography and space exploration navigation and timing (e.g., on the Moon). "GPS-equivalent" navigation (i.e., by measuring time of flight of coded signals) outside of the near-Earth environment, such as the Moon and Mars, can be greatly simplified

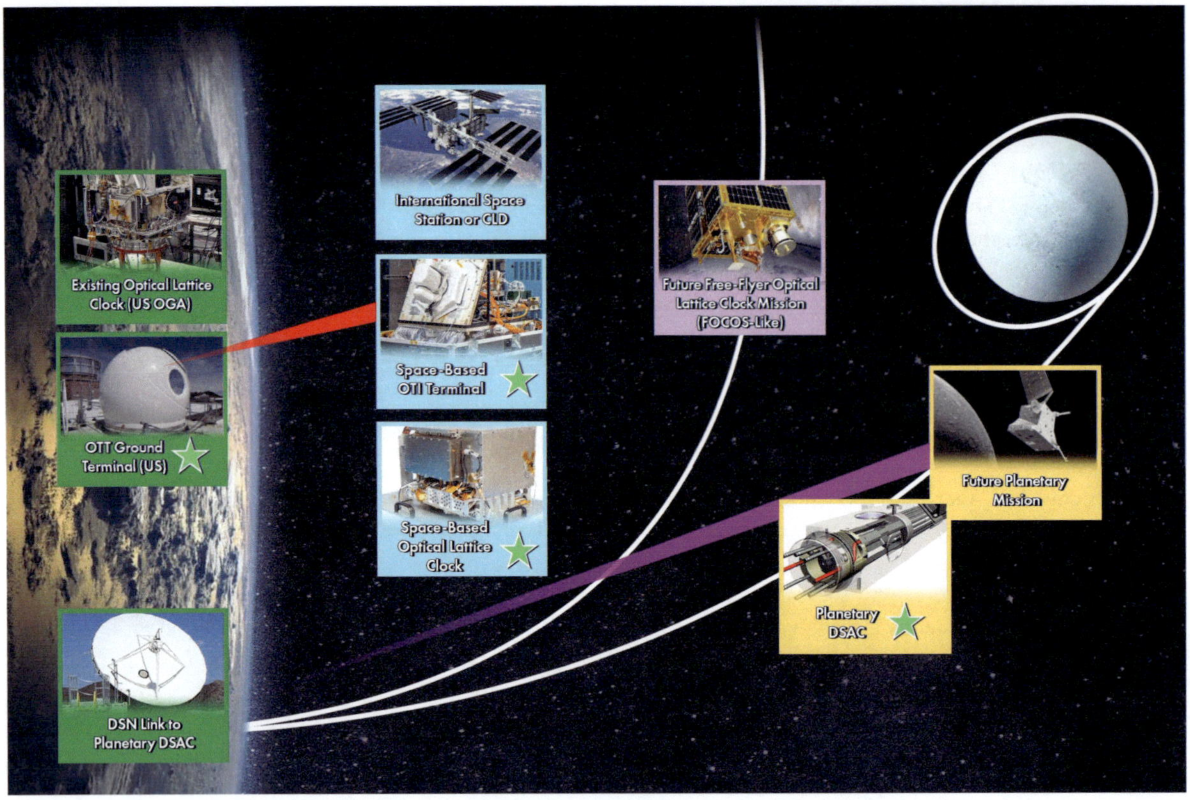

FIGURE 6-12 Overview of locations for the PFaST (Probing the Fabric of Spacetime) elements with key hardware elements denoted with a green star; notional diagram. SOURCES: Provided by The Aerospace Corporation under contract with the National Academies of Sciences, Engineering, and Medicine. Composed by Tim Warchocki. Image credits: Most images courtesy of NASA. *Future Free-Flyer Optical Lattice Clock Mission (FOCOS-Like)*: NASA/MSFC. *Uranus*: NASA/JPL. *Future Planetary Mission*: ESA/ATG medialab/NASA/JHU APL/Carnegie Institution of Washington.

using high-precision, low-drift clocks on the navigating spacecraft. Specifically, the need for multiple linearly independent measurements to estimate local clock bias can be reduced by implementing true one-way ranging and precision Doppler, which, combined with a suitable gravitational model, would allow autonomous, on-board, real-time position and velocity estimation in the cis-lunar, Mars, and solar system environments for future space missions. For example, Figure 6-12 shows an overview of locations for the PFaST elements with key hardware elements denoted with a green star.

Broad Considerations of PFaST, Including the Associated Facilities and Platforms

While the DSAC-2 elements and needed technology developments were assessed as low to medium risk, those associated with the OTT link and high-performance OLC were mostly assessed as medium to high. While the fundamental concepts support BPS science, development of the technological capabilities and system integration required for flying a DSAC-2 on a deep space mission is more properly within the realm of one of the other NASA SMD divisions or STMD. The schedule risks and challenges of developing OLCs and the OTT link to the required performance and durability level lend themselves better to a mostly ground focus in the near term, coupled with in-space experiments to be done on the ISS or a CLD, depending on timing. For example, overcoming the effects of atmospheric disturbances and Doppler closing rates of kilometers per second are serious challenges, and it is difficult to predict how quickly they will be overcome. Other government agencies, notably the Air Force Research

Laboratory (AFRL) and NIST, have long-standing R&D programs relevant to their own needs. In addition, there are several potential international collaborations.

Opportunities exist today for collaboration and mission cost-sharing within the international scientific community and between U.S. federal research programs, and they may substantially reduce the costs of the recommended fundamental physics research campaign to place OLCs in space. Because the results of fundamental physics research is typically published in journals with an international distribution, and because the results of fundamental physics research involves international scientific collaborations already, and builds human knowledge that spans across national boundaries, a strong commitment exists to cooperate and share expenses between nations in support of fundamental physics research.

The Space Optical Clocks (SOC) project is one example of such international collaboration, which is funded by the European Space Agency (ESA) and the German government. This program includes subcomponents of OLC that are suitable for later space use, such as all-solid-state lasers, low-power consumption, and compact dimensions. These components have been developed and validated. This included a demonstration of laser-cooling and magneto-optical trapping of strontium atoms in a compact breadboard apparatus and demonstration of a transportable clock laser with 1 Hz linewidth. Non-destructive detection of atom excitations and minimization of decoherence effects have achieved a fractional frequency stability of 10^{-16}. Furthermore, ESA operates the SOC project with the goal of installing and operating an OLC on the ISS with improved frequency stability of at least one order of magnitude over more conventional cesium-based atomic clocks. The payload is planned to include an OLC complete with an integrated frequency comb and the additional equipment and infrastructure necessary to support Two-Way Optical Time and Frequency Transfer (TWOTFT) to enable accurate comparisons of the ISS clock with ground clocks located in several countries and continents. To this end, the EU-FP7-SPACE-2010-1 Project No. 263500 (SOC2) operated from 2011 to 2015. It developed accurate transportable OLC demonstrators on the ground, having fractional relative frequency stability of 1×10^{-15} in a 1-second integration time, and relative fractional accuracies better than 5×10^{-17}. The devices were based on trapped neutral ytterbium and strontium atoms. This ESA development demonstrates the importance of an extensive terrestrial technology effort to ensure that OLC ground-based technologies are subjected to extensive environmental testing early in their development cycle, to ensure that they can continue to perform at these advanced levels following launch loads, and in the radiation environments and weightless conditions of space. It is much easier to de-risk the technical design and to contain costs if such flight qualification is done well before the new OLC technologies enter the flight campaign.[16]

Another opportunity for collaborative cost reduction for a future NASA-based space OLC program may be achieved through joint project agreements with DoD and the Departments of Energy and Commerce. Many government agencies have been effective in developing commercial off-the-shelf (COTS) technologies that are delivered through industrial and academic consortiums. One such consortium, called the QED-C, was mandated in the National Quantum Initiative Act, and was established with support from NIST within the Department of Commerce.[17] Another is Q-NEXT, which is a Department of Energy (DOE) National Quantum Information Science Research Center led by DOE's Argonne National Laboratory. The AFRL has led the development of programs in photonic integrated circuits to miniaturize and ruggedize these once huge and vulnerable photonic benches for flight operations, in cooperation with national laboratories, industrial partners, and universities. Given the potential cost savings from such collaborations, it appears that it is worth exploring this possibility.

NOTIONAL CONCEPT: POLAR RADIATION OF MODEL ORGANISMS

The committee also considered PROMO, a notional concept that provides a capability to conduct research on biological and physical samples to better understand their response to the harsh space environment of microgravity and giant cosmic ray space radiation for extended periods of time not suitable for crewed scientific research.

At present, the plans for Artemis missions have few opportunities for research prior to dispatching crew into extended beyond-LEO missions. The research opportunities that are envisioned to exist within cis-lunar space are

[16] See Schiller (2015).

[17] See QED-C, 2020, "The Quantum Consortium," https://quantumconsortium.org, for more details.

expected to be severely limited in volume and frequency. This sets an interesting conundrum where some critical research cannot be met with the current deep space platforms, yet they would richly inform human exploration beyond LEO during the Artemis missions. Of particular concern in deep space are radiation effects from high atomic number and high-energy cosmic rays that are both highly damaging to biological systems and difficult to shield (Cucinotta et al. 2011; Giovanetti et al. 2020). Moreover, extremely limited data exist on the combined effects of radiation and microgravity. Therefore, there exists a strong need for laboratory science with animals, organs on chips, plants, and other model systems in vehicles that sample both microgravity and space radiation for extended periods of time.

Missions that experience high levels of radiation and microgravity by traveling completely outside the Van Allen belts are extremely expensive and well beyond current funding concepts. However, there exists a tractable approach that builds on the notional concept known as the "Rodent Mission of Unusual Size," a concept studied by NASA (Robinson 2017). In this concept, a crew capsule, such as Dragon or Starliner that had passed its rated life for human flight, would be retrofitted with suitable rodent habitats and could potentially support tens of rats for up to a 90-day mission in a polar orbit similar to that used for weather satellites (800 km, ~90-degree inclination). Such an orbit passes through the polar regions effectively unshielded by Earth's magnetic field and thus with the higher radiation flux. One central goal would be to get larger numbers of intact mammals, including genetically diverse populations, into the combination of high radiation and microgravity to gauge the response of intact mammals in the deep space environment. Figure 6-13 shows the polar orbit of the proposed notional concept freeflyer that accesses a wider range of radiation and microgravity than typical orbits used currently for BPS research.

Preliminary assessment indicates that the avionics and reentry systems are adequate with minimal changes. Return of fixed samples and living mammals, recovered on timelines similar to returning crews, would provide a rich set of samples and data for assessment.

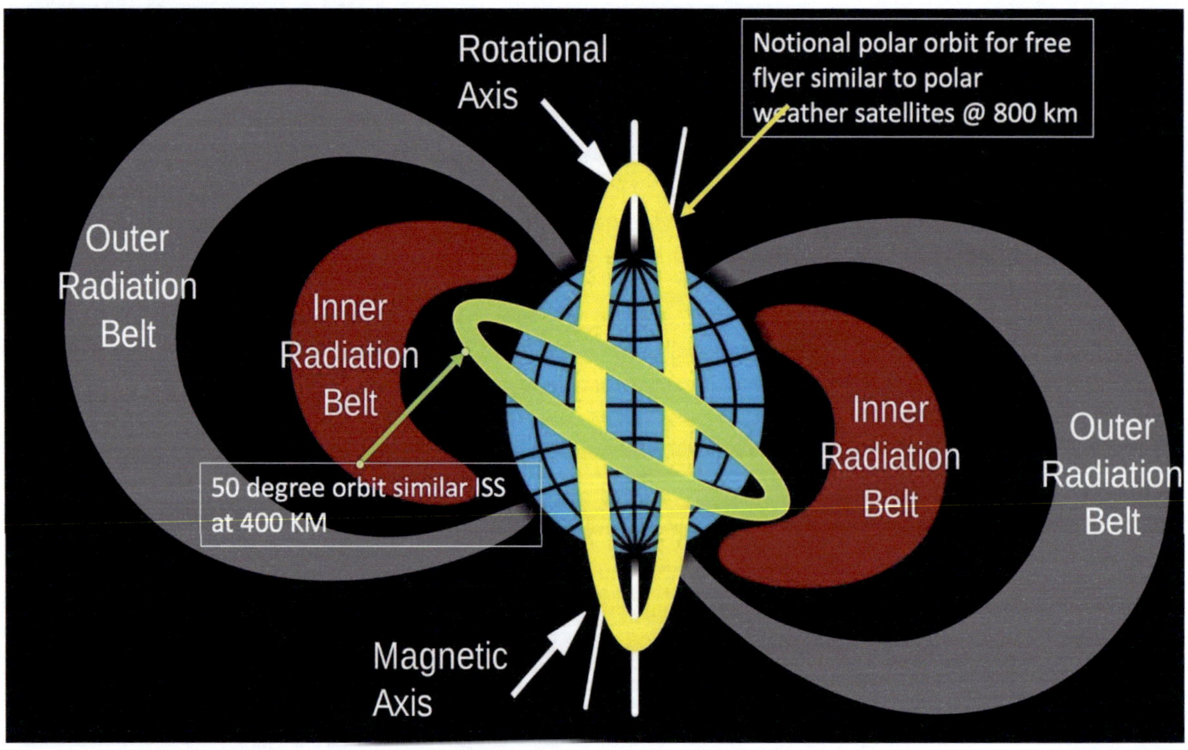

FIGURE 6-13 Polar orbit of the proposed notional concept freeflyer accesses a wider range of radiation and microgravity than typical orbits used currently for biological and physical sciences research.

The costs of a repurposed crew capsule polar freeflyer mission are highly dependent on acquisition strategies and the extent to which NASA and the potential providers, including those with strategic objectives of interplanetary flight, can find common interest and approaches to achieving the goals at low cost. Acquisition with traditional models shows potential cost of approximately $1 billion. However, there are good reasons to believe that acquisition approaches finding common interest between NASA and the potential systems partners, and flexibilities to leverage low-cost approaches and manage risks, could result in a mission cost at 25 to 50 percent of the traditional approaches. As more vehicles become available, either dedicated science vehicles or potentially after serving as human-rated systems, the opportunities for innovative deployments will expand.

SUMMARY

The proposed research campaigns, multi-agency opportunity, and notional concept for BPS research platforms offer transformative opportunities for sustained, coordinated science that advance the knowledge needed for extended space missions and return benefits to society. Such ambitious efforts will require investment in the human capital—including broadened participation by more talented scientists inspired by such team-based, multi-year, mission-focused research—and in the physical capital and collaboration within and beyond reach of U.S. federal agencies (see Figures 6-14 to 6-16). Chapter 7 closes this decadal survey report with recommendations to maintain strategy and address challenges for disciplinary balance, infrastructure, and access, and the vibrantly sustained space science community required to realize the promise of this research and to achieve answers to the KSQs in BPS over the coming decade.

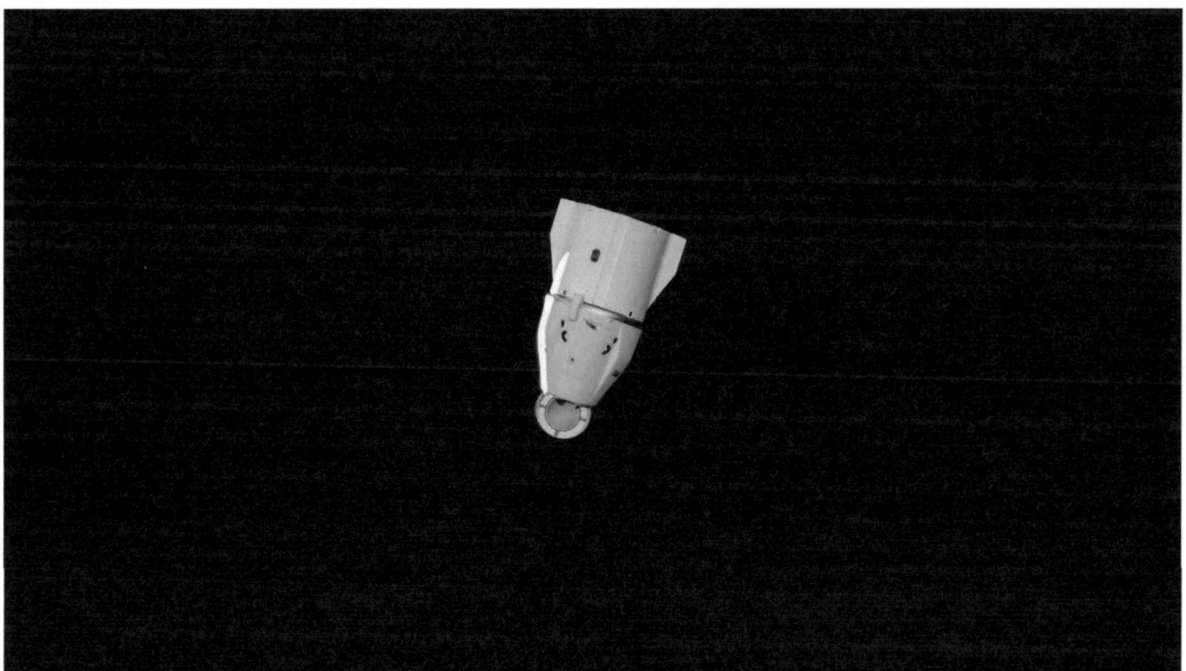

FIGURE 6-14 The United States has an increasing number of space vehicles available for innovative science research. Here the SpaceX Dragon resupply ship, packed with more than 4,000 pounds of return cargo and science experiments for analysis, is pictured departing the vicinity of the International Space Station several minutes after undocking from the Harmony module's forward port. SOURCE: Courtesy of NASA, https://www.flickr.com/photos/nasa2explore/52303935825, CC BY-NC 2.0.

FIGURE 6-15 Boeing's CST-100 Starliner crew ship approaches the International Space Station on the company's Orbital Flight Test-2 mission before automatically docking to the Harmony module's forward port. The orbiting laboratory was flying 268 miles above the South Pacific at the time of this photograph. SOURCE: Courtesy of NASA, https://www.flickr.com/photos/nasa2explore/52095126237, CC BY-NC 2.0.

FIGURE 6-16 The Cygnus space freighter from Northrop Grumman, with its prominent cymbal-shaped UltraFlex solar arrays, is pictured moments after its release from the Canadarm2 robotic arm, ending its 83-day stay at the International Space Station. SOURCE: Courtesy of NASA, https://www.flickr.com/photos/nasa2explore/49887215571, CC BY-NC 2.0.

7

Strategy and Challenges for Disciplinary Balance, Infrastructure, and Access and a Vibrantly Sustained Space Science Community

With the development of advanced reusable rockets and the resulting significant decrease of payload launch costs over the past decade (Roberts 2022), together with anticipated future reductions in cost, the United States has an unprecedented opportunity to be the leader in crewed and uncrewed missions to the Moon, Mars, and beyond. (See Figure 7-1.) To realize this opportunity, the United States needs to also be among the international leaders in the innovation of science and technology in space. Over the next decade, this unique opportunity could enable transformative advances in both the biological and physical sciences (BPS)—from successfully sustaining deep-space missions and lunar habitats to discovering the origin of the dark matter and energy that make up most of the universe. At the same time, society faces numerous current challenges on Earth arising from the impacts of climate change and population growth on both the environment and civil organization, including a rising occurrence of pandemics and maintaining resilient and sustainable food and manufacturing and supply chains in the face of volatile access to resources. Across the world, these instances can exacerbate related challenges in managing socioeconomic disparities, realizing the underutilized potential in our full human diversity, and ensuring equitable impact of collective investments. Research in space exploration supports all of these challenges and opportunities. However, over this same timeframe, increasing international competition is weakening U.S. leadership in space exploration, the underlying science and technology talent that supports that national effort, and the associated workforce development. Therefore, it is timely and important that the United States, including the National Aeronautics and Space Administration (NASA), seizes the opportunity of the coming decade to ensure a future for space exploration by addressing these challenges through fundamental discovery and technology innovation in this always aspirational and inspirational endeavor.

In response to the prior decadal survey, *Recapturing a Future for Space Exploration: Life and Physical Sciences Research for a New Era* (NRC 2011), NASA stood up the Division of Space Life and Physical Sciences Research and Applications (SLPSRA) in the Human Exploration and Operations (HEO) Mission Directorate. This NASA unit is now named the Division of Biological and Physical Sciences (BPS Division) and is relocated in the Science Mission Directorate (SMD). These moves by NASA resulted in a scientifically engaged division that is moving rapidly to accommodate the science-intensive needs of space exploration. In lockstep, the science community—including principal investigator faculty, students, and postdoctoral researchers funded by NASA—has developed a level of accomplishment and stability. This effort and engagement between the federal government, U.S. research community, and general public—as the current taxpayer supporters of and future beneficiaries of

FIGURE 7-1 Astronaut Gene Cernan on the Moon during the 1972 Apollo 17 mission. NASA is planning to return astronauts to the Moon in this decade, for the first time in over half a century. SOURCE: Courtesy of NASA, https://www.flickr.com/photos/nasacommons/52476555977/in/album-72157634974000238.

and the workforce enabling BPS research—is now in need of significant expanded investment if the science is to thrive and contribute to ambitious future space exploration missions and Earth benefits of research outcomes.

Finding 7-1: NASA responded to the 2011 decadal survey by standing up what is now the BPS Division within SMD. As a result, the scientific community in these important and diverse fields has begun to be rebuilt. Importantly, however, much work remains to establish a healthy and sustainable BPS community with resources commensurate with its long-term scientific and technological mission. Much of this work lies in the area of funding beyond simple inflationary adjustments toward large investments that scale to the work needed, similar to the levels that existed during the space shuttle era.

As described in Chapter 2, many of the impressive research outcomes in BPS research have come from the gradual rebuilding, resourcing, and coordination of the U.S. academic space science research community with NASA-employed scientists; divisions of other government agencies (e.g., National Institutes of Health [NIH] and National Science Foundation [NSF]) that share scientific goals; industry entities (typically for-profit commercial entities) that serve as technical solution providers, research collaborators, or sponsors; and the international community of space explorers, including those whose partnerships have sustained International Space Station (ISS) operations and access for BPS experiments. This rebuilding phase over the past decade has been steady but slow. However, the rapid changes in world events, the government-anchored low Earth orbit (LEO) research infrastructure, the nascent standup of the U.S. Space Force, and the role of the private sector are such that BPS research progress will not maintain momentum without increased attention and investment. Rather, lessons of the past decade of retreading the space science research community, infrastructure, and priorities need to inform strategy to ensure U.S. leadership in BPS research and its applications in the coming decade. Some of these lessons are captured in Finding 7-2, which summarizes the successful elements of change over the prior decade.

Finding 7-2: A robust and resilient BPS program requires

- A healthy and regular cadence of proposal calls and grant dollar awards that are consistent with sustaining a diverse and productive BPS community over the course of the next decade, including the necessity of training a diverse scientific workforce of sufficient size and caliber to maintain the BPS community over a generational timescale;
- Broadened and more inclusive participation in the U.S. BPS community, including diversity of both scientific expertise and lived socioeconomic experience, recognizing the slow progress in attracting and retaining women and underrepresented minorities into graduate and postgraduate research roles;
- A total science budget sufficient to meet current national needs and international competitor/collaborator challenges;
- Interactions with other U.S. government and non-U.S. space agencies necessary for optimal BPS community productivity in science and technology development; and
- Significant awareness and collaboration with the emerging commercial space science, platforms, and activities, as appropriate for BPS program goals.

Throughout this report, the critical role of the ISS as a research platform has been recognized as critical to this community of science and the resulting body of work. The largest portion of the current state of work reviewed in Chapter 2 is derived from ISS experiments. Many of the key scientific questions (KSQs) demand experiments that require the extended microgravity exposure afforded by the ISS. Yet the transition from the ISS era to the era of multiple commercial LEO destinations (CLDs) is not yet clearly stated. NASA sees itself as a consumer, along with commercial entities seeking orbital laboratories, of scientific capabilities in CLDs. While NASA has published resource needs, actual design requirements are not yet available (NASA 2023a).

Finding 7-3: The private sector is engaged in development of CLDs, on which the nation's research in BPS will depend. However, science-design requirements have yet to be published. This delay may result in an unintended consequence that CLD companies develop revenue sources to focus on commercial markets, deemphasizing government-funded or fundamental research for public benefit.

As discussed below, the BPS Division at NASA and BPS research community across the nation are severely underfunded with respect to the scope and technical challenge of science that is required to meet the national needs for space exploration, as well as other U.S. national competitiveness and national security interests that depend on advancement of the same science.

Recommendation 7-1: Because the nation benefits from global leadership in space science and technology, and given the emergence of commercial platforms that can be tasked to the nation's science, NASA should:

- **Seek significant funding increases for biological and physical sciences with new monies or through rebalancing the portfolio across the Science Mission Directorate, and in coordination with other U.S. government agencies, as the community needs to grow significantly in size to reach the science goals of the nation;**
- **Actively engage commercial spaceflight firms, using science funding as a driver and with all due haste, to ensure that science needs are met with clear priority, guaranteeing that national science needs are enabled along with those of potential commercial customers using those platforms; and**
- **Ensure that the funded science community fully engages diversity and inclusivity in the pursuit of the nation's space exploration science priorities.**

Commercial firms may have their own interests that are not compatible with those of government customers. They may also lack the long-term stability and sustainability of government projects. Based on recent and current experience, navigating these engagements will not be easy.

GRAND OPPORTUNITIES FOR BIOLOGICAL AND PHYSICAL SCIENCES IN SPACE

Strategic Overview

The next decade will provide unparalleled prospects for BPS in space. The extraterrestrial environment—with its widely varied temperatures, pressures, toxicities, gravitational forces, cosmic radiation, and vast distances—represent extraordinary challenges for exploration, as well as grand opportunities for advancing fundamental science and technologies to benefit humankind. The strategy for BPS scientific priorities—and the underlying technology and workforce development required to meet those priorities—needs to be grounded in the reality that major advances require interdisciplinary expertise, sustained capital investment, and international partnerships in the public and private sector.

Past decades—including the prior decade outlined in Chapter 2—provide ample examples of NASA's contributions in BPS, ranging from medical studies of aging and bone loss and development of plant growth platforms to enhance astronaut health, to tests of general relativity that led to the now invaluable and ubiquitous global positioning system (GPS). BPS discoveries also led to understanding and development of new transport and processing techniques for liquids, soft matter (foams, gels, granular systems), optics (lasers), and pastes for additive manufacturing (AM).

Over the next decade, the stage has been set for transformative progress in BPS research and applications by the wide variety of modern space vehicles, along with powerful new scientific tools, measurement instruments, and analytical techniques enabled by access to BPS "big data," machine learning, and artificial intelligence (AI). These rapidly expanding capabilities fundamentally change the way that science and technological development will be done, allowing vast quantities of high-precision data to be acquired and analyzed autonomously.

An important example can be found in the revolutionary advances in the biological subfields of "-omics"—originally genomics, but which now includes transcriptomics, epigenomics, proteomics, and metabolomics, as well as many subfields within these domains. Taken together, these tools are creating a comprehensive picture of the reference genome, physiological state, and dynamic responses of an organism, including the response to the unique environments in space. These observational technologies that support prediction of the response of organisms to space are complemented by new technologies that enable control and design of organisms for the space mission environment. For example, tools for genetic manipulation, including the adaptation of Clustered Regularly Interspaced Short Palindromic Repeats (CRISPR)-associated systems, bring capabilities to far more easily adapt plants and microbes for space applications, which include engineering increased resilience to the conditions of space, augmented metabolisms for biosynthesis of key nutrients and high-value chemicals and materials, and bioprocesses that convert in situ resources and valorize waste streams for use by the mission. Space-based genetics could also enable response to mission surprise wherein drugs, nutrients, or materials not originally shipped to space are found to be needed. Modifying plants and microbes to biosynthesize these resources where logistical delivery of these needs is impossible or too expensive could be both economical and the only possible route to respond to scarcity when far from Earth. Concurrently, there have been significant advances in robotics, laboratory automation, and human-machine interaction for both measurement and production processes that enable AI-supported laboratories of increasing automation that can perform critical tasks and experiments. Development of efficient and sustainable bioprocesses for space has significant implications for development of sustainable biotechnologies for Earth. The time is therefore ripe for development of an integrated biotechnological research and development program for space.

Equally exciting are the frontiers in the physical sciences that are accelerated in recent years by highly sensitive quantum and precision instruments, including enhanced microscopy and optical atomic clocks, and again aided by big data and AI techniques. The physical science in space community is now ready to answer some of the most fundamental questions about dark matter and energy, non-equilibrium statistical mechanics, quantum information, and gravity. Much of the required instrumentation can also be used to further technology with diverse and practical utility for space exploration, including materials with properties accessible only from non-equilibrium processing, materials made with responsive active matter, materials and methods for additive and digital manufacturing, and quantum sensors and clocks to aid autonomous spacecraft and vehicle navigation. A deeper understanding of transport phenomena in the space environment, building on progress over recent decades, is essential for enabling expanded space exploration far from Earth, as well as for providing fundamental insights that are valuable for manufacturing and resource processes on Earth. Similarly, deepening understanding of space-based combustion science will inform fire safety and supercritical water oxidation that may sustain life beyond Earth's surface. Last, advances in design and manufacturing, including design-for-disassembly and the recovery and reprocessing of materials from products and natural resources, have received a renewed U.S. focus in the past decade that couples to societal needs for Earth resource stewardship and workforce opportunities. These advances are not unique to BPS research in space environments, but fortunately can be leveraged to support operations more autonomously to increase reliability, reproducibility, and defraying the cost of astronaut time.

Key Scientific Questions and Research Campaigns

The grand BPS opportunities over the next decade emerged in this report to encompass three themes and 11 KSQs spanning, and often linking, the biological and physical sciences. (See Table S-1.) These questions underpin a vision and comprehensive strategy for international leadership in transformational BPS and high-impact technological applications and serve as the predictably resourced base effort recommended for BPS in the coming decade. A focus on this set of KSQs will help the United States establish leadership in knowledge across three themes: adapting to space during the travel phase, living in space for extended durations, and leveraging space to reveal phenomena that are otherwise obscured by features of Earth's environment, including its gravitational forces.

These 11 KSQs were identified through broad community input, expert panel discussion, and steering committee deliberation—on consideration of many worthy, ambitious scientific questions and topics that were articulated and advocated for by both seasoned and emerging space science experts in the BPS research community. This decadal report could have summarized 110 interesting and important scientific avenues of discovery to pursue in the coming decade, which would have produced admittedly more narrow questions aligned with academic discipline boundaries and funding lines. Indeed, the selection criteria described in Chapter 3 and the resulting KSQs signal a strong level of intellectual cohesion, convergence, and maturation in BPS space research. That progress in prioritization and discipline integration is owed to the hard work emerging from the prior decadal and mid-decadal recommendations; to insights provided by the Committee on Biological and Physical Sciences in Space (CBPSS) to the community and NASA on implementing those recommendations; and to NASA in collaboration with other government agencies, the ISS National Laboratory, and the private sector as champions of space exploration.

A strategy that presents sufficiently focused, persistent, and resourced pursuit of these KSQs will enable space exploration in LEO, on the Moon, and on the way to Mars. Such a strategy will maintain the interest and endorsement of the U.S. public because the research outcomes of this BPS research will confer benefits to society on Earth in addition to the inspirational images and accomplishments attained in those space environments. Together, these KSQs synergistically advance human knowledge about the universe; facilitate extraterrestrial survival, exploration, and thriving in sustained missions in orbit and on extraterrestrial surfaces; and enable profound terrestrial impact. Progress on each of these KSQs is critical. Successful implementation of this strategy will necessarily require engagement of other government agencies and the private sector, as well as international partnerships. These KSQs offer the basis for creation of numerous possible calls for proposals to which individual investigators, small teams, and even a few substantial centers could submit transformative ideas.

An implicit aspect of this strategy is for NASA and the community of BPS investigators to intentionally and deliberately sunset the space-based research efforts of questions for which the answers are sufficiently clear to enable space exploration, or that fail to rise to a level of transformative discovery that demands use of our limited space research resources. This term "sunset" is a specific although gentle indication that projects will end, and NASA BPS funding for those projects will cease, so that the community can apply the valuable resources of taxpayer funds and space-based crews and resources to other, now more pressing questions. The research intensity and resource of at least some of these KSQs need to be answered sufficiently well within the decade in order to be sunset.

Recommendation 7-2: To maintain research campaign momentum, NASA should require external advisory committees to evaluate research campaign team progress and emergent technologies annually.

The majority of these KSQs provide the knowledge development that is necessary for adapting to and living in space. The other KSQs are equally important to transformational science that probes phenomena that are simply inaccessible on Earth. However, to truly move the needle for human exploration and understanding of our universe, there is also a need for more-coordinated, larger-scale campaigns with a strong mission objective that is greater than any one scientific question. Not all KSQs require a campaign-level effort to address forcefully and completely; however, all research campaigns need to enable answering at least one KSQ among other benefits. Campaigns are aimed at the most important aspects of the KSQs that would require more integrated planning and team coordination, and that would include ideally reusable, extensible capabilities and infrastructure beyond the central campaign goals to enable the broader community to accelerate space science inquiries. Research campaigns need to both inspire and enable space exploration, as well as both capture the imagination of Earth-bound citizens and return benefits of knowledge, products, or technological capabilities to society. Two notional research campaigns were recommended in Chapter 6. Their selection was informed by evaluation for technical risk and cost estimation (TRACE)—an effort new to the BPS space research community and to this decadal survey series. (See Appendix E.) To move from these two notional research campaigns—informed by broad community input and 2 years of study and evaluation by diverse experts comprising the committee and panels—to coordinated teams and scopes, further workshopping and requests for proposals would be required. Additionally presented are one whole-of-government initiative with national objectives broader than BPS research and one research mission concept that can be de-risked and developed in this decade to enable BPS research outcomes in the following decade.

These research campaigns are a new and thus higher risk element of NASA BPS support and programming and represent a collectively higher cost than single-investigator projects that utilize existing research infrastructure. Thus, a regular process of evaluation including external (non-NASA) advisory committees for each research campaign is warranted. These bodies would be advisory to NASA and also serve as a valuable resource to the research campaign leadership, similar to external advisory committees of major Earth-based interdisciplinary, multi-site research centers. It is possible but not necessary that these external advisory committees comprise members of the CBPSS administered by the Space Studies Board of the National Academies of Science, Engineering, and Medicine and/or members of the nascent Biological and Physical Advisory Committee (BPAC), which was chartered to provide recommendations to NASA on BPS scope and priorities.

Table 7-1 summarizes the themes, notional research campaigns, initiatives, and KSQs, along with the specific space environments hosting such BPS investigations. This table summarizes the recommended programmatic

TABLE 7-1 Biological and Physical Science (BPS) Research Themes, Notional Campaigns, and Key Scientific Questions in Space Environments

Research Campaigns, Multi-Agency Opportunity, or Platform Concept[a]	Key Scientific Questions (may be addressed in part by inclusion in coordinated research campaigns)[b]	Space Environment(s) (would be accessed during the research campaign, initiative, or platform use)
Adapting to Space		
Prioritized BPS research program effort independent of notional research campaigns, with subsequent investigations on Polar Radiation of Model Organisms (PROMO) platform	• How does the space environment influence biological mechanisms required for organisms to survive the transitions to and from space, and thrive while off Earth? • How do genetic diversity and life history influence physiological adaptation to the space environment? • How does the space environment alter interactions between organisms? • What are the mechanisms by which organisms sense and respond to physical properties of surroundings, and to applied mechanical forces, including gravitational force?	Low Earth orbit (LEO), including the International Space Station (ISS) and commercial LEO destinations (CLDs); transit to lunar and martian surfaces Future PROMO: polar orbit, uncrewed with months-long cumulative exposure to microgravity and cosmic ray radiation Short suborbital experiments and technology maturation
Living and Traveling in Space		
Biogenerative Life Support Systems (BLiSS)	• What are the important multi-generational effects of the space environment on growth, development, and reproduction? • What principles guide the integration of biological and abiotic systems to create sustainable and functional extraterrestrial habitats?	LEO, including the ISS and CLDs
Manufacturing and Materials Processes for Sustainability in Space (MATRICES)	• What principles enable identification, extraction, processing, and use of materials found in extraterrestrial environments to enable long-term, sustained human and robotic space exploration? • What are the relevant chemical and physical properties and phenomena that govern the behavior of fluids in space environments?	LEO, including the ISS and CLDs; lunar surface and martian samples Short suborbital experiments and technology maturation
Probing Phenomena Hidden by Gravity or Terrestrial Limitations		
MATRICES	• What are the fundamental principles that organize the structure and functionality of materials, including but not limited to soft and active matter?	LEO, including the ISS and CLDs
Probing the Fabric of Spacetime (PFaST)	• What are the fundamental laws that govern the behavior of systems that are far from equilibrium? • What new physics, including particle physics, general relativity, and quantum mechanics, can be discovered with experiments that can only be carried out in space?	Leveraging deep-space destinations, including planetary mission to Uranus

[a] See Chapter 6.
[b] See Table 3-1 and Chapters 4 and 5 for fuller discussion of these KSQs.

activities of BPS research for the coming decade. Fuller discussion of the KSQs in Chapters 4 and 5 outline the aspirational benefits of those efforts and investments to society. This latter aspect of broader understanding of and appreciation for such BPS research is the answer to the natural "so what?" question that space science experts will seek to convey to non-experts in their specialized field, including fellow scientists and non-scientist citizens.

Such research campaigns supplement and augment the steady funding of the recommended KSQs. The two recommended NASA-resourced campaigns are important new efforts to design, resource, and evaluate mission-based BPS research with an anticipated envelope of growing, then sustained, and then reduced reliance on NASA support as the campaign objectives are achieved over approximately a decade. The KSQs, in contrast, require steady and sustained effort as a set over the course of the entire decade; some may be addressed in whole or in part as an element of a research campaign, and others will be addressed entirely outside of a research campaign although perhaps using the same or related ground-based or space-based research infrastructure. Note that the majority of KSQs include explicit consideration of living organisms, including the potential for organisms in the animal kingdom as model systems that may predict answers for human responses during adaptation and sustained living durations in space. The two recommended, notional research campaigns do not explicitly include use of animal models, and thus addressing KSQs related to mammalian cells and to animal organisms remains part of the critical base effort of BPS research.

These campaigns need to be adaptable but at the same time be resilient to scientific and technological change and surprise, and also nurture a U.S. talent base to drive researchers into the next decade sustainably and thus maintain leadership in space-based biological and physical sciences. Importantly, multiple campaigns (at least two) over the coming decade will be essential given the diverse scientific disciplines in BPS.

Challenges

Realizing the grand opportunities for BPS in space over the next decade will require systematic processes to recognize, address, and manage many significant challenges. Any ambition with this scope has, as its central challenge, the prioritization and ultimately right-sizing and scheduling of investment in the critical questions and campaigns. The critical aims of this BPS management challenge are to

1. Balance an agile and effective ability to assess the most critical areas for new investment, further investments, and ramp-down as research and development progresses with investment in longer-term programs, such as research campaigns, whose large size and persistence are justified by the significance of their deliverables but whose effectiveness can be tracked and improved over time as well.
2. Ensure that the funding levels and their investments are sufficient to train, engage, and maintain a sizable and interactive scientific and technological workforce spanning multiple disciplines, career stages, and socioeconomic backgrounds while managing the goals of (1).
3. Develop the essential ground- and space-based technologies (instruments, vehicles, and data analysis tools); reusable infrastructures (i.e., physics and biology laboratories and new "cassettes" that hold different classes of experimental types that can operate within them), driven by the highest-priority science identified in (1); and properly subsidized protocols for ensuring their usability by both the core BPS and research campaign scientists.
4. Ensure alliance and cooperation across national and international public and private organizations engaged in space-based research, operations, and industry such that the above goals can be met efficiently and with maximum benefit to the U.S. public.
5. Communicate regularly and effectively to the U.S. public and all stakeholders about the value provided by the BPS program through advances in fundamental scientific knowledge, new technologies critical to expanded space exploration, and capabilities that can improve life on Earth.

A challenge internal to the scientific community, including both NASA-employed scientists as well as NASA-funded researchers in academia and industry, is the discipline to end research pursuits deliberately and gracefully. Important to the framing of this decadal survey, well-posed questions are anticipated to have useful answers (as well as to beget distinct questions). Challenges external to the BPS community are numerous in the coming decade. Among the most important external factors that provide challenges to a successful BPS program are

- Scientific, political, and national priorities and investment within and beyond NASA;
- The rapid pace of scientific and technological change and emerging scientific opportunities;
- International collaboration and competition on Earth and in the space environment;
- Private investment in space activity, both domestic and international, and including private crews and missions;
- NASA mission continuity and coordination across directorates and centers; and
- Platform capabilities and associated uncertainties with space vehicle access and launch, including orbital mechanics that can affect launch cadence (critical to some BPS space science).

Success of the BPS program over the next decade will require successful management of the internal and external challenges outlined above. Such management will require regular evaluation of scientific, technology, and programmatic progress, including an evolving assessment of the key scientific questions, research campaigns, ground-based programs, and recommendations in this and future decadal surveys and mid-decadal assessments. It will also be essential to have expert assessment of and input on non-technical challenges such as managing the public–private partnerships, competitions, and transition of publicly funded invention to industry and ensuring a vibrant, diverse, and sustained BPS community. Decision rules that establish a priori sensibilities for response to changes in resource levels and prioritization based on favorable and unfavorable scenarios with respect to today (2023) provide a pathway to predictable, smoother organizational approaches to external disruptions. (See Box 7-1.) This approach favors attention to outcomes (i.e., actually answering prioritized scientific questions deemed critical for space exploration and use) rather than to a process that dilutes focus as resources contract or expand. Practices established in other recent societal disruptions at the global scale (e.g., COVID-19 pandemic) and local scale (e.g., emergency responses in publicly accessible buildings) indicate the value of table-top exercises to practice the cadence and communication of such decisions under stressful circumstances.

The decision-making framework described in Box 7-1 is powerful in its simplicity and brevity; intentional in its reliance on qualitative annual comparisons rather than specific resource thresholds; and demanding of its implementation team. When the decision rules are this few in number and this clear in consequence, the full community (of researchers and the public, private, and international parties) can read and prepare for the signals that will maintain momentum in answering KSQs and competing for multi-year research campaigns. When the measure is relative in comparison to the prior year, the guideposts are spaced closely enough to reduce uncertainty in continuity of progress for only demonstrably impactful efforts. When incentives for broadening representation and perspectives of BPS researchers are applied to both NASA as the key sponsor and the community of research leaders who select and build their research teams, a sense of shared values and responsibility for growing the inclusive U.S. talent pipeline without apology is clear at the outset. When the choices are laid out so starkly, it is imperative that the decision team becomes practiced in evaluating external conditions regularly and building internal and community trust for sound responses to scenarios. The sense of urgency and laser focus on multiple prioritized outcomes is a natural by-product of such a framework, appropriate to the ambitious and competitive/collaborative uses of the space environment that the United States has the opportunity to lead with principles of inclusion, excellence, and sustainability. Again, this type of emergency preparedness may seem obvious coming out of a global pandemic as are the visible failures of those teams that do not prioritize table-top exercises and timely communication to all stakeholders.

While management of these various challenges will primarily be the responsibility of NASA BPS Division leadership, momentum and accountability will continue to benefit from a body independent of NASA, such as the current CBPSS administered by the Space Studies Board of the National Academies and the nascent BPAC that was chartered to provide recommendations on BPS scope and priorities, with broad discussion of that role within and outside NASA. In particular, the attention by those committees to a non-monetary return on investment (in terms of science published, technologies invented, workforce developed, missions completed, infrastructure built, industry created, etc.) would help to identify growth opportunities as well as sunset opportunities for KSQs, research campaign elements, and entire research campaigns.

BOX 7-1
Decision Rules to Guide the NASA Biological and Physical Sciences Division (BPS Division) Response to Favorable and Unfavorable Changes in External Circumstances Affecting Progress on Key Science Questions (KSQs) or Research Campaigns, Relative to Prior Year

1. **NASA is appropriated *more* or *less* federal funding for the BPS Division:** If the NASA BPS Division receives *more* federal funding than anticipated, it is allocated to research programs and teams that documented at least one published milestone that was demonstrated to the general public, and that documented maintenance or increase in diversity of research project participants per NASA's definition and strategic plan. If the NASA BPS Division receives *less* federal funding than anticipated, maintaining funding levels on KSQs that hit research milestones in the prior budget year is prioritized.
2. **NASA-sponsored researchers are granted *more* or *less* access to the International Space Station (ISS):** If researchers are granted *more* crew time or upmass on the ISS, experiments that serve as development or validation of commercial low Earth orbit (LEO) destination-planned experiments are prioritized. If researchers are granted *less* crew time or upmass on ISS, technical/biological replicate experiments are prioritized.
3. **BPS researchers have *more* or *less* access to commercial LEO destinations (CLDs) or payload service providers:** If researchers have *more* access to CLDs, projects focused on KSQs representing all three themes and research campaign elements are prioritized. If researchers have *less* access to CLDs, projects focused on KSQs representing at least the adapting to space theme and probing hidden phenomena theme are prioritized until answered.
4. **NASA gains *more* or *less* U.S. interagency cooperation and co-funding of BPS research:** If NASA gains *more* cooperation and co-funding from other U.S. agencies, KSQs and research campaigns with synergy with those agencies are increased in scope, participation, or duration. If NASA anticipates *less* cooperation and co-funding, KSQs that enable space exploration are prioritized.
5. **The United States enjoys *more* or *less* international cooperation with launch, crew time for research, or infrastructure and mission co-development:** If international collaboration *increases*, research campaigns are expanded in scope and participation. If international collaboration *contracts*, KSQs and research campaigns that include industry cooperation are prioritized to enable continued progress on these research programs.

Stimulating Opportunities for Cross-Disciplinary Research

It is hard to imagine a more opportune time to leverage technological advances across BPS in space, including the prospects for transformative discoveries in basic science, facilitating space travel and exploration, and providing benefits to life on Earth. This exciting BPS effort will necessarily require extensive cross-disciplinary research, which will require NASA and its partners to step outside of comfort zones/silos and increase partnerships. These partnerships include those with federal agencies that include a science mission, such as NIH and NSF, as well as with academia and non-governmental organizations (NGOs) and for-profit industry.

Individual examples of such partnerships even within the federal government exist and have been fruitful (e.g., NIH-NASA for tissue chips), but these have been limited in scope and duration of co-investment. NASA could more broadly leverage advances in tissue and cell science, as well as genomics and other -omics methods, and biotechnology, across their relevant programs, through partnerships with NIH, NSF, the National Institute of Standards and Technology (NIST), the Department of Energy (DOE), and other partners. In the physical sciences, cross disciplinary partnership opportunities include the realization of space-based, state-of-the-art precision measurements using optical atomic clocks, quantum sensors, and other emerging tools; here, NASA could partner with other government agencies—for example, the Department of Defense (DoD), DOE, and NIST—to accelerate advances for the national interest. Research on climate, materials science, and manufacturing on Earth are also poised to make substantial advances by leveraging the opportunities provided by space-based research facilities.

Furthermore, there is an absolute need for data sharing and hubs for data processing for public and private data sets. NASA BPS Division has been a champion of Open Science and sharing of data generated by NASA-sponsored research; the Life Sciences Data Archive (LSDA), GeneLab, and Physical Sciences Informatics (PSI) databases are the principal data repositories, and PSI also now includes physical artifact archives. Similarly, it is essential for NASA to enact substantial outreach to the broader scientific community to integrate these data with other data sets (e.g., in biology sciences, data managed by NCBI, EBI, NMDC, DOE Kbase, UK Biobank, or AllofUs) (Arkin et al. 2018; Eloe-Fadrosh et al. 2021), and to provide funding outside of groups generating the data, to enable more broad and complete analyses of these unique data sets generated in space.

Parallel with a more integrative, open, multi-disciplinary approach to research and data sharing—including broader adoption of findable, accessible, interoperable, and reusable (FAIR) principles (Wilkinson et al. 2016)—there is a need to build a more robust multidisciplinary space research training ecosystem through data access and exchange with other research training environments and programs. One step in this regard would be to develop a BPS Training Grant program that is not siloed by one government agency, with internships at public or private parts of the space science research ecosystem that could combine small-scale, "start-to-finish" experiences of an individual trainee's project to provide a more rounded, real-world translatable experience. In this way, BPS programs will develop well-rounded technologists and scientists that are less siloed, more adaptable, and creative.

Managing Choices Within the BPS Research Portfolio

Even in the absence of sudden or sharp changes that prompt use of the decision rules outlined above, the resources available for NASA in any given fiscal year may not be adequate to support the full array of multi-disciplinary BPS KSQs, research campaigns, and other activities recommended by the present decadal survey. One approach, suggested by the most recent mid-decadal review, is that each research direction and associated technology being developed or considered for BPS activities (for example, as found in the evolving NASA Technology Taxonomy and NASA Strategic Technology Integration framework) be classified as falling into one of four categories: *lead*, *collaborate*, *watch*, or *park*. These four categories can help NASA determine the level of cooperative development with other entities (within and outside the federal government) and thus how best to optimize their research and technology development expenditures and efforts.

- *Lead*. NASA's needs and timing for a BPS research area and enabling technology are predominant and essential for the core NASA BPS mission—for example, in subject areas addressing key questions or necessary for research campaigns outlined in this decadal survey. In this category, advancing the research and technology will require NASA investment without substantial shared investments by others. Efforts in this category will typically include (1) maintaining substantial intramural (within NASA and NASA-affiliated entities) expertise and infrastructure for the BPS research area; and (2) ongoing support of a robust community of extramural (outside NASA) BPS researchers.
- *Collaborate*. NASA determines that a BPS research area and/or BPS technology development activity is best pursued via an interdependent partnership with other organizations (government, industry, academia, or international partners) using shared investments. Such collaborations can include but are not limited to subject areas addressing key questions or necessary for research campaigns outlined in this decadal survey. An example is NASA and another government agency coordinating research, possibly also issuing joint calls for research proposals, and communicating project results to each other through means such as joint workshops. Another form is a public–private partnership in which NASA provides part of the funding with cost sharing by the industry partner. With academia, NASA could fund university investigators for

direct collaboration with NASA intramural researchers on pilot projects relevant to downstream scientific opportunities (e.g., informing possible new research directions and space missions relevant to the next decade of BPS). NASA can also provide its partners with access to unique infrastructure, research advances, and in-house expertise that significantly influence the direction of the collaboration. Collaboration allows NASA's in-house technical experts to develop technology and perform research that they may not have otherwise been afforded the opportunity to conduct.

- *Watch*. NASA maintains high vigilance in the monitoring of emerging research and corresponding technology development efforts within government, industry, academia, and potential international partners. Activities in this category will most likely be research and technology development that is not unique to core NASA BPS research areas and campaigns but that may be relevant to downstream scientific opportunities. It is important that NASA stay actively engaged in the national and international scientific dialog to remain poised to react to developments that meet NASA needs in the near and longer term. One means of staying actively engaged in the national and international scientific dialog is attendance at and participation in scientific conferences by NASA researchers, as well as through NASA selectively funding external pilot projects in emerging scientific areas of potential but not yet clear BPS relevance.
- *Park*. NASA decides that pursuing research and corresponding technology development requires better definition of scientific goals, mission or operational requirements, and/or technical viability (even over a 10- to 20-year timescale) before proceeding with any significant effort. NASA maintains a minimal effort for research and technology development in this category until better definition is achieved.

CBPSS and BPAC could provide feedback to NASA on its position in this lead/collaborate/watch/park trade-off space, either annually or at the mid-decadal timepoint.

INFRASTRUCTURE FOR BPS RESEARCH FROM EARTH TO LEO, THE MOON, AND BEYOND TO MARS

Overview of the Current, Known Infrastructure for Ground-Based Validation and Support for LEO, the Lunar Surface, and on the Way to Mars

NASA and commercial partners are currently flying or developing a wide variety of research platforms, as illustrated in Figure 7-2. These include the following platforms: (1) ground, (2) suborbital, (3) LEO, (4) beyond the Van Allen belts but not on the lunar surface, (5) on the lunar surface, and (6) on the way to the Martian orbit and surface. NASA BPS Division science is required to effectively develop and engage the vehicles in support of lunar and martian exploration, yet it is unclear whether BPS capability and utilization are being included in the designs and operations plans for all opportunities beyond LEO.

Finding 7-4: The BPS community has access to a much wider range of relevant space-related vehicles than ever before. However, BPS does not regularly offer access to those vehicles as a dedicated part of its grant portfolio. Instead, NASA BPS Division in SMD relies on the space science community to fund such efforts through separate applications to the NASA Space Technology Mission Directorate (STMD).

Recommendation 7-3: Because key questions identified in this study benefit from access to multiple spaceflight-related platforms, the Biological and Physical Sciences Program should

- **Coordinate funding opportunities with the Space Technology Mission Directorate such that access to the range of spaceflight and spaceflight-related platforms is efficiently employed to answer key science questions, especially those questions that inform technology development for space exploration; and**
- **Maintain a foundational approach to science, building through a strong, vibrant program of ground-based, suborbital, orbital, lunar, martian, and beyond missions.**

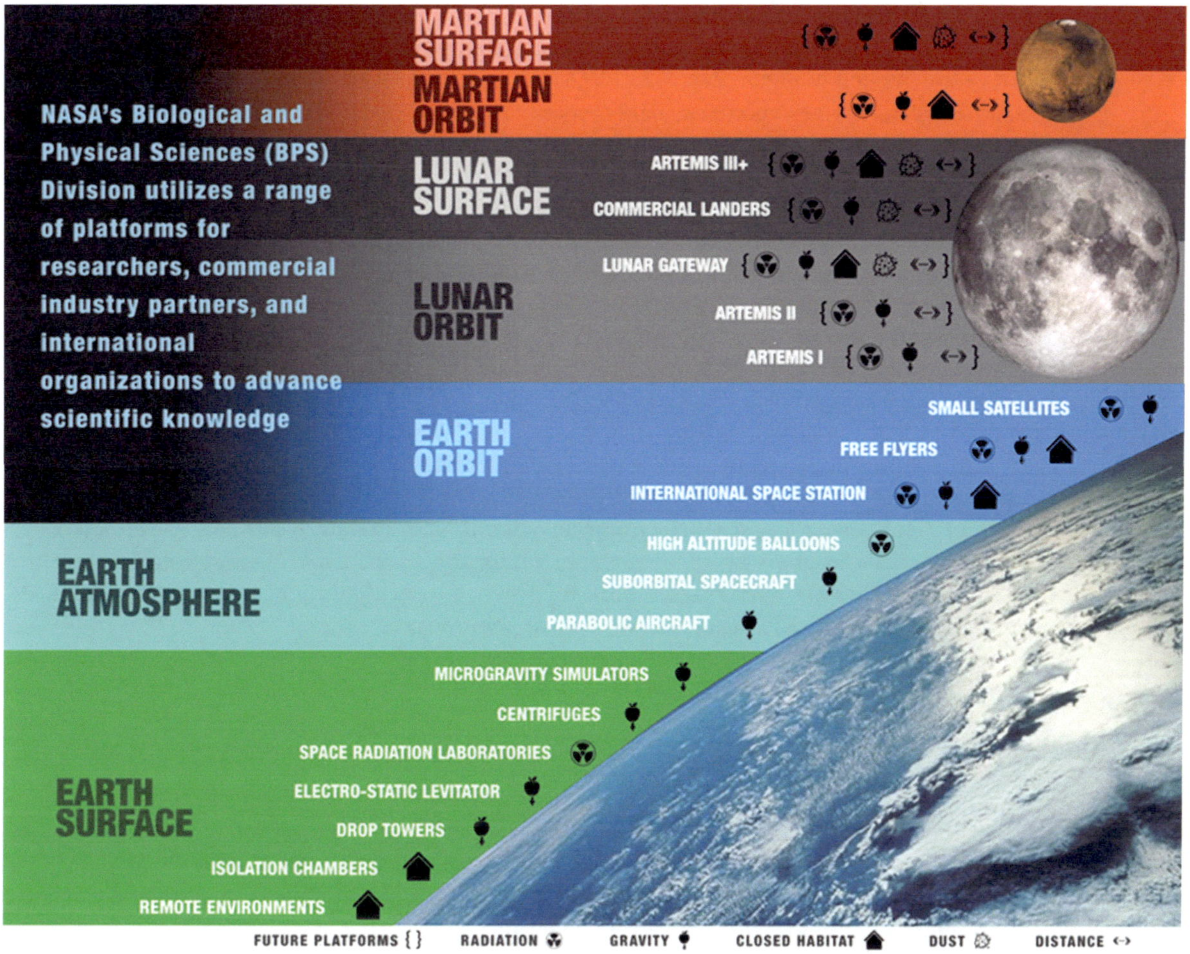

FIGURE 7-2 NASA research platforms anticipated for biological and physical sciences (BPS) research in the coming decade. These research platforms enable the pioneering of scientific discovery. SOURCE: Courtesy of NASA, https://science.nasa.gov/biological-physical/focus-areas.

Precedent for the coordination prompted in Recommendation 7-3 includes an example within the current decade between these two NASA mission directorates (BPS in SMD and STMD). By defining roles and knowledge transfer opportunities for cryogenic propellant tank microgravity data and simulations, the ZBOT experiment series supported by BPS[1] included ISS-based measurements on simulant fluids, from which models of fundamental fluid physics could be based, and potentially later transitioned to STMD's planned technology demonstration flight missions.

Infrastructure for BPS research arguably includes the structural, computational, logistical, and mission control prerequisites for the successful support of the scientific study or research campaign.

- *Structural infrastructure* includes ground-based laboratories on Earth, on space stations/ships, or on extraterrestrial surfaces, including vehicles, communications hardware, computational hardware, roads, power, and so on. These are indicated in Figure 7-2.

[1] See NASA Science, "Space Biology Program," https://science.nasa.gov/biological-physical/programs/space-biology.

- *Computational infrastructure* is the common access of a well-maintained environment for storage, analysis, and transport of data where needed at the appropriate timescales as missions and institutional responsibilities require. These are implicit in Figure 7-2.
- *Logistical infrastructure* is the expected support to manage general operations; maintenance; supply chains including food and waste management with associated upmass and downmass limit considerations, power management, water management, air management, fuel management, medical, and repair supplies; basic search, retrieval, and placement of materials and data from and into storage; maintaining server performance and accessibility; and ensuring minimal staffing of mission control and other infrastructure. These are typically not represented in renderings of destinations such as in Figure 7-2.
- *Mission control* includes the operations managers and decision support to launch and maintain communication with sites of structural infrastructure. This is generally implicit in renderings of space destinations such as in Figure 7-2.

Any mission addressing KSQs, a research campaign, or an integrated service will use a subset of these types of infrastructure and the platforms in Figure 7-2.

Recommendation 7-4: Because key scientific questions identified in this study support the effective utilization of, and benefit from access to, deep-space exploration platforms, NASA should ensure that scientific opportunities are maximized within the range of spaceflight and spaceflight-related platforms intended for lunar, cislunar and Mars transit solutions.

Ground-Based Research Facilities

General Biological, Chemical, and Physical Laboratory Facilities

There is a need for more ground-based facilities to examine subsets of spaceflight-associated stressors, either singly or in combination. Controlled facilitates permitting exposure of model organisms to elevated CO_2 levels, altered partial pressure, and radiation will be needed. In addition, increased availability of ground-based facilities that replicate orbiting and lunar facilities will be needed for preflight testing and for generation of ground control data. Controlled and automated facilities that can support small animal models, including *C. elegans*, *Drosophila*, and rodents, will also be needed to carry out comparative and multi-generational studies. Last, controlled plant and microbial growth chambers that can accommodate microgravity simulators (such as random positioning machines) and bioprocess controllers with the properly constrained form factors to also facilitate exposing organisms to increasing CO_2 levels, altered partial pressure, and radiation would be useful. Genetic manipulation (e.g., with CRISPR) to optimize organism resilience and productivity could be another strategy that is developed and investigated on the ground for deployment in space. However, another aspect that needs to be explored is simulating how such engineering could be done given the constraints of space laboratories. Ground-based facilities for biological, chemical, physical, and/or topically integrated experiments are necessary for a variety of reasons. Some allow investigators to interrogate individual components of the spaceflight environment in a more controlled and less expensive manner, while others provide preliminary evidence for future flight studies. Still others allow investigators to validate flight hardware. In any case, the need for such facilities will only increase as flight opportunities grow.

Radiation Facilities

While the radiation environment beyond the Van Allen belts is ever-changing and challenging to replicate on Earth, access to particle accelerators and similar facilities need be continued and expanded, including through partnerships with international institutions. Use of on-Earth-based beam lines for simulation of deep-space radiation and beam line harnesses for material and biological samples are essential for preliminary testing and validation. Advantages of such facilities include acute exposures to space-relevant high linear energy transfer radiations, both singly and in combination (e.g., carbon, iron, and high-energy protons). Limitations include limited access and the inability to perform prolonged low-dose/low-dose-rate studies.

Large Centrifuge Facilities

Although models for gravitational/inertial environments below the nominal 1 *g* are somewhat limited on Earth, large centrifuge facilities allow investigators to interrogate the impact of chronic exposure to increases in the gravitational environment (e.g., hypergravity). Furthermore, these facilities could, conceivably, be used to examine the impact of combinations of space-relevant stressors, such as low-dose radiation combined with hypergravity. However, while not as expensive as flight experiments, these facilities and the necessary support personnel are not inexpensive to maintain and support.

Habitat Analogs

While there are several isolation and confinement analogs, their utility for BPS research might be somewhat limited for animal-based research, except perhaps for method development and training purposes. In general, most of these facilities are suited for Human Research Program (HRP)-related research. However, some facilities, such as the Antarctic stations, can be useful for studying changes in circadian rhythm and extreme environments on plant growth.

High-Altitude Balloons

High-altitude balloons close to the magnetic poles allow chronic exposures (days to months) to radiation environments closer to what is seen in deep space. This includes galactic cosmic ray (GCR)-like radiation profiles.

Parabolic Flight Facilities

Parabolic flights can provide repeated, approximately 20-second exposures to microgravity. These flights generally go through 20–40 cycles of "micro" to 1.8 *g* levels over a 1–2 hour period. The acceleration level in the low-g period can also be adjusted—for example, providing "martian parabolas" or "lunar parabolas." Higher acceleration in the low-g part provides longer duration, as the change in altitude and speed of the aircraft is the limiting factor. These flight profiles allow the opportunity to study biological and physical phenomena that respond very quickly to changes in the gravitational environment. Furthermore, hardware can be tested to see if it performs as expected in the microgravity environment prior to actual launch to LEO and beyond. (See Figure 7-3.)

Suborbital Flights

Suborbital spaceflight platforms now include researcher-tended and untended payload flights within human-rated vehicles, as well as payloads on sounding rockets. Similar to parabolic flight on airplanes, these opportunities provide exposure to microgravity (~10^{-4} *g*). However, the length of exposure is for considerably longer, ranging from 3–15 minutes (Elgindi et al. 2021). Increasing access to crewed suborbital research flights is expected through the coming decade.

Drop Towers

Drop towers can provide very short but very high-quality microgravity (10^{-6} *g*) exposures in the range of less than 10 s. These facilities are likely more suitable for specific physical sciences experiments requiring short observation timescales.

Ground Control and Mission Support Facilities

Many flight experiments, particularly those that involve sample return, require access to ground facilities for tissue processing and animal housing. For example, rodent studies require access to local Animal Care Facilities and laboratory space for tissue collection and preparation. These could be near the launch site (Kennedy Space Center or Vandenberg Space Force Base), near the landing site (local universities), on the capsule recovery vehicle itself, or in "mobile" laboratories. Last, because much of the hardware for biological studies is "open" to the cabin environment of the launch vehicle or space platform, ground controls are typically maintained in environment simulation chambers that closely match parameters of the vehicle, typically on a 1- to 2-day delay. These parameters include

FIGURE 7-3 Northwestern University researchers gathered data for their foam experiment during parabolic flight. SOURCE: Courtesy of NASA, https://www.nasa.gov/sites/default/files/thumbnails/image/jsc2015e065650.jpg.

day/night cycles, temperature, humidity, and carbon dioxide and oxygen levels affecting both physical and biological systems. These complexities and present limitations reflect dependence on experimental formats that in turn depend on long-term, variable prestorage on Earth with uncertain launch times. Some experiments are flown only once, and those that anticipate sample return to address questions fully grapple with uncertain storage and return times.

Finding 7-5: With the increase in launch facilities and destination platforms, experiment cadence will accelerate in the coming decade.

Finding 7-6: The BPS community cannot continue to rely on experimental formats that depend on long-term, variable prestorage on Earth with uncertain launch times; single-pass "treatments" in space; and sample return with uncertain storage and return times.

Recommendation 7-5: The U.S. government, including NASA, should develop and maintain sufficient ground-based infrastructure to validate and support biological and physical sciences missions. Some of these facilities already exist and simply need to be upgraded, while others have yet to be conceived and built.

A Space-Oriented Data Science Ecosystem

This decadal survey concludes in the Year of Open Science, but the NASA BPS Division has dedicated resources to the managed yet open sharing of BPS research data well before this federal government proclamation.

There has been, and will continue to be, massive amounts of data generated from federally and commercially funded space-relevant research. For biology, for example, these data include -omics, cell population distributions, functional performance assessment data, and images. In BPS studies, multi-terabyte data sets comprising image and video data, with associated metadata and data analytics, are now common. Furthermore, there is a considerable amount of metadata associated with these studies that are unique to the spaceflight environment. Indeed, making data acquired in past and future space missions and ground-based analogs more accessible through data set curation and user-friendly interfaces can promote education and talent recruitment in space science. To ensure that these data are stored, organized, curated, and openly available to the public, data repositories need to be maintained by and for the broader research community.

> Finding 7-7: The ability to manage and interpret the large amounts of data in BPS, both within and across data sets, remains an immature practice and science.

Currently, there are two such databases directly associated with NASA. Biological (non-astronaut) data are typically stored and managed by the GeneLab database at Ames Research Center. Physical sciences data are typically curated in the PSI database associated most closely with Glenn Research Center. In addition to storing and distributing data, members of these facilities also develop tools and analysis pipelines to analyze the data, as well as generate analysis manuscripts of their own.

> Finding 7-8: Spaceflight experiments have unique sets of metadata associated with each experiment. These metadata are critically important to the usefulness of the data for additional analysis. Experiments flown aboard private, commercial, international, and potential U.S. defense-managed vehicles and platforms will generate similarly important data and may require data/IP agreements.

> **Recommendation 7-6: NASA should continue to expand the investment in open and shared computational infrastructure (CI) to support storage, analysis, and dissemination of its biological and physical data, while ensuring linkage to the original and archived samples.**
>
> - **For biological sciences, GeneLab should be continued and efforts made to ensure findable, accessible, interoperable, and reusable access from other critical international biological resource CIs.**
> - **NASA should recognize the need for long-term investment to maintain, update, and improve such community-serving CI and physical repositories over time.**

Sample Repositories

Space-related experiments are incredibly unique and valuable. Furthermore, there are often samples (biological or physical material artifacts) generated as part of such studies that are not directly related to the goals of the primary study or can be utilized for companion studies. For example, studies involving characterization of space effects on bone may not need tissues and organs involved in other physiological systems (such as the central nervous system or the gut microbiome), and rapidly solidified metals analyzed for structural parameters by the principal research team could also later be analyzed for functional properties by another research team. Ground-based studies, such as those involving radiation exposures at NASA Space Radiation Laboratory (at Brookhaven National Laboratory) or hypergravity exposures at the Ames Research Center, also generate a plethora of samples (biological tissues or engineered materials). Rather than discarding these resources, tissues are often collected, processed, and stored by NASA, and eventually distributed to interested investigators not associated with the primary science; physical samples can also be stored and redistributed.

Currently, NASA has several mechanisms for this. For biological samples, this includes the Biospecimen Sharing Program (BSP) and the NASA Biological Institutional Scientific Collection (NBISC) at Ames Research Center, as well as the Life Sciences Data Archive (LSDA) and Animal Life Sciences Data Archive (ALSDA). For physical sciences, samples are curated as part of PSI. As lunar, asteroid, and Mars missions advance, there will likely be opportunities to collect and store regolith for future ground-based study.

Finding 7-9: Spaceflight experiments can generate additional tissues and materials that are useful beyond the original scientific objectives of primary studies. Additional facilities, support personnel, and user protocols are needed for storage, handling, and availability of spaceflight-specific samples and tissues to maximize use of these materials for research.

Coordination with Commercial Destinations and International Facilities

With the changing space research landscape, policies and standards are needed for interacting with these new entities. Private-sector and international partners will have different modes of operation and requirements. These considerations range from using international laboratories and facilities (e.g., the European Space Agency [ESA] Environmental Control and Life Support System [ECLSS]) to launch vehicles (e.g., SpaceX and Northrop Grumman) and CLDs (e.g., Axiom and Blue Origin).

Because commercial companies are engaged in development of CLDs on which the exploration of national science needs will depend, Finding 7-3 and Recommendation 7-1 regarding the development of specifications are critical.

Coordination Between Other Government Agencies Within the United States

Many federally funded research organizations within the United States have goals that are synergistic to varying degrees with the mission of the BPS community. The most obvious of these are within NASA (including HRP, HEO, and STMD). Peripheral to these, but outside NASA, are the Center for the Advancement of Science in Space (CASIS) management of the ISS National Laboratory (ISSNL), the Translational Institute for Space Health (TRISH) and DoD's Space Force. And, last, many basic and applied research units of DoD and DOE, and well as NIH and NSF, could benefit from or contribute capabilities to BPS-related research. For example, the impact of low-dose radiation exposure is a component of DOE, DoD, and NIH's National Institute of Allergy and Infectious Diseases research portfolios. Sustainable technologies for waste stream recycling, in situ resource utilization (ISRU), biomanufacturing, and agriculture are also in the remit of NSF, DOE, DoD, and the U.S. Department of Agriculture (USDA).

A lack of coordination between NASA and other federal agencies (owing to granular and/or uncoordinated funding) is a persistent challenge that was also noted in the previous decadal survey. NASA needs to establish mechanisms of long-term interagency cooperation and resource planning to sustain a robust, long-term inter- and multi-disciplinary BPS research program where the individual agency objectives are subordinate to common unifying scientific goals in the national interest. Increasingly, this coordination reaches into the commercial sector, to suppliers of spaceflight hardware and facilities. A clear example is the development and space-based realization of state-of-the-art optical atomic clocks and clock networks. This challenging scientific (and supporting technological) goal will require coordinated effort both on the ground and in space, with no single agency having a complete capability to meet the overall goal. Thus, efforts at NASA need to be coordinated with complementary work at federal agencies such as NSF, NIST, and DoD. The body coordinating research between agencies could also be empowered to issue cross-agency calls for proposals, especially for long-term grants, on an annual or a biannual basis. Best practices developed by NSF and NIH could be used to ensure competent and transparent peer review of these proposals. In addition to proposal and grant coordination, challenges will exist in coordinating all phases of science, including policies and procedures for experiment compliance such as biosafety, handling of intellectual property, and other agreements that reach across platforms, providers, and agencies.

Recommendation 7-7: NASA should work with the other appropriate U.S. government agencies with a goal to establish an office or a mechanism for commercial sponsorship and collaboration with non-profit organizations, including academia and government research agencies. That office/mechanism should have the primary focus of

- **Coordinating the work between these commercial sectors and government agencies;**
- **Providing guidance on or facilitating research compliance, data security, and material transfer agreements, including prototype agreements;**

- **Representing multiple space environments and destinations (e.g., not only the International Space Station in low Earth orbit); and**
- **Communicating these opportunities to the research community.**

Beyond the U.S. government and industry, there is a historic reliance on international collaborations in the BPS space research community. Spanning decades concurrent with tumultuous world events, those collaborations have been maintained with space science and technology perceived as a collective effort and public good. However, this history cannot be taken to ensure future cooperation and friendly competition on the currently strained global stage. Thus, it is important that NASA continues to facilitate appropriate international engagements on behalf of the BPS community that requires access to space.

Recommendation 7-8: NASA should work with appropriate government agencies to establish clear guidelines for international collaborations within the biological and physical sciences—in particular, for support of non-U.S. students and scholars, to balance two goals:

- **Sustain and advance the U.S. leadership in the relevant areas of research, possibly by attracting the best and brightest globally; and**
- **Support a robust global research community and information exchange, fostering partnerships with other space programs and U.S. access to other nations' ground-based and space assets.**

Communication of Technoeconomics of Space Science with the Research Community

This decadal survey grappled with the costs of BPS research that have been less understood by the research community during the ISS and shuttle launch eras. Namely, launch and related costs of BPS-sponsored research have not been included in the costs borne by the sponsored research team; these costs are borne by NASA and essentially cost-shared by the taxpayer as part of the larger launch event of which the research is only a part. At this stage of commercial LEO destination development and payload services, the division between the total cost of research and the researcher-aware costs are also unclear. When and how these shared costs are differently distributed is among the uncertainties of the coming BPS research decade, and lack of clarity on this point is a threat to a thriving future for BPS research in space.

Ensuring that analyses of cost evaluation—for projects addressing KSQs or for modules of research campaigns—are transparent and available will also help in comparison of different technological approaches to solve the same scientific or technological problem in space. Growing a community well-informed about the practical limitations or bottlenecks in the implementation of its proposed plans and technologies is increasingly important as the BPS community and broader U.S. space exploration community plan longer and more complicated missions.

PROVIDING AND MAINTAINING SPACE ENVIRONMENT ACCESS FOR BPS RESEARCH

Overall Decline in Funding and Community Needs to Be Remedied

One measure of commitment to space life and physical sciences is the overall NASA programmatic budget associated with microgravity science. The funding for the BPS program also suffered dramatic cuts after about 2005, when a decision was made to redirect research funding to Project Constellation. Figure 7-4 shows BPS task book and funding data, with the BPS budget decreasing by more than 30-fold from 1996 to 2010, followed by a modest rebound in support to the present. In fiscal year (FY) 2016, the funding had been compared to FY 2010. This increase demonstrates NASA's efforts to restore space life and physical science research following ISS assembly completion. However, as shown in Figure 7-4, the funding environment has not been restored to its earlier levels of ~ $700 million in inflation-adjusted dollars.

A year-by-year look at the number of tasks and principal investigators supported by the BPS Division (Figure 7-4) reveals that between 2004 and 2009, both decreased by a factor of 7. This trend suggests that it is highly unlikely that increased efficiency can cover the loss of research output. The funding trends are consistent

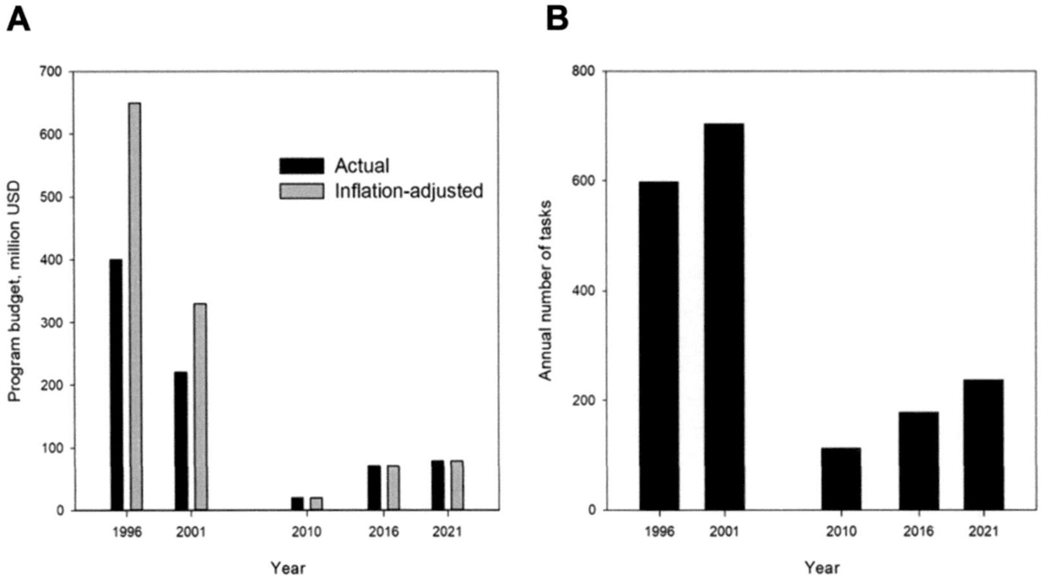

FIGURE 7-4 (A) Annual Division of Biological and Physical Sciences (BPS Division) budget in actual and inflation-corrected (for 2022) U.S. dollars; and (B) annual number of BPS Division–funded tasks. SOURCE: Committee created based on NASA data.

for both biological and physical sciences, as Figure 7-4 shows. Figures 7-4 and 7-5 also show that, although the BPS Division budget was slowly increasing at least for some time since 2006, it has plateaued or decreased since 2020. The trends appear consistent for funding of BPS. Moreover, the grant funding available to external (non-NASA) U.S. researchers appears to represent only a modest ~25 percent of the overall BPS Division budget.

This historical perspective joins with the TRACE cost analyses of Chapter 6 and the political and social realities expressed in this chapter to arrive at a recommended funding level that is a significant increase from the current BPS level. (See Box 7-2.)

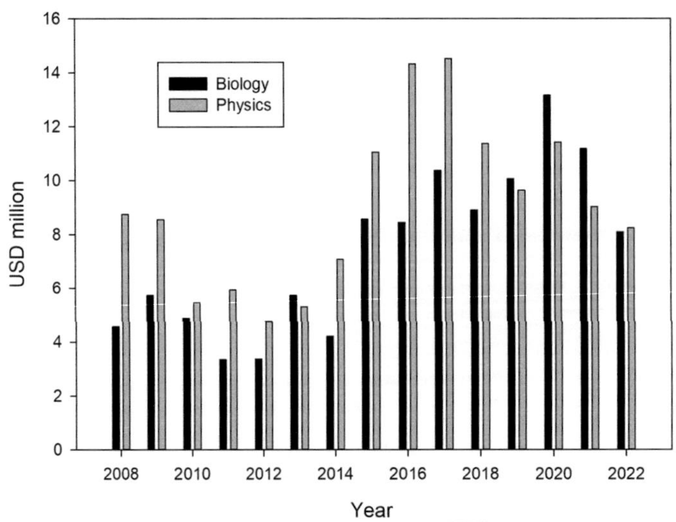

FIGURE 7-5 Division of Biological and Physical Sciences annual totals for grant budgets in 2008–2022 for biological (labeled here simply as biology) and physical sciences (labeled here inaccurately as physics). SOURCE: Committee created based on NASA data.

BOX 7-2
Defining the Billion-Dollar Need

This decadal survey argues for significantly increased funding for the Division of Biological and Physical Sciences (BPS Division) to create a stable, sustainable, and visible program. The recommended funding increase (Recommendation 7-9) is based on three principles that all drive toward the inescapable conclusion that to provide sustainability and visibility for the program, while funding the science needed to support the expansion of space activities prioritized by this survey, the budget must rise by a factor of 10 well before the end of the decade. This increase is supported by the intersection of the principles of cost analysis, political reality, and historical precedence.

- *Cost of driving science questions to conclusions within the decade.* The technical risk and cost evaluation (TRACE) analyses of the two research campaigns that were developed to move several key scientific questions (KSQs) toward practical solutions within the decade each require $100 million to $400 million per year of the decade. However, the campaigns cannot even be considered without the recommended increase in funding (see Figures 6-5 and 6-10). The campaigns are one major new concept of achieving closure of certain KSQs within the coming decade, yet they cannot be conducted at the loss of the broader core program of science for *all* the KSQs. The KSQs are scaled to be largely addressable in the coming decade. At the *current* budget level, all KSQs can be addressed, but few will be driven to solution. At the recommended funding increase, many can be *retired* within the decade through a robust core budget increase and the consideration of the recommended campaigns.
- *Political, scientific, and societal reality.* The increased annual expenditure not only enables the BPS research community to regain breadth and pace required of U.S. leadership to thrive in the space economy, but it is also a budget that citizen taxpayers and members of Congress will recognize and pay attention to within a program as large as the NASA budget. This funding level can weather annual fluctuations of a few tens of millions of dollars without major damage to ongoing research projects or the stability of the research community and stabilizes the core funding as large enough to allow reprogramming of funds within the budget. Moreover, a budget nearer to $1 billion raises the profile of BPS science into the realm of the other divisions within NASA's Science Mission Directorate (SMD), tangibly recognizing the tremendous human and commercial interest in space exploration, the expansion of the BPS program beyond low Earth orbit (LEO), and the development of the space economy. Last, this level is also required to cooperate or compete with other nation's investments in space-based research and resources. For example, during this decadal study period China has made substantial progress toward establishing a new space station in LEO and plans for crewed moon missions. (See Chapter 1, Box 1-1.) These are investments by China in people and infrastructure that will address similar scientific questions and opportunities as those presented by this study for the United States, but currently are not anticipated to include U.S. collaboration or access.
- *Historical precedence and higher occupancy of space.* In the 2000s, the BPS budget at NASA was cut severely—not because the science was completed, but because of the emergent need to fund the build-out of the Constellation program to the Moon. In 2010, Constellation was canceled, but the funding for BPS was not restored and many researchers left the field. A return to funding levels of the space shuttle era would target ~$700 million inflation-adjusted dollars in 2023. (See Figure 7-4.) Given the dramatic increase in the person-days in space now compared to the space shuttle era, and the concomitant increase in space science both needed and possible, $700 million is clearly less than a restoration of a similar relative funding level as that which existed in the space shuttle era. Moreover, NASA has begun making substantial progress toward the goal of returning humans to the Moon. The success of the Artemis I mission in fall 2022 and the debut of the first test launch of the Starship rocket in 2023 are both major steps on the way to this goal. These exciting new capabilities for human spaceflights to the Moon strongly argue that the budget cuts

from Constellation are no longer needed for rocket development, and therefore could be available for restoration of the BPS budget.

- *Key assumption.* This budget-level recommendation assumes that transportation costs and mission integration and operations costs will continue to be borne outside the BPS Division, just as they are now. Currently, neither the BPS Division nor the International Space Station National Laboratory (ISSNL) pay the actual costs for the launches, use of the space station and its facilities, or astronaut time. These costs for the conduct of science on the ISS dwarf the budget for BPS. (See Box 1-1 and Figures 6-5 and 6-10.) Put another way, the BPS Division cannot be charged for launch costs or mission integration and operations costs within this recommended funding level. The BPS Division need fund only the science. Any change from this assumption would require a drastic reconsideration of the BPS Division budget.

SOURCE: Data from Thomas (2023).

Finding 7-10: The BPS program is severely underfunded relative to current need, essentially preventing the development of a truly robust and resilient program that can meet the space exploration science needs of the nation.

Recommendation 7-9: To retire many of the key scientific questions by the end of the decade, NASA should establish support for the Biological and Physical Sciences Program to levels that reflect the current national need and to build the science community in size, diversity of technical expertise and lived experience, and capability to reach the science goals of the nation, toward levels that are an order of magnitude above the current funding and well before the end of the decade.

NASA considered the KSQs as the prioritized guidance for science in the coming decade. These prioritized questions certainly consume the current funding levels. The funding strategies can include larger team projects, individual grants and smaller exploratory de-risking, preliminary data, high-risk/high-reward pilot projects, and/or data mining studies. These prioritized KSQs also scale into much larger projects and many can be addressed partially in the research campaigns, but those campaigns can only be considered after a significant increase in funding.

- *Current funding levels:* All funding of BPS directly addresses the KSQs that are set out in this report as the priority for research in the next decade.
- *Recommended funding levels:* Research campaigns can be pursued only with funding provided well beyond the current funding levels, and research campaigns can be funded outside of and in addition to what is the current science funding strategy within the BPS Division. The KSQs scale into and are addressed by the research campaigns.

Recommendation 7-10: To maintain a viable scientific community, the numerical majority of supported principal investigators (i.e., fraction of research team leaders) should be extramural (i.e., not NASA employees) and funding levels should be commensurate with addressing the key scientific questions.

Recommendation 7-11: NASA should establish periodic reviews of selected research campaigns to ensure coordinated access to the space environment, publicly communicated progress on research milestones, and facilitation of collaborations and public–private partnerships as required to meet these ambitious goals.

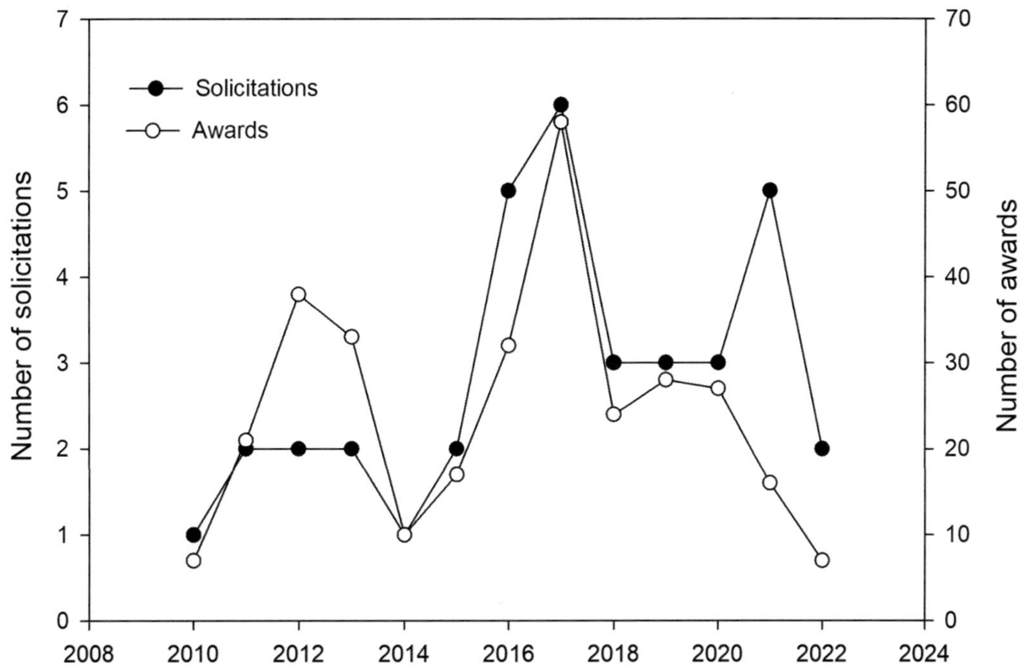

FIGURE 7-6 Number of solicitations (left axis, solid circles) and number of awards (right axis, open circles) from the BPS Division by year. SOURCE: Committee created based on NASA data.

Improving the Reliability of Funding to Improve the BPS Community

Schedule of Funding Opportunity Announcements Need Improvement

Federal agencies sponsoring scientific research have different approaches to proposal submissions. NIH has very specific and regular dates for proposal release, due dates, and awards (three times per year). For NSF, proposals can be submitted year-round. For NASA BPS, a fixed announcement schedule with annual calls would allow investigators to plan and reduce the uncertainty surrounding the current funding model. (See Figure 7-6.) The funding opportunity cycle could also be consistent and realistic to ensure a reasonable interval between opportunity announcement and proposal due date.

Reliability and Resilience to External Setback or Surprise Are Necessary

Multiple funding sources support BPS research. For transformational research, it is critically important not just to have funding, but to have *sustained* funding, with funding mechanisms appropriate for supporting research spanning multiple years and accommodating uncertainties that can lead to delays and setbacks. Delays can be owing to relatively benign reasons (e.g., flight scheduling problems or supply-chain issues), but it is important to develop contingency plans for major disruptions because of external events adversely affecting global research, such as pandemics, disruptive climate events, and war. Decision rules (see Box 7-1) and table-top exercises in contingency planning aid this management practice and can also inform communication plans with the BPS community, whose careers and educations can be disrupted inadvertently without such proactive communication.

As a result of declining overall funds, an unpredictable call for proposal schedule, and extreme sensitivity to delays and setbacks out of the control of investigators, funding for BPS-relevant science is subject to stressors for both lead investigators and more junior members of those potential research teams—particularly compared with other venues and sponsors of BPS research. The funding environment needs to be consistent and adaptable to these stressors to maintain and grow the BPS workforce. Without this consistency, it becomes increasingly difficult to operate a laboratory in this field and, even more importantly, attract new graduate students and postdoctoral researchers to the BPS research community. Stressors well known to the current BPS community that affect continuity of research and researchers, beyond the above findings of historically decreased research expenditures and irregular funding opportunities, include the following:

- *Mission/flight delays can dissolve research teams and dampen scientific impact.* Flight delays can have drastic effects on the funding environment in individual laboratories, particularly when these delays occur over years rather than weeks to months. Students graduate, postdoctorates leave for faculty positions, and investigators retire. Technologies improve and scientific theories change. If these kinds of unintended delays occur, budgets need to be adaptable enough to respond. The BPS community is small, and institutional knowledge is lost when funding requires team members to move on.
- *"Moonshot" projects induce sudden shifts in broader funding.* Drastic changes in research priorities driven by political aspirations can leave entire research disciplines completely out of contention for funding for years. Every federal administration's desire to have its "moonshot" project is understandable and encouraging, as it refreshes engagement with the nation's leaders and the public. A rendezvous with an asteroid, a very fast return to the Moon, or a shift to a Mars mission are all wonderful ideas and often reflect current public interests in space science. But the moonshot ideas historically tend to cause drastic shifts in BPS research priorities, making long-term strategic planning exceptionally difficult, particularly for the investigative teams doing the work. There needs to be enough resources for both moonshot projects and long-term programs essential to sustaining a viable BPS community that can survive these short-term shifts in priorities.
- *Flight experiments can be overly constrained.* Despite the time, effort, and funding put into studies including flight experiments, most studies are "one and done" without opportunity to show repeatability or expand on previous results. This is a significant weakness in the current scientific approach and portfolio, as compared to Earth-based research, where replicate experiments, trials, and peer comparisons are considered necessary to separate spurious from robust findings. Anticipated increases in launch vehicles and flight opportunities can address this constraint, primarily by pursuing ground-based research with a minority of studies pursuing space-based opportunities at any given time to ensure de-risking with ground-based preparation as well as bandwidth for repeated and incrementally expanding flight experiments.
- *Ground-based BPS activities can mitigate stressors.* Ground-based research both complements and enables space-based investigations. For example, detailed studies often need to be performed on the ground to provide calibrations or comparisons to space-based research. Ground-based research could also frequently precede space-based experiments—for example, to establish proof of concept of new scientific ideas or approaches, and to develop and then mature new technologies to the point where they can be considered for later space-based implementation. Ground-based research is also essential to maintaining the healthy cohort of research leaders and trainees needed to generate and execute successful and relevant space missions. These facilities are not as dependent on flight scheduling or partitioning of crew time and attention. These facilities also provide opportunities for synergistic and dual-use research with other agencies wherein shared infrastructure accelerates both sides and shared research might help space and Earth.

SUSTAINING PUBLIC AWARENESS AND SUPPORT FOR BPS RESEARCH AND ITS SOCIETAL BENEFITS

Relative Public Awareness and Investment in BPS Compared with Other NASA and U.S. Government–Funded Research in Space

NASA's BPS program has two key missions. The first is to use the space environment to advance outstanding fundamental science, which will allow us to both survive and thrive in space. The second is to leverage the space environment to perform experiments that cannot be done on Earth to advance fundamental discoveries in science, technology, and space exploration. These dual missions include the general goal of enhancing education, innovation, and economic vitality, and to execute on this the BPS Division currently administers three essential programs: NASA's Space Biology Program,[2] Physical Science Program,[3] and Commercially Enabled Rapid Space Science project (CERISS).[4] This raises the question of the level of relative public awareness and investment in BPS and these programs compared with other NASA (e.g., HRP, STMD, HERO, and HEOMD) and U.S. government–funded (e.g., CASIS, TRISH, and DoD/Space Force) organizations that are engaged in space related activities.

Initiatives such as the GeneLab Data System and LSDA, which promote open science, have been particularly successful in engaging the public and raising awareness of the BPS Division's mission. The Space Biology Program has also focused on understanding how to mitigate adverse changes caused by the space environment by developing novel treatment approaches to promote healing and regeneration (NASA 2023b; White House 2023).[5]

The BPS Division's Physical Science Program has promoted open science via the PSI data repository for experiments performed on the ISS, space shuttle flights, and free-flyers. The PSI data repository, which is accessible and open to the public, provides the opportunity for researchers to further the BPS Physical Science Program mission through informatics.

CERISS is aimed at working with the commercial space industry to dramatically increase the efficacy and pace of research supported by the BPS Division's Research Programs. The mandate of this project is rapidly developing and expanding, including long-term goals such as the development of enhanced capabilities for research on the ISS as well as in commercial LEO and lunar surface projects. Updates regarding ongoing initiatives planned by the commercial space industry (e.g., SpaceX, Blue Origin, and Virgin Galactic) have garnered recent media attention in the popular press.

> Finding 7-11: Articles in newspapers and magazines—including well-known publications such as *The New York Times*, *The Washington Post*, and *Time Magazine*—often focus on the contribution of the commercial space industry to fascinating NASA-based research, rather than the research process and findings.

The BPS Division has launched numerous exciting initiatives resulting in essential research progress over the past decade. It has also made a strong commitment to open science, which is key to improving public awareness. In comparison to the BPS Division, the research discoveries of other NASA research institutions, such as NASA's planetary science and astrophysics programs, are more readily and regularly communicated by the popular press to the public. Furthermore, the expanding commercial space industry also is more readily and regularly reported by the popular press. Indeed, there is relatively limited news coverage of BPS Division–funded research by the press beyond that distributed by NASA's own news service.[6] The increasing level of press focus on commercial space flight is expected given recent growth and progress in this area but can pose a threat to public awareness and investment in BPS Division–sponsored research. Greater public awareness of BPS accomplishments is a shared responsibility of NASA and the research community, which can leverage the positive coverage of NASA by providing BPS-specific content and clear "so what?" descriptions of the societal implications. Social media channels such as LinkedIn, Facebook, Twitter, and Instagram are underutilized as vehicles for sharing outcomes using a focused branding strategy that elevates BPS community achievements.

National Leaders and Inspirers/Influencers Who Participated in or Use Outcomes of BPS Research

Invested and knowledgeable individuals who have participated in and/or use outcomes of BPS research in areas of space biology and physical science research can engage the public in the excitement and significance of BPS

[2] See NASA Science, "Space Biology Program," https://science.nasa.gov/biological-physical/programs/space-biology.

[3] See NASA Science, "Physical Sciences Program," https://science.nasa.gov/biological-physical/programs/physical-sciences.

[4] See NASA Science, "Commercially Enabled Rapid Space Science Project (CERISS)," https://science.nasa.gov/biological-physical/commercial.

[5] See NASA, "2023: Year of Open Science," https://nasa.github.io/Transform-to-Open-Science/year-of-open-science.

[6] NASA science news can be found at https://science.nasa.gov/science-news.

research and discoveries. While space biology research can potentially be more controversial, owing to bioethical considerations (discussed later in this chapter), there are numerous opportunities to engage the public. Individuals who can serve in advocacy roles are essential for sustaining and growing programs and missions such as those of BPS because they serve as conduits between the scientific community and the public at large.

Finding 7-12: Well-known public champions for BPS-focused research are relatively scarce. Internet searches for "space biophysicist," "space biologist," or "space physicist" retrieve very few names of persons who are nationally recognized for their NASA BPS research and who currently serve as prominent advocates for the program.

To sustain public awareness and support for BPS research and its societal benefits, it will be vital to tap into the diversity of individuals who directly participate in the research and mission of BPS, as well as other committed individuals who can serve as champions and influencers to inspire the next generation. To optimally engage the public, champions could include diverse individuals from all career pathways and stages of training such as principal investigators, postdoctoral fellows, PhD and master's graduate students, and undergraduate students who participate in or use outcomes of BPS research.[7] These individuals as well as those leading the research teams can effectively contribute as influencers to advance public awareness of the mission of BPS.

Scientific Meetings and Government Outreach

NASA has historically supported research meetings so that the BPS community can meet to discuss its research and explore new ideas. This is also a common activity for new students to quickly learn about different aspects of the field or to identify future postdoctoral positions. These meetings, such as American Society for Gravitational and Space Research and the NASA Space Radiation Workshop, are critical components of the BPS community.

Finding 7-13: BPS-relevant scientific meetings do not typically engage members of the government in the exchange of information and ideas. This participation gap highlights a missed opportunity for interaction between scientists at all career pathways and government leaders that fosters dissemination of findings and permits real-time response to questions and insights.

Suggested Modes of Local, National, and Global Building of Awareness and Support, Beyond the BPS Research Community

The BPS Division has a strong track record of public outreach through citizen science activities and contributions to building capacity for science, technology, engineering, and mathematics (STEM) education that can garner favorable public support at local, national, and global levels. The BPS Division currently participates in several educationally oriented programs that are designed to broaden participation and knowledge of space sciences. For example, awareness and support is furthered through the Science Activation (SciAct) program, a competitively selected network of teams from across the United States who focus on bringing NASA science to learners of all ages, backgrounds, and communities.[8] The BPS Division participated in the first edition of NASA's new space biology bootcamp in which educators from Historically Black Colleges and Universities (HBCUs) and Minority Serving Institutions (MSIs) were taught the basics of space biology, experimental design, and the technology used for collecting data. Educators also were given instruction and tools for the analysis of real data from GeneLab (Tabor 2022a). These types of activities and practices are essential for building public awareness and support.

NASA is transitioning to an operational model where the agency will rely more on industry and international collaborators rather than directing spaceflight programs. In essence, NASA will become a "customer" and BPS

[7] See NASA Taskbook, "The NASA Task Book," https://taskbook.nasaprs.com/tbp/welcome.cfm.

[8] See NASA Science, "Science Activation (SciAct) Program," https://science.nasa.gov/learners.

researchers will need to adapt to this new scenario by strengthening ties with global partners and communities (NASA 2022a). The Biological and Physical Sciences Open Data Initiative is a major strength of the BPS research portfolio around which to grow global support.[9] BPS crop research in space is a major attractant of global attention and support and is highly valued by the United Nations, in part because of its relevance for agriculture in resource-limited environments (Agronet 2022; Ripples Nigeria 2019; *Times of India* 2021; United Nations 2023).

Finding 7-14: BPS research elicits national and global awareness through participation in NASA's educational outreach activities and its citizen science projects, in which interested members of the public around the world can collaborate with BPS researchers. While the BPS national engagement portfolio already comprises student learning opportunities, internships, and fellowships, these could be better communicated to the public. BPS programs are valued globally and have benefited from international collaborations. BPS is well positioned to adapt to NASA's new strategic approach by leveraging the strengths of commercial and international partners to maximize outcomes of funded research.

NASA BPS can build on this public outreach foundation to cultivate global awareness and support for research activities. The Biological and Physical Sciences Open Data Initiative can be advertised widely and tapped for programmatic impact. GeneLab's reach can be further expanded by engaging more of the public through initiatives such as citizen science, crowdsourcing, observatory science, and real-time science. The Spaceflight Technology, Applications, and Research (STAR) program for principal investigators, senior research scientists, and postdoctoral scholars has a global reach and can serve as a model for engaging international scientists in BPS research.[10] This can also encourage participation of international scholars and citizens in BPS research needs in an appropriate, mutually beneficial, and cooperative manner. Last, this public outreach over the first 5 years can build momentum and partnerships for future BPS research campaigns at the mid-decadal timepoint. Like the decision rules (see Box 7-1) that anticipate surprise and manage outcomes in the research program itself, the next 5 years of this decadal time horizon must build momentum in public awareness and appreciation of practical impact provided by BPS research, and build sensitivity to external shocks that affect the BPS research community.

Recommendation 7-12: NASA should identify mechanisms to compete new or additional research campaigns within 5 years, in light of anticipated changes to access to low Earth orbit and the inevitable but unknown changes in research, technology, funding, and space mission directives that will ensue after this report is issued.

Stability of the Scientific Community, Including Stability of Disciplines or Subdisciplines

According to the NSF 2022 Science and Engineering Indicators, the U.S. share of global research and development (R&D) remained high at 27 percent in 2019 (NSF 2022c). The United States is a global leader in the publication of high-impact articles, as measured by citations (NSF 2022b, Figure 22). Between 2000 and 2018, U.S. authors contributed almost double the number of highly cited articles when compared with their overall publication output. U.S. publications with international authors numbered 40 percent in 2020, a twofold change from 2000 and underscoring the collaborative spirit of U.S. S&E (NSF 2022b). Patents are indicators of innovation, and the United States is strong in this area. In 2020, U.S. inventors were granted 164,000 patents by the U.S. Patent and Trademark Office, a 53 percent increase from 2010 (NSF 2022b, Figure 25). The United States accounts for 23 percent of the world's scientific R&D services industry, whose purpose is to increase basic knowledge and to apply research findings to develop products in engineering, physics, and life sciences. However, between 2010 to 2020, the U.S. share of international patents declined from 15 percent to 10 percent, while China's share of international patents increased to 49 percent, a threefold gain from their share in 2010 (NSF 2022b, Figure 25). Moreover, the U.S. proportion of R&D growth has been declining over the past 2 decades. (See Figure 7-7A.)

[9] See NASA Science, "Data: Biological and Physical Sciences Open Data," https://science.nasa.gov/biological-physical/data.
[10] See NASA Science, "NASA's STAR Program," https://science.nasa.gov/biological-physical/programs/star.

FIGURE 7-7 (A) Shares of worldwide research and development (R&D) expenditures, by selected region, country, or economy: 2000, 2010, and 2019. (B) National R&D intensity, expressed as gross R&D investment as a percentage of gross domestic product, whereby the United States currently ranks fourth. NOTE: OECD, Organisation for Economic Co-operation and Development. SOURCES: (A) Courtesy of National Science Foundation; NCSES, National Patterns of R&D Resources; OECD, MSTI March 2021 release; UNESCO, UIS, R&D dataset. (B) Courtesy of American Association for the Advancement of Science, https://www.aaas.org/sites/default/files/2023-05/AAAS%20Global%20RD%20Update%20April%202023.pdf.

As of this report, the United States ranks only fourth among nations in R&D intensity, expressed as percentage of R&D investment relative to gross domestic product (Zimmermann 2023). According to Organisation for Economic Co-operation and Development (OECD) data reported by the American Academy of Arts and Sciences (AAAS), the annual growth in U.S. R&D expenditures since 2000 has been eclipsed significantly by several nations, and that rate of increase is notable in Korea, Israel, and China. (See Figure 7-7B.)

STEM Degrees and the BPS Workforce

The BPS research workforce will continue to require advanced education in STEM, and must compete with other inspiring fields and careers for a limited pool of U.S. STEM talent. In 2021, more than 90 percent of 25- to 64-year-olds in the United States had completed a high school degree, making the United States among the top 10 countries in educational attainment.[11] Overall, it is encouraging that there has been strong growth in the STEM workforce in the United States in the past decade. Notably, following the COVID-19 pandemic, the Bureau of Labor Statistics predicts continued growth in the next decade in disciplinary sectors such as biochemistry/biophysics (15 percent), physics/astronomy (8 percent), and data science (36 percent) that will contribute new talent to the BPS enterprise.[12] However, it is notable that in terms of the research workforce, the United States has the third largest workforce globally, but as of 2023 ranks only 18th in researchers as a share of the total workforce compared with other countries.

This STEM pool will be increasingly diverse in ethnic/racial demographic composition in the coming decade. While this is a statement of fact based on recent census data (e.g., showing increasing representation of Hispanic White and Black residents in the under age 15 population in the United States [Brookings 2019]; with Hispanics accounting for more than a quarter of this demographic and projected to comprise 35 percent of U.S. children by 2050 [Passel and Cohn 2008]), BPS can use this awareness to prepare intentionally to broaden its community of perspectives. Inclusive teams of diverse lived experiences are documented to also be creative, problem-solving teams in many work contexts. The associate's (AA) degree conferred by community colleges is an increasingly sought after and affordable educational option for those who seek entry to STEM professions and especially for those who come from under-resourced and historically underrepresented groups in STEM (NSF 2022d). Student achievement of bachelor's, master's, and doctoral degrees in the natural sciences and mathematics disciplines central to the BPS mission has remained steady over the past decade for doctoral degrees, and increased for bachelor's and master's degrees (Department of Education 2022).

Although these trends in STEM-seeking students in the United States across wide socioeconomic and other population descriptor backgrounds are promising, they obscure underlying disparities in the achievement of STEM degrees and entry to STEM professions by women and historically marginalized populations (Department of Education 2022). For example, a snapshot analysis of persons earning STEM degrees and certificates in 2020–2021 that considers race and ethnicity shows that men outnumbered women across all racial and ethnic groups, with Black women achieving closest parity and White women least as compared with their male peers. (See Table 7-2; nonbinary gender was not a reported category in that statistical digest.) Moreover, Blacks, Hispanics, Native Americans, and Pacific Islanders were underrepresented in bachelor's degree achievement in comparison to their proportion of the population (13.6 percent, 18.9 percent, 1.3 percent, 0.3 percent) (U.S. Census Bureau 2023).

For decades, the United States has led the world in international education and served as a magnet for the more than 900,000 international students seeking U.S. degrees (Department of State 2023). In 2019, about 36 percent of master's of science and engineering degrees and 33 percent of science and engineering doctorates were earned by students on temporary visas (NSF 2022d). However, the appeal of a U.S. degree for international students lessened since 2019, owing not only to the pandemic transient but also to the rising cost of a U.S. degree in the context of greater study options in other countries.[13]

Short-term prospects for a well-educated BPS workforce are promising. Long term, the projections for a BPS-ready STEM workforce are more uncertain, owing to persistent gaps in STEM degree achievement by historically underrepresented groups. Graduate programs have relied on a strong international presence that is expected to decline, leaving graduate programs with fewer students to sustain research and teaching in S&E academic departments.

---

[11] See U.S. Census Bureau release of educational attainment data (2022), https://www.census.gov/newsroom/press-releases/2022/educational-attainment.html, and World Population Review of most education countries (2023), https://worldpopulationreview.com/country-rankings/most-educated-countries.

[12] See Bureau of Labor Statistics, "Occupational Outlook Handbook," https://www.bls.gov/ooh.

[13] See K. Fisher and S. Aslanian, "Fading Beacon," https://www.apmreports.org/episode/2021/08/03/fading-beacon-why-america-is-losing-international-students.

TABLE 7-2 Percentage Distribution of Science, Technology, Engineering, and Mathematics (STEM) Degrees (All) Certificates Conferred by Postsecondary Institutions, by Race/Ethnicity and Sex of Student: Academic Year 2020–2021

Race	Males	Females
White	33.0%	15.7%
Black	4.5%	3.1%
Hispanic	8.2%	4.7%
Asian/Pacific Islander	7.0%	4.7%
American Indian/Alaska Native	0.3%	0.1%
Two or more races	2.2%	1.4%
Nonresident	10.0%	5.2%

SOURCE: Based on data from Institute of Education Science, National Center for Education Statistics, 2022, "Table 318.45. Number and Percentage Distribution of Science, Technology, Engineering, and Mathematics (STEM) Degrees/Certificates Conferred by Postsecondary Institutions, by Race/Ethnicity, Level of Degree/Certificate, and Sex of Student: Academic Years 2011–2012 Through 2020–2021," https://nces.ed.gov/programs/digest/d22/tables/dt22_318.45.asp.

BUILDING AND SUSTAINING A DIVERSE AND ETHICAL U.S. SCIENCE COMMUNITY IN GLOBAL CONTEXT

National Competitiveness and Advancement of Human Knowledge for Society, as Motivators of Diverse and Ethical U.S. BPS Community

Sustaining national S&E leadership and advancement of human knowledge for a thriving and economically robust society serve as key motivators for the growth of a diverse and ethical U.S. BPS community. As the BPS mission continues to expand, many rapidly developing considerations pose ethical challenges that need to be addressed. While space exploration is motivated by scientific discovery, it also is motivated by resource extraction and allocation, geopolitical and strategic considerations, and the need to ensure the public trust. These issues are the focus of *space ethics*—a field that studies the ethical, legal, and societal implications of space exploration and addresses questions such as: "Even if we can do certain things in space, should we?"

> In a highly competitive global science and engineering environment, the U.S. must stay on the leading edge of the practice of S&E, improving the research environment and setting the standard for ethics and values.
>
> —Dr. Ellen Ochoa[14]

International in scope, space ethics encompasses perspectives from diverse disciplines such as philosophy, astronomy, environmental studies, history, law, psychology, religious studies, and sociology. Space ethics is a field of study that deals with the moral and ethical issues that arise in the exploration and utilization of outer-space exploration with the goal of ensuring that activities are conducted responsibly and in a sustainable manner and guided by ethical principles and values. Some of the key issues that are addressed in space ethics include

- *Resource utilization:* How to use and distribute the resources that are found through space exploration fairly and ethically.
- *Environmental protection:* Considerations of the potential impact of human space exploration on the environment in space and on other planets and the potential impact of bringing extraterrestrial materials back to Earth.

[14] Quote from Ellen Ochoa, former NASA astronaut and former director of Johnson Space Center; also first Hispanic woman to go to space, on Space Shuttle *Discovery*, National Science Foundation, https://www.nsf.gov/nsb/publications/2021/nsbct042821.pdf.

- *International cooperation:* How to fairly and ethically share the benefits and responsibilities of cooperative endeavors among many countries.
- *Property rights:* Who has the right to claim ownership of resources, and what criteria would be used to determine these claims.
- *Human rights:* What rights and protections would be afforded as humans begin to travel, live, and work in space and on other planets.
- *Benefits to humanity:* How to ensure that communities from all over the world can use and benefit from space exploration research, technology, and applications.

Space exploration in the next decade will be undertaken in a landscape of conflicting national and corporate interests. More than 70 space agencies exist around the world, dominated by nations that are technologically advanced or with a well-financed military presence, and more are predicted to join this group. Moreover, private space entrepreneurship is expected to grow, and companies such as SpaceX and Blue Origin are developing space tourism and venturing into R&D. The legal aspects of space exploration are fueling the emergence of the field of *space law*, which comprises international treaties and principles governing outer space such as those developed under the aegis of the United Nations and overseen by the United Nations Office for Outer Space Affairs (UNOOSA). While there are existing institutions that consider ethical, legal, and social implications of space exploration, their abilities to enforce compliance are currently limited. Ethical conduct will depend on international cooperation and cultivating dialog, discussion, and articulation of shared values and mutual respect.

In light of recent economic and social upheaval and growing economic difficulties, public support for space exploration is mixed (Foust 2019; Konicki and Pethokoukis 2022; Sabin 2021). It is critical that BPS (and NASA) practices and policies be in place that increase the public trust and ensure that space exploration proceeds in a transparent manner where the benefits are prominently shared. To this end, the public needs to be engaged well in advance on potentially controversial matters such as the use of animals in space research, the environmental consequences of LEO satellites and associated space debris, and the alteration of habitats on other planets via mining and natural resource extraction.

An ethical BPS workforce is essential to fulfill the BPS mission in support of national competitiveness and advancement of human knowledge for society. However, BPS Division guidelines for responsible conduct in research were not easily accessible on the BPS/NASA website.[15]

Workforce or Collaboration Weaknesses at Present or Anticipated

A robust BPS STEM workforce is vital to develop innovative breakthroughs and engage in space-related activities that will grow the economy and uphold national leadership during the next decade. Global circumstances such as the COVID-19 pandemic, climate change, and economic instability pose challenges to the growth and strength of the STEM workforce. The U.S. STEM enterprise is global and benefits from the contributions of foreign-born workers who comprised almost 20 percent of all STEM workers in the United States.

Increased unemployment and the chance of a global recession have caused economic disruptions and raise uncertainty about prospects for STEM enrollment in higher education and the development of a thriving STEM business sector (Langin 2022; OECD 2021). During the COVID-19 pandemic years between 2019 and 2021, the unemployment rates for STEM and non-STEM occupations increased for most demographic groups. Although unemployment for the STEM workforce remained lower than for non-STEM workers during the COVID-19 pandemic, the economic impact was felt more keenly among historically underrepresented groups. Black and Hispanic STEM workers experienced the highest rate of pandemic unemployment, while White and Asian STEM workers experienced the lowest unemployment rates.[16]

[15] This paragraph was changed after release of the report to the sponsor to remove a description of guidelines for which no source could be found.

[16] National Science Foundation, "STEM Unemployment," https://ncses.nsf.gov/pubs/nsf23315/report/stem-unemployment.

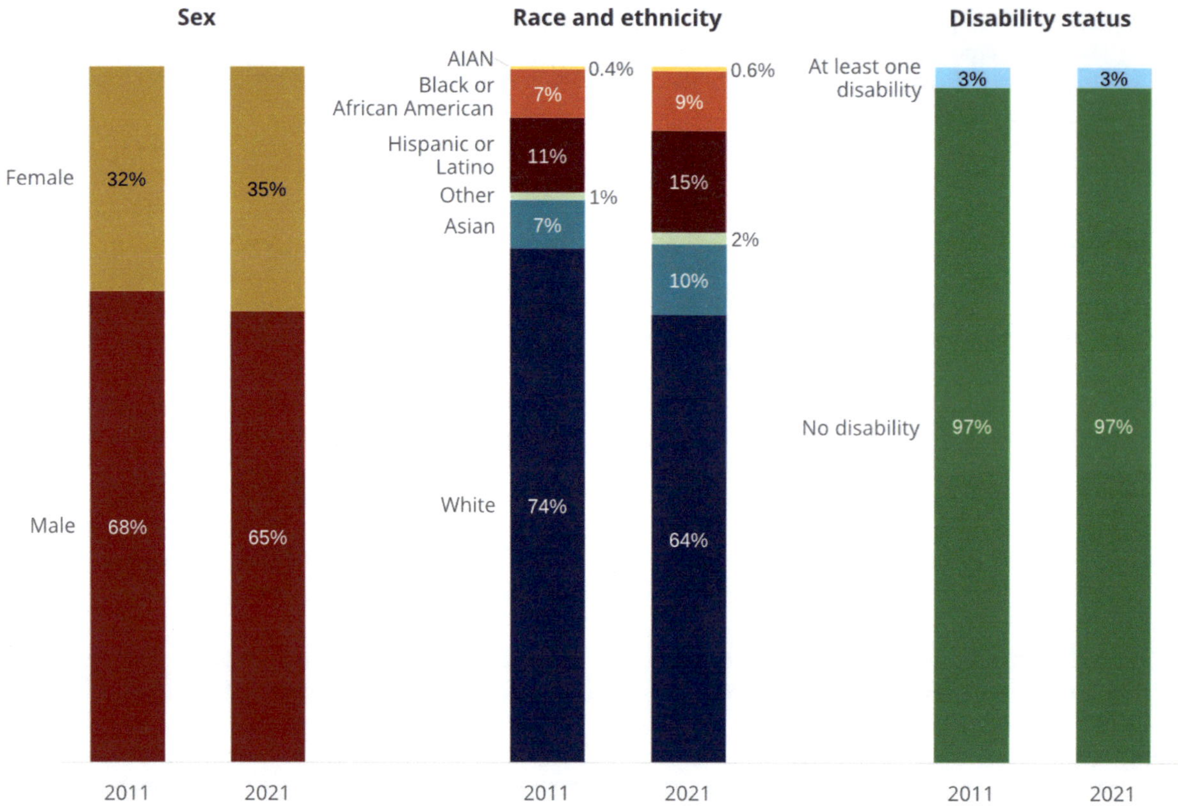

FIGURE 7-8 Characteristics of the STEM workforce ages 18–74: 2011 and 2021. SOURCE: Courtesy of Census Bureau, Current Population Survey, Annual Social and Economic Supplement, 2021.

Disparities in Participation and Earnings in the STEM Workforce

In 2021, as the global population emerged from the pandemic, nearly a quarter (24 percent) of the U.S. workforce was employed in STEM occupations.[17] Between 2011 and 2021, the number of women and historically underrepresented minorities increased within the STEM workforce but lagged in comparison to their representation. Even though women are half (51 percent) of the total U.S. population, they comprise only a third (35 percent) of persons employed in STEM. (See Figure 7-8.) Combined, historically underrepresented minorities held about 25 percent of STEM jobs but were disproportionately a higher share of the skilled technical workforce than of STEM workers with at least a bachelor's degree.[18] Persons with disabilities comprise 26 percent of U.S. adults (CDC 2023) but only 3 percent of the 2021 STEM workforce (Figure 7-8). The STEM workforce also is characterized by pronounced demographic disparities. In 2020, women in the STEM workforce had lower median earnings than men in S&E, S&E-related, and middle-skill occupations, and historically underrepresented minorities had lower median earnings as compared with White or Asian STEM workers.[19] Across all groups, persons with a bachelors' degree or higher earned significantly higher wages. (See Figure 7-9.)

[17] National Science Foundation, "The STEM Workforce," https://ncses.nsf.gov/pubs/nsf23315/report/the-stem-workforce.

[18] National Science Foundation, "STEM Occupation," Figure 3-3, https://ncses.nsf.gov/pubs/nsf23315/report/stem-occupations#educational-attainment-of-the-stem-workforce.

[19] National Science Foundation, "STEM Median Wage and Salary Earnings," Figure 4-1, https://ncses.nsf.gov/pubs/nsf23315/report/stem-median-wage-and-salary-earnings.

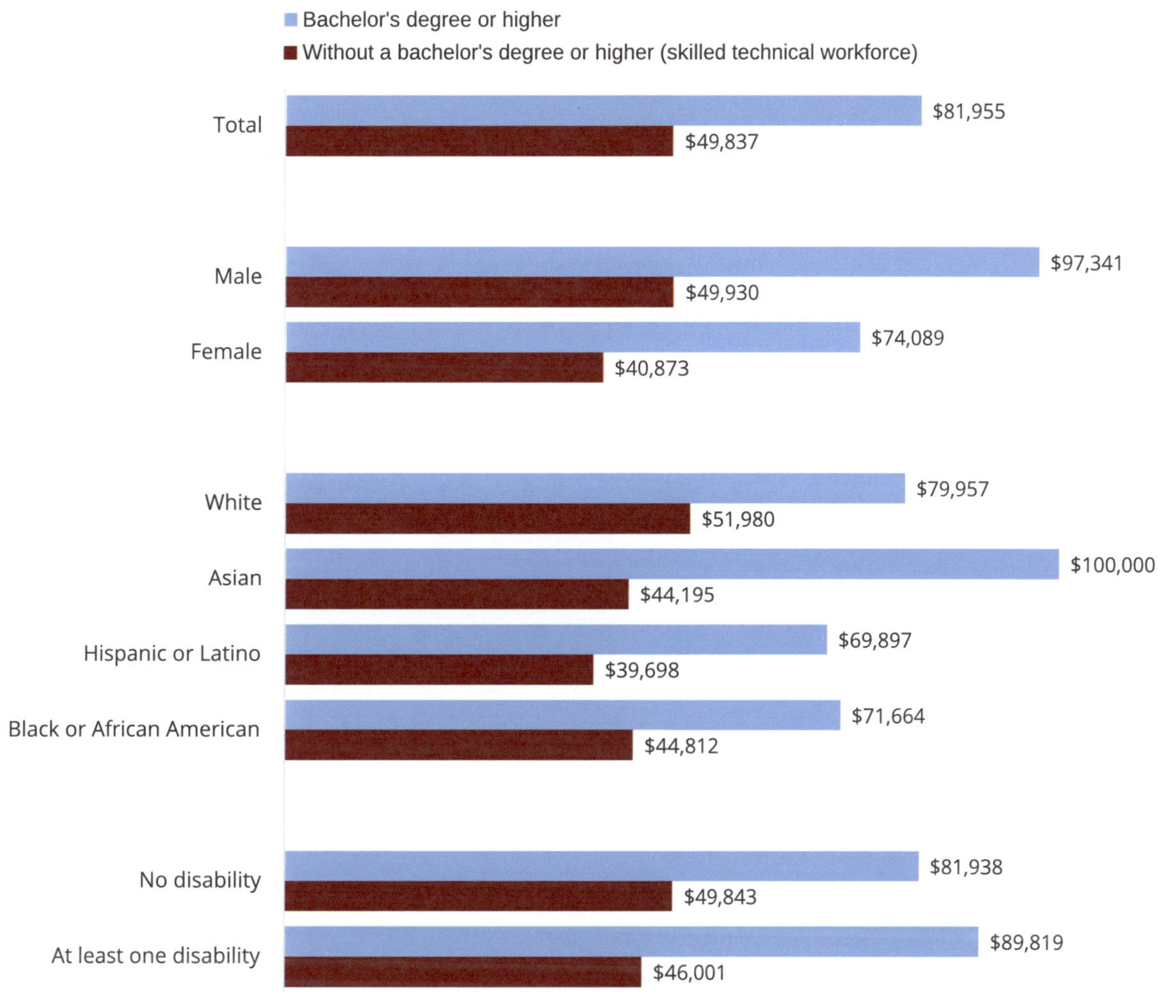

FIGURE 7-9 Median wage and salary earnings of the science, technology, engineering, and mathematics (STEM) workforce ages 18–74 by education and by sex, ethnicity, race, and disability status: 2020. SOURCE: Courtesy of NSF NCES; Census Bureau, Current Population Survey, Annual Social and Economic Supplement, 2021.

The U.S. STEM workforce is characterized by decades of persistent disparities in STEM degree achievement among demographic groups.[20] In 2019, historically underrepresented groups received a disproportionally lower share of all degrees with one exception: Hispanics achieved a greater share of AA degrees in comparison to their representation in the general population. (See Figure 7-10.)

Although women earn half or more of overall higher education degrees at each degree level, their degree achievement in S&E fields is lower in comparison with male peers, with very large disparities between men and women in fields such as engineering, computer science, math, and physical science (NSF 2022a). S&E degree achievement is influenced by the complex intersectionality of discipline, race/ethnicity, and gender. (See Table 7-2.) For example, across disciplines, women earn a lesser share of STEM bachelor's degrees in fields such as mathematics and computer sciences, while within a discipline such as physical sciences, women of

[20] See HED-16, https://ncses.nsf.gov/pubs/nsb20223/demographic-attributes-of-s-e-degree-recipients.

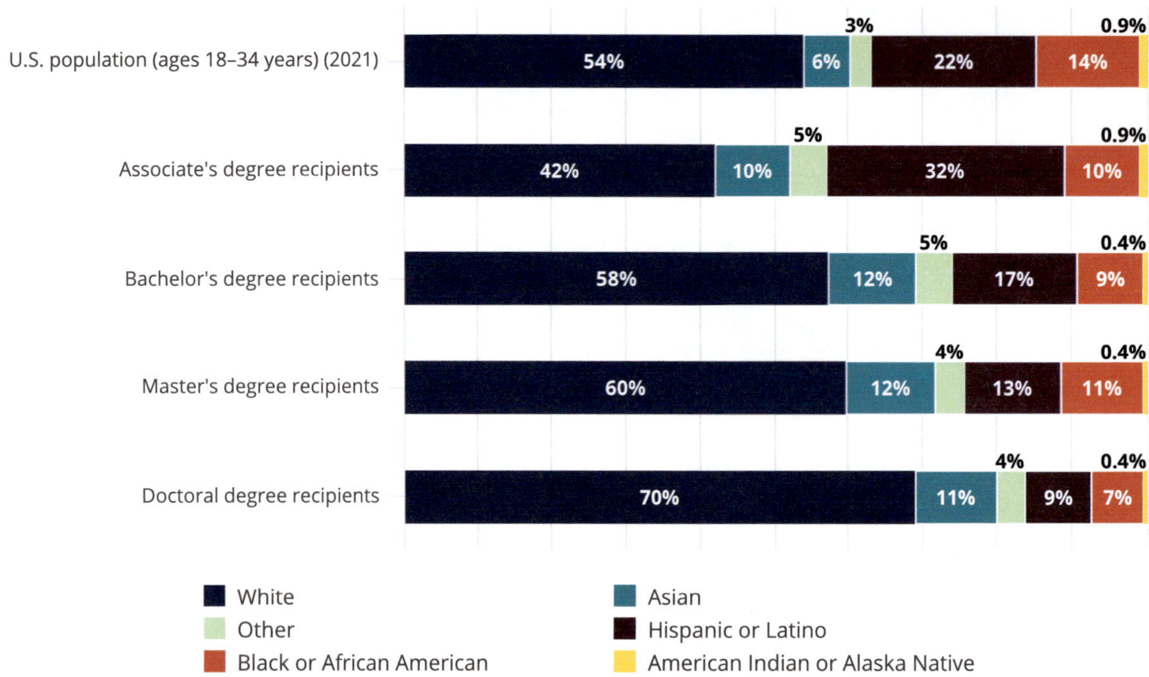

FIGURE 7-10 U.S. population ages 18–34 and science and engineering degree recipients: 2020. SOURCE: Courtesy of National Center for Science and Engineering Statistics, special tabulations (2022, Table Builder) of the National Center for Education Statistics, Integrated Postsecondary Education Data System, Completions Survey, provisional release data. Population data from Census Bureau, Current Population Survey, Annual Social and Economic Supplement, 2021.

certain race/ethnic backgrounds may earn a greater or lesser share of degrees than male peers or members of other racial/ethnic groups.

Growing the Next Generation of Space Science Researchers

A strong research training program for students at all levels (for flight- and ground-based research) would increase research productivity and also provide a trained workforce for the future. In the mid-1990s, the plant space biology program provided such training grants—the NASA Specialized Centers of Research and Training (Spooner and Guikema 1992). These large grants provided inter-, multi-disciplinary approaches to fundamental questions of interest to BPS. They provided funding for multiple principal investigators (6 to 12) and supported multiple postdoctoral researchers, graduate students, and undergraduates. The awards also required significant outreach to K–12 teachers and students and the general public. NIH provides an example of a tiered training grant program that strengthens and supports the next generation of biomedical researchers. (See Table 7-3.) Such approaches for multi-year, mentored support of graduate students and postdoctorates is considered best practice for momentum, quality, and mentorship of junior researchers; other models such as NSF Research Training Groups were considered (NRC 2009) and found beneficial to education goals but also less aligned to the needs of growing an inclusive research community sustainably.

The need for professionals with STEM expertise will remain high during the next decade, and the advancement of a well-educated space science workforce in support of the BPS mission presents challenges. Space exploration requires well-trained scientists, engineers, and technicians. Demand may outpace the ability of the current STEM workforce to fill positions, and those already employed in STEM and non-STEM fields may need to reskill or specialize to remain competitive. It is essential to eliminate the demographic disparities in STEM wages and education that

TABLE 7-3 Example of Tiered National Institutes of Health Training Grants

Training Grants		Amount per Recipient	Duration
F30	Individual predoctoral National Research Service Award for MD/PhD fellowships. Individual fellowships for predoctoral training that leads to the combined MD/PhD degrees.	Award budgets composed of stipends, tuition and fees, and institutional allowance	4 years
F32	Provides postdoctoral research training to individuals, to broaden their scientific background and extend their potential for research in specified health-related areas.	$30,000–$100,000 annually	2–4 years
T32	Enables institutions to recruit individuals for predoctoral and postdoctoral research training in specified shortage areas. Prepares qualified predoctoral and/or postdoctoral trainees for careers that have a significant impact on the related research needs of the nation.	$30,000–$100,000 annually	Up to 5 years, renewable

SOURCE: Based on data from the National Institutes of Health.

presently characterize the U.S. STEM workforce. Moreover, the long-term impact of the current political landscape on the nation's ability to retain and recruit foreign-born workers who are vital contributors to science and engineering innovation is uncertain, and their continued participation needs to be ensured. BPS needs individuals with the necessary skills and expertise to fill positions and undertake research in BPS-relevant space science fields. For this reason, it is essential to recruit, train, and retain diverse talent within all career pathways and degree levels. NASA is a participating agency in the *OSTP's Interagency Roadmap to Support Space-Related STEM Education and Workforce* (NSTC 2022), which provides guidance for strengthening space-related STEM education and developing a workforce strategy in support of space R&D. This momentum is supported by the National Academies' consensus study report published in 2023, *Foundations of a Healthy and Vital Research Community for NASA Science*.

Recommendation 7-13: NASA should ensure diversity, equity, inclusivity, and accessibility in the pursuit of the nation's space exploration science priorities, including instituting a requirement of documented progress in diversity among NASA-sponsored research teams seeking multi-year funding or multiple sponsorship requests over the coming decade. This inclusivity should be intentionally broad in concept, with respect to visible and less visible characteristics of historically underrepresented groups in biological and physical sciences research and leadership.

Recommendation 7-14: Project grants should be funded at levels and duration consistent with the project aims with full support for trainees (postdoctorates, graduate students, and undergraduates), including travel for trainees and principal investigators to support the mission and participate in scientific meetings. Full funding representing the total costs of research (direct and indirect) is imperative to be inclusive of participation by all trainees.

Diversity in Many Forms, Including Participation in BPS Research and in Space Missions That Carry Out BPS Experiments or Space Exploration

Inclusion, Diversity, Equity, and Accessibility at NASA

Diversity of representation and of thinking is an undisputed foundation for innovation and a recognizable feature of a resilient, equitable, inclusive, and accessible STEM environment (Campbell et al. 2013; Freeman and Huang 2014a; NASEM 2023). Indeed, the National Academies have issued many reports advocating for more diverse, equitable, and inclusive STEM that are maintained as a collection of resources.[21] NASA maintains an

[21] National Academies of Sciences, Engineering, and Medicine, "Diversity, Equity, and Inclusion Collection of Reports," https://nap.nationalacademies.org/collection/97/diversity-equity-and-inclusion.

BOX 7-3
Inclusion, Diversity, Equity, and Accessibility Working Group

SMD [NASA Science Mission Directorate] is actualizing an environment where IDEA principles are integrated into our ways of working and being. We recognize the barriers and obstacles that historically marginalized groups face societally, in the workplace, and in science. We are committed to relieving these inequities with boldness and forward-learning practices—for systemic and lasting change. We value every SMD team member for their diversity of thought, background, and whole selves. And, larger still, we desire that all science funded by NASA, regardless of location, is comprised of teams that reflect the nation.

—Tresa Mitchell and Eric Holmes, IDEA Working Group Co-Chairs

Office of Diversity and Equal Opportunity that oversees diversity and civil rights policies and services and supports several special emphasis programs.

In 2022, NASA leadership participated in a national taskforce that issued a roadmap with goals for interagency-wide action to build STEM education and the STEM workforce (NSTC 2022). The roadmap specified as the third goal to: "[e]mploy a diverse workforce to bring the benefits of space to more communities." This same year, NASA issued policies and practices as part of a strategic plan that stipulates NASA's aspirations for mission directorates to embed diversity, equity, inclusion, and accessibility (DEIA, used interchangably within and beyond NASA as IDEA; see Box 7-3) into all interactions with their workforce and contractors (NASA 2022d,e). (See Figure 7-10.) The NASA strategic plan specifically defines DEIA (see Table 7-4) and the term "underserved communities" that is also referred to in the economic development sphere as socially or economically disadvantaged individuals

TABLE 7-4 NASA Definitions for Diversity, Equity, Inclusion, and Accessibility for a Shared Understanding of These Terms

Term	NASA Definition
Diversity	The entire universe of differences and similarities.
Equity	The consistent and systematic fair, just, and impartial treatment of all individuals, including individuals who belong to underserved communities.[a]
Inclusion	The full participation, belonging, and contribution of organizations and individuals.
Accessibility	Providing accommodations and modifications to ensure equal access to employment and participation in activities, eliminating and reducing physical barriers to promote equitable opportunities, and ensuring that every outward-facing and internal activity or electronic space can be accessed by every person independently.

[a] The term "underserved communities" refers to populations sharing a particular characteristic, as well as geographic communities, that have been systematically denied a full opportunity to participate in aspects of economic, social, and civic life. Underserved communities include the following: people with disabilities; LGBTQIA+ individuals; individuals with limited English proficiency; older adults; people of color; individuals in rural communities; first-generation professionals, college students, and immigrants; formerly incarcerated individuals; persons adversely impacted by persistent poverty, discrimination, or inequality; women; individuals facing religious discrimination; veterans and military spouses; and parents, caregivers, and individuals facing pregnancy discrimination. They have been historically denied full opportunity including Blacks and African Americans; Hispanics and Latinos; Indigenous, Native American, and Native Alaskan persons; Asian Americans, Native Hawaiians, and Other Pacific Islanders; other persons of color; members of religious groups; lesbian, gay, bisexual, transgender, queer/questioning, intersex, and asexual (LGBTQIA+) persons; persons with disabilities; persons who live in rural areas; and people otherwise adversely affected by persistent poverty or inequality.

SOURCE: NASA, 2021, *Fiscal Years 2022–2026, NASA Strategic Plan for Diversity, Equity, Inclusion, & Accessibility*, Washington, DC: Office of Diversity and Equal Opportunity, https://www.nasa.gov/sites/default/files/atoms/files/nasa_deia_strategic_plan-fy22-fy26-final_tagged.pdf.

(SEDI); this plan focuses on four goals that are consistent with this decadal survey's recommendations regarding (1) workforce diversity; (2) workforce equity and inclusion (employee experience); (3) accessibility and accommodation; and (4) DEIA integration into the NASA mission.

Ensuring Achievement of a Goal Requires Accountability Standards and Progress Measurements

To this end, NASA provides a yearly open access report that summarizes agency-wide progress toward workforce diversity using the federal STEM workforce and Relevant Civilian Labor Force (RCLF) benchmarks as a comparison group (NASA 2021). The report, however, does not provide disaggregate data for SMD nor for BPS.

Although the NASA 2021 report provided evidence that NASA met or exceeded the RCLF averages for most STEM occupations, the report also flagged a need to ensure that underrepresented and underserved communities have the opportunity and are encouraged to pursue careers in STEM, as well as experience equity and inclusion in the NASA and STEM workforce. The report noted that Asian Americans and Pacific Islanders (AAPI) and Hispanics account for a lower percentage of the Senior Executive Service (SES) compared to their overall representation in the NASA workforce, and that Blacks, Hispanics, and women make up a smaller proportion of both ST (scientific or professional) and SL (senior level) positions than their overall representation in the workforce.

Diversity Equity, Inclusion, and Accessibility at SMD/BPS

SMD, which oversees the BPS Division, maintains a diversity unit that produced SMD's own IDEA-focused strategic plan in February 2022 and produced a report that is openly accessible.[22] SMD proposes to promote DEIA through many potentially transformational activities and practices such as the initiation of an SMD Bridge program to partner with MSIs; a PI Launchpad that will support first-time proposers; adoption of a dual anonymous peer review; codes of conduct for mission teams; and collection, evaluation, and publication of demographics of Research Opportunities in Earth and Science Program (ROSES) proposers and awardees.[23] SMD also is promoting GEM Fellowships for underrepresented minorities in STEM graduate programs, and maintains a page that disseminates space research opportunities for the high school students, graduates, and postdoctorates.[24] However, the DEIA strategic plan does not state whether diversity and representation in SMD research will be tracked, nor does the Science by the Numbers portal provide disaggregated information specific for BPS research. (See Figure 7-11.)

Finding 7-15: The BPS program and the community it supports are increasingly diverse and inclusive, yet work still needs to be done to ensure momentum and broader participation at all levels of the research enterprise.

Finding 7-16: Supporting workforce development is necessary to sustain a strong research community and to maintain a vigorous research program. Historical disparities in the achievement of STEM degrees propagate throughout the STEM enterprise and lead to persistent representation gaps in research and organizational leadership by women, historically marginalized groups, and persons with disabilities. The gaps are likely to be greater than currently perceived because data have not been consistently collected for many groups; there are little or no data for groups such as those who self-identify as LGBQT+, nor do data consistently capture how intersectionality affects participation and outcomes. Military veterans and returning adult learners are untapped higher education constituencies who often are overlooked and who would bring valuable life and work experiences to space exploration missions.

The 2022 National Academies report *Advancing Diversity, Equity, Inclusion, and Accessibility in the Leadership of Competed Space Missions* (NASEM 2022) provides a deep analysis and offers specific recommendation

[22] NASA, "NASA Inclusion, Diversity, Equity, and Accessibility at SMD," https://science.nasa.gov/about-us/idea, and NASA Science Mission Directorate Inclusion, Diversity, Equity, and Accessibility, https://science.nasa.gov/science-red/s3fs-public/atoms/files/SMD_IDEA_Annual%20Report_2022.pdf.

[23] NASA, "NASA Inclusion, Diversity, Equity, and Accessibility at SMD," https://science.nasa.gov/about-us/idea.

[24] See NASA Science Continuing Opportunities, https://science.nasa.gov/learners/learner-opportunities.

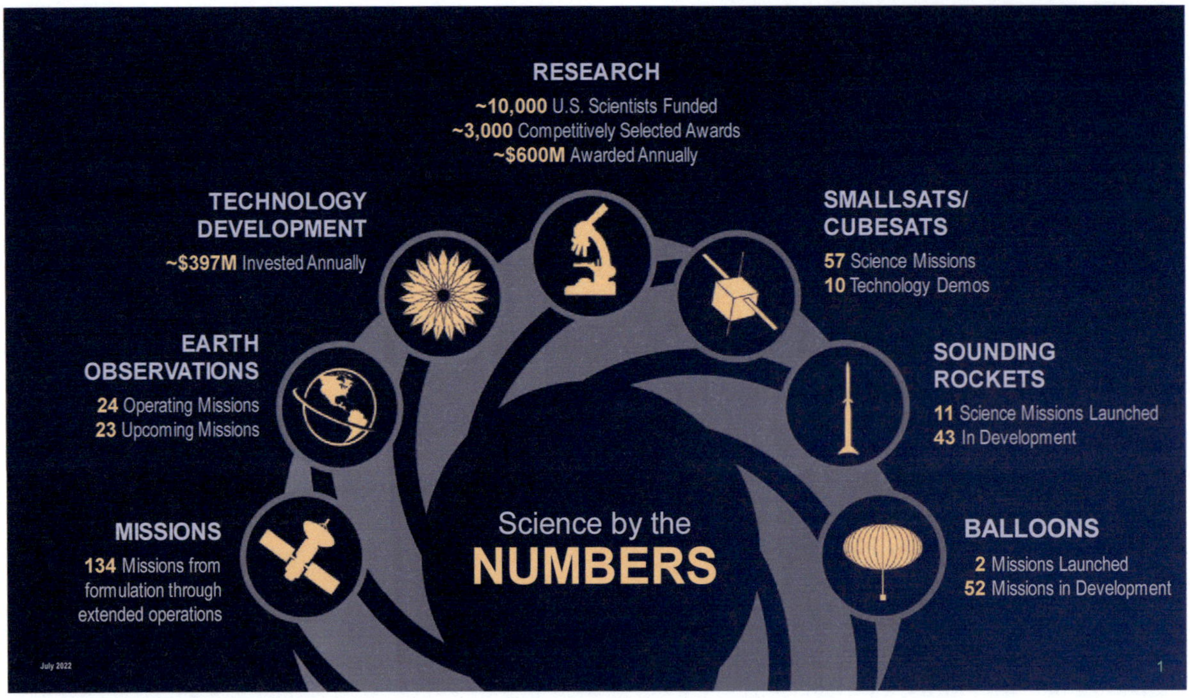

FIGURE 7-11 Science by the numbers. SOURCE: Courtesy of NASA Science Mission Directorate.

to SMD on how to make leadership of space missions more accessible, inclusive, and equitable. Although focused on the Heliophysics, Astrophysics, Planetary Science, and Earth Science divisions, the report recognized that most of the recommendations could be applied throughout the agency to lead to a more diverse workforce in all NASA-funded research and training programs.

> Finding 7-17: BPS is a DEIA leader within NASA owing to its record of embedding DEIA into its policies and programs. Noteworthy examples include an established commitment to open science, a broad portfolio of outreach and training activities such as GeneLab and the STAR Program, and development of its own DEIA strategic plan. As NASA's flagship program in basic sciences with an interdisciplinary focus, BPS is well-positioned to recruit from a wide range of individuals with broad interests in STEM who can be recruited to space exploration enterprises in academic and commercial sectors. Outcomes of such commitment can be substantiated by longitudinal data collection, regular periodic and public sharing of those data, and leadership-supported celebration of more inclusive research workplaces and communities.

OUTLOOK

The coming decade of biological and physical science research in space environments is one in which the United States has the opportunity to thrive, lead, and foster positive collaborations with industry and international partners. This future is not guaranteed, especially given the decade of tenuous rebuilding of BPS research teams, platforms, and protocols. The investment in the BPS research community and infrastructure within and beyond NASA that is recommended in this decadal survey is necessary to ensure that position. By the end of the decade, the ISS will be a relic of past partnerships, and several missions will have been completed that take research questions well beyond the hostile environment of LEO. The U.S. BPS community has an amazing decade of discovery, transformation, and translation ahead—if we seize it. (See Figure 7-12.)

FIGURE 7-12 In the 1970s, NASA commissioned space artists to depict the far future of humans in space. Although this future is still a long way off, biological and physical sciences research is taking steps today that will enable people to live and work in space for increasingly long durations. SOURCES: Courtesy of NASA Ames Research Center. Art work by Don Davis.

References

Abdel-Hafiz, M., P. Ablewski, A. Al-Masoudi, H. Álvarez Martínez, P. Balling, G. Barwood, E. Benkler, et al. 2019. "Guidelines for Developing Optical Clocks with 10^{-18} Fractional Frequency Uncertainty." arXiv:1906.11495 [physics.atom-Ph]. https://doi.org/10.48550/arXiv.1906.11495.

Abe, M., P. Adamson, M. Borcean, D. Bortoletto, K. Bridges, S.P. Carman, S. Chattopadhyay, et al. 2021. "Matter-wave Atomic Gradiometer Interferometric Sensor (MAGIS-100)." *Quantum Science and Technology* 6:044003.

Abels, E.R., and X.O. Breakefield. 2016. "Introduction to Extracellular Vesicles: Biogenesis, RNA Cargo Selection, Content, Release, and Uptake." *Cellular and Molecular Neurobiology* 36(3):301–312.

Acres, J.M., M.J. Youngapelian, and J. Nadeau. 2021. "The Influence of Spaceflight and Simulated Microgravity on Bacterial Motility and Chemotaxis." *npj Microgravity* 7(1):7.

Afshinnekoo, E., R.T. Scott, M.J. MacKay, E. Pariset, E. Cekanaviciute, R. Barker, S. Gilroy, et al. 2020. "Fundamental Biological Features of Spaceflight: Advancing the Field to Enable Deep-Space Exploration." *Cell* 183(5):1162–1184.

Afshinnekoo, E., R.T. Scott, M.J. MacKay, E. Pariset, E. Cekanaviciute, R. Barker, S. Gilroy, et al. 2021. "Fundamental Biological Features of Spaceflight: Advancing the Field to Enable Deep-Space Exploration." *Cell* 184(24):6002.

Agrawal, A.F., and M.C. Whitlock. 2012. "Mutation Load: The Fitness of Individuals in Populations Where Deleterious Alleles Are Abundant." *Annual Review of Ecology, Evolution, and Systematics* 43(1):115–135.

Agresti, J.J., E. Antipov, A.R. Abate, K. Ahn, A.C. Rowat, J.C. Baret, M. Marquez, A.M. Klibanov, A.D. Griffiths, and D.A. Weitz. 2010. "Ultrahigh-Throughput Screening in Drop-Based Microfluidics for Directed Evolution." *Proceedings of the National Academy of Sciences* 107(9):4004–4009.

Agronet. 2022. "Así es Como la NASA Cultiva Chiles en El Espacio." https://www.agronet.gov.co/Noticias/Paginas/As%C3%AD-es-como-la-NASA-cultiva-chiles-en-el-espacio.aspx.

Agui, J.H., and R.A. Wilkinson. 2010. "Granular Flow and Dynamics of Lunar Simulants in Excavating Implements." Pp. 84–94 in *Earth and Space 2010: Engineering, Science, Construction, and Operations in Challenging Environments*, G. Song and R.B. Malla, eds. 12th Biennial International Conference on Engineering, Construction, and Operations in Challenging Environments; and Fourth NASA/ARO/ASCE Workshop on Granular Materials in Lunar and Martian Exploration. March 14. Honolulu, Hawaii. Reston, VA: American Society of Civil Engineers.

Aguilera, D.N., H. Ahlers, B. Battelier, A. Bawamia, A. Bertoldi, R. Bondarescu, K. Bongs, et al. 2014. "STE-QUEST—Test of the Universality of Free Fall Using Cold Atom Interferometry." *Classical and Quantum Gravity* 31(11):115010.

Akiyama, T., K. Horie, E. Hinoi, M. Hiraiwa, A. Kato, Y. Maekawa, A. Takahashi, and S. Furukawa. 2020. "How Does Spaceflight Affect the Acquired Immune System?" *npj Microgravity* 6(1):14.

Albers, H., A. Herbst, L.L. Richardson, H. Heine, D. Nath, J. Hartwig, C. Schubert, et al. 2020. "Quantum Test of the Universality of Free Fall Using Rubidium and Potassium." *The European Physical Journal D* 74(7):145.

Allam, R., S. Martin, B. Forrest, J. Fetvedt, X. Lu, D. Freed, G.W. Brown, T. Sasaki, M. Itoh, and J. Manning. 2017. "Demonstration of the Allam Cycle: An Update on the Development Status of a High Efficiency Supercritical Carbon Dioxide Power Process Employing Full Carbon Capture." *Energy Procedia* 114:5948–5966.

Allen, J.P., M. Nelson, and A. Alling. 2003. "The Legacy of Biosphere 2 for the Study of Biospherics and Closed Ecological Systems." *Advances in Space Research* 31(7):1629–1639.

Alonso, I., C. Alpigiani, B. Altschul, H. Araujo, G. Arduini, J. Arlt, L. Badurina, et al. 2022. "Cold Atoms in Space: Community Workshop Summary and Proposed Road-Map." *EPJ Quantum Technology* 9(30). https://doi.org/10.1140/epjqt/s40507-022-00147-w.

Alwood, J.S., K. Yumoto, R. Mojarrab, C.L. Limoli, E.A. Almeida, N.D. Searby, and R.K. Globus. 2010. "Heavy Ion Irradiation and Unloading Effects on Mouse Lumbar Vertebral Microarchitecture, Mechanical Properties and Tissue Stresses." *Bone* 47(2):248–255.

Amselem, S. 2019. "Remote Controlled Autonomous Microgravity Lab Platforms for Drug Research in Space." *Pharmaceutical Research* 36(12):183.

Angelos, E., D.K. Ko, S. Zemelis-Durfee, and F. Brandizzi. 2021. "Relevance of the Unfolded Protein Response to Spaceflight-Induced Transcriptional Reprogramming in Arabidopsis." *Astrobiology* 21(3):367–380.

Antypas, D., A. Banerjee, C. Bartram, M. Baryakhtar, J. Betz, J.J. Bollinger, C. Boutan, et al. 2022. "New Horizons: Scalar and Vector Ultralight Dark Matter." arXiv:2203.14915 [hep-ex]. https://doi.org/10.48550/arXiv.2203.14915.

Araújo, P.A., J. Malheiro, I. Machado, F. Mergulhão, L. Melo, and M. Simões. 2016. "Influence of Flow Velocity on the Characteristics of *Pseudomonas Fluorescens* Biofilms." *Journal of Environmental Engineering* 142(7):04016031.

Arcadi, G., M. Dutra, P. Ghosh, M. Lindner, Y. Mambrini, M. Pierre, S. Profumo and F.S. Queiroz. 2018. "The Waning of the WIMP? A Review of Models, Searches, and Constraints, Giorgio." *The European Physical Journal C* 78(203).

Arkin, A.P., R.W. Cottingham, C.S. Henry, N.L. Harris, R.L. Stevens, S. Maslov, P. Dehal, et al. 2018. "KBase: The United States Department of Energy Systems Biology Knowledgebase." *Nature Biotechnology* 36(7):566–569.

Artusio, F., A. Castellví, A. Sacristán, R. Pisano, and J.A. Gavira. 2020. "Agarose Gel as a Medium for Growing and Tailoring Protein Crystals." *Crystal Growth & Design* 20(8):5564–5571.

Asenbaum, P., C. Overstreet, M. Kim, J. Curti, and M.A. Kasevich. 2020. "Atom-Interferometric Test of the Equivalence Principle at the 10^{-12} Level." *Physical Review Letters* 125(19):191101.

Ashby, N. 2003. "Relativity in the Global Positioning System." *Living Reviews in Relativity* 6(1):1.

Ashida, A. 1994. "Recycling of Trace Elements Required for Humans in CELSS." *Advances in Space Research* 14(11):177–187.

Atkinson, C. 1997. "NASA Tests Composters for Space." *Biocycle* 38(3):47–48.

Attinger, D., C. Frankiewicz, A.R. Betz, T.M. Schutzius, R. Ganguly, A. Das, C.-J. Kim, and C.M. Megaridis. 2014. "Surface Engineering for Phase Change Heat Transfer: A Review." *MRS Energy & Sustainability* 1:E4.

Aubret, A., M. Youssef, S. Sacanna, and J. Palacci. 2018. "Targeted Assembly and Synchronization of Self-Spinning Microgears." *Nature Physics* 14(11):1114–1118.

Augustine, C., and J.W. Tester. 2009. "Hydrothermal Flames: From Phenomenological Experimental Demonstrations to Quantitative Understanding." *Journal of Supercritical Fluids* 47(3):415–430.

Averesch, N.J., A.J. Berliner, S.N. Nangle, S. Zezulka, G.L. Vengerova, D. Ho, C.A. Casale, et al. 2022. "Biomanufacturing for Space-Exploration—What to Take and When to Make." *Preprints* 2022070329.

Averesch, N.J., V.E. Pane, F. Kracke, M. Ziesack, S.N. Nangle, P.A. Silver, R.W. Waymouth, and C.S. Criddle. 2021. "Biocatalytic Formation of Novel Polyesters with *para*-Hydroxyphenyl Groups in the Backbone—Engineering *Cupriavidus Necator* for Production of High-Performance Materials from CO_2 and Electricity." *bioRxiv* 2021.12.12.472320, preprint.

Aviles-Gaxiola, S., G. Olivo-Vázquez, L. Cabanillas-Bojórquez, E. Gutiérrez-Grijalva, and J. Heredia. 2020. "Plants as Biofactories for Phenolic Compounds." In *Plant Phenolics in Sustainable Agriculture*, R. Lone, R. Shuab and A. Kamili, eds. Singapore: Springer.

Bacci, S., and D. Bani. 2022. "The Epidermis in Microgravity and Unloading Conditions and Their Effects on Wound Healing." *Frontiers in Bioengineering Biotechnology* 10:666434.

Baek, M., F. DiMaio, I. Anishchenko, J. Dauparas, S. Ovchinnikov, G.R. Lee, J. Wang, et al. 2021. "Accurate Prediction of Protein Structures and Interactions Using a Three-Track Neural Network." *Science* 373(6557):871–876.

Baetz, U., and E. Martinoia. 2014. "Root Exudates: The Hidden Part of Plant Defense." *Trends in Plant Science* 19(2):90–98.

Bai, P., Y. Li, J. Bai, and H. Xu. 2022. "Markedly Decreased Growth Rate and Biofilm Formation Ability of Acinetobacter Schindleri After a Long-Duration (64 Days) Spaceflight." *European Review for Medical and Pharmacological Sciences* 26(11):4001–4015.

Balasubramaniam, R. 2021. "It's All About Fluids." Input paper submitted to the Committee for the Decadal Survey on Biological and Physical Sciences Research in Space 2023–2032.

Ballantine, K.E., and J. Ruostekoski. 2020. "Optical Magnetism and Huygens' Surfaces in Arrays of Atoms Induced by Cooperative Responses." *Physical Review Letters* 125(14):143604.

Bandstra, E.R., R.W. Thompson, G.A. Nelson, J.S. Willey, S. Judex, M.A. Cairns, E.R. Benton, M.E. Vazquez, J.A. Carson, and T.A. Bateman. 2009. "Musculoskeletal Changes in Mice from 20–50 cGy of Simulated Galactic Cosmic Rays." *Radiation Research* 172(1):21–29.

Baran, R., S. Marchal, S. Garcia Campos, E. Rehnberg, K. Tabury, B. Baselet, M. Wehland, D. Grimm, and S. Baatout. 2021. "The Cardiovascular System in Space: Focus on in Vivo and in Vitro Studies." *Biomedicines* 10(1).

Barker, R., S.V. Costes, J. Miller, S.G. Gebre, J. Lombardino, and S. Gilroy. 2021. "Rad-Bio-App: A Discovery Environment for Biologists to Explore Spaceflight-Related Radiation Exposures." *npj Microgravity* 7(1):15.

Barker, R., J. Lombardino, K. Rasmussen, and S. Gilroy. 2020. "Test of *Arabidopsis* Space Transcriptome: A Discovery Environment to Explore Multiple Plant Biology Spaceflight Experiments." *Frontiers in Plant Science* 11:147.

Barrila, J., S.F. Sarker, N. Hansmeier, S. Yang, K. Buss, N. Briones, J. Park, et al. 2021. "Evaluating the Effect of Spaceflight on the Host–Pathogen Interaction Between Human Intestinal Epithelial Cells and Salmonella Typhimurium." *npj Microgravity* 7(1):9.

Barrila, J., J. Yang, K.P. Franco Meléndez, S. Yang, K. Buss, T.J. Davis, B.J. Aronow, et al. 2022. "Spaceflight Analogue Culture Enhances the Host-Pathogen Interaction Between Salmonella and a 3-D Biomimetic Intestinal Co-Culture Model." *Frontiers in Cellular and Infection Microbiology* 12:705647.

Basu, D., and E.S. Haswell. 2017. "Plant Mechanosensitive Ion Channels: An Ocean of Possibilities." *Current Opinion in Plant Biology* 40:43–48.

Bataglieri, M., A. Belloni, A. Chou, P. Cushman, B. Echenard, R. Essig, J. Estrada, et al. 2017. "US Cosmic Visions: New Ideas in Dark Matter 2017: Community Report." arXiv:1707.04591 [hep-ph]. https://doi.org/10.48550/arXiv.1707.04591.

Baule, A., F. Morone, H.J. Herrmann, and H.A. Makse. 2018. "Edwards Statistical Mechanics for Jammed Granular Matter." *Reviews of Modern Physics* 90(1):015006.

Be, N.A., A. Avila-Herrera, J.E. Allen, N. Singh, A. Checinska Sielaff, C. Jaing, and K. Venkateswaran. 2017. "Whole Metagenome Profiles of Particulates Collected from the International Space Station." *Microbiome* 5(1):81.

Beheshti, A., K. Chakravarty, H. Fogle, H. Fazelinia, W.A.D. Silveira, V. Boyko, S.L. Polo, et al. 2019. "Multi-Omics Analysis of Multiple Missions to Space Reveal a Theme of Lipid Dysregulation in Mouse Liver." *Scientific Reports* 9(1):19195.

Beheshti, A., J. Miller, Y. Kidane, D. Berrios, S.G. Gebre, and S.V. Costes. 2018. "NASA Genelab Project: Bridging Space Radiation Omics with Ground Studies." *Radiation Research* 189(6):553–559.

Beisel, N.S., J. Noble, W.B. Barbazuk, A.-L. Paul, and R.J. Ferl. 2019. "Spaceflight-Induced Alternative Splicing During Seedling Development in Arabidopsis Thaliana." *npj Microgravity* 5(1):9.

Beloy, K., M.I. Bodine, T. Bothwell, S.M. Brewer, S.L. Bromley, J.-S. Chen, J.-D. Deschênes, et al. 2021. "Frequency Ratio Measurements at 18-Digit Accuracy Using an Optical Clock Network." *Nature* 591(7851):564–569.

Bennett, W.R., P.J. Smith, and I.J. Jakupca. 2020. *Analysis of 100-W Regenerative Fuel Cell Demonstration*. Glenn Research Center, National Aeronautics and Space Administration. Cleveland, OH. https://ntrs.nasa.gov/api/citations/20205000357/downloads/TM-20205000357.pdf.

Benoit, M.R., and D.M. Klaus. 2007. "Microgravity, Bacteria, and the Influence of Motility." *Advances in Space Research* 39(7):1225–1232.

Berger, T., D. Matthiä, S. Burmeister, C. Zeitlin, R. Rios, N. Stoffle, N.A. Schwadron, et al. 2020. "Long Term Variations of Galactic Cosmic Radiation on Board the International Space Station, on the Moon and on the Surface of Mars." *Journal of Space Weather and Space Climate* 10:34.

Berliner, A.J., J.M. Hilzinger, A.J. Abel, M.J. McNulty, G. Makrygiorgos, N.J.H. Averesch, S. Sen Gupta, et al. 2021. "Towards a Biomanufactory on Mars." Perspective. *Frontiers in Astronomy and Space Sciences* 8.

Berliner, A.J., I. Lipsky, D. Ho, J.M. Hilzinger, G. Vengerova, G. Makrygiorgos, M.J. McNulty, et al. 2022. "Space Bioprocess Engineering on the Horizon." *Communications Engineering* 1(1):13.

Berns, M.W. 2020. "Laser Scissors and Tweezers to Study Chromosomes: A Review." Review. *Frontiers in Bioengineering and Biotechnology* 8.

Berthier, L., and G. Biroli. 2011. "Theoretical Perspective on the Glass Transition and Amorphous Materials." *Reviews of Modern Physics* 83(2):587–645.

Bertone, G., and D. Hooper. 2018. "History of Dark Matter." *Reviews of Modern Physics* 90(4):045002.

Bigley, A.B., N.H. Agha, F.L. Baker, G. Spielmann, H.E. Kunz, P.L. Mylabathula, B.V. Rooney, et al. 2019. "NK Cell Function Is Impaired During Long-Duration Spaceflight." *Journal of Applied Physiology (1985)* 126(4):842–853.

Bijlani, S., C. Parker, N.K. Singh, M.A. Sierra, J. Foox, C.C.C. Wang, C.E. Mason, and K. Venkateswaran. 2022. "Genomic Characterization of the Titan-Like Cell Producing Naganishia Tulchinskyi, the First Novel Eukaryote Isolated from the International Space Station." *Journal of Fungi (Basel)* 8(2).

Bijlani, S., N.K. Singh, V.V.R. Eedara, A.R. Podile, C.E. Mason, C.C.C. Wang, and K. Venkateswaran. 2021. "*Methylobacterium Ajmalii* sp. nov., Isolated from the International Space Station." Original Research. *Frontiers in Microbiology* 12.

Bijlani, S., E. Stephens, N.K. Singh, K. Venkateswaran, and C.C.C. Wang. 2021. "Advances in Space Microbiology." *iScience* 24(5):102395.

Biteen, J.S., P.C. Blainey, Z.G. Cardon, M. Chun, G.M. Church, P.C. Dorrestein, S.E. Fraser, et al. 2016. "Tools for the Microbiome: Nano and Beyond." *ACS Nano* 10(1):6–37.

Blaber, E.A., M.J. Pecaut, and K.R. Jonscher. 2017. "Spaceflight Activates Autophagy Programs and the Proteasome in Mouse Liver." *International Journal of Molecular Sciences* 18(10).

Bland, R., R. Brukardt, W. Gangware, and D. Swartz. 2022. "A Different Space Race: Raising Capital and Accelerating Growth." https://www.mckinsey.com/industries/aerospace-and-defense/our-insights/a-different-space-race-raising-capital-and-accelerating-growth-in-space.

Blaser, J.P., M. Bye, G. Cavallo, T. Damour, C.W.F. Everitt, A. Hedin, R.W. Hellings, Y. Jafry, R. Laurance, and M. Lee. 1993. *STEP: Satellite Test of the Equivalence Principle. Report on the Phase a Study*. NASA. https://ntrs.nasa.gov/citations/19950019764.

Blaustein, R.A., A.G. McFarland, S. Ben Maamar, A. Lopez, S. Castro-Wallace, and E.M. Hartmann. 2019. "Pangenomic Approach to Understanding Microbial Adaptations Within a Model Built Environment, the International Space Station, Relative to Human Hosts and Soil." *mSystems* 4(1).

Bloch, I., J. Dalibard, and S. Nascimbène. 2012. "Quantum Simulations with Ultracold Quantum Gases." *Nature Physics* 8(4):267–276.

Bloomfield, S.A., D.A. Martinez, R.D. Boudreaux, and A.V. Mantri. 2016. "Microgravity Stress: Bone and Connective Tissue." *Comprehensive Physiology* 6(2):645–686.

Blount, P., and I. Iscla. 2020. "Life with Bacterial Mechanosensitive Channels, from Discovery to Physiology to Pharmacological Target." *Microbiology and Molecular Biology Reviews* 84(1):e00055-00019.

Blue Origin. 2023. "Blue Alchemist Technology Powers Our Lunar Future." Blog. Blue Origin for the Benefit of Earth. https://www.blueorigin.com/news/blue-alchemist-powers-our-lunar-future.

Boerma, M., C.G. van der Wees, H. Vrieling, J.P. Svensson, J. Wondergem, A. van der Laarse, L.H. Mullenders, and A.A. van Zeeland. 2005. "Microarray Analysis of Gene Expression Profiles of Cardiac Myocytes and Fibroblasts After Mechanical Stress, Ionising or Ultraviolet Radiation." *BMC Genomics* 6:6.

Boonsirichai, K., C. Guan, R. Chen, and P.H. Masson. 2002. "Root Gravitropism: An Experimental Tool to Investigate Basic Cellular and Molecular Processes Underlying Mechanosensing and Signal Transmission in Plants." *Annual Review of Plant Biology* 53:421–447.

Boonstra, J. 1999. "Growth Factor-Induced Signal Transduction in Adherent Mammalian Cells Is Sensitive to Gravity." *FASEB Journal* 13(Suppl):S35–S42.

Botelho, J., F. Grosso, and L. Peixe. 2019. "Antibiotic Resistance in Pseudomonas Aeruginosa—Mechanisms, Epidemiology and Evolution." *Drug Resistance Updates* 44:100640.

Bothwell, T., C.J. Kennedy, A. Aeppli, D. Kedar, J.M. Robinson, E. Oelker, A. Staron, and J. Ye. 2022. "Resolving the Gravitational Redshift Across a Millimetre-Scale Atomic Sample." *Nature* 602(7897):420–424.

Bouchet, F., and J. Sommeria. 2002. "Emergence of Intense Jets and Jupiter's Great Red Spot as Maximum-Entropy Structures." *Journal of Fluid Mechanics* 464:165–207.

Bourgoin, M., R. Kervil, C. Cottin-Bizonne, F. Raynal, R. Volk, and C. Ybert. 2020. "Kolmogorovian Active Turbulence of a Sparse Assembly of Interacting Marangoni Surfers." *Physical Review X* 10(2):021065.

Boyer, F., É. Guazzelli, and O. Pouliquen. 2011. "Unifying Suspension and Granular Rheology." *Physical Review Letters* 107(18):188301.

Braak, H., U. Rüb, W.P. Gai, and K. Del Tredici. 2003. "Idiopathic Parkinson's Disease: Possible Routes by Which Vulnerable Neuronal Types May Be Subject to Neuroinvasion by an Unknown Pathogen." *Journal of Neural Transmission (Vienna)* 110(5):517–536.

Braam, J. 2005. "In Touch: Plant Responses to Mechanical Stimuli." *New Phytologist* 165:373–389.

Bracker, G.P., S. Schneider, R. Wunderlich, H. Fecht, J. Zhao, and R.W. Hyers. 2020. "Confirmation of Anomalous Nucleation in Zirconium." *Journal of The Minerals, Metals and Materials Society* 72:3140–3146.

Bragg, J.G., M.A. Supple, R.L. Andrew, and J.O. Borevitz. 2015. "Genomic Variation Across Landscapes: Insights and Applications." *New Phytologist* 207(4):953–967.

Branco, S., A. Schauster, H.-L. Liao, and J. Ruytinx. 2022. "Mechanisms of Stress Tolerance and Their Effects on the Ecology and Evolution of Mycorrhizal Fungi." *New Phytologist* 235(6):2158–2175.

Brandizzi, F. 2022. *Life Beyond Earth: Effect of Space Flight on Seeds with Improved Nutritional Value*. https://taskbook.nasaprs.com/Tbp/index.cfm?action=public_query_taskbook_content&TASKID=14849.

Brendel, L.P.M., S.L. Caskey, M.K. Ewert, D. Hengeveld, J.E. Braun, and E.A. Groll. 2021. "Review of Vapor Compression Refrigeration in Microgravity Environments." *International Journal of Refrigeration* 123:169–179.

Brenner, M., and V.J. Hearing. 2008. "The Protective Role of Melanin Against UV Damage in Human Skin." *Photochemistry and Photobiology* 84(3):539–549.

Brewer, S.M., J.-S. Chen, A.M. Hankin, E.R. Clements, C.W. Chou, D.J. Wineland, D.B. Hume, and D.R. Leibrandt. 2019. "An $^{27}Al^{+}$ Quantum-Logic Clock with Systematic Uncertainty Below 10^{-18}." *Physical Review Letters* 123(3):033201.

Brinckerhoff, W.B., P.A. Willis, A.J. Ricco, D.A. Kaplan, R.M. Danell, A. Grubisic, M.F. Mora, et al. 2022. "Europan Molecular Indicators of Life Investigation (EMILI) for a Future Europa Lander Mission." Original Research. *Frontiers in Space Technologies* 2.

Britten, R.A., L.L. Wellman, and L.D. Sanford. 2021. "Progressive Increase in the Complexity and Translatability of Rodent Testing to Assess Space-Radiation Induced Cognitive Impairment." *Neuroscience and Biobehavioral Reviews* 126:159–174.

Brookings. 2019. "Less than Half of U.S. Children Under 15 are White, Census Shows." https://www.brookings.edu/research/less-than-half-of-us-children-under-15-are-white-census-shows.

Broutin, I., M. Riès-Kautt, and A. Ducruix. 1997. "Crystallographic Analyses of Lysozyme and Collagenase Microgravity Grown Crystals Versus Ground Controls." *Journal of Crystal Growth* 181(1):97–108.

Brown, D.R., T.B. Stout, J.A. Dirks, and N. Fernandez. 2012. "The Prospects of Alternatives to Vapor Compression Technology for Space Cooling and Food Refrigeration Applications." *Energy Engineering* 109(6):7–20.

Brown, E., N. Rodenberg, J. Amend, A. Mozeika, E. Steltz, M.R. Zakin, H. Lipson, and H.M. Jaeger. 2010. "Universal Robotic Gripper Based on the Jamming of Granular Material." *Proceedings of the National Academy of Sciences* 107(44):18809–18814.

Bryan, J.E., and J.S. Yagoobi. 1997. "Heat Transport Enhancement of Monogroove Heat Pipe with Electrohydrodynamic Pumping." *AIAA Journal of Thermophysics and Heat Transfer* 11(3):454–460.

Bubenheim, D., and K. Wignarajah. 1997. "Recycling of Inorganic Nutrients for Hydroponic Crop Production Following Incineration of Inedible Biomass." *Advances in Space Research* 20(10):2029–2035.

Buckner, T.L., R.A. Bilodeau, S.Y. Kim, and R. Kramer-Bottiglio. 2020. "Roboticizing Fabric by Integrating Functional Fibers." *Proceedings of the National Academy of Sciences* 117(41):25360–25369.

Burgner, S.E., K. Nemali, G.D. Massa, R.M. Wheeler, R.C. Morrow, and C.A. Mitchell. 2020. "Growth and Photosynthetic Responses of Chinese Cabbage (Brassica rapa L. cv. Tokyo Bekana) to Continuously Elevated Carbon Dioxide in a Simulated Space Station 'Veggie' Crop-Production Environment." *Life Sciences in Space Research* 27:83–88.

Burt, E.A., J.D. Prestage, R.L. Tjoelker, D.G. Enzer, D. Kuang, D.W. Murphy, D.E. Robison, J.M. Seubert, R.T. Wang, and T.A. Ely. 2021. "Demonstration of a Trapped-Ion Atomic Clock in Space." *Nature* 595(7865):43–47.

Buyel, J.F., E. Stöger, and L. Bortesi. 2021. "Targeted Genome Editing of Plants and Plant Cells for Biomanufacturing." *Transgenic Research* 30(4):401–426.

Califar, B., N.J. Sng, A. Zupanska, A.-L. Paul, and R.J. Ferl. 2020. "Root Skewing-Associated Genes Impact the Spaceflight Response of Arabidopsis Thaliana." Original Research. *Frontiers in Plant Science* 11.

Campbell, L., S. Mehtani, M. Dozier, and J. Rinehart. 2013. "Gender-Heterogeneous Working Groups Produce Higher Quality Science." *PLOS One* 8(10):e79147.

Canepa, A. 2019. "Searches for Supersymmetry at the Large Hadron Collider." *Reviews in Physics* 4:100033.

Carattoli, A. 2013. "Plasmids and the Spread of Resistance." *International Journal of Medical Microbiology* 303(6):298–304.

Carney, D., H. Müller, and J.M. Taylor. 2021. "Using an Atom Interferometer to Infer Gravitational Entanglement Generation." *PRX Quantum* 2(3):030330.

Carriot, J., M. Jamali, and K.E. Cullen. 2015. "Rapid Adaptation of Multisensory Integration in Vestibular Pathways." Mini Review. *Frontiers in Systems Neuroscience* 9.

Carriot, J., I. Mackrous, and K.E. Cullen. 2021. "Challenges to the Vestibular System in Space: How the Brain Responds and Adapts to Microgravity." Review. *Frontiers in Neural Circuits* 15.

Casaburi, G., I. Goncharenko-Foster, A.A. Duscher, and J.S. Foster. 2017. "Transcriptomic Changes in an Animal-Bacterial Symbiosis Under Modeled Microgravity Conditions." *Scientific Reports* 7(1):46318.

Casey, T., O.V. Patel, and K. Plaut. 2015. "Transcriptomes Reveal Alterations in Gravity Impact Circadian Clocks and Activate Mechanotransduction Pathways with Adaptation Through Epigenetic Change." *Physiological Genomics* 47(4):113–128.

Cassaro, A., C. Pacelli, M. Baqué, B. Cavalazzi, G. Gasparotto, R. Saladino, L. Botta, et al. 2022. "Investigation of Fungal Biomolecules After Low Earth Orbit Exposure: A Testbed for the Next Moon Missions." *Environmental Microbiology* 24(7):2938–2950.

Castaneda, A.J., N.J. O'Connor, J.S. Yagoobi, J.R. Didion, M.S. Martins, and M.M. Hasan. 2023. "Dielectrophoretically-Assisted EHD-Driven Liquid Film Flow Boiling in the Presence and Absence of Gravity." *ASME Journal of Heat Transfer* 145(3):031601.

Castelein, S.M., T.F. Aarts, J. Schleppi, R. Hendrikx, A.J. Böttger, D. Benz, M. Marechal, et al. 2021. "Iron Can Be Microbially Extracted from Lunar and Martian Regolith Simulants and 3D Printed into Tough Structural Materials." *PLOS One* 16(4):e0249962.

Castro, V.A., A.N. Thrasher, M. Healy, C.M. Ott, and D.L. Pierson. 2004. "Microbial Characterization During the Early Habitation of the International Space Station." *Microbial Ecology* 47(2):119–126.

Castro-Wallace, S.L., C.Y. Chiu, K.K. John, S.E. Stahl, K.H. Rubins, A.B.R. McIntyre, J.P. Dworkin, et al. 2017. "Nanopore DNA Sequencing and Genome Assembly on the International Space Station." *Scientific Reports* 7(1):18022.

Cavanagh, P.R., A.J. Rice, A.A. Licata, et al. 2013. "A Novel Lunar Bed Rest Analogue." *Aviation, Space, and Environmental Medicine* 84(11):1191–1195.

CDC (Centers for Disease Control and Prevention). 2023. Disability Impacts All of Us. Infographic at https://www.cdc.gov/ncbddd/disabilityandhealth/infographic-disability-impacts-all.html.

Cervantes, J.L., and B.Y. Hong. 2016. "Dysbiosis and Immune Dysregulation in Outer Space." *International Reviews of Immunology* 35(1):67–82.

Cestellos-Blanco, S., S. Friedline, K.B. Sander, A.J. Abel, J.M. Kim, D.S. Clark, A.P. Arkin, and P. Yang. 2021. "Production of PHB from CO_2-Derived Acetate with Minimal Processing Assessed for Space Biomanufacturing." Original Research. *Frontiers in Microbiology* 12:700010.

Chancellor, J.C., R.S. Blue, K.A. Cengel, S.M. Auñón-Chancellor, K.H. Rubins, H.G. Katzgraber, and A.R. Kennedy. 2018. "Limitations in Predicting the Space Radiation Health Risk for Exploration Astronauts." *npj Microgravity* 4(1):8.

Chandler, J.O., F.B. Haas, S. Khan, L. Bowden, M. Ignatz, E.M.A. Enfissi, F. Gawthrop, et al. 2020. "Rocket Science: The Effect of Spaceflight on Germination Physiology, Ageing, and Transcriptome of Eruca Sativa Seeds." *Life (Basel)* 10(4).

Chatterjee, A., J.L. Plawsky, P.C.W. Jr., D.F. Chao, R.J. Sicker, T. Lorik, L. Chestney, R. Margie, J. Eustace, and J. Zoldak. 2013. "Constrained Vapor Bubble Heat Pipe Experiment Aboard the International Space Station." *Journal of Thermophysics and Heat Transfer* 27(2):309–319.

Chegg. 2023. "Adult Life Cycle of Mice." Biology Questions and Answers. 2023. https://www.chegg.com/homework-help/questions-and-answers/adult-life-cycle-mice-fertilized-egg-polar-bodies-zona-pellucida-cleavage-blastocyst-birth-q70738289.

Chen, B.G.-g., and C.D. Santangelo. 2018. "Branches of Triangulated Origami Near the Unfolded State." *Physical Review X* 8(1):011034.

Chen, Q., R.S. Amano, and M. Xin. 2006. "Experimental Study of Flow Patterns and Regimes of Condensation in Horizontal Three-Dimensional Micro-Fin Tubes." *Heat and Mass Transfer* 43(2):201–206.

Chen, Y., X. Li, C. Scheibner, V. Vitelli, and G. Huang. 2021. "Realization of Active Metamaterials with Odd Micropolar Elasticity." *Nature Communications* 12(1):5935.

Chen, Z., S. Stanbouly, N.C. Nishiyama, X. Chen, M.D. Delp, H. Qiu, X.W. Mao, and C. Wang. 2021. "Spaceflight Decelerates the Epigenetic Clock Orchestrated with a Global Alteration in DNA Methylome and Transcriptome in the Mouse Retina." *Precision Clinical Medicine* 4(2):93–108.

Chevalier, Y., and M.-A. Bolzinger. 2013. "Emulsions Stabilized with Solid Nanoparticles: Pickering Emulsions." *Colloids and Surfaces A: Physicochemical and Engineering Aspects* 439:23–34.

Chiaramonte, F.P., and J.A. Joshi. 2004. *Workshop on Critical Issues in Microgravity Fluids, Transport, and Reaction Processes in Advanced Human Support Technology*. NASA. https://ntrs.nasa.gov/citations/20040035652.

Chien, Y.-C., D.P. Stocker, U.G. Hegde, and D. Dunn-Rankin. 2022. "Electric-Field Effects on Methane Coflow Flames Aboard the International Space Station (ISS): ACME E-FIELD Flames." *Combustion and Flame* 246:112443.

Choi, W.G., R.J. Barker, S.H. Kim, S.J. Swanson, and S. Gilroy. 2019. "Variation in the Transcriptome of Different Ecotypes of Arabidopsis Thaliana Reveals Signatures of Oxidative Stress in Plant Responses to Spaceflight." *American Journal of Botany* 106(1):123–136.

Cialdai, F., A. Colciago, D. Pantalone, A.M. Rizzo, S. Zava, L. Morbidelli, F. Celotti, D. Bani, and M. Monici. 2020. "Effect of Unloading Condition on the Healing Process and Effectiveness of Platelet Rich Plasma as a Countermeasure: Study on in Vivo and in Vitro Wound Healing Models." *International Journal of Molecular Sciences* 21(2).

Cichos, F., K. Gustavsson, B. Mehlig, and G. Volpe. 2020. "Machine Learning for Active Matter." *Nature Machine Intelligence* 2(2):94–103.

Clark, B.C., and T.M. Manini. 2012. "What Is Dynapenia?" *Nutrition* 28(5):495–503.

Clark, N., J. Maclennan, M. Glaser, C. Park, R. Stannarius, P. Tin, and N. Hall. 2011. "Observation and Analysis of Smectic Islands in Space (OASIS)." In *49th AIAA Aerospace Sciences Meeting Including the New Horizons Forum and Aerospace Exposition*. January 4–7. Orlando, FL.

Clarke, A.C. 1962. *Profiles of the Future: An Enquiry into the Limits of the Possible*. New York: Harper and Row.

Clauwaert, P., M. Muys, A. Alloul, J.D. Paepe, A. Luther, X. Sun, C. Ilgrande, et al. 2017. "Nitrogen Cycling in Bioregenerative Life Support Systems: Challenges for Waste Refinery and Food Production Processes." *Progress in Aerospace Sciences* 91:87–98.

Cockell, C.S., P. Rettberg, E. Rabbow, and K. Olsson-Francis. 2011. "Exposure of Phototrophs to 548 Days in Low Earth Orbit: Microbial Selection Pressures in Outer Space and on Early Earth." *The ISME Journal* 5(10):1671–1682.

Cockell, C.S., R. Santomartino, K. Finster, A.C. Waajen, L.J. Eades, R. Moeller, P. Rettberg, et al. 2020. "Space Station Biomining Experiment Demonstrates Rare Earth Element Extraction in Microgravity and Mars Gravity." *Nature Communications* 11(1):5523.

Cockell, C.S., R. Santomartino, K. Finster, A.C. Waajen, N. Nicholson, C.-M. Loudon, L.J. Eades, et al. 2021. "Microbially-Enhanced Vanadium Mining and Bioremediation Under Micro- and Mars Gravity on the International Space Station." Original Research. *Frontiers in Microbiology* 12.

Cohen, B., M. Dai, S.B. Yakushin, and C. Cho. 2019. "The Neural Basis of Motion Sickness." *Journal of Neurophysiology* 121(3):973–982.

Coker, J. 2019. "Recent Advances in Understanding Extremophiles [Version 1; Peer Review: 2 Approved]." *F1000Research* 8(1917).

Collicott, S.H., and A. Alexeenko. 2019. "Propulsion Testing in Commercial Re-Usable Sub-Orbital Rockets." AIAA 2019-4354. *AIAA Propulsion and Energy 2019 Forum*.

Collins, P.J., R.N. Grugel, and A. Radlińska. 2021. "The Influence of Variable Gravity on the Microstructural Development of Tricalcium Silicate Pastes." *Earth and Space 2021* 59–66.

Colombo, M., G. Raposo, and C. Théry. 2014. "Biogenesis, Secretion, and Intercellular Interactions of Exosomes and Other Extracellular Vesicles." *Annual Review of Cell and Developmental Biology* 30:255–289.

Conrath, M., P.J. Canfield, P.M. Bronowicki, M.E. Dreyer, M.M. Weislogel, and A. Grah. 2013. "Capillary Channel Flow Experiments Aboard the International Space Station." *Physical Review. E: Statistical, Nonlinear, and Soft Matter Physics* 88(6):063009.

Correll, M.J., T.P. Pyle, K.D. Millar, Y. Sun, J. Yao, R.E. Edelmann, and J.Z. Kiss. 2013. "Transcriptome Analyses of Arabidopsis Thaliana Seedlings Grown in Space: Implications for Gravity-Responsive Genes." *Planta* 238(3):519–533.

Cox, S.E. 2013. "Energy Metabolism." Pp. 177–185 in *Encyclopedia of Human Nutrition*, 3rd ed. Cambridge, MA: Academic Press.

Crisp, P.A., D. Ganguly, S.R. Eichten, J.O. Borevitz, and B.J. Pogson. 2016. "Reconsidering Plant Memory: Intersections Between Stress Recovery, RNA Turnover, and Epigenetics." *Science Advances* 2(2):e1501340.

Crucian, B.E., A. Choukèr, R.J. Simpson, S. Mehta, G. Marshall, S.M. Smith, S.R. Zwart, et al. 2018. "Immune System Dysregulation During Spaceflight: Potential Countermeasures for Deep Space Exploration Missions." *Frontiers in Immunology* 9:1437.

Crucian, B., R.J. Simpson, S. Mehta, R. Stowe, A. Chouker, S.A. Hwang, J.K. Actor, A.P. Salam, D. Pierson, and C. Sams. 2014. "Terrestrial Stress Analogs for Spaceflight Associated Immune System Dysregulation." *Brain, Behavior, and Immunity* 39:23–32.

Crucian, B., R.P. Stowe, S. Mehta, H. Quiriarte, D. Pierson, and C. Sams. 2015. "Alterations in Adaptive Immunity Persist During Long-Duration Spaceflight." *npj Microgravity* 1(1):15013.

Cucinotta, F.A., and M. Durante. 2006. "Cancer Risk from Exposure to Galactic Cosmic Rays: Implications for Space Exploration by Human Beings." *Lancet Oncology* 7(5):431–435.

Cucinotta, F.A., I. Plante, A.L. Ponomarev, and M.H. Kim. 2011. "Nuclear Interactions in Heavy Ion Transport and Event-Based Risk Models." *Radiation Protection Dosimetry* 143(2–4):384–390.

Cucinotta, F.A., H. Wu, M.R. Shavers, and K. George. 2003. "Radiation Dosimetry and Biophysical Models of Space Radiation Effects." *Gravitational and Space Biology Bulletin* 16(2):11–18.

Cui, C., Y. Li, S. Wang, M. Ren, C. Yang, Z. Jiang, and J. Zhang. 2020. "Review on an Advanced Combustion Technology. Supercritical Hydrothermal Combustion." *Applied Sciences* 10(5):1645.

Curran, H.J. 2019. "Developing Detailed Chemical Kinetic Mechanisms for Fuel Combustion." *Proceedings of the Combustion Institute* 37(1):57–81.

da Silveira, W.A., H. Fazelinia, S.B. Rosenthal, E.C. Laiakis, M.S. Kim, C. Meydan, Y. Kidane, et al. 2020. "Comprehensive Multi-Omics Analysis Reveals Mitochondrial Stress as a Central Biological Hub for Spaceflight Impact." *Cell* 183(5):1185–1201.

Dabrowska, A., J. Kumar, and C. Rallis. 2022. "Nutrient-Response Pathways in Healthspan and Lifespan Regulation." *Cells* 11(9):1568.

Dailey, C., C. Bradley, D.F. Jackson Kimball, I.A. Sulai, S. Pustelny, A. Wickenbrock, and A. Derevianko. 2021. "Quantum Sensor Networks as Exotic Field Telescopes for Multi-Messenger Astronomy." *Nature Astronomy* 5(2):150–158.

Damkiær, S., L. Yang, S. Molin, and L. Jelsbak. 2013. "Evolutionary Remodeling of Global Regulatory Networks During Long-Term Bacterial Adaptation to Human Hosts." *Proceedings of the National Academy of Sciences* 110(19):7766–7771.

Danko, D., D. Bezdan, E.E. Afshin, S. Ahsanuddin, C. Bhattacharya, D.J. Butler, K.R. Chng, et al. 2021. "A Global Metagenomic Map of Urban Microbiomes and Antimicrobial Resistance." *Cell* 184(13):3376–3393.

Darch, S.E., C.B. Ibberson, and M. Whiteley. 2017. "Evolution of Bacterial 'Frenemies.' " *mBio* 8(3).

De Groh, K.K., and B.A. Banks. 2021. *Misse-Flight Facility Polymers and Composites Experiment 1-4 (PCE 1-4)*. NASA. https://ntrs.nasa.gov/citations/20205008863.

de Groot, R.P., P.J. Rijken, J. den Hertog, J. Boonstra, A.J. Verkleij, S.W. de Laat, and W. Kruijer. 1991. "Nuclear Responses to Protein Kinase C Signal Transduction Are Sensitive to Gravity Changes." *Experimental Cell Research* 197(1):87–90.

De Marco, L., G. Valtolina, K. Matsuda, W.G. Tobias, J.P. Covey, and J. Ye. 2019. "A Degenerate Fermi Gas of Polar Molecules." *Science* 363(6429):853–856.

Dee. 2018. "Farmbot Has Arrived!" The Payload Blog, Kennedy Space Center Visitor Complex. https://www.kennedyspacecenter.com/blog/04/farmbot-has-arrived.

Delp, M.D., J.M. Charvat, C.L. Limoli, R.K. Globus, and P. Ghosh. 2016. "Apollo Lunar Astronauts Show Higher Cardiovascular Disease Mortality: Possible Deep Space Radiation Effects on the Vascular Endothelium." *Scientific Reports* 6(1):29901.

Department of Education. 2022. "Table 318.45. Number and Percentage Distribution of Science, Technology, Engineering, and Mathematics (STEM) Degrees/Certificates Conferred by Postsecondary Institutions, by Race/Ethnicity, Level of Degree/Certificate, and Sex of Student: Academic Years 2011–12 through 2020–21." Institute of Education Sciences, National Center for Education Statistics, Integrated Postsecondary Education Data System. https://nces.ed.gov/programs/digest/d22/tables/dt22_318.45.asp.

Department of State. 2023. "Enrollment Trends." *Open Doors*. https://opendoorsdata.org/data/international-students/enrollment-trends.

Deppner, C., W. Herr, M. Cornelius, P. Stromberger, T. Sternke, C. Grzeschik, A. Grote, et al. 2021. "Collective-Mode Enhanced Matter-Wave Optics." *Physical Review Letters* 127(10):100401.

Derevianko, A., K. Gibble, L. Hollberg, N.R. Newbury, C. Oates, M.S. Safronova, L.C. Sinclair, and N. Yu. 2022. "Fundamental Physics with a State-of-the-Art Optical Clock in Space." *Quantum Science and Technology* 7(4):044002.

Derevianko, A., and M. Pospelov. 2014. "Hunting for Topological Dark Matter with Atomic Clocks." *Nature Physics* 10(2014): 933–936.

Detrell, G. 2021. "Chlorella Vulgaris Photobioreactor for Oxygen and Food Production on a Moon Base—Potential and Challenges." Review. *Frontiers in Astronomy and Space Sciences* 8.

Devahdhanush, V.S., I. Mudawar, H.K. Nahra, R. Balasubramaniam, M.M. Hasan, and J.R. Mackey. 2022. "Experimental Heat Transfer Results and Flow Visualization of Vertical Upflow Boiling in Earth Gravity with Subcooled Inlet Conditions—in Preparation for Experiments Onboard the International Space Station." *International Journal of Heat and Mass Transfer* 188:122603.

Dhir, V.K., G.R. Warrier, E. Aktinol, D. Chao, J. Eggers, W. Sheredy, and W. Booth. 2012. "Nucleate Pool Boiling Experiments (NPBX) on the International Space Station." *Microgravity Science and Technology* 24(5):307–325.

Diedrich, A., S.Y. Paranjape, and D. Robertson. 2007. "Plasma and Blood Volume in Space." *American Journal of the Medical Sciences* 334(1):80–85.

Dietrich, D.L., V. Nayagam, M.C. Hicks, P.V. Ferkul, F.L. Dryer, T. Farouk, B.D. Shaw, et al. 2014. "Droplet Combustion Experiments Aboard the International Space Station." *Microgravity Science and Technology* 26(2):65–76.

Discher, D.E., L. Smith, S. Cho, M. Colasurdo, A.J. García, and S. Safran. 2017. "Matrix Mechanosensing: From Scaling Concepts in 'Omics' Data to Mechanisms in the Nucleus, Regeneration, and Cancer." *Annual Review of Biophysics* 46:295–315.

Dissanayake, L., and L.N. Jayakody. 2021. "Engineering Microbes to Bio-Upcycle Polyethylene Terephthalate." Mini Review. *Frontiers in Bioengineering and Biotechnology* 9.

Dong, C.Y. Fu, B. Xie, M. Wang and H. Liu. 2017. "Element Cycling and Energy Flux Responses in Ecosystem Simulations Conducted at the Chinese Lunar Palace-1." *Astrobiology* 17(1):78–86.

Douglas, G.L., R.M. Wheeler, and R.F. Fritsche. 2021. "Sustaining Astronauts: Resource Limitations, Technology Needs, and Parallels Between Spaceflight Food Systems and Those on Earth." *Sustainability* 13(16):9424.

Dowling, T.E. 2020. "Jupiter-Style Jet Stability." *The Planetary Science Journal* 1(1):1–9.

Dryer, F.L. 2015. "Chemical Kinetic and Combustion Characteristics of Transportation Fuels." *Proceedings of the Combustion Institute* 35(1):117–144.

Drysdale, A., M. Ewert, and A. Hanford. 1999. "Equivalent System Mass Studies of Missions and Concepts," SAE Technical Paper 1999-01-2081. https://doi.org/10.4271/1999-01-2081.

Drysdale, A., S. Maxwell, M. Ewert, and A. Hanford. 2000. "Systems Analysis of Life Support for Long-Duration Missions." SAE Technical Paper 2000-01-2394.

Du, H.M., C. Liu, X.W. Jin, C.F. Du, Y. Yu, S. Luo, W.Z. He, and S.Z. Zhang. 2022. "Overexpression of the Aldehyde Dehydrogenase Gene Zmaldh Confers Aluminum Tolerance in Arabidopsis Thaliana." *International Journal of Molecular Sciences* 23(1).

Dufrêne, Y.F., and A. Persat. 2020. "Mechanomicrobiology: How Bacteria Sense and Respond to Forces." *Nature Reviews Microbiology* 18(4):227–240.

Dukler, A.E., J.A. Fabre, J.B. McQuillen, and R. Vernon. 1988. "Gas-Liquid Flow at Microgravity Conditions: Flow Patterns and Their Transitions." *International Journal of Multiphase Flow* 14(4):389–400.

Dunn, J.O., M.G. Mythen, and M.P. Grocott. 2016. "Physiology of Oxygen Transport." *BJA Education* 16(10):341–348.

Durante, M. 2014. "Space Radiation Protection: Destination Mars." *Life Sciences in Space Research* 1:2–9.

Durante, M., and F.A. Cucinotta. 2008. "Heavy Ion Carcinogenesis and Human Space Exploration." *Nature Reviews: Cancer* 8(6):465–472.

Duri, L.G., A.G. Caporale, Y. Rouphael, S. Vingiani, M. Palladino, S. De Pascale, and P. Adamo. 2022. "The Potential for Lunar and Martian Regolith Simulants to Sustain Plant Growth: A Multidisciplinary Overview." Review. *Frontiers in Astronomy and Space Sciences* 8.

Duri, L.G., A. Pannico, S.A. Petropoulos, A.G. Caporale, P. Adamo, G. Graziani, A. Ritieni, S. De Pascale, and Y. Rouphael. 2022. "Bioactive Compounds and Antioxidant Activity of Lettuce Grown in Different Mixtures of Monogastric-Based Manure with Lunar and Martian Soils." Original Research. *Frontiers in Nutrition* 9.

El Karoui, M., M. Hoyos-Flight, and L. Fletcher. 2019. "Future Trends in Synthetic Biology-A Report." *Frontiers in Bioengineering and Biotechnology* 7:175.

Elgart, S.R., M.P. Little, L.J. Chappell, C.M. Milder, M.R. Shavers, J.L. Huff, and Z.S. Patel. 2018. "Radiation Exposure and Mortality from Cardiovascular Disease and Cancer in Early NASA Astronauts." *Scientific Reports* 8(1):8480.

Elgindi, M., J. Sapudom, I.H. Ibrahim, M. Al-Sayegh, W. Chen, A. Garcia-Sabaté, and J.C.M. Teo. 2021. "May the Force Be with You (or Not): The Immune System Under Microgravity." *Cells* 10(8).

Elliott, E.R., M.C. Krutzik, J.R. Williams, R.J. Thompson, and D.C. Aveline. 2018. "NASA's Cold Atom Lab (CAL): System Development and Ground Test Status." *npj Microgravity* 4(1):16.

Eloe-Fadrosh, E.A., F. Ahmed, A., M. Babinski, J. Baumes, M. Borkum, L. Bramer, et al. 2021. "The National Microbiome Data Collaborative Data Portal: An Integrated Multi-Omics Microbiome Data Resource." *Nucleic Acids Research* 50(D1):D828–D836.

Ely, T.A., J. Seubert, N. Bradley, T. Drain, and S. Bhaskaran. 2021. "Radiometric Autonomous Navigation Fused with Optical for Deep Space Exploration." *The Journal of the Astronautical Sciences* 68:300–325.

English, K.L., J.J. Bloomberg, A.P. Mulavara, and L.L. Ploutz-Snyder. 2019. "Exercise Countermeasures to Neuromuscular Deconditioning in Spaceflight." *Comprehensive Physiology* 10(1):171–196.

Erbil, H.Y. 2014. "The Debate on the Dependence of Apparent Contact Angles on Drop Contact Area or Three-Phase Contact Line: A Review." *Surface Science Reports* 69(4):325–365.

ESA (European Space Agency). 2020. "Orbital Space Aging Tests Offered for Space Age Materials." https://www.esa.int/Enabling_Support/Space_Engineering_Technology/Orbital_space_ageing_tests_offered_for_space_age_materials.

Escobar, C., N. Altaf, R. Barker, M. Bhuiyan, M. Correll, R. Fritsche, S. Humphrey, et al. "Research Campaign—Artificial Intelligence for Autonomous Space Plant Production." Input paper submitted to the Committee on Decadal Survey on Biological and Physical Sciences Research in Space 2023–2032. https://www.nationalacademies.org/our-work/decadal-survey-on-life-and-physical-sciences-research-in-space-2023-2032.

Espinosa-Hoyos, D., H. Du, N.X. Fang, and K.J. Van Vliet. 2017. "Poly(HDDA)-Based Polymers for Microfabrication and Mechanobiology." *MRS Advances* 2(24):1315–1321.

Espinosa-Hoyos, D., A. Jagielska, K.A. Homan, H. Du, T. Busbee, D.G. Anderson, N.X. Fang, J.A. Lewis, and K.J. Van Vliet. 2018. "Engineered 3D-Printed Artificial Axons." *Scientific Reports* 8(1):478.

Everett, H. 2022. "Astronauts Test 3D Bioprinted Skin Bandages in Space." 3D Printing Industry. https://3dprintingindustry.com/news/astronauts-test-3d-bioprinted-skin-bandages-in-space-202445.

Fajardo-Cavazos, P., and W.L. Nicholson. 2021. "Mechanotransduction in Prokaryotes: A Possible Mechanism of Spaceflight Adaptation." *Life (Basel)* 11(1).

Falkowski, P.G., T. Fenchel, and E.F. Delong. 2008. "The Microbial Engines That Drive Earth's Biogeochemical Cycles." *Science* 320(5879):1034–1039.

Farouk, T.I., D. Dietrich, F.E. Alam, and F.L. Dryer. 2017. "Isolated N-Decane Droplet Combustion—Dual Stage and Single Stage Transition to 'Cool Flame' Droplet Burning." *Proceedings of the Combustion Institute* 36(2):2523–2530.

Fedderke, M.A., P.W. Graham, and S. Rajendran. 2022. "Asteroids for μHz Gravitational-Wave Detection." *Physical Review D* 105(10):103018.

Ferkul, P. 2017. "Centrifuge in Free Fall: Combustion at Partial Gravity." Paper presented at the Annual Meeting American Society for Gravitational and Space Research, Seattle, WA.

Ferl, R.J., J. Koh, F. Denison, and A.L. Paul. 2015. "Spaceflight Induces Specific Alterations in the Proteomes of Arabidopsis." *Astrobiology* 15(1):32–56.

Fernander, M.C., P.K. Parsons, B. Khaled, A. Bradley, J.L. Graves, and M.D. Thomas. 2022. "Adaptation to Simulated Microgravity in Streptococcus Mutans." *npj Microgravity* 8(1):17.

Fernandez-Gonzalo, R., S. Baatout, and M. Moreels. 2017. "Impact of Particle Irradiation on the Immune System: From the Clinic to Mars." *Frontiers in Immunology* 8:177.

Fester, D.A., R.N. Eberhardt, and J.R. Tegart. 1975. "Space Shuttle Reaction Control Subsystem Propellant Acquisition Technology." *Journal of Spacecraft and Rockets* 12(12):705–710.

Feynman, R.P. 1982. "Simulating Physics with Computers." *International Journal of Theoretical Physics* 21(6):467–488.

Fitts, R.H., S.W. Trappe, D.L. Costill, P.M. Gallagher, A.C. Creer, P.A. Colloton, J.R. Peters, J.G. Romatowski, J.L. Bain, and D.A. Riley. 2010. "Prolonged Space Flight-Induced Alterations in the Structure and Function of Human Skeletal Muscle Fibres." *Journal of Physiology* 588(Pt 18):3567–3592.

Flynn-Evans, E.E., L.K. Barger, A.A. Kubey, J.P. Sullivan, and C.A. Czeisler. 2016. "Circadian Misalignment Affects Sleep and Medication Use Before and During Spaceflight." *npj Microgravity* 2(1):15019.

Fonseca, A.P., and J.C. Sousa. 2007. "Effect of Shear Stress on Growth, Adhesion and Biofilm Formation of Pseudomonas Aeruginosa with Antibiotic-Induced Morphological Changes." *International Journal of Antimicrobial Agents* 30(3):236–241.

Foox, J., S.W. Tighe, C.M. Nicolet, J.M. Zook, M. Byrska-Bishop, W.E. Clarke, M.M. Khayat, et al. 2021. "Performance Assessment of DNA Sequencing Platforms in the ABRF Next-Generation Sequencing Study." *Nature Biotechnology* 39(9):1129–1140.

Foster, J.S., C.L. Khodadad, S.R. Ahrendt, and M.L. Parrish. 2013. "Impact of Simulated Microgravity on the Normal Developmental Time Line of an Animal-Bacteria Symbiosis." *Scientific Reports* 3:1340.

Foust, J. 2019. "How Much Does the Public Care About Returning to the Moon?" *SpaceNews*. https://spacenews.com/foust-forward-how-much-does-the-public-care-about-returning-to-the-moon.

Francoz, E., P. Ranocha, H. Nguyen-Kim, E. Jamet, V. Burlat, and C. Dunand. 2015. "Roles of Cell Wall Peroxidases in Plant Development." *Phytochemistry* 112:15–21.

Fredriksson, H. 2022. "Analysis of the Solidification Process of Metal Alloys Under Microgravity Conditions." Original Research. *Frontiers in Metals and Alloys* 1.

Freeman, R.B., and W. Huang. 2014a. "Collaboration: Strength in Diversity." *Nature* 513(305).

Freitas-Dias, R., T.I. Lima, J.M. Costa-Junior, L.M. Gonçalves, H.N. Araujo, F.M.M. Paula, G.J. Santos, et al. 2022. "Offspring from Trained Male Mice Inherit Improved Muscle Mitochondrial Function Through PPAR Co-Repressor Modulation." *Life Sciences* 291:120239.

Freundlich, A.A. Ignatiev, C. Horton, M. Duke, P. Curreri, and L. Sibille. 2005. "Manufacture of Solar Cells on the Moon." Pp. 794–797 in *IEEE Photovoltaic Specialists Conference, 2005*. Lake Buena Vista, FL.

Friedman, M.A., A. Abood, B. Senwar, Y. Zhang, C.R. Maroni, V.L. Ferguson, C.R. Farber, and H.J. Donahue. 2020. "Genetic Variability Affects the Skeletal Response to Unloading." *bioRxiv* 2020.06.26.174326, preprint.

Friedman, M.A., A. Abood, B. Senwar, Y. Zhang, C.R. Maroni, V.L. Ferguson, C.R. Farber, and H.J. Donahue. 2021. "Genetic Variability Affects the Skeletal Response to Immobilization in Founder Strains of the Diversity Outbred Mouse Population." *Bone Reports* 15:101140.

Friedman, M.A., Y. Zhang, J.S. Wayne, C.R. Farber, and H.J. Donahue. 2019. "Single Limb Immobilization Model for Bone Loss from Unloading." *Journal of Biomechanics* 83:181–189.

Friedman, R. 1993. *Risks and Issues in Fire Safety on the Space Station*. NASA. Cleveland, OH. https://ntrs.nasa.gov/api/citations/19940022933/downloads/19940022933.pdf.

Friedman, R. 1996. "Fire Safety in Spacecraft." *Fire and Materials* 20(5):235–243.

Frieman, J.A., M.S. Turner, and D. Huterer. 2008. "Dark Energy and the Accelerating Universe." *Annual Review of Astronomy and Astrophysics* 46:385–432.

Frydenlund Michelsen, C., S.M. Hossein Khademi, H. Krogh Johansen, H. Ingmer, P.C. Dorrestein, and L. Jelsbak. 2016. "Evolution of Metabolic Divergence in Pseudomonas Aeruginosa During Long-Term Infection Facilitates a Proto-Cooperative Interspecies Interaction." *The ISME Journal* 10(6):1323–1336.

Frye, K., S. Abend, W. Bartosch, A. Bawamia, D. Becker, H. Blume, C. Braxmaier, et al. 2021. "The Bose-Einstein Condensate and Cold Atom Laboratory." *EPJ Quantum Technology* 8(1).

Fu, Y., L. Li, B. Xie, C. Dong, M. Wang, B. Jia, L. Shao, et al. 2016. "How to Establish a Bioregenerative Life Support System for Long-Term Crewed Missions to the Moon or Mars." *Astrobiology* 16(12):925–936.

Fujita, O. 2015. "Solid Combustion Research in Microgravity as a Basis of Fire Safety in Space." *Proceedings of the Combustion Institute* 35(3):2487–2502.

Fujita, S.I., L. Rutter, Q. Ong, and M. Muratani. 2020. "Integrated RNA-Seq Analysis Indicates Asynchrony in Clock Genes Between Tissues Under Spaceflight." *Life (Basel)* 10(9).

Furukawa, S., A. Nagamatsu, M. Nenoi, A. Fujimori, S. Kakinuma, T. Katsube, B. Wang, et al. 2020. "Space Radiation Biology for 'Living in Space.' " *BioMed Research Intitute* 2020:4703286.

Galhardo, R.S., P.J. Hastings, and S.M. Rosenberg. 2007. "Mutation as a Stress Response and the Regulation of Evolvability." *Critical Reviews in Biochemistry and Molecular Biology* 42(5):399–435.

Gallo, C., L. Ridolfi, and S. Scarsoglio. 2020. "Cardiovascular Deconditioning During Long-Term Spaceflight Through Multiscale Modeling." *npj Microgravity* 6(1):27.

Gambara, G., M. Salanova, S. Ciciliot, S. Furlan, M. Gutsmann, G. Schiffl, U. Ungethuem, P. Volpe, H.C. Gunga, and D. Blottner. 2017. "Microgravity-Induced Transcriptome Adaptation in Mouse Paraspinal Longissimus Dorsi Muscle Highlights Insulin Resistance-Linked Genes." *Frontiers in Physiology* 8:279.

Ganesan, V., R. Patel, J. Hartwig, and I. Mudawar. 2021. "Universal Critical Heat Flux (CHF) Correlations for Cryogenic Flow Boiling in Uniformly Heated Tubes." *International Journal of Heat and Mass Transfer* 166:120678.

Garland, J.L., K.L. Cook, M. Johnson, R. Sumner, and N. Fields. 1997. "Density and Composition of Microorganisms During Long-Term (418 Day) Growth of Potato Using Biologically Reclaimed Nutrients from Inedible Plant Biomass." *Advances in Space Research* 20(10):1931–1937.

Garrett-Bakelman, F.E., M. Darshi, S.J. Green, R.C. Gur, L. Lin, B.R. Macias, M.J. McKenna, et al. 2019. "The NASA Twins Study: A Multidimensional Analysis of a Year-Long Human Spaceflight." *Science* 364(6436). https://doi.org/10.1126/science.aau8650.

Gaskill, M. 2019. "Solving the Challenges of Long Duration Space Flight with 3D Printing." https://www.nasa.gov/mission_pages/station/research/news/3d-printing-in-space-long-duration-spaceflight-applications.

Gatens, R. 2022. "The International Space Station: The Decade of Results." Presentation to the Committee on Biological and Physical Sciences in Space (March 23, 2022). https://www.nationalacademies.org/event/03-21-2022/committee-on-biological-and-physical-sciences-in-space-space-science-week-2022.

Gavira, J.A., F. Otálora, L.A. González-Ramírez, E. Melero, A.E.S.v. Driessche, and J.M. García-Ruíz. 2020. "On the Quality of Protein Crystals Grown Under Diffusion Mass-Transport Controlled Regime (I)." *Crystals* 10(2):68.

Georgescu, I. 2020. "25 Years of BEC." *Nature Reviews Physics* 2(8):396.

Giannone, M. 2017. "From 2D to 3D, Space Station Microscope Gets an Upgrade." https://www.nasa.gov/feature/from-2d-to-3d-space-station-microscope-gets-an-upgrade.

Gilbert, R., N. Tanenbaum, and S. Bhattacharya. 2022. "Asparagine Biosynthesis as a Mechanism of Increased Host Lethality Induced by Serratia Marcescens in Simulated Microgravity Environments." *Heliyon* 8(5):e09379.

Gilbert, R., M. Torres, R. Clemens, S. Hateley, R. Hosamani, W. Wade, and S. Bhattacharya. 2020. "Spaceflight and Simulated Microgravity Conditions Increase Virulence of Serratia Marcescens in the Drosophila Melanogaster Infection Model." *npj Microgravity* 6(1):4.

Giorgini, S., L.P. Pitaevskii, and S. Stringari. 2008. "Theory of Ultracold Atomic Fermi Gases." *Reviews of Modern Physics* 80(4):1215–1274.

Giovanetti, A., F. Tortolici, and S. Rufini. 2020. "Why Do the Cosmic Rays Induce Aging?" *Frontiers in Physiology* 11:955.

Glaubitz, B., K. Kullack, W. Dreier, A. Seidel, W. Soellner, A. Diefenbach, and S. Schneider. 2015. "The Electro-Magnetic Levitator (EML) on Board the ISS—an Overview and Outlook." *Deutscher Luft- und Raumfahrtkongress* 1–9.

Gòdia, F., J. Albiol, J. Montesinos, J. Pérez, N. Creus, F. Cabello, X. Mengual, A. Montras and C. Lasseur. 2002. "MELiSSA: A Loop of Interconnected Bioreactors to Develop Life Support in Space." *Journal of Biotechnology* 99(3):319–330.

Goel, N., T.L. Bale, C.N. Epperson, S.G. Kornstein, G.R. Leon, L.A. Palinkas, J.W. Stuster, and D.F. Dinges. 2014. "Effects of Sex and Gender on Adaptation to Space: Behavioral Health." *Journal of Womens Health (Larchmt)* 23(11):975–986.

Gold, W.M., and L.L. Koth. 2016. "Pulmonary Function Testing." *Murray and Nadel's Textbook of Respiratory Medicine* 407–435.

Golla, B.R., A. Mukhopadhyay, B. Basu, and S.K. Thimmappa. 2020. "Review on Ultra-High Temperature Boride Ceramics." *Progress in Materials Science* 111:100651.

Good, B.H., M.J. McDonald, J.E. Barrick, R.E. Lenski, and M.M. Desai. 2017. "The Dynamics of Molecular Evolution Over 60,000 Generations." *Nature* 551(7678):45–50.

Gordon, V.D., and Wang, L. 2019. "Bacterial Mechanosensing: The Force Will Be with You, Always." *Journal of Cell Science* 132(7):jcs227694.

Graham, T., R. Scorza, R. Wheeler, B. Smith, C. Dardick, A. Dixit, D. Raines, et al. 2015. "Over-Expression of FT1 in Plum (*Prunus domestica*) Results in Phenotypes Compatible with Spaceflight: A Potential New Candidate Crop for Bioregenerative Life Support Systems." *Gravitational and Space Research* 3(1):39–50.

Grant, K.C., C.L.M. Khodadad, and J.S. Foster. 2014. "Role of Hfq in an Animal–Microbe Symbiosis Under Simulated Microgravity Conditions." *International Journal of Astrobiology* 13(1):53–61.

Green, M.J., J.W. Aylott, P. Williams, A.M. Ghaemmaghami, and P.M. Williams. 2021. "Immunity in Space: Prokaryote Adaptations and Immune Response in Microgravity." *Life* 11(2):112.

Greiner, M., O. Mandel, T. Esslinger, T.W. Hänsch, and I. Bloch. 2002. "Quantum Phase Transition from a Superfluid to a Mott Insulator in a Gas of Ultracold Atoms." *Nature* 415(6867):39–44.

Groover, A. 2016. "Gravitropisms and Reaction Woods of Forest Trees—Evolution, Functions and Mechanisms." *New Phytologist* 211:790–802.

Guéguinou, N., C. Huin-Schohn, M. Bascove, J.L. Bueb, E. Tschirhart, C. Legrand-Frossi, and J.P. Frippiat. 2009. "Could Spaceflight-Associated Immune System Weakening Preclude the Expansion of Human Presence Beyond Earth's Orbit?" *Journal of Leukocyte Biology* 86(5):1027–1038.

Guerges, M. 2021. "NASA Additively Manufactured Rocket Engine Hardware Passes Cold Spray, Hot Fire Tests." https://www.nasa.gov/centers/marshall/news/releases/2021/nasa-additively-manufactured-rocket-engine-hardware-passes-cold-spray-hot-fire-tests.html.

Guibaud, A., G. Legros, J.-L. Consalvi, and J. Torero. 2022. "Fire Safety in Spacecraft: Past Incidents and Deep Space Challenges." *Acta Astronautica* 195:344–354.

Guillamat, P., Z. Kos, J. Hardoüin, J. Ignés-Mullol, M. Ravnik, and F. Sagués. 2018. "Active Nematic Emulsions." *Science Advances* 4(4):eaao1470. https://www.science.org/doi/10.1126/sciadv.aao1470.

Guimarces, A.R., I. Correia, I. Sousa, C. Oliveira, G. Moura, A.R. Bezerra, and M.A.S. Santos. 2021. "tRNAs as a Driving Force of Genome Evolution in Yeast." *Frontiers in Microbiology* 12:634004.

Guntur, S.V., C. Mackowiak, and R.M. Wheeler. 1999. "Recycling of Na in Advanced Life Support: Strategies Based on Crop Production Systems." *Life Support & Biosphere Science* 6(2):153–160.

Guzman, A. 2023. "ISS Research Results Resources Library." NASA. https://www.nasa.gov/mission_pages/station/research/results_category.

Häder, D.P., M. Braun, D. Grimm, and R. Hemmersbach. 2017. "Gravireceptors in Eukaryotes—a Comparison of Case Studies on the Cellular Level." *npj Microgravity* 3:13.

Hadler, K., D.J.P. Martin, J. Carpenter, J.J. Cilliers, A. Morse, S. Starr, J.N. Rasera, K. Seweryn, P. Reiss, and A. Meurisse. 2020. "A Universal Framework for Space Resource Utilisation (SRU)." *Planetary and Space Science* 182:104811.

Haines, S.R., A. Bope, J.M. Horack, M.E. Meyer, and K.C. Dannemiller. 2019. "Quantitative Evaluation of Bioaerosols in Different Particle Size Fractions in Dust Collected on the International Space Station (ISS)." *Applied Microbiology and Biotechnology* 103(18):7767–7782.

Hall, N., K. Logsdon, and K. Magee. 2006. "Shear History Extensional Rheology Experiment: A Proposed ISS Experiment." *44th AIAA Aerospace Sciences Meeting and Exhibit*.

Hall, N., G. McKinley, P. Erni, J. Soulages, and K. Magee. 2009. "Preliminary Findings from the SHERE ISS Experiment." *47th AIAA Aerospace Sciences Meeting Including the New Horizons Forum and Aerospace Exposition*.

Hamant, O., M.G. Heisler, H. Jönsson, P. Krupinski, M. Uyttewaal, P. Bokov, F. Corson, et al. 2008. "Developmental Patterning by Mechanical Signals in Arabidopsis." *Science (New York)* 322(5908):1650–1655.

Handakumbura, P., K. Hixson, and A. Rivas-Ubach. 2022. *C4 Photosynthesis in Space*. https://taskbook.nasaprs.com/tbp/index.cfm?action=public_query_taskbook_content&TASKID=14806.

Hangarter, R.P. 1997. "Gravity, Light and Plant Form." *Plant, Cell & Environment* 20(6):796–800.

Hansen, H.C., W.L. Johnson, M. Meyer, A. Werkheiser, and J.R. Stephens. 2020. "Cryogenic Fluid Management Technologies Enabling for the Artemis Program and Beyond." Paper presented at ASCEND. https://ntrs.nasa.gov/citations/20205008297.

Hao, Z., L. Li, Y. Fu, and H. Liu. 2018. "The Influence of Bioregenerative Life-Support System Dietary Structure and Lifestyle on the Gut Microbiota: A 105-Day Ground-Based Space Simulation in Lunar Palace 1." *Environmental Microbiology* 20(10):3643–3656.

Hargens, A.R., and S. Richardson. 2009. "Cardiovascular Adaptations, Fluid Shifts, and Countermeasures Related to Space Flight." *Respiratory Physiology & Neurobiology* 169(Suppl 1):S30–S33.

Hargens, A.R., and D.E. Watenpaugh. 1996. "Cardiovascular Adaptation to Spaceflight." *Medicine & Science in Sports & Exercise* 28(8):977–982.

Harm, D.L., R.T. Jennings, J.V. Meck, M.R. Powell, L. Putcha, C.P. Sams, S.M. Schneider, L.C. Shackelford, S.M. Smith, and P.A. Whitson. 2001. "Invited Review: Gender Issues Related to Spaceflight: A NASA Perspective." *Journal of Applied Physiology* 91(5):2374–2383.

Hartwig, J., S. Darr, and A. Asencio. 2016. "Assessment of Existing Two Phase Heat Transfer Coefficient and Critical Heat Flux Correlations for Cryogenic Flow Boiling in Pipe Quenching Experiments." *International Journal of Heat and Mass Transfer* 93:441–463.

Hasterok, R., P. Catalan, S.P. Hazen, A.C. Roulin, J.P. Vogel, K. Wang, and L.A.J. Mur. 2022. "Brachypodium: 20 Years as a Grass Biology Model System; the Way Forward?" *Trends in Plant Science* 27(10):1002–1016.

Haveman, N.J., C.L.M. Khodadad, A.R. Dixit, A.S. Louyakis, G.D. Massa, K. Venkateswaran, and J.S. Foster. 2021. "Evaluating the Lettuce Metatranscriptome with Minion Sequencing for Future Spaceflight Food Production Applications." *npj Microgravity* 7(1):22.

Haveman, N.J.S., and A.C. Schuerger. 2022. "Diagnosing an Opportunistic Fungal Pathogen on Spaceflight-Grown Plants Using the Minion Sequencing Platform." *Astrobiology* 22(1):1–6.

Haveman, N., M. Settles, A. Zupanska, T. Graham, B. Califar, J. Callaham, D. Jha, et al. 2023. "Elevating the Use of Genetic Engineering to Support Sustainable Plant Agriculture for Human Space Exploration." Input paper submitted to the Committee on Decadal Survey on Biological and Physical Sciences Research in Space 2023–2032. https://www.nationalacademies.org/our-work/decadal-survey-on-life-and-physical-sciences-research-in-space-2023-2032.

He, M., J.P. Gales, É. Ducrot, Z. Gong, G. Yi, S. Sacanna, and D.J. Pine. 2020. "Colloidal Diamond." *Nature* 585:524–529.

Heard, E., and R.A. Martienssen. 2014. "Transgenerational Epigenetic Inheritance: Myths and Mechanisms." *Cell* 157(1): 95–109.

Heinse, R., S.B. Jones, M. Tuller, G.E. Bingham, I.G. Podolski and D. Or. 2009. "Providing Optimal Root-Zone Fluid Fluxes: Effects of Hysteresis on Capillary-Dominated Water Distributions in Reduced Gravity," Technical Paper 2009-01-2360. SAE International.

Heinse, R., O. Monje, J. Richards, N. Dufour, and N. Lewis. 2022. "Root-Zone Water Stress Assessment for the Advanced Plant Habitat (APH) Inaugural Mission on the International Space Station (ISS)." Paper presented at the 44th COSPAR Scientific Assembly, Athens, Greece, July 16–24.

Heisler, M.G., O. Hamant, P. Krupinski, M. Uyttewaal, C. Ohno, H. Jönsson, et al. 2010. "Alignment Between PIN1 Polarity and Microtubule Orientation in the Shoot Apical Meristem Reveals a Tight Coupling Between Morphogenesis and Auxin Transport." *PLOS Biology* 8(10):e1000516.

Hendrickx, L., H. De Wever, V. Hermans, F. Mastroleo, N. Morin, A. Wilmotte, P. Janssen, and M. Mergeay. 2006. "Microbial Ecology of the Closed Artificial Ecosystem MELiSSA (Micro-Ecological Life Support System Alternative): Reinventing and Compartmentalizing the Earth's Food and Oxygen Regeneration System for Long-Haul Space Exploration Missions." *Research in Microbiology* 157(1):77–86.

Hernández-Coronado, M., P.C. Dias Araujo, P.-L. Ip, C.O. Nunes, R. Rahni, M.M. Wudick, M.A. Lizzio, J.A. Feijó, and K.D. Birnbaum. 2022. "Plant Glutamate Receptors Mediate a Bet-Hedging Strategy Between Regeneration and Defense." *Developmental Cell* 57(4):451–465.

Herranz, R., J.P. Vandenbrink, A. Villacampa, A. Manzano, W.L. Poehlman, F.A. Feltus, J.Z. Kiss, and F.J. Medina. 2019. "RNAseq Analysis of the Response of *Arabidopsis thaliana* to Fractional Gravity Under Blue-Light Stimulation During Spaceflight." Original Research. *Frontiers in Plant Science* 10.

Herzog, D., V. Seyda, E. Wycisk, and C. Emmelmann. 2016. "Additive Manufacturing of Metals." *Acta Materialia* 117:371–392.

Heveran, C.M., S.L. Williams, J. Qiu, J. Artier, M.H. Hubler, S.M. Cook, J.C. Cameron, and W.V. Srubar. 2020. "Biomineralization and Successive Regeneration of Engineered Living Building Materials." *Matter* 2(2):481–494. https://doi.org/10.1016/j.matt.2019.11.016.

Hewezi, T., V. Pantalone, M. Bennett, C. Neal Stewart, Jr., and T.M. Burch-Smith. 2018. "Phytopathogen-Induced Changes to Plant Methylomes." *Plant Cell Reports* 37(1):17–23.

Higashitani, A., T. Hashizume, M. Takiura, N. Higashitani, M. Teranishi, R. Oshima, S. Yano, K. Kuriyama, and A. Higashibata. 2021. "Histone Deacetylase HDA-4-Mediated Epigenetic Regulation in Space-Flown C. Elegans." *npj Microgravity* 7(1):33.

Hilzinger, J.M., S. Freidline, D. Sivanandan, Y.-F. Cheng, S. Yamazaki, D.S. Clark, J.M. Skerker, and A.P. Arkin. 2022. "Acetaminophen Production in the Edible, Filamentous Cyanobacterium *Arthrospira Platensis*." *bioRxiv* 2022.06.30.498297, preprint.

Ho, D., G. Makrygiorgos, A. Hill, and A.J. Berliner. 2022. "Towards an Extension of Equivalent System Mass for Human Exploration Missions on Mars." *npj Microgravity* 8(1):30.

Hofstra, B., V.V. Kulkarni, S. Munoz-Najar Galvez, B. He, D. Jurafsky, and D.A. McFarland. 2020. "The Diversity–Innovation Paradox in Science." *Proceedings of the National Academy of Sciences* 117(17):9284–9291.

Holtze, S., E. Gorshkova, S. Braude, A. Cellerino, P. Dammann, T.B. Hildebrandt, A. Hoeflich, et al. 2021. "Alternative Animal Models of Aging Research." Review. *Frontiers in Molecular Biosciences* 8.

Hołyńska, M., A. Tighe, and C. Semprimoschnig. 2018. "Coatings and Thin Films for Spacecraft Thermo-Optical and Related Functional Applications." *Advanced Materials Interfaces* 5(11):1701644.

Honda, Y., S. Honda, M. Narici, and N.J. Szewczyk. 2014. "Spaceflight and Ageing: Reflecting on Caenorhabditis Elegans in Space." *Gerontology* 60(2):138–142.

Hopkins, S.R., A.C. Henderson, D.L. Levin, K. Yamada, T. Arai, R.B. Buxton, and G.K. Prisk. 2007. "Vertical Gradients in Regional Lung Density and Perfusion in the Supine Human Lung: The Slinky Effect." *Journal of Applied Physiology* 103(1):240–248.

Horneck, G., D.M. Klaus, and R.L. Mancinelli. 2010. "Space Microbiology." *Microbiology and Molecular Biology Reviews* 74(1):121–156.

Horneck, G., R. Moeller, J. Cadet, T. Douki, R.L. Mancinelli, W.L. Nicholson, C. Panitz, et al. 2012. "Resistance of Bacterial Endospores to Outer Space for Planetary Protection Purposes—Experiment Protect of the Expose-E Mission." *Astrobiology* 12(5):445–456.

Hoson, T., and K. Wakabayashi. 2015. "Role of the Plant Cell Wall in Gravity Resistance." *Phytochemistry* 112:84–90.

Hughes-Fulford, M. 2003. "Function of the Cytoskeleton in Gravisensing During Spaceflight." *Advances in Space Research* 32(8):1585–1593.

Hughes-Fulford, M., T.T. Chang, E.M. Martinez, and C.F. Li. 2015. "Spaceflight Alters Expression of Microrna During T-Cell Activation." *FASEB Journal* 29(12):4893–4900.

Hughson, R.L., A.D. Robertson, P. Arbeille, J.K. Shoemaker, J.W. Rush, K.S. Fraser, and D.K. Greaves. 2016. "Increased Postflight Carotid Artery Stiffness and Inflight Insulin Resistance Resulting from 6-Mo Spaceflight in Male and Female Astronauts." *American Journal Physiology Heart and Circulatory Physiology* 310(5):H628–H638.

Hupfeld, K.E., H.R. McGregor, V. Koppelmans, N.E. Beltran, I.S. Kofman, Y.E. De Dios, R.F. Riascos, et al. 2022. "Brain and Behavioral Evidence for Reweighting of Vestibular Inputs with Long-Duration Spaceflight." *Cerebral Cortex* 32(4):755–769.

Hupfeld, K.E., H.R. McGregor, P.A. Reuter-Lorenz, and R.D. Seidler. 2021. "Microgravity Effects on the Human Brain and Behavior: Dysfunction and Adaptive Plasticity." *Neuroscience and Biobehavioral Reviews* 122:176–189.

Hurlbert, K.M., L.C. Witte, F.R. Best, and C. Kurwitz. 2004. "Scaling Two-Phase Flows to Mars and Moon Gravity Conditions." *International Journal of Multiphase Flow* 30(4):351–368.

Hutcheson, J.D., and E. Aikawa. 2018. "Extracellular Vesicles in Cardiovascular Homeostasis and Disease." *Current Opinion in Cardiology* 33(3):290–297.

Hyman, A.A., and K. Simons. 2012. "Cell Biology. Beyond Oil and Water—Phase Transitions in Cells." *Science* 337(6098): 1047–1049.

Ibáñez, S., E. Carneros, P.S. Testillano, and J.M. Pérez-Pérez. 2020. "Advances in Plant Regeneration: Shake, Rattle and Roll." *Plants (Basel)* 9(7).

Ijioma, E.R., H. Izuhara, M. Mimura, and T. Ogawa. 2015. "Homogenization and Fingering Instability of a Microgravity Smoldering Combustion Problem with Radiative Heat Transfer." *Combustion and Flame* 162(10):4046–4062.

Ikeuchi, M., B. Rymen, and K. Sugimoto. 2020. "How Do Plants Transduce Wound Signals to Induce Tissue Repair and Organ Regeneration?" *Current Opinion in Plant Biology* 57:72–77.

Inguscio, M., W. Ketterle, and C. Salomon. 2008. "Ultra-Cold Fermi Gases." *Proceedings of the International School of Physics "Enrico Fermi"* 164. Amsterdam, Netherlands: IOS Press.

Inoue, K., H. Ohta, H. Asano, O. Kawanami, R. Imai, K. Suzuki, Y. Shinmoto, T. Kurimoto, and S. Matsumoto. 2021a. "Heat Loss Analysis of Flow Boiling Experiments Onboard International Space Station with Unclear Thermal Environmental Conditions (2nd Report: Liquid-Vapor Two-Phase Flow Conditions at Test Section Inlet)." *Microgravity Science and Technology* 33(5):57.

Inoue, K., H. Ohta, Y. Toyoshima, H. Asano, O. Kawanami, R. Imai, K. Suzuki, Y. Shinmoto, and S. Matsumoto. 2021b. "Heat Loss Analysis of Flow Boiling Experiments Onboard International Space Station with Unclear Thermal Environmental Conditions (1st Report: Subcooled Liquid Flow Conditions at Test Section Inlet)." *Microgravity Science and Technology* 33(2):28.

Isachenkov, M., S. Chugunov, I. Akhatov, and I. Shishkovsky. 2021. "Regolith-Based Additive Manufacturing for Sustainable Development of Lunar Infrastructure—an Overview." *Acta Astronautica* 180:650–678.

ISSNL (ISS National Laboratory). 2017. "The ISS National Lab and NCATS Announce Five Projects Selected from International Space Station Funding Opportunity Focused on Human Physiology Research." http://www.iss-casis.org/iss360/casis-and-ncats-announce-five-projects-selected-from-international-space-station-funding-opportunity-focused-on-human-physiology-research.

ISSNL. 2018. *ISS U.S. National Lab 2018 Annual Report*. Melbourne, FL. https://www.issnationallab.org/ar2018.

ISSNL. 2023. "In-Space Production Applications: Advanced Manufacturing and Materials." Center for the Advancement of Science in Space. https://www.issnationallab.org/ispa-materials-manufacturing.

Ivanova, T., S. Sapunova, P. Kostov, and I. Dandolov. 2001. "First Successful Space Seed-to-Seed Plant Growth Experiment in the SVET-2 Space Greenhouse in 1997." *Aerospace Research in Bulgaria* (16):12–23.

Jaalouk, D.E., and J. Lammerding. 2009. "Mechanotransduction Gone Awry." *Nature Reviews: Molecular Cell Biology* 10(1):63–73.

Jackson, N.E., M.A. Webb, and J.J. de Pablo. 2019. "Recent Advances in Machine Learning Towards Multiscale Soft Materials Design." *Current Opinion in Chemical Engineering* 23:106–114.

Jain, M.R., M. Li, W. Chen, T. Liu, S.M. de Toledo, B.N. Pandey, H. Li, B.M. Rabin, and E.I. Azzam. 2011. "In Vivo Space Radiation-Induced Non-Targeted Responses: Late Effects on Molecular Signaling in Mitochondria." *Current Molecular Pharmacology* 4(2):106–114.

Jaishankar, A., S. Haward, N.R. Hall, K. Magee, and G. McKinley. 2012. "Shear History Extensional Rheology Experiment II (SHERE II) Microgravity Rheology with Non-Newtonian Polymeric Fluids." Poster No. GRC-E-DAA-TN7182. https://ntrs.nasa.gov/api/citations/20130011410/downloads/20130011410.pdf?attachment=true.

James, J.T., and N. Kahn-Mayberry. 2009. "Human Health and Performance Risks of Space Exploration Missions." Ch. 13 in *Risk of Adverse Health Effects from Lunar Dust Exposure*. NASA Johnson Space Center.

Jayawardena, S.S., V. Balakotaiah, and L.C. Witte. 1997. "Flow Pattern Transition Maps for Microgravity Two-Phase Flows." *AIChE Journal* 43(6):1637.

Jemison, M., and R. Olabisi. 2021. "Biomaterials for Human Space Exploration: A Review of Their Untapped Potential." *Acta Biomaterialia* 128:77–99.

Jerde, E.A. 2021. "Chapter 2—the Apollo Program." Pp. 9–36 in *Sample Return Missions*, A. Longobardo, ed. Amsterdam, Netherlands: Elsevier.

Jiang, P., S.J. Green, G.E. Chlipala, F.W. Turek, and M.H. Vitaterna. 2019. "Reproducible Changes in the Gut Microbiome Suggest a Shift in Microbial and Host Metabolism During Spaceflight." *Microbiome* 7(1):113.

John, S., F. Abou-Issa, and K.H. Hasenstein. 2021. "Space Flight Cultivation for Radish in the Advanced Plant Habitat." *Gravitational and Space Research* 9(1):121–132.

Johnson, C.M., H.O. Boles, L.E. Spencer, L. Poulet, M. Romeyn, J.M. Bunchek, R. Fritsche, G.D. Massa, A. O'Rourke, and R.M. Wheeler. 2021. "Supplemental Food Production with Plants: A Review of NASA Research." Mini Review. *Frontiers in Astronomy and Space Sciences* 8.

Johnson, C.M., A. Subramanian, S. Pattathil, M.J. Correll, and J.Z. Kiss. 2017. "Comparative Transcriptomics Indicate Changes in Cell Wall Organization and Stress Response in Seedlings During Spaceflight." *American Journal of Botany* 104(8):1219–1231.

Johnson, M. 2022. "Publication Metrics from the International Space Station Results." https://www.nasa.gov/mission_pages/station/research/news/b4h-3rd/sv-publication-metrics.

Johnson, W.L., and J.R. Stephens. 2018. "NASA's Cryogenic Fluid Management Technology Development Roadmaps." Paper presented at the JANNAF In-Space Chemical Propulsion TIM, Huntsville, AL, August 27. https://ntrs.nasa.gov/citations/20190000305.

Jonscher, K.R., A. Alfonso-Garcia, J.L. Suhalim, D.J. Orlicky, E.O. Potma, V.L. Ferguson, M.L. Bouxsein, et al. 2016. "Correction: Spaceflight Activates Lipotoxic Pathways in Mouse Liver." *PLOS One* 11(5):e0155282.

Jordan, S.P., K.S.M. Lee, and J. Preskill. 2012. "Quantum Algorithms for Quantum Field Theories." *Science* 336(6085):1130–1133.

Joseph, C., and D. Wood. 2019. "Understanding Socio-Technical Issues Affecting the Current Microgravity Research Marketplace." Paper presented at the 2019 IEEE Aerospace Conference, 2–9 March 2019.

Joseph, C., and D. Wood. 2021. "Analysis of the Microgravity Research Ecosystem and Market Drivers of Accessibility." *New Space* 9(2):123–138.

Ju, Y. 2021. "Understanding Cool Flames and Warm Flames." *Proceedings of the Combustion Institute* 38(1):83–119.

Juhl, O.J., E.G. Buettmann, M.A. Friedman, R.C. DeNapoli, G.A. Hoppock, and H.J. Donahue. 2021. "Update on the Effects of Microgravity on the Musculoskeletal System." *npj Microgravity* 7(1):28.

Jumper, J., R. Evans, A. Pritzel, T. Green, M. Figurnov, O. Ronneberger, K. Tunyasuvunakool, et al. 2021. "Highly Accurate Protein Structure Prediction with Alphafold." *Nature* 596(7873):583–589.

Just, G.H., K. Smith, K.H. Joy, and M.J. Roy. 2020. "Parametric Review of Existing Regolith Excavation Techniques for Lunar in Situ Resource Utilisation (ISRU) and Recommendations for Future Excavation Experiments." *Planetary and Space Science* 180:104746.

Kakoulidou, I., E.V. Avramidou, M. Baránek, S. Brunel-Muguet, S. Farrona, F. Johannes, E. Kaiserli, et al. 2021. "Epigenetics for Crop Improvement in Times of Global Change." *Biology* 10(8):766.

Kamada, M., M. Oka, K. Miyamoto, E. Uheda, C. Yamazaki, T. Shimazu, H. Sano, et al. 2020. "Microarray Profile of Gene Expression in Etiolated Pisum Sativum Seedlings Grown Under Microgravity Conditions in Space: Relevance to the International Space Station Experiment 'Auxin Transport.' " *Life Sciences in Space Research* 26:55–61.

Karl, D., F. Kamutzki, A. Zocca, O. Goerke, J. Guenster, and A. Gurlo. 2018. "Towards the Colonization of Mars by in-Situ Resource Utilization: Slip Cast Ceramics from Martian Soil Simulant." *PLOS One* 13(10):e0204025.

Kennedy, A.R., B. Crucian, J.L. Huff, S.L. Klein, D. Morens, D. Murasko, C.A. Nickerson, and G. Sonnenfeld. 2014. "Effects of Sex and Gender on Adaptation to Space: Immune System." *Journal of Women's Health (Larchmt)* 23(11):956–958.

Kenyon, C.J. 2010. "The Genetics of Ageing." *Nature* 464(7288):504–512.

Khadka, U., V. Holubec, H. Yang, and F. Cichos. 2018. "Active Particles Bound by Information Flows." *Nature Communications* 9(1):3864.

Khalil, A.S., and J.J. Collins. 2010. "Synthetic Biology: Applications Come of Age." *Nature Reviews: Genetics* 11(5):367–379.

Kharangate, C.R., C. Konishi, and I. Mudawar. 2016. "Consolidated Methodology to Predicting Flow Boiling Critical Heat Flux for Inclined Channels in Earth Gravity and for Microgravity." *International Journal of Heat and Mass Transfer* 92:467–482.

Kharlamova, A., A. Proshchina, V. Gulimova, Y. Krivova, P. Soldatov, and S. Saveliev. 2021. "Cerebellar Morphology and Behavioural Correlations of the Vestibular Function Alterations in Weightlessness." *Neuroscience and Biobehavioral Reviews* 126:314–328.

Khodadad, C.L.M., M.E. Hummerick, L.E. Spencer, A.R. Dixit, J.T. Richards, M.W. Romeyn, T.M. Smith, R.M. Wheeler, and G.D. Massa. 2020. "Microbiological and Nutritional Analysis of Lettuce Crops Grown on the International Space Station." Original Research. *Frontiers in Plant Science* 11:199.

Khodadad, C.L.M., C.M. Oubre, V.A. Castro, S.M. Flint, M.C. Roman, C.M. Ott, C.J. Spern, et al. 2021. "A Microbial Monitoring System Demonstrated on the International Space Station Provides a Successful Platform for Detection of Targeted Microorganisms." *Life* 11(6):492.

Khossravi, E.A., and A.R. Hargens. 2021. "Visual Disturbances During Prolonged Space Missions." *Current Opinion in Ophthalmology* 32(1):69–73.

Kiffer, F., T. Alexander, J. Anderson, T. Groves, T. McElroy, J. Wang, V. Sridharan, M. Bauer, M. Boerma, and A. Allen. 2020. "Late Effects of ^{1}H + ^{16}O on Short-Term and Object Memory, Hippocampal Dendritic Morphology and Mutagenesis." Original Research. *Frontiers in Behavioral Neuroscience* 14.

Kim, H.N., K.K. Richardson, K.J. Krager, W. Ling, P. Simmons, A.R. Allen, and N. Aykin-Burns. 2021. "Simulated Galactic Cosmic Rays Modify Mitochondrial Metabolism in Osteoclasts, Increase Osteoclastogenesis and Cause Trabecular Bone Loss in Mice." *International Journal of Molecular Sciences* 22(21).

Kim, S.-M., and I. Mudawar. 2013. "Universal Approach to Predicting Saturated Flow Boiling Heat Transfer in Mini/Micro-Channels—Part II. Two-Phase Heat Transfer Coefficient." *International Journal of Heat and Mass Transfer* 64:1239–1256.

Kim, W., F.K. Tengra, J. Shong, N. Marchand, H.K. Chan, Z. Young, R.C. Pangule, et al. 2013. "Effect of Spaceflight on Pseudomonas Aeruginosa Final Cell Density Is Modulated by Nutrient and Oxygen Availability." *BMC Microbiology* 13:241.

Kim, W., F.K. Tengra, Z. Young, J. Shong, N. Marchand, H.K. Chan, R.C. Pangule, et al. 2013. "Spaceflight Promotes Biofilm Formation by Pseudomonas Aeruginosa." *PLOS One* 8(4):e62437.

Kirke, J., X.-L. Jin, and X.-H. Zhang. 2020. "Expression of a Tardigrade Dsup Gene Enhances Genome Protection in Plants." *Molecular Biotechnology* 62:563–571.

Kiss, J.Z., E. Brinckmann, and C. Brillouet. 2000. "Development and Growth of Several Strains of Arabidopsis Seedlings in Microgravity." *International Journal of Plant Sciences* 161(1):55–62.

Kiss, J.Z., K.D. Millar, and R.E. Edelmann. 2012. "Phototropism of Arabidopsis Thaliana in Microgravity and Fractional Gravity on the International Space Station." *Planta* 236(2):635–645.

Kliss, M. 2016. "Understanding the NASA TA6: Human Health, Life Support, and Habitation." Paper for the 46th International Conference on Environmental Systems, Vienna. https://ttu-ir.tdl.org/handle/2346/67489.

Klopp, C., T. Trittel, A. Eremin, K. Harth, R. Stannarius, C.S. Park, J.E. Maclennan, and N.A. Clark. 2019. "Structure and Dynamics of a Two-Dimensional Colloid of Liquid Droplets." *Soft Matter* 15(40):8156–8163.

Knox, B.P., A. Blachowicz, J.M. Palmer, J. Romsdahl, A. Huttenlocher, C.C.C. Wang, N.P. Keller, and K. Venkateswaran. 2016. "Characterization of Aspergillus Fumigatus Isolates from Air and Surfaces of the International Space Station." *mSphere* 1(5).

Kobayashi, H., Y.N. Graven, W.J. Broughton, and X. Perret. 2004. "Flavonoids Induce Temporal Shifts in Gene-Expression of Nod-Box Controlled Loci in Rhizobium sp. NGR234." *Molecular Microbiology* 51(2):335–347.

Kodger, T.E., P.J. Lu, G.R. Wiseman, and D.A. Weitz. 2017. "Stable, Fluorescent Polymethylmethacrylate Particles for the Long-Term Observation of Slow Colloidal Dynamics." *Langmuir* 33(25):6382–6389.

Kohn, F.P.M., and R. Ritzmann. 2018. "Gravity and Neuronal Adaptation, in Vitro and in Vivo—from Neuronal Cells Up to Neuromuscular Responses: A First Model." *European Biophysics Journal* 47(2):97–107.

Kolkowitz, S., I. Pikovski, N. Langellier, M.D. Lukin, R.L. Walsworth, and J. Ye. 2016. "Gravitational Wave Detection with Optical Lattice Atomic Clocks." *Physical Review D* 94(12):124043.

Konicki, J., and J. Pethokoukis. 2022. "Do Americans Care About Space?" *The New Atlantis*. https://www.aei.org/articles/do-americans-care-about-space.

Konishi, C., H. Lee, I. Mudawar, M.M. Hasan, H.K. Nahra, N.R. Hall, J.D. Wagner, R.L. May, and J.R. Mackey. 2015a. "Flow Boiling in Microgravity: Part 1—Interfacial Behavior and Experimental Heat Transfer Results." *International Journal of Heat and Mass Transfer* 81:705–720.

Konishi, C., H. Lee, I. Mudawar, M.M. Hasan, H.K. Nahra, N.R. Hall, J.D. Wagner, R.L. May, and J.R. Mackey. 2015b. "Flow Boiling in Microgravity: Part 2—Critical Heat Flux Interfacial Behavior, Experimental Data, and Model." *International Journal of Heat and Mass Transfer* 81:721–736.

Kono, M., K. Ito, T. Niioka, T. Kadota, and J.i. Sato. 1996. "Current State of Combustion Research in Microgravity." *Symposium (International) on Combustion* 26(1):1189–1199.

Konstantinova, N., B. Korbei, and C. Luschnig. 2021. "Auxin and Root Gravitropism: Addressing Basic Cellular Processes by Exploiting a Defined Growth Response." *International Journal of Molecular Sciences* 22(5):2749.

Koo, H., R.N. Allan, R.P. Howlin, P. Stoodley, and L. Hall-Stoodley. 2017. "Targeting Microbial Biofilms: Current and Prospective Therapeutic Strategies." *Nature Reviews Microbiology* 15(12):740–755.

Koos, E., and N. Willenbacher. 2011. "Capillary Forces in Suspension Rheology." *Science* 331(6019):897–900.

Krebs, R.A., and V. Loeschcke. 1994. "Response to Environmental Change: Genetic Variation and Fitness in Drosophila Buzzatii Following Temperature Stress." Pp. 309–321 in *Conservation Genetics*, V. Loeschcke, S.K. Jain, and J. Tomiuk, eds. Basel, Switzerland: Birkhäuser Basel.

Kruse, C.P.S., P. Basu, D.R. Luesse, and S.E. Wyatt. 2017. "Transcriptome and Proteome Responses in RNAlater Preserved Tissue of Arabidopsis Thaliana." *PLOS One* 12(4):e0175943.

Kruse, C.P.S., A.D. Meyers, P. Basu, S. Hutchinson, D.R. Luesse, and S.E. Wyatt. 2020. "Spaceflight Induces Novel Regulatory Responses in Arabidopsis Seedling as Revealed by Combined Proteomic and Transcriptomic Analyses." *BMC Plant Biology* 20(1):237.

Kruse, C.P.S., and S.E. Wyatt. 2022. "Nitric Oxide, Gravity Response, and a Unified Schematic of Plant Signaling." *Plant Science* 314:111105.

Kuang, A., A. Popova, G. McClure, and M.E. Musgrave. 2005. "Dynamics of Storage Reserve Deposition During Brassica Rapa L. Pollen and Seed Development in Microgravity." *International Journal of Plant Sciences* 166(1):85–96.

Kuang, A., A. Popova, Y. Xiao, and M.E. Musgrave. 2000. "Pollination and Embryo Development in Brassica Rapa L. in Microgravity." *International Journal of Plant Sciences* 161(2):203–211.

Kültz, D. 2020a. "Defining Biological Stress and Stress Responses Based on Principles of Physics." *Journal of Experimental Zoology Part A: Ecological and Integrative Physiology* 333(6):350–358.

Kültz, D. 2020b. "Evolution of Cellular Stress Response Mechanisms." *Journal of Experimental Zoology Part A: Ecological and Integrative Physiology* 333(6):359–378.

Kumar, S., A.S. Beena, M. Awana, and A. Singh. 2017. "Physiological, Biochemical, Epigenetic and Molecular Analyses of Wheat (Triticum Aestivum) Genotypes with Contrasting Salt Tolerance." *Frontiers in Plant Science* 8:1151.

Kumar, S.P., and M.B. Plenio. 2020. "On Quantum Gravity Tests with Composite Particles." *Nature Communications* 11(1):3900.

Kundrot, C.E., R.A. Judge, M.L. Pusey, and E.H. Snell. 2001. "Microgravity and Macromolecular Crystallography." *Crystal Growth & Design* 1(1):87–99.

Kwon, T., J.A. Sparks, J. Nakashima, S.N. Allen, Y. Tang, and E.B. Blancaflor. 2015. "Transcriptional Response of Arabidopsis Seedlings During Spaceflight Reveals Peroxidase and Cell Wall Remodeling Genes Associated with Root Hair Development." *American Journal of Botany* 102(1):21–35.

La Spina, M., P.S. Contreras, A. Rissone, N.K. Meena, E. Jeong, and J.A. Martina. 2021. "MiT/TFE Family of Transcription Factors: An Evolutionary Perspective." Review. *Frontiers in Cell and Developmental Biology* 8.

Lackner, J.R., and P. Dizio. 2006. "Space Motion Sickness." *Experimental Brain Research* 175(3):377–399.

Laiakis, E.C., I. Shuryak, A. Deziel, Y.W. Wang, B.L. Barnette, Y. Yu, R.L. Ullrich, A.J. Fornace, Jr., and M.R. Emmett. 2021. "Effects of Low Dose Space Radiation Exposures on the Splenic Metabolome." *International Journal of Molecular Sciences* 22(6).

Lam, C.W., V. Castranova, P.C. Zeidler-Erdely, R. Renne, R. Hunter, R. McCluskey, R.R. Scully, et al. 2022. "Comparative Pulmonary Toxicities of Lunar Dusts and Terrestrial Dusts (TiO_2 & SiO_2) in Rats and an Assessment of the Impact of Particle-Generated Oxidants on the Dusts' Toxicities." *Inhalation Toxicology* 34(3–4):51–67.

Lam, C.W., R.R. Scully, Y. Zhang, R.A. Renne, R.L. Hunter, R.A. McCluskey, B.T. Chen, et al. 2013. "Toxicity of Lunar Dust Assessed in Inhalation-Exposed Rats." *Inhalation Toxicology* 25(12):661–678.

Lang, C., and M. Bamsay. "International Partnership Opportunities in a Lunar Surface Plant Production Demonstrator, Research Campaign White Paper." Input paper submitted to the Committee on Decadal Survey on Biological and Physical Sciences Research in Space 2023–2032. https://www.nationalacademies.org/our-work/decadal-survey-on-life-and-physical-sciences-research-in-space-2023-2032.

Langevin, D., and M. Vignes-Adler. 2014. "Microgravity Studies of Aqueous Wet Foams." *The European Physical Journal E* 37(3):16.

Langin, K. 2022. "Pandemic Led to Historic Drop in U.S. STEM Ph.D. Graduates, New Data Suggest." *Science*. https://www.science.org/content/article/pandemic-led-historic-drop-u-s-stem-ph-d-graduates-new-data-suggest.

Larson, M., C. Talso, A. Croonquist, W. Holmes, D. Langford, L. Feng-Chuan, and J. Pensinger. 2002. "A Low Temperature Facility for Experiments on the International Space Station." Conference paper. P. 1-1 in *Proceedings, IEEE Aerospace Conference*. Big Sky, MT: Institute of Electrical and Electronics Engineers.

Law, C.K., and G.M. Faeth. 1994. "Opportunities and Challenges of Combustion in Microgravity." *Progress in Energy and Combustion Science* 20(1):65–113.

Leach, J.E., M. Ryba-White, Q. Sun, C.J. Wu, E. Hilaire, C. Gartner, O. Nedukha, et al. 2001. "Plants, Plant Pathogens, and Microgravity—a Deadly Trio." *Gravitational and Space Biology Bulletin* 14(2):15–23.

Leblanc, A., T. Matsumoto, J. Jones, J. Shapiro, T. Lang, L. Shackelford, S.M. Smith, et al. 2013. "Bisphosphonates as a Supplement to Exercise to Protect Bone During Long-Duration Spaceflight." *Osteoporosis International* 24(7):2105–2114.

Lee, H., I. Mudawar, and M.M. Hasan. 2013a. "Experimental and Theoretical Investigation of Annular Flow Condensation in Microgravity." *International Journal of Heat and Mass Transfer* 61:293–309.

Lee, H., I. Mudawar, and M.M. Hasan. 2013b. "Flow Condensation in Horizontal Tubes." *International Journal of Heat and Mass Transfer* 66:31–45.

Lee, H.S., H. Merte Jr., and F. Chiaramonte. 1997. "Pool Boiling Curve in Microgravity." *Journal of Thermophysics and Heat Transfer* 11(2):216–222.

Lee, J., L.E. O'Neill, S. Lee, and I. Mudawar. 2019. "Experimental and Computational Investigation on Two-Phase Flow and Heat Transfer of Highly Subcooled Flow Boiling in Vertical Upflow." *International Journal of Heat and Mass Transfer* 136:1199–1216.

Lee, J., L.E. O'Neill, and I. Mudawar. 2020. "3-D Computational Investigation and Experimental Validation of Effect of Shear-Lift on Two-Phase Flow and Heat Transfer Characteristics of Highly Subcooled Flow Boiling in Vertical Upflow." *International Journal of Heat and Mass Transfer* 150:119291.

Lee, M.C., and U.E. Israelsson. 2003. "Fundamental Physics Research Aboard the International Space Station." *Physica B: Condensed Matter* 329–333:1649–1650.

Lee, M.D., A. O'Rourke, H. Lorenzi, B.M. Bebout, C.L. Dupont, and R.C. Everroad. 2021. "Reference-Guided Metagenomics Reveals Genome-Level Evidence of Potential Microbial Transmission from the ISS Environment to an Astronaut's Microbiome." *iScience* 24(2):102114.

Lee, S.H., I. Mudawar, and M.M. Hasan. 2016. "Thermal Analysis of Hybrid Single-Phase, Two-Phase and Heat Pump Thermal Control System (TCS) for Future Spacecraft." *Applied Thermal Engineering* 100:190–214.

Lee, S.-J., A. Lehar, J.U. Meir, C. Koch, A. Morgan, L.E. Warren, R. Rydzik, et al. 2020. "Targeting Myostatin/Activin a Protects Against Skeletal Muscle and Bone Loss During Spaceflight." *Proceedings of the National Academy of Sciences* 117(38):23942–23951.

Lee, W.C., H. Shi, Z. Poon, L.M. Nyan, T. Kaushik, G.V. Shivashankar, J.K.Y. Chan, C.T. Lim, J. Han, and K.J. Van Vliet. 2014. "Multivariate Biophysical Markers Predictive of Mesenchymal Stromal Cell Multipotency." *Proceedings of the National Academy of Sciences* 111(42):E4409–E4418.

Lehner, B.A.E., J. Schlechten, A. Filosa, A. Canals Pou, D.G. Mazzotta, F. Spina, L. Teeney, et al. 2019. "End-to-End Mission Design for Microbial ISRU Activities as Preparation for a Moon Village." *Acta Astronautica* 162:216–226.

Lenski, R.E., and M. Travisano. 1994. "Dynamics of Adaptation and Diversification: A 10,000-Generation Experiment with Bacterial Populations." *Proceedings of the National Academy of Sciences* 91(15):6808–6814.

Levernier, N., O. Pouliquen, and Y. Forterre. 2021. "An Integrative Model of Plant Gravitropism Linking Statoliths Position and Auxin Transport." *Frontiers in Plant Science* 12:651928.

Levine, H.G. 2010. "The Influence of Microgravity on Plants" Presentation at NASA ISS Research Academy and Pre-Application Meeting. August 3–5. League City, TX. https://www.nasa.gov/pdf/478076main_Day1_P03c_Levine_Plants.pdf.

Lewis, N.G., L. Davin, and M. Kahn. 2022. "Dissecting Beneficial Plant-Microbe Interactions and Their Efficacy in the ISS Spaceflight Environment, A Model Study." https://taskbook.nasaprs.com/tbp/index.cfm?action=public_query_taskbook_content&TASKID=15156.

Leyser, O. 2005. "Auxin Distribution and Plant Pattern Formation: How Many Angels Can Dance on the Point of PIN?" *Cell* 121(6):819–822.

Li, B., L. Sun, J. Huang, C. Göschl, W. Shi, J. Chory, and W. Busch. 2019. "GSNOR Provides Plant Tolerance to Iron Toxicity Via Preventing Iron-Dependent Nitrosative and Oxidative Cytotoxicity." *Nature Communications* 10(1):3896.

Li, F., Y. Liu, and J. Leng. 2019. "Progress of Shape Memory Polymers and Their Composites in Aerospace Applications." *Smart Materials and Structures* 28(10):103003.

Li, N., C. Wang, S. Sun, C. Zhang, D. Lü, Q. Chen, and M. Long. 2018. "Microgravity-Induced Alterations of Inflammation-Related Mechanotransduction in Endothelial Cells on Board SJ-10 Satellite." Original Research. *Frontiers in Physiology* 9.

Liao, J., G. Liu, O. Monje, G.W. Stutte, and D.M. Porterfield. 2004. "Induction of Hypoxic Root Metabolism Results from Physical Limitations in O_2 Bioavailability in Microgravity." *Advances in Space Research* 34(7):1579–1584.

Limbert, G. 2017. "Mathematical and Computational Modelling of Skin Biophysics: A Review." *Proceedings of the Royal Society A: Mathematical, Physical and Engineering Sciences* 473(2203):20170257.

Limoli, D.H., G.B. Whitfield, T. Kitao, M.L. Ivey, M.R. Davis, Jr., N. Grahl, D.A. Hogan, et al. 2017. "Pseudomonas Aeruginosa Alginate Overproduction Promotes Coexistence with Staphylococcus Aureus in a Model of Cystic Fibrosis Respiratory Infection." *mBio* 8(2).

Lin, H., F. Rosenberger, J.I.D. Alexander, and A. Nadarajah. 1995. "Convective-Diffusive Transport in Protein Crystal Growth." *Journal of Crystal Growth* 151(1):153–162.

Lindsay, J. 1976. "The Lunar Soil." In *Lunar Stratigraphy and Sedimentology*. Amsterdam, Netherlands: Elsevier.

Link, B.M., S.J. Durst, W. Zhou, and B. Stankovic. 2003. "Seed-to-Seed Growth of Arabidopsis Thaliana on the International Space Station." *Advances in Space Research* 31(10):2237–2243.

Lithgow, G.J., T.M. White, S. Melov, and T.E. Johnson. 1995. "Thermotolerance and Extended Life-Span Conferred by Single-Gene Mutations and Induced by Thermal Stress." *Proceedings of the National Academy of Sciences* 92(16):7540–7544.

Liu, D., B. Xie, Y. Dong, and H. Liu. 2020. "Semi-Continuous Fermentation of Solid Waste in Closed Artificial Ecosystem: Microbial Diversity, Function Genes Evaluation." *Life Sciences in Space Research* 25:136–142.

Liu, H., L. Tian, and D. Wang. 2021. "Notch Receptor GLP-1 Regulates Toxicity of Simulated Microgravity Stress by Activating Germline-Intestine Communication of Insulin Signaling in C. elegans." *Biochemical and Biophysical Research Communications* 534:248–253.

Liu, Y., W. Rosikiewicz, Z. Pan, N. Jillette, P. Wang, A. Taghbalout, J. Foox, et al. 2021. "DNA Methylation-Calling Tools for Oxford Nanopore Sequencing: A Survey and Human Epigenome-Wide Evaluation." *Genome Biology* 22(1):295.

Llorente, B., T.C. Williams, and H.D. Goold. 2018. "The Multiplanetary Future of Plant Synthetic Biology." *Genes (Basel)* 9(7).

Lloyd, S.A., G.S. Lewis, Y. Zhang, E.M. Paul, and H.J. Donahue. 2012. "Connexin 43 Deficiency Attenuates Loss of Trabecular Bone and Prevents Suppression of Cortical Bone Formation During Unloading." *Journal of Bone and Mineral Research* 27(11):2359–2372.

Lloyd, S.A., A.E. Loiselle, Y. Zhang, and H.J. Donahue. 2013. "Connexin 43 Deficiency Desensitizes Bone to the Effects of Mechanical Unloading Through Modulation of Both Arms of Bone Remodeling." *Bone* 57(1):76–83.

Lo, Y.M.D., D.S.C. Han, P. Jiang, and R.W.K. Chiu. 2021. "Epigenetics, Fragmentomics, and Topology of Cell-Free DNA in Liquid Biopsies." *Science* 372(6538):eaaw3616.

Locey, K.J., and J.T. Lennon. 2016. "Scaling Laws Predict Global Microbial Diversity." *Proceedings of the National Academy of Sciences* 113(21):5970–5975.

Lombardino, J., S. Bijlani, N.K. Singh, J.M. Wood, R. Barker, S. Gilroy, C.C.C. Wang, and K. Venkateswaran. 2022. "Genomic Characterization of Potential Plant Growth-Promoting Features of *Sphingomonas* Strains Isolated from the International Space Station." *Microbiology Spectrum* 10(1):e01994-01921.

Lopatkin, N.A., V.P. Kozlov, and M.A. Grishin. 1992. "[Cancer of the Kidney: Nephrectomy or Resection?]." *Urologiia i Nefrologiia* (4–6):3–6.

Lu, H., G. Yuan, S.H. Strauss, T.J. Tschaplinski, G.A. Tuskan, J.-G. Chen, and X. Yang. 2020. "Reconfiguring Plant Metabolism for Biodegradable Plastic Production." *BioDesign Research*.

Lu, Z., J. Cui, L. Wang, N. Teng, S. Zhang, H.-M. Lam, Y. Zhu, et al. 2021. "Genome-Wide DNA Mutations in Arabidopsis Plants After Multigenerational Exposure to High Temperatures." *Genome Biology* 22(1):160.

Lubensky, T.C., C.L. Kane, X. Mao, A. Souslov, and K. Sun. 2015. "Phonons and Elasticity in Critically Coordinated Lattices." *Reports on Progress in Physics* 78(7):073901.

Ludlow, A.D., M.M. Boyd, J. Ye, E. Peik, and P.O. Schmidt. 2015. "Optical Atomic Clocks." *Reviews of Modern Physics* 87(2): 637–701.

Ludtka, C., J. Silberman, E. Moore, and J.B. Allen. 2021. "Macrophages in Microgravity: The Impact of Space on Immune Cells." *npj Microgravity* 7(1):13.

Lüthi, M.N., A.E. Berardi, T. Mandel, L.B. Freitas, and C. Kuhlemeier. 2022. "Single Gene Mutation in a Plant MYB Transcription Factor Causes a Major Shift in Pollinator Preference." *Current Biology* 32(24):5295–5308.

Mackowiak, C.L., J.L. Garland, R.F. Strayer, B.W. Finger, and R.M. Wheeler. 1996. "Comparison of Aerobically-Treated and Untreated Crop Residue as a Source of Recycled Nutrients in a Recirculating Hydroponic System." *Advances in Space Research* 18(1–2):281–287.

Mackowiak, C.L., R.M. Wheeler, G.W. Stutte, N.C. Yorio, and J.C. Sager. 1997. "Use of Biologically Reclaimed Minerals for Continuous Hydroponic Potato Production in a CELSS." *Advances in Space Research* 20(10):1815–1820.

MacLean, R.C., C. Torres-Barceló, and R. Moxon. 2013. "Evaluating Evolutionary Models of Stress-Induced Mutagenesis in Bacteria." *Nature Reviews: Genetics* 14(3):221–227.

Mader, T.H., C.R. Gibson, A.F. Pass, L.A. Kramer, A.G. Lee, J. Fogarty, W.J. Tarver, et al. 2011. "Optic Disc Edema, Globe Flattening, Choroidal Folds, and Hyperopic Shifts Observed in Astronauts After Long-Duration Space Flight." *Ophthalmology* 118(10):2058–2069.

Malkani, S., C.R. Chin, E. Cekanaviciute, M. Mortreux, H. Okinula, M. Tarbier, A.S. Schreurs, et al. 2020. "Circulating Mirna Spaceflight Signature Reveals Targets for Countermeasure Development." *Cell Reports* 33(10):108448.

Mallory, S.A., C. Valeriani, and A. Cacciuto. 2018. "An Active Approach to Colloidal Self-Assembly." *Annual Review of Physical Chemistry* 69(1):59–79.

Manian, V., J. Orozco, H. Gangapuram, H. Janwa, and C. Agrinsoni. 2021. "Network Analysis of Gene Transcriptions of Arabidopsis Thaliana in Spaceflight Microgravity." *Genes* 12(3):337.

Mann, V., A. Sundaresan, S.K. Mehta, B. Crucian, M.F. Doursout, and S. Devakottai. 2019. "Effects of Microgravity and Other Space Stressors in Immunosuppression and Viral Reactivation with Potential Nervous System Involvement." *Neurology India* 67(Suppl):S198–S203.

Manzano, A., E. Carnero-Diaz, R. Herranz, and F.J. Medina. 2022. "Recent Transcriptomic Studies to Elucidate the Plant Adaptive Response to Spaceflight and to Simulated Space Environments." *iScience* 25(8):104687.

Mao, Q., A. Acharya, A. Rodríguez-delaRosa, F. Marchiano, B. Dehapiot, Z. Al Tanoury, J. Rao, et al. 2022. "Tension-Driven Multi-Scale Self-Organisation in Human iPSC-Derived Muscle Fibers." *Elife* 11.

Mao, X.W., M. Boerma, D. Rodriguez, M. Campbell-Beachler, T. Jones, S. Stanbouly, V. Sridharan, A. Wroe, and G.A. Nelson. 2018. "Acute Effect of Low-Dose Space Radiation on Mouse Retina and Retinal Endothelial Cells." *Radiation Research* 190(1):45–52.

Mao, X.W., S. Byrum, N.C. Nishiyama, M.J. Pecaut, V. Sridharan, M. Boerma, A.J. Tackett, et al. 2018. "Impact of Spaceflight and Artificial Gravity on the Mouse Retina: Biochemical and Proteomic Analysis." *International Journal of Molecular Sciences* 19(9).

Mao, X.W., N.C. Nishiyama, S.D. Byrum, S. Stanbouly, T. Jones, J. Holley, V. Sridharan, et al. 2020. "Spaceflight Induces Oxidative Damage to Blood-Brain Barrier Integrity in a Mouse Model." *FASEB Journal* 34(11):15516–15530.

Marchetti, M.C., J.F. Joanny, S. Ramaswamy, T.B. Liverpool, J. Prost, M. Rao, and R.A. Simha. 2013. "Hydrodynamics of Soft Active Matter." *Reviews of Modern Physics* 85(3):1143–1189.

Marfia, G., S.E. Navone, L. Guarnaccia, R. Campanella, M. Locatelli, M. Miozzo, P. Perelli, et al. 2022. "Space Flight and Central Nervous System: Friends or Enemies? Challenges and Opportunities for Neuroscience and Neuro-Oncology." *Journal of Neuroscience Research* 100(9):1649–1663.

Mark, S., G.B. Scott, D.B. Donoviel, L.B. Leveton, E. Mahoney, J.B. Charles, and B. Siegel. 2014. "The Impact of Sex and Gender on Adaptation to Space: Executive Summary." *Journal of Women's Health (Larchmt)* 23(11):941–947.

Marre, S., and K.F. Jensen. 2010. "Synthesis of Micro and Nanostructures in Microfluidic Systems." 10.1039/B821324K. *Chemical Society Reviews* 39(3):1183–1202.

Marshall-Goebel, K., S.S. Laurie, I.V. Alferova, P. Arbeille, S.M. Auñón-Chancellor, D.J. Ebert, S.M.C. Lee, et al. 2019. "Assessment of Jugular Venous Blood Flow Stasis and Thrombosis During Spaceflight." *JAMA Network Open* 2(11):e1915011.

Martin, N., J. Singh, and A. Aballay. 2017. "Natural Genetic Variation in the Caenorhabditis Elegans Response to Pseudomonas Aeruginosa." *G3 (Bethesda)* 7(4):1137–1147.

Martinovich, S. 2016. "Collegiate Inventors Awarded 2016 Lemelson-MIT Student Prize." MIT News, Massachusetts Institute of Technology. https://news.mit.edu/2016/collegiate-inventors-awarded-lemelson-mit-student-prize-0412.

Mashinsky, A., I. Ivanova, T. Derendyaeva, G. Nechitailo, and F. Salisbury. 1994. " 'From Seed-to-Seed' Experiment with Wheat Plants Under Space-Flight Conditions." *Advances in Space Research* 14(11):13–19.

Masica, W.J., and J.A. Salzman. 1967. *Experimental Investigation of Liquid-Propellant Reorientation*. NASA TN D-3789. Washington, DC: NASA.

Masini, M.A., V. Bonetto, M. Manfredi, A. Pastò, E. Barberis, S. Timo, V.V. Vanella, et al. 2022. "Prolonged Exposure to Simulated Microgravity Promotes Stemness Impairing Morphological, Metabolic and Migratory Profile of Pancreatic Cancer Cells: A Comprehensive Proteomic, Lipidomic and Transcriptomic Analysis." *Cellular and Molecular Life Sciences* 79(5):226.

Massa, G.D., N.F. Dufour, J.A. Carver, M.E. Hummerick, R.M. Wheeler, R.C. Morrow, and T.M. Smith. 2017. "VEG-01: Veggie Hardware Validation Testing on the International Space Station." *Open Agriculture* 2(1):33–41.

Massa, G., G. Newsham, M.E. Hummerick, J.L. Caro, G.W. Stutte, R.C. Morrow, and R.M. Wheeler. 2013. "Preliminary Species and Media Selection for the Veggie Space Hardware." *Gravitational and Space Research* 1(1):95–106.

Massa, G.D., G. Newsham, M.E. Hummerick, R.C. Morrow, and R.M. Wheeler. 2017. "Plant Pillow Preparation for the Veggie Plant Growth System on the International Space Station." *Gravitational and Space Research* 5(1):24–34.

Masson, P.H. 2019. *Using Brachypodium Distachyon to Investigate Monocot Plant Adaptation to Spaceflight*. https://taskbook.nasaprs.com/Tbp/index.cfm?action=public_query_taskbook_content&TASKID=12402.

Masson, P.H., and S.-H. Su. 2022. *Can Polyamines Mitigate Plant Stress Response Under Microgravity Conditions?* https://taskbook.nasaprs.com/Tbp/index.cfm?action=public_query_taskbook_content&TASKID=15023.

Mathelié-Guinlet, M., F. Viela, D. Alsteens, and Y.F. Dufrêne. 2021. "Stress-Induced Catch-Bonds to Enhance Bacterial Adhesion." *Trends in Microbiology* 29(4):286–288.

Matsumura, T., T. Noda, M. Muratani, R. Okada, M. Yamane, A. Isotani, T. Kudo, S. Takahashi, and M. Ikawa. 2019. "Male Mice, Caged in the International Space Station for 35 Days, Sire Healthy Offspring." *Scientific Reports* 9(1):13733.

Maurer, M., and J. Lammerding. 2019. "The Driving Force: Nuclear Mechanotransduction in Cellular Function, Fate, and Disease." *Annual Review of Biomedical Engineering* 21:443–468.

May, S. 2012. "What Is Microgravity?" *NASA Knows! (Grades 5–8)*. https://www.nasa.gov/audience/forstudents/5-8/features/nasa-knows/what-is-microgravity-58.html.

Mazars, C., C. Brière, S. Grat, C. Pichereaux, M. Rossignol, V. Pereda-Loth, B. Eche, et al. 2014. "Microsome-Associated Proteome Modifications of Arabidopsis Seedlings Grown on Board the International Space Station Reveal the Possible Effect on Plants of Space Stresses Other Than Microgravity." *Plant Signaling & Behavior* 9(9):e29637.

Mazzolai, B., F. Tramacere, I. Fiorello, and L. Margheri. 2020. "The Bio-Engineering Approach for Plant Investigations and Growing Robots. A Mini-Review." Mini Review. *Frontiers in Robotics and AI* 7.

McIntyre, A.B.R., N. Alexander, K. Grigorev, D. Bezdan, H. Sichtig, C.Y. Chiu, and C.E. Mason. 2019. "Single-Molecule Sequencing Detection of N6-Methyladenine in Microbial Reference Materials." *Nature Communications* 10(1):579.

McLean, R.J.C., and T.J. Beveridge. 1990. "Metal Binding Capacity of Bacterial Surfaces and Their Ability to Form Mineralized Aggregates." Pp. 185–222 in *Microbial Mineral Recovery*, H.L. Ehrlich and C.L. Brierley, eds. New York: McGraw-Hill.

McPherson, A., and L.J. DeLucas. 2015. "Microgravity Protein Crystallization." *npj Microgravity* 1(1):15010.

McSweeney, K. 2021. "Diversity and Inclusion Create Better Outcomes: A Look at the Science." Now powered by Northrop Grumman. https://now.northropgrumman.com/diversity-and-inclusion-create-better-outcomes-a-look-at-the-science.

Mehlstäubler, T., G. Grosche, C. Lisdat, P. Schmidt, and H. Denker. 2018. "Atomic Clocks for Geodesy." *Reports on Progress in Physics* 81(6):064401.

Mehta, S.K., D.C. Bloom, I. Plante, R. Stowe, A.H. Feiveson, A. Renner, A. Dhummakupt, et al. 2018. "Reactivation of Latent Epstein-Barr Virus: A Comparison After Exposure to Gamma, Proton, Carbon, and Iron Radiation." *International Journal of Molecular Sciences* 19(10).

Mehta, S.K., M.L. Laudenslager, R.P. Stowe, B.E. Crucian, C.F. Sams, and D.L. Pierson. 2014. "Multiple Latent Viruses Reactivate in Astronauts During Space Shuttle Missions." *Brain, Behavior, and Immunity* 41:210–217.

Mendes, R., P. Garbeva, and J.M. Raaijmakers. 2013. "The Rhizosphere Microbiome: Significance of Plant Beneficial, Plant Pathogenic, and Human Pathogenic Microorganisms." *FEMS Microbiology Reviews* 37(5):634–663.

Menezes, A.A., J. Cumbers, J.A. Hogan, and A.P. Arkin. 2015. "Towards Synthetic Biological Approaches to Resource Utilization on Space Missions." *Journal of The Royal Society Interface* 12(102):20140715.

MetaSUB International Consortium. 2016. "The Metagenomics and Metadesign of the Subways and Urban Biomes (MetaSUB) International Consortium Inaugural Meeting Report." *Microbiome* 4(1):24.

Meyer, C., and W. Schneider. 2018. "NASA Advanced Explorations Systems: 2018 Advancements in Life Support Systems." Paper presented at the AIAA SPACE and Astronautics Forum and Exposition. https://ntrs.nasa.gov/citations/20180006596.

Meyers, A., and S.E. Wyatt. 2022. "Plant Space Biology in the Genomics Age." *Annual Plant Reviews* 5:123–150.

Mezzenga, R. 2021. "Grand Challenges in Soft Matter." *Frontiers in Soft Matter* 1(December).

Mhatre, S.D., J. Iyer, S. Puukila, A.M. Paul, C.G.T. Tahimic, L. Rubinstein, M. Lowe, et al. 2022. "Neuro-Consequences of the Spaceflight Environment." *Neuroscience and Biobehavioral Reviews* 132:908–935.

Mickens, M.A., M. Torralba, S.A. Robinson, L.E. Spencer, M.W. Romeyn, G.D. Massa, and R.M. Wheeler. 2019. "Growth of Red Pak Choi Under Red and Blue, Supplemented White, and Artificial Sunlight Provided by LEDs." *Scientia Horticulturae* 245:200–209.

Miles II, J.D., and G. Lunn. 2013. "Separation and Purification of Mineral Salts from Spacecraft Wastewater Processing via Electrostatic Beneficiation." NASA Report KSC-2013-213.

Millar, K.D.L., P. Kumar, M.J. Correll, J.L. Mullen, R.P. Hangarter, R.E. Edelmann, and J.Z. Kiss. 2010. "A Novel Phototropic Response to Red Light Is Revealed in Microgravity." *New Phytologist* 186(3):648–656.

Mironova, V., and J. Xu. 2019. "A Single-Cell View of Tissue Regeneration in Plants." *Current Opinion in Plant Biology* 52:149–154.

Mishra, B., and U. Luderer. 2019. "Reproductive Hazards of Space Travel in Women and Men." *Nature Reviews Endocrinology* 15(12):713–730.

Modica, K.J., Y. Xi, and S.C. Takatori. 2022. "Porous Media Microstructure Determines the Diffusion of Active Matter: Experiments and Simulations." Original Research. *Frontiers in Physics* 10.

Monje, O., J.T. Richards, J.A. Carver, D.I. Dimapilis, H.G. Levine, N.F. Dufour, and B.G. Onate. 2020. "Hardware Validation of the Advanced Plant Habitat on ISS: Canopy Photosynthesis in Reduced Gravity." Original Research. *Frontiers in Plant Science* 11.

Monje, O., G.W. Stutte, G.D. Goins, D.M. Porterfield, and G.E. Bingham. 2003. "Farming in Space: Environmental and Biophysical Concerns." *Advances in Space Research* 31(1):151–167.

Moore, A.D., Jr., M.E. Downs, S.M. Lee, A.H. Feiveson, P. Knudsen, and L. Ploutz-Snyder. 2014. "Peak Exercise Oxygen Uptake During and Following Long-Duration Spaceflight." *Journal of Applied Physiology (1985)* 117(3):231–238.

Moosavi, D., D. Wolovsky, A. Depompeis, D. Uher, D. Lennington, R. Bodden, and C.E. Garber. 2021. "The Effects of Spaceflight Microgravity on the Musculoskeletal System of Humans and Animals with an Emphasis on Exercise as a Countermeasure: A Systematic Scoping Review." *Physiological Research* 70:119–151.

Mora, M., A. Perras, T.A. Alekhova, L. Wink, R. Krause, A. Aleksandrova, T. Novozhilova, and C. Moissl-Eichinger. 2016. "Resilient Microorganisms in Dust Samples of the International Space Station—Survival of the Adaptation Specialists." *Microbiome* 4(1):65.

Moreno-Villanueva, M., M. Wong, T. Lu, Y. Zhang, and H. Wu. 2017. "Interplay of Space Radiation and Microgravity in DNA Damage and DNA Damage Response." *npj Microgravity* 3(1):14.

Morris, J.F. 2020. "Shear Thickening of Concentrated Suspensions: Recent Developments and Relation to Other Phenomena." *Annual Review of Fluid Mechanics* 52(1):121–144.

Morrison, M.D., J.B. Thissen, F. Karouia, S. Mehta, C. Urbaniak, K. Venkateswaran, D.J. Smith, and C. Jaing. 2021. "Investigation of Spaceflight Induced Changes to Astronaut Microbiomes." Original Research. *Frontiers in Microbiology* 12.

Morrow, M.A. 2006. "Clinorotation Differentially Inhibits T-Lymphocyte Transcription Factor Activation." *In Vitro Cellular & Developmental Biology—Animal* 42(5):153–158.

Mortreux, M., and M.E. Rosa-Caldwell. 2020. "Approaching Gravity as a Continuum Using the Rat Partial Weight-Bearing Model." https://doi.org/10.3390/life10100235.

Mota, F.L., Y. Song, J. Pereda, B. Billia, D, Tourret, J. -M. Debierre, R. Trivedi, A. Karma, and N. Bergeon. 2017. "Convection Effects During Bulk Transparent Alloy Solidification in DECLIC-DSI and Phase-Field Simulations in Diffusive Conditions." *Journal of The Minerals, Metals and Materials Society* 69:1280–1288.

Motil, B.J., E. Ramé, P. Salgi, M. Taghavi, and V. Balakotaiah. 2021. "Gas–Liquid Flows Through Porous Media in Microgravity: The International Space Station Packed Bed Reactor Experiment." *AIChE Journal* 67(1):e17031.

Mudawar, I. 2017. "Chapter Five—Flow Boiling and Flow Condensation in Reduced Gravity." Pp. 225–306 in *Advances in Heat Transfer*, E.M. Sparrow, J.P. Abraham, and J.M. Gorman, eds. Amsterdam, Netherlands: Elsevier. Reprint.

Mulavara, A.P., T. Ruttley, H.S. Cohen, B.T. Peters, C. Miller, R. Brady, L. Merkle, and J.J. Bloomberg. 2012. "Vestibular-Somatosensory Convergence in Head Movement Control During Locomotion After Long-Duration Space Flight." *Journal of Vestibular Research* 22(2):153–166.

Musacchia, X.J., and R.E. Barr. 1968. "Survival of Whole-Body-Irradiated Hibernating and Active Ground Squirrels; Citellus Tridecemlineatus." *Radiation Research* 33(2):348–356.

Musgrave, M.E. 2002. "Seeds in Space." *Seed Science Research* 12(1):1–17.

Musgrave, M.E., A. Kuang, and S.W. Matthews. 1997. "Plant Reproduction During Spaceflight: Importance of the Gaseous Environment." *Planta* 203(Suppl):S177–S184.

Musgrave, M.E., A. Kuang, Y. Xiao, S.C. Stout, G.E. Bingham, L.G. Briarty, M.A. Levinskikh, V.N. Sychev, and I.G. Podolski. 2000. "Gravity Independence of Seed-to-Seed Cycling in Brassica Rapa." *Planta* 210(3):400–406.

Nagayama, K., and T. Fukuei. 2020. "Cyclic Stretch-Induced Mechanical Stress to the Cell Nucleus Inhibits Ultraviolet Radiation-Induced DNA Damage." *Biomechanics and Modeling in Mechanobiology* 19(2):493–504.

Najrana, T., and J. Sanchez-Esteban. 2016. "Mechanotransduction as an Adaptation to Gravity." Review. *Frontiers in Pediatrics* 4.

Nakashima, J., J.A. Sparks, J.A. Carver, S.D. Stephens, T. Kwon, and E.B. Blancaflor. 2014. "Delaying Seed Germination and Improving Seedling Fixation: Lessons Learned During Science and Payload Verification Tests for Advanced Plant Experiments (Apex) 02-1 in Space." *Gravitational and Space Research* 2(1):54–67.

Nangle, S.N., M.Y. Wolfson, L. Hartsough, N.J. Ma, C.E. Mason, M. Merighi, V. Nathan, et al. 2020. "The Case for Biotech on Mars." *Nature Biotechnology* 38(4):401–407.

Napoli, A., D. Micheletti, M. Pindo, S. Larger, A. Cestaro, J.-P. de Vera, and D. Billi. 2022. "Absence of Increased Genomic Variants in the Cyanobacterium Chroococcidiopsis Exposed to Mars-Like Conditions Outside the Space Station." *Scientific Reports* 12(1):8437.

NASA (National Aeronautics and Space Administration). 2013. *A Researcher's Guide to the International Space Station—Microbial Research*. Houston, TX: Johnson Space Center. http://www.nasa.gov/sites/default/files/files/Microbial-Observatory-Mini-Book-04-28-14-508.pdf.

NASA. 2017. "Advanced Plant Habitat." https://www.nasa.gov/sites/default/files/atoms/files/advanced-plant-habitat.pdf.

NASA. 2019. "Spectrum." https://www.nasa.gov/sites/default/files/atoms/files/iss_spectrum_flyer_9091020191.pdf.

NASA. 2020a. "Commercial Lunar Payload Services." https://www.nasa.gov/content/commercial-lunar-payload-services.

NASA. 2020b. "Veggie Fact Sheet 508." https://www.nasa.gov/sites/default/files/atoms/files/veggie_fact_sheet_508.pdf.

NASA. 2021. *National Aeronautics and Space Administration (NASA) Model Equal Employment Opportunity Program Status Report: FY 2021*. Washington, DC: Office of Diversity and Equal Opportunity. https://www.nasa.gov/sites/default/files/atoms/files/odeo-fy21_model_715_report_tagged.pdf.

NASA. 2022a. *Aerospace Safety Advisory Panel Annual Report*. Washington, DC: NASA Aerospace Safety Advisory Panel. https://oiir.hq.nasa.gov/asap/documents/ASAP_2022_Annual_Report.pdf.

NASA. 2022b. *International Space Station Transition Report*. Washington, DC. https://www.nasa.gov/sites/default/files/atoms/files/2022_iss_transition_report-final_tagged.pdf.

NASA. 2022c. *Moon to Mars Objectives*. Washington, DC. https://www.nasa.gov/sites/default/files/atoms/files/m2m-objectives-exec-summary.pdf.

NASA. 2022d. *NASA Strategic Plan for Diversity, Equity, Inclusion, and Accessibility*. Washington, DC. https://www.nasa.gov/sites/default/files/atoms/files/nasa_deia_strategic_plan-fy22-fy26-final_tagged.pdf.

NASA. 2022e. "National Aeronautics and Space Administration Policy Statement on Diversity, Equity, Inclusion, and Accessibility for NASA's Workforce and Workplaces." https://www.nasa.gov/sites/default/files/atoms/files/2022_administrator_policy_statement_on_deia_at_nasa_tagged.pdf.

NASA. 2023a. "Commercial Low-Earth Orbit Destination (CLD) Capabilities of Interest and Resource Needs." Washington, DC. https://ntrs.nasa.gov/api/citations/20230003013/downloads/CLD%20Utilization%20Cap_STH%20White%20Paper%20v12b%20%20FINAL%200208_2023.pdf.

NASA. 2023b. "Transform to Open Science (TOPS)." https://science.nasa.gov/open-science/transform-to-open-science.

NASA GeneLab. "Homepage." https://genelab.nasa.gov.

NASEM (National Academies of Sciences, Engineering, and Medicine). 2018. *A Midterm Assessment of Implementation of the Decadal Survey on Life and Physical Sciences Research at NASA*. Washington, DC: The National Academies Press.

NASEM. 2019. *Science Breakthroughs to Advance Food and Agricultural Research by 2030*. Washington, DC: The National Academies Press.

NASEM. 2020. *Manipulating Quantum Systems: An Assessment of Atomic, Molecular, and Optical Physics in the United States*. Washington, DC: The National Academies Press.

NASEM. 2022a. *Advancing Diversity, Equity, Inclusion, and Accessibility in the Leadership of Competed Space Missions*. Washington, DC: The National Academies Press.

NASEM. 2022b. *Physics of Life*. Washington, DC: The National Academies Press.

NASEM. 2023. *Advancing Antiracism, Diversity, Equity, and Inclusion in STEMM Organizations: Beyond Broadening Participation*. Washington, DC: The National Academies Press.

Nath, S., G.M. Aron, G.M. Southard, and R.J. McLean. 2010. "Potential for Largemouth Bass Virus to Associate with and Gain Protection from Bacterial Biofilms." *Journal of Aquatic Animal Health* 22(2):95–101.

Nauman, E.A., C.M. Ott, E. Sander, D.L. Tucker, D. Pierson, J.W. Wilson, and C.A. Nickerson. 2007. "Novel Quantitative Biosystem for Modeling Physiological Fluid Shear Stress on Cells." *Applied and Environmental Microbiology* 73(3): 699–705.

Nayagam, V., D.L. Dietrich, P.V. Ferkul, M.C. Hicks, and F.A. Williams. 2012. "Can Cool Flames Support Quasi-Steady Alkane Droplet Burning?" *Combustion and Flame* 159(12):3583–3588.

Neelam, S., B. Richardson, R. Barker, C. Udave, S. Gilroy, M.J. Cameron, H.G. Levine, and Y. Zhang. 2020. "Changes in Nuclear Shape and Gene Expression in Response to Simulated Microgravity Are Linc Complex-Dependent." *International Journal of Molecular Sciences* 21(18).

Nelson, E.S., L. Mulugeta, and J.G. Myers. 2014. "Microgravity-Induced Fluid Shift and Ophthalmic Changes." *Life (Basel)* 4(4):621–665.

Nguyen, H.P., P.H. Tran, K.-S. Kim, and S.-G. Yang. 2021. "The Effects of Real and Simulated Microgravity on Cellular Mitochondrial Function." *npj Microgravity* 7(1):44.

Ni, Z., S. Zhou, S. Li, L. Kuang, H. Chen, X. Luo, J. Ouyang, M. He, X. Du, and L. Chen. 2020. "Exosomes: Roles and Therapeutic Potential in Osteoarthritis." *Bone Research* 8(1):25.

Nicholson, W.L., and A.J. Ricco. 2020. "Nanosatellites for Biology in Space: In Situ Measurement of Bacillus Subtilis Spore Germination and Growth After 6 Months in Low Earth Orbit on the O/Oreos Mission." *Life* 10(1):1.

Nickerson, C.A., A.A. Medina-Colorado, J. Barrila, G. Poste, and C.M. Ott. 2022. "A Vision for Spaceflight Microbiology to Enable Human Health and Habitat Sustainability." *Nature Microbiology* 7(4):471–474.

Nickerson, C.A., C.M. Ott, S.J. Mister, B.J. Morrow, L. Burns-Keliher, and D.L. Pierson. 2000. "Microgravity as a Novel Environmental Signal Affecting Salmonella Enterica Serovar Typhimurium Virulence." *Infection and Immunity* 68(6):3147–3152.

Nickerson, C.A., C.M. Ott, J.W. Wilson, R. Ramamurthy, and D.L. Pierson. 2004. "Microbial Responses to Microgravity and Other Low-Shear Environments." *Microbiology and Molecular Biology Reviews* 68(2):345–361.

Nickerson, C.A., N.R. Pellis, and C.M. Ott. 2016. *Effect of Spaceflight and Spaceflight Analogue Culture on Human and Microbial Cells: Novel Insights into Disease Mechanisms*. New York: Springer.

NIH (National Institutes of Health). 2018. "2018 Tissue Chips in Space Projects." https://ncats.nih.gov/tissuechip/projects/space2018.

Ninarello, A., L. Berthier, and D. Coslovich. 2017. "Models and Algorithms for the Next Generation of Glass Transition Studies." *Physical Review X* 7(2):021039.

Nonić, M., and M. Šijačić-Nikolić. 2019. "Genetic Diversity: Sources, Threats, and Conservation." Pp. 1–15 in *Life on Land*, W.L. Filho, A.M. Azul, L. Brandli, et al., eds. Cham, Switzerland: Springer International Publishing.

NRC (National Research Council). 2000. *Future Biotechnology Research on the International Space Station*. Washington, DC: National Academy Press.

NRC. 2009. *Evaluation of NSF's Program of Grants for Vertical Integration of Research and Education in the Mathematical Sciences (VIGRE)*. Washington, DC: The National Academies Press.

NRC. 2011. *Recapturing a Future for Space Exploration: Life and Physical Sciences Research for a New Era*. Washington, DC: The National Academies Press.

NSF (National Science Foundation). 2022a. "Demographic Attributes of S&E Degree Recipients." Science and Engineering Indicators. https://ncses.nsf.gov/pubs/nsf23315/report/science-and-engineering-degrees-earned#degrees-earned-by-women-in-broad-s-e-fields.

NSF. 2022b. "U.S. and Global Science and Technology Capabilities." Science and Engineering Indicators. https://ncses.nsf.gov/pubs/nsb20221/u-s-and-global-science-and-technology-capabilities#research-publications.

NSF. 2022c. "U.S. and Global Research and Development." Science and Engineering Indicators. https://ncses.nsf.gov/pubs/nsb20221/u-s-and-global-research-and-development.

NSF. 2022d. "U.S. and Global STEM Education and Labor Force." Science and Engineering Indicators. https://ncses.nsf.gov/pubs/nsb20221/u-s-and-global-stem-education-and-labor-force.

NSTC (National Science and Technology Council). 2022. *Interagency Roadmap to Support Space-Related STEM Education and Workforce*. Office of Science and Technology Policy. https://www.whitehouse.gov/wp-content/uploads/2022/09/09-2022-Interagency-Roadmap-to-Support-Space-Related-STEM-Education-and-Workforce.pdf.

NSTC. 2023. *National Low Earth Orbit Research and Development Strategy*. Low Earth Orbit Science and Technology Interagency Working Group. Washington, DC: Office of Science and Technology Policy. https://www.whitehouse.gov/wp-content/uploads/2023/03/NATIONAL-LEO-RD-STRATEGY-033123.pdf.

NTSS (NASA Technical Standards System). 2011. "Flammability, Offgassing, and Compatibility Requirements and Test Procedures." NASA-STD-6001, version B. Office of the NASA Chief Engineer. https://standards.nasa.gov/standard/NASA/NASA-STD-6001.

NTSS. 2022. "NASA Space Flight Human-System Standard Volume 2: Human Factors, Habitability, and Environmental Health." NASA-STD-3001 Vol 2, version C. Office of the NASA Chief Engineer. https://standards.nasa.gov/standard/NASA/NASA-STD-3001-VOL-2.

Nussle, R.C., E.W. Otto, and D.A. Petrash. 1963. *Effect of Contact Angle and Tank Geometry on the Configuration of the Liquid-Vapor Interface During Weightlessness*. NASA. https://ntrs.nasa.gov/citations/19630013804.

OECD (Organisation for Economic Co-operation and Development). 2021. "How Will COVID-19 Reshape Science, Technology and Innovation?" OECD Policy Responses to Coronavirus (COVID-19). https://www.oecd.org/coronavirus/policy-responses/how-will-covid-19-reshape-science-technology-and-innovation-2332334d.

Ogneva, I.V., and M.A. Usik. 2021. "Mitochondrial Respiration in Drosophila Ovaries After a Full Cycle of Oogenesis Under Simulated Microgravity." *Current Issues in Molecular Biology* 43(1):176–186.

Ogneva, I.V., M.A. Usik, M.V. Burtseva, N.S. Biryukov, Y.S. Zhdankina, V.N. Sychev, and O.I. Orlov. 2020. "Drosophila Melanogaster Sperm under Simulated Microgravity and a Hypomagnetic Field: Motility and Cell Respiration." *International Journal of Molecular Sciences* 21(17).

Ogneva, I.V., M.A. Usik, S.S. Loktev, Y.S. Zhdankina, N.S. Biryukov, O.I. Orlov, and V.N. Sychev. 2019. "Testes and Duct Deferens of Mice During Space Flight: Cytoskeleton Structure, Sperm-Specific Proteins and Epigenetic Events." *Scientific Reports* 9(1):9730.

Ogneva, I.V., Y.S. Zhdankina, and O.V. Kotov. 2022. "Sperm of Fruit Fly Drosophila Melanogaster Under Space Flight." *International Journal of Molecular Sciences* 23(14):7498.

Ohmae, N., M. Takamoto, Y. Takahashi, M. Kokubun, K. Araki, A. Hinton, I. Ushijima, et al. 2021. "Transportable Strontium Optical Lattice Clocks Operated Outside Laboratory at the Level of 10–18 Uncertainty." *Advanced Quantum Technologies* 4(8):2100015.

Ohta, H., H. Asano, O. Kawanami, K. Suzuki, R. Imai, Y. Shinmoto, S. Matsumoto, et al. 2016. "Development of Boiling and Two-Phase Flow Experiments on Board ISS (Research Objectives and Concept of Experimental Setup)." *International Journal of Microgravity Science and Application* 33(1):330102.

Okuno, T., H. Nakamura, T. Tezuka, S. Hasegawa, K. Takase, M. Katsuta, M. Kikuchi, and K. Maruta. 2016. "Study on the Combustion Limit, Near-Limit Extinction Boundary, and Flame Regimes of Low-Lewis-Number $CH_4/O_2/CO_2$ Counterflow Flames Under Microgravity." *Combustion and Flame* 172:13–19.

Okutani, T. 2002. "Materials Synthesis Using Microgravity Circumstances." *Journal of Advanced Science* 14(4). https://www.jstage.jst.go.jp/article/jsas1989/14/4/14_4_143/_pdf/-char/ja.

Onorato, G., E. Di Schiavi, and F. Di Cunto. 2020. "Understanding the Effects of Deep Space Radiation on Nervous System: The Role of Genetically Tractable Experimental Models." Review. *Frontiers in Physics* 8.

Onyak, J.R., M.N. Vergara, and J.M. Renna. 2022. "Retinal Organoid Light Responsivity: Current Status and Future Opportunities." *Translational Research: The Journal of Laboratory and Clinical Medicine* 250:98–111.

Orazi, G., and G.A. O'Toole. 2019. " 'It Takes a Village': Mechanisms Underlying Antimicrobial Recalcitrance of Polymicrobial Biofilms." *Journal of Bacteriology* 202(1):e00530-00519.

O'Rourke, A., M.D. Lee, W.C. Nierman, R.C. Everroad, and C.L. Dupont. 2020. "Genomic and Phenotypic Characterization of Burkholderia Isolates from the Potable Water System of the International Space Station." *PLOS One* 15(2):e0227152.

Ott, C.M., A. Crabbé, J.W. Wilson, J. Barrila, S.L. Castro-Wallace, and C.A. Nickerson. 2020. "Microbial Stress: Spaceflight-Induced Alterations in Microbial Virulence and Infectious Disease Risks for the Crew." Pp. 327–355 in *Stress Challenges and Immunity in Space: From Mechanisms to Monitoring and Preventive Strategies*, A. Choukèr, ed. Cham: Springer International Publishing. Reprint.

Otto, C. 2016. "Fluid Shifts in Microgravity: The Visual Impairment and Intracranial Pressure Syndrome in U.S. Astronauts." *Investigative Ophthalmology and Visual Science* 57(12).

Paez, M.Y., L.I. Mudie, and P.S. Subramanian. 2020. "Spaceflight Associated Neuro-Ocular Syndrome (Sans): A Systematic Review and Future Directions." *Eye Brain* 12:105–117.

Pagnotti, G.M., M. Styner, G. Uzer, V.S. Patel, L.E. Wright, K.K. Ness, T.A. Guise, J. Rubin, and C.T. Rubin. 2019. "Combating Osteoporosis and Obesity with Exercise: Leveraging Cell Mechanosensitivity." *Nature Reviews: Endocrinology* 15(6):339–355.

Palacci, J., S. Sacanna, A.P. Steinberg, D.J. Pine, and P.M. Chaikin. 2013. "Living Crystals of Light-Activated Colloidal Surfers." *Science* 339(6122):936–940.

Panda, C.D., M. Tao, J. Egelhoff, M. Ceja, V. Xu, and H. Müller. 2022. "Minute-Scale Gravimetry Using a Coherent Atomic Spatial Superposition." arXiv:2210.07289 [physics.atom-ph]. https://doi.org/10.48550/arXiv.2210.07289.

Pandey, G., C.B. Yadav, P.P. Sahu, M. Muthamilarasan, and M. Prasad. 2017. "Salinity Induced Differential Methylation Patterns in Contrasting Cultivars of Foxtail Millet (Setaria Italica L.)." *Plant Cell Reports* 36(5):759–772.

Papadopoulos, E., F. Aghili, O. Ma, and R. Lampariello. 2021. "Robotic Manipulation and Capture in Space: A Survey." Review. *Frontiers in Robotics and AI* 8.

Papadopoulos, L., M.A. Porter, K.E. Daniels, and D.S. Bassett. 2018. "Network Analysis of Particles and Grains." *Journal of Complex Networks* 6(4):485–565.

Papike, J.J., S.B. Simon, and J.C. Laul. 1982. "The Lunar Regolith: Chemistry, Mineralogy, and Petrology." *Reviews of Geophysics* 20(4):761–826.

Paradise, R.K., D.A. Lauffenburger, and K.J. Van Vliet. 2011. "Acidic Extracellular pH Promotes Activation of Integrin $\alpha_v\beta_3$." *PLOS One* 6(1):e15746. https://doi.org/10.1371/journal.pone.0015746.

Paradise, R.K., M.J. Whitfield, D.A. Lauffenburger, and K.J. Van Vliet. 2013. "Directional Cell Migration in an Extracellular pH Gradient: A Model Study with an Engineered Cell Line and Primary Microvascular Endothelial Cells." *Experimental Cell Research* 319(4):487–497. https://doi.org/https://doi.org/10.1016/j.yexcr.2012.11.006.

Paradiso, R., R. Buonomo, M. Dixon, G. Barbieri, and S. De Pascale. 2014. "Soybean Cultivation for Bioregenerative Life Support Systems (BLISSs): The Effect of Hydroponic System and Nitrogen Source." *Advances in Space Research* 53(3):574–584.

Paradiso, R., V. De Micco, R. Buonomo, G. Aronne, G. Barbieri, and S. De Pascale. 2014. "Soilless Cultivation of Soybean for Bioregenerative Life-Support Systems: A Literature Review and the Experience of the Melissa Project—Food Characterisation Phase I." *Plant Biology (Stuttgart, Germany)* 16(Suppl 1):69–78.

París, R., M.M. Vazquez, M. Graziano, M.C. Terrile, N.D. Miller, E.P. Spalding, M.S. Otegui, and C.A. Casalongué. 2018. "Distribution of Endogenous NO Regulates Early Gravitropic Response and PIN2 Localization in Arabidopsis Roots." Original Research. *Frontiers in Plant Science* 9.

Parker, C.W., N. Singh, S. Tighe, A. Blachowicz, J.M. Wood, A. Seuylemezian, P. Vaishampayan, et al. 2020. "End-to-End Protocol for the Detection of SARS-CoV-2 from Built Environments." *mSystems* 5(5).

Parkes, T.L., A.J. Elia, D. Dickinson, A.J. Hilliker, J.P. Phillips, and G.L. Boulianne. 1998. "Extension of *Drosophila* Lifespan by Overexpression of Human SOD1 in Motorneurons." *Nature Genetics* 19(2):171–174.

Parkhitko, A.A., E. Filine, S.E. Mohr, A. Moskalev, and N. Perrimon. 2020. "Targeting Metabolic Pathways for Extension of Lifespan and Healthspan Across Multiple Species." *Ageing Research Reviews* 64:101188.

Parmar, N., K.H. Singh, D. Sharma, L. Singh, P. Kumar, J. Nanjundan, Y.J. Khan, D.K. Chauhan, and A.K. Thakur. 2017. "Genetic Engineering Strategies for Biotic and Abiotic Stress Tolerance and Quality Enhancement in Horticultural Crops: A Comprehensive Review." *3 Biotech* 7(4):239.

Particle Data Group, R.L. Workman, V.D. Burkert, V. Crede, E. Klempt, U. Thoma, L. Tiator, et al. 2022. "Review of Particle Physics." *Progress of Theoretical and Experimental Physics* (8).

Passel, J.S., and Cohn, D. 2008. "U.S. Population Projections: 2005–2050." Pew Research Center. https://www.pewresearch.org/hispanic/2008/02/11/us-population-projections-2005-2050.

Patel, V.K., F. Robinson, J.S. Yagoobi, and J. Didion. 2013. "Terrestrial and Micro-Gravity Experimental Study of Micro-Scale Heat Transport Device Driven by Electrohydrodynamic Conduction Pumping." *IEEE Transactions on Industry Applications* 49(6):2397–2401.

Patel, V.K., J. Seyed-Yagoobi, F. Robinson, and J.R. Didion. 2016. "Effect of Gravity on Electrohydrodynamic Conduction Driven Liquid Film Flow Boiling." *AIAA Journal of Thermophysics and Heat Transfer* 30(2):429–437.

Paul, A.-L., C.E. Amalfitano, and R.J. Ferl. 2012. "Plant Growth Strategies Are Remodeled by Spaceflight." *BMC Plant Biology* 12(1):232.

Paul, A.-L., S.M. Elardo, and R. Ferl. 2022. "Plants Grown in Apollo Lunar Regolith Present Stress-Associated Transcriptomes That Inform Prospects for Lunar Exploration." *Communications Biology* 5(1):382.

Paul, A.-L., N. Haveman, B. Califar, and R.J. Ferl. 2021. "Epigenomic Regulators Elongator Complex Subunit 2 and Methyltransferase 1 Differentially Condition the Spaceflight Response in Arabidopsis." *Frontiers in Plant Science* 12:691790.

Paul, A.-L., N.J. Sng, A.K. Zupanska, A. Krishnamurthy, E.R. Schultz, and R.J. Ferl. 2017. "Genetic Dissection of the Arabidopsis Spaceflight Transcriptome: Are Some Responses Dispensable for the Physiological Adaptation of Plants to Spaceflight?" *PLOS One* 12(6):e0180186.

Paul, A.-L., A.K. Zupanska, D.T. Ostrow, Y. Zhang, Y. Sun, J.-L. Li, S. Shanker, W.G. Farmerie, C.E. Amalfitano, and R.J. Ferl. 2012. "Spaceflight Transcriptomes: Unique Responses to a Novel Environment." *Astrobiology* 12(1):40–56.

Paul, A.-L., A.K. Zupanska, E.R. Schultz, and R.J. Ferl. 2013. "Organ-Specific Remodeling of the Arabidopsis Transcriptome in Response to Spaceflight." *BMC Plant Biology* 13(1):112.

Pecaut, M.J., D.S. Gridley, and G.A. Nelson. 2003. "Long-Term Effects of Low-Dose Proton Radiation on Immunity in Mice: Shielded vs. Unshielded." *Aviation, Space, Environmental Medicine* 74(2):115–124.

Pelfrey, J., and L. Jordan. 2008. "An EXPRESS Rack Overview and Support for Microgravity Research on the International Space Station (ISS)." Paper for the 46th AIAA Aerospace Sciences Meeting and Exhibit.

Perez-Garcia, P., and M.A. Moreno-Risueno. 2018. "Stem Cells and Plant Regeneration." *Developmental Biology* 442(1): 3–12.

Perrin, E., G. Bacci, L. Garrelly, F. Canganella, G. Bianconi, R. Fani, and A. Mengoni. 2018. "Furnishing Spaceship Environment: Evaluation of Bacterial Biofilms on Different Materials Used Inside International Space Station." *Research in Microbiology* 169(6):289–295.

Persat, A., C.D. Nadell, M.K. Kim, F. Ingremeau, A. Siryaporn, K. Drescher, N.S. Wingreen, B.L. Bassler, Z. Gitai, and H.A. Stone. 2015. "The Mechanical World of Bacteria." *Cell* 161(5):988–997.

Phillips, K.W. 2014. "How Diversity Makes Us Smarter." *Scientific American* 311(4):42–47.

Pierson, D.L., R.P. Stowe, T.M. Phillips, D.J. Lugg, and S.K. Mehta. 2005. "Epstein-Barr Virus Shedding by Astronauts During Space Flight." *Brain, Behavior, and Immunity* 19(3):235–242.

Piofczyk, T., G. Jeena, and A. Pecinka. 2015. "Arabidopsis Natural Variation Reveals Connections Between UV Radiation Stress and Plant Pathogen-Like Defense Responses." *Plant Physiology and Biochemistry* 93:34–43.

Piper, M.D.W., and L. Partridge. 2018. "Drosophila as a Model for Ageing." *Biochimica et Biophysica Acta Molecular Basis of Disease* 1864(9 Pt A):2707–2717.

Pippia, P., L. Sciola, M. Cogoli-Greuter, M.A. Meloni, A. Spano, and A. Cogoli. 1996. "Activation Signals of T Lymphocytes in Microgravity." *Journal of Biotechnology* 47(2–3):215–222.

Pisharody, S., B. Borchers, and G. Schlick. 1996. "Solid Waste Processing in a CELSS: Nitrogen Recovery." *Life Support & Biosphere Science* 3(1–2):61–65.

Ploutz-Snyder, L., S. Bloomfield, S.M. Smith, S.K. Hunter, K. Templeton, and D. Bemben. 2014. "Effects of Sex and Gender on Adaptation to Space: Musculoskeletal Health." *Journal of Women's Health (Larchmt)* 23(11):963–966.

Popova, A., A. Kuang, G. McClure, and M. Musgrave. 2002. "Reserve Nutrient Substance Accumulation in Brassica Rapa L. Seeds in Microgravity Conditions (STS-87)." *Journal of Gravitational Physiology* 9(1):P237–P238.

Popova, N.K., A.V. Kulikov, and V.S. Naumenko. 2020. "Spaceflight and Brain Plasticity: Spaceflight Effects on Regional Expression of Neurotransmitter Systems and Neurotrophic Factors Encoding Genes." *Neuroscience and Biobehavioral Reviews* 119:396–405.

Poughon, L., C. Laroche, C. Creuly, C.-G. Dussap, C. Paille, C. Lasseur, P. Monsieurs, et al. 2020. "Limnospira Indica PCC8005 Growth in Photobioreactor: Model and Simulation of the ISS and Ground Experiments." *Life Sciences in Space Research* 25:53–65.

Poulet, L., K. Engeling, T. Hatch, S. Stahl-Rommel, Y.-A. Velez Justiniano, S. Castro-Wallace, J. Bunchek, et al. 2022. "Large-Scale Crop Production for the Moon and Mars: Current Gaps and Future Perspectives." Perspective. *Frontiers in Astronomy and Space Sciences* 8.

Poulet, L., J.P. Fontaine, and C.G. Dussap. 2018. "A Physical Modeling Approach for Higher Plant Growth in Reduced Gravity Environments." *Astrobiology* 18(9):1093–1100.

Poulet, L., C. Zeidler, J. Bunchek, P. Zabel, V. Vrakking, D. Schubert, G. Massa, and R. Wheeler. 2021. "Crew Time in a Space Greenhouse Using Data from Analog Missions and Veggie." *Life Sciences in Space Research* 31:101–112.

Prakash, B., B.M. Veeregowda, and G. Krishnappa. 2003. "Biofilms: A Survival Strategy of Bacteria." *Current Science* 85(9):1299–1307.

Prisk, G.K. 2014. "Microgravity and the Respiratory System." *European Respiratory Journal* 43(5):1459–1471.

Prosperetti, A. 2017. "Vapor Bubbles." *Annual Review of Fluid Mechanics* 49(1):221–248.

Raj, R., J. Kim, and J. McQuillen. 2009. "Subcooled Pool Boiling in Variable Gravity Environments." *ASME Journal of Heat and Mass Transfer* 131(9).

Raj, R., J. Kim, and J. McQuillen. 2012. "Pool Boiling Heat Transfer on the International Space Station: Experimental Results and Model Verification." *Journal of Heat Transfer* 134(10).

Rapin, G., N. Caballero, I. Gaponenko, B. Ziegler, A. Rawleigh, E. Moriggi, T. Giamarchi, S.A. Brown, and P. Paruch. 2021. "Roughness and Dynamics of Proliferating Cell Fronts as a Probe of Cell–Cell Interactions." *Scientific Reports* 11(1):8869.

Rasera, J.N., J.J. Cilliers, J.A. Lamamy, and K. Hadler. 2020. "The Beneficiation of Lunar Regolith for Space Resource Utilisation: A Review." *Planetary and Space Science* 186:104879.

Reagan, S. 2017. *MSRR Rack Materials Science Research Rack*. ID: 20170012377. Paper at the Annual Meeting American Society for Gravitational and Space Research (ASGSR 2017).

Rechavi, O., L. Houri-Ze'evi, S. Anava, W.S.S. Goh, S.Y. Kerk, G.J. Hannon, and O. Hobert. 2014. "Starvation-Induced Transgenerational Inheritance of Small RNAs in C. Elegans." *Cell* 158(2):277–287.

Rechavi, O., G. Minevich, and O. Hobert. 2011. "Transgenerational Inheritance of an Acquired Small RNA-Based Antiviral Response in C. Elegans." *Cell* 147(6):1248–1256.

Redwire. 2020. "Redwire Acquires Made in Space, the Leader in On-Orbit Space Manufacturing Technologies." https://redwirespace.com/newsroom/redwire-acquires-made-in-space-the-leader-in-on-orbit-space-manufacturing-technologies/?rdws=nnn.xffxcv.tfd&rdwj=120.

Regal, C.A., C. Ticknor, J.L. Bohn, and D.S. Jin. 2003. "Creation of Ultracold Molecules from a Fermi Gas of Atoms." *Nature* 424(6944):47–50.

Reichert, P., W. Prosise, T.O. Fischmann, G. Scapin, C. Narasimhan, A. Spinale, R. Polniak, et al. 2019. "Pembrolizumab Microgravity Crystallization Experimentation." *npj Microgravity* 5:28.

Reichhardt, T. 1998. "Biologists Recommend Scrapping NASA's Research on Crystals." *Nature* 394(6690):213.

Reid, J.B., and J.J. Ross. 2011. "Regulation of Tissue Repair in Plants." *Proceedings of the National Academy of Sciences* 108(42):17241–17242.

Reitz, B., C. Lotz, N. Gerdes, S. Linke, E. Olsen, K. Pflieger, S. Sohrt, et al. 2021. "Additive Manufacturing Under Lunar Gravity and Microgravity." *Microgravity Science and Technology* 33(2):25.

Reitz, R.D., and G. Duraisamy. 2015. "Review of High Efficiency and Clean Reactivity Controlled Compression Ignition (RCCI) Combustion in Internal Combustion Engines." *Progress in Energy and Combustion Science* 46:12–71.

Reynolds, R.J., I.V. Bukhtiyarov, G.I. Tikhonova, S.M. Day, I.B. Ushakov, and T.Y.U. Gorchakova. 2019. "Contrapositive Logic Suggests Space Radiation Not Having a Strong Impact on Mortality of US Astronauts and Soviet and Russian Cosmonauts." *Scientific Reports* 9(1):8583.

Ripples Nigeria. 2019. "NASA Set to Grow First Fruits in Space." Ripples Nigeria. https://www.ripplesnigeria.com/nasa-set-to-grow-first-fruits-in-space.

Ritchie, H., M. Roser, and P. Rosado. 2022. "Energy Mix." https://ourworldindata.org/energy-mix.

Roberts, T.G. "Space Launch to Low Earth Orbit: How Much Does It Cost?" 2022. Aerospace Security. Center for Strategic & International Studies. https://aerospace.csis.org/data/space-launch-to-low-earth-orbit-how-much-does-it-cost.

Robinson, J. 2017. Rodent Research Workshop. Presentation at the 33rd Annual Meeting of the American Society for Gravitational and Space Research.

Robling, A.G., and C.H. Turner. 2009. "Mechanical Signaling for Bone Modeling and Remodeling." *Critical Reviews in Eukaryotic Gene Expression* 19(4):319–338.

Roger, F., A. Godhe, and L. Gamfeldt. 2012. "Genetic Diversity and Ecosystem Functioning in the Face of Multiple Stressors." *PLOS One* 7(9):e45007.

Rojas-Alva, U., and G. Jomaas. 2022. "A Historical Overview of Experimental Solid Combustion Research in Microgravity." *Acta Astronautica* 194:363–375.

Ronca, A.E., J.S. Alwood, J.R. Alberts, J.G. Steller, L.K. Christenson, Y. Shirazi, and A.-L. Paul. 2021. "Mammalian Multi-Generational Studies in Space." Input paper submitted to the Committee on Biological and Physical Sciences Research in Space 2023–2032. https://www.nationalacademies.org/our-work/decadal-survey-on-life-and-physical-sciences-research-in-space-2023-2032.

Ronca, A.E., E.L. Moyer, Y. Talyansky, M. Lowe, S. Padmanabhan, S. Choi, C. Gong, S.M. Cadena, L. Stodieck, and R.K. Globus. 2019. "Behavior of Mice Aboard the International Space Station." *Scientific Reports* 9(1):4717.

Ronney, P.D. 1998. "Understanding Combustion Processes Through Microgravity Research." *Symposium (International) on Combustion* 27(2):2485–2506.

Rosenzweig, J.A., S. Ahmed, J. Eunson, Jr., and A.K. Chopra. 2014. "Low-Shear Force Associated with Modeled Microgravity and Spaceflight Does Not Similarly Impact the Virulence of Notable Bacterial Pathogens." *Applied Microbiology and Biotechnology* 98(21):8797–8807.

Royal Museums Greenwich. 2023. "What Was the First Animal Sent into Space?" https://www.rmg.co.uk/stories/topics/what-was-first-animal-space.

Roy-O'Reilly, M., A. Mulavara, and T. Williams. 2021. "A Review of Alterations to the Brain During Spaceflight and the Potential Relevance to Crew in Long-Duration Space Exploration." *npj Microgravity* 7(1):5.

Rubenstein, M., A. Cornejo, and R. Nagpal. 2014. "Programmable Self-Assembly in a Thousand-Robot Swarm." *Science* 345(6198):795–799.

Rutter, L., R. Barker, D. Bezdan, H. Cope, S.V. Costes, L. Degoricija, K.M. Fisch, et al. 2020. "A New Era for Space Life Science: International Standards for Space Omics Processing." *Patterns* 1(9):100148.

Saad, M.M., A.A. Eida, and H. Hirt. 2020. "Tailoring Plant-Associated Microbial Inoculants in Agriculture: A Roadmap for Successful Application." *Journal of Experimental Botany* 71(13):3878–3901.

Sabin, S. 2021. "Nearly Half the Public Wants the U.S. to Maintain Its Space Dominance. Appetite for Space Exploration Is a Different Story." *Morning Consult*. https://pro.morningconsult.com/articles/space-force-travel-exploration-poll.

Sacco, E., and S.K. Moon. 2019. "Additive Manufacturing for Space: Status and Promises." *The International Journal of Advanced Manufacturing Technology* 105(10):4123–4146.

Safronova, M., and D. Budker. 2022. "Focus Issue on Quantum Sensors for New-Physics Discoveries." *Quantum Science and Technology*. Collection of papers. https://iopscience.iop.org/journal/2058-9565/page/Focus-on-Quantum-Sensors-for-New-Physics-Discoveries.

Safronova, M.S., and D. Budker. 2021. "Quantum Technologies and the Elephants." *Quantum Science and Technology* 6(4):040401.

Safronova, M.S., D. Budker, D. DeMille, D.F. Jackson Kimball, A. Derevianko, and C.W. Clark. 2018. "Search for New Physics with Atoms and Molecules." *Reviews of Modern Physics* 90(2):025008.

Saletore, Y., K. Meyer, J. Korlach, I.D. Vilfan, S. Jaffrey, and C.E. Mason. 2012. "The Birth of the Epitranscriptome: Deciphering the Function of RNA Modifications." *Genome Biology* 13(10):175.

Salisbury, F.B. 1992. "Some Challenges in Designing a Lunar, Martian, or Microgravity CELSS." *Acta Astronautica* 27:211–217.

Salisbury, F.B. 1997. "Growing Super-Dwarf Wheat in Space Station Mir." *Life Support & Biosphere Science* 4(3–4):155–166.

Salisbury, F.B. 1999. "Growing Crops for Space Explorers on the Moon, Mars, or in Space." *Advances in Space Biology and Medicine* 7:131–162.

Salzman, J.A., W.J. Masica, and R.F. Lacovic. 1973. *Low Gravity Reorientation in a Scale-Model Centaur Liquid-Hydrogen Tank*. NASA. https://ntrs.nasa.gov/citations/19730007525.

Sampathkumar, A., A. Yan, P. Krupinski, and E.M. Meyerowitz. 2014. "Physical Forces Regulate Plant Development and Morphogenesis." *Current Biology* 24(10):R475–R483.

Sangsuwan, T., M. Mannervik, and S. Haghdoost. 2022. "Transgenerational Effects of Gamma Radiation Dose and Dose Rate on Drosophila Flies Irradiated at an Early Embryonal Stage." *Mutation Research—Genetic Toxicology Environmental Mutagenesis* 881:503523.

Sanner, C., N. Huntemann, R. Lange, C. Tamm, E. Peik, M.S. Safronova, and S.G. Porsev. 2019. "Optical Clock Comparison for Lorentz Symmetry Testing." *Nature* 567(7747):204–208.

Santomartino, R., L. Zea, and C.S. Cockell. 2022. "The Smallest Space Miners: Principles of Space Biomining." *Extremophiles* 26(1):7.

Sarkar, R., and F. Pampaloni. 2022. "In Vitro Models of Bone Marrow Remodelling and Immune Dysfunction in Space: Present State and Future Directions." *Biomedicines* 10(4):766.

Sarraf, D.B., and W.G. Anderson. 2007. "Heat Pipes for High Temperature Thermal Management." Paper presented at the ASME 2007 InterPACK Conference collocated with the ASME/JSME 2007 Thermal Engineering Heat Transfer Summer Conference. https://doi.org/10.1115/IPACK2007-33984.

Saulmon, M.M., K.F. Reardon, and W.Z. Sadeh. 1996. "A Bioreactor System for the Nitrogen Loop in a Controlled Ecological Life Support System." *Advances in Space Research* 18(1):289–292.

Scherer, G.F., and P. Pietrzyk. 2014. "Gravity-Dependent Differentiation and Root Coils in Arabidopsis Thaliana Wild Type and Phospholipase-a-I Knockdown Mutant Grown on the International Space Station." *Plant Biology (Stuttgart, Germany)* 16(Suppl 1):97–106.

Schermelleh, L., A. Ferrand, T. Huser, C. Eggeling, M. Sauer, O. Biehlmaier, and G.P.C. Drummen. 2019. "Super-Resolution Microscopy Demystified." *Nature Cell Biology* 21(1):72–84.

Schiller, S. 2015. *Final Report Summary—SOC2 (Towards Neutral-Atom Space Optical Clocks: Development of High-Performance Transportable and Breadboard Optical Clocks and Advanced Subsystems*. Luxembourg: EU Publications. https://cordis.europa.eu/project/id/263500/reporting.

Schiwon, K., K. Arends, K.M. Rogowski, S. Fürch, K. Prescha, T. Sakinc, R. Van Houdt, G. Werner, and E. Grohmann. 2013. "Comparison of Antibiotic Resistance, Biofilm Formation and Conjugative Transfer of Staphylococcus and Enterococcus Isolates from International Space Station and Antarctic Research Station Concordia." *Microbial Ecology* 65(3):638–651.

Schkolnik, V., D. Budker, O. Fartmann, V. Flambaum, L. Hollberg, T. Kalaydzhyan, S. Kolkowitz, et al. 2022. "Optical Atomic Clock Aboard an Earth-Orbiting Space Station (OACESS): Enhancing Searches for Physics Beyond the Standard Model in Space." *Quantum Science and Technology* 8(2023):014003.

Schmoch, T., J.H. Westhoff, S.O. Decker, A. Skarabis, G.F. Hoffmann, C. Dohna-Schwake, U. Felderhoff-Müser, et al. 2021. "Next-Generation Sequencing Diagnostics of Bacteremia in Pediatric Sepsis." *Medicine (Baltimore)* 100(25):e26403.

Schuerger, A.C., B.S. Amaradasa, N.S. Dufault, M.E. Hummerick, J.T. Richards, C.L. Khodadad, T.M. Smith, and G.D. Massa. 2021. "Fusarium Oxysporum as an Opportunistic Fungal Pathogen on Zinnia Hybrida Plants Grown on Board the International Space Station." *Astrobiology* 21(9):1029–1048.

Schwartzkopf, S.H. 1997. "Human Life Support for Advanced Space Exploration." *Advances in Space Biology and Medicine* 6:231–253.

Schwarzenbach, H., D.S.B. Hoon, and K. Pantel. 2011. "Cell-Free Nucleic Acids as Biomarkers in Cancer Patients." *Nature Reviews Cancer* 11(6):426–437.

Scott, T.J., and N.S. Vonortas. 2017. *An Economic Appraisal of Microgravity Protein Crystallization for Drug Development*. Washington, DC: NASA. https://www.nasa.gov/sites/default/files/atoms/files/microg_protein_crystallization_scott-vonortas_09152017.pdf.

Seawright, J.W., V. Sridharan, R.D. Landes, M. Cao, P. Singh, I. Koturbash, X.-W. Mao, et al. 2019. "Effects of Low-Dose Oxygen Ions and Protons on Cardiac Function and Structure in Male C57BL/6J Mice." *Life Sciences in Space Research* 20:72–84.

Sedda, M.A., C.P.L. Berry, K. Jani, P. Amaro-Seoane, P. Auclair, J. Baird, T. Baker, et al. 2021. "The Missing Link in Gravitational-Wave Astronomy: A Summary of Discoveries Waiting in the Decihertz Range." *Experimental Astronomy (Dordr)* 51(3):1427–1440.

Senatore, G., F. Mastroleo, N. Leys, and G. Mauriello. 2018. "Effect of Microgravity & Space Radiation on Microbes." *Future Microbiology* 13(7):831–847.

Sertoglu, K. 2020. "International Space Station Bioprinting Facility Receives a 3D Printed Upgrade." https://3dprintingindustry.com/news/international-space-station-bioprinting-facility-receives-a-3d-printed-upgrade-168402.

Shackelford, L.C., A.D. LeBlanc, T.B. Driscoll, H.J. Evans, N.J. Rianon, S.M. Smith, E. Spector, D.L. Feeback, and D. Lai. 2004. "Resistance Exercise as a Countermeasure to Disuse-Induced Bone Loss." *Journal of Applied Physiology (1985)* 97(1):119–129.

Shang, L., Y. Cheng, and Y. Zhao. 2017. "Emerging Droplet Microfluidics." *Chemical Reviews* 117(12):7964–8040.

Shevtsov, J., C. Escobar, A.C. Schuerger, C.M. Trouillefou, L.K. Tuominen, R. Wheeler, N. O'Neil, D.G. Zaharescu, O. Monje, and M. Gül. *Bioregenerative Life Support Systems: Coordinated Research into Organisms, Technology and Systems Integration*. Input paper submitted to the Committee on Biological and Physical Sciences Research in Space 2023–2032. https://www.nationalacademies.org/our-work/decadal-survey-on-life-and-physical-sciences-research-in-space-2023-2032.

Shi, Q., J. Sun, C. Hou, Y. Li, Q. Zhang, and H. Wang. 2019. "Advanced Functional Fiber and Smart Textile." *Advanced Fiber Materials* 1(1):3–31.

Shigenaga, A.M., and C.T. Argueso. 2016. "No Hormone to Rule Them All: Interactions of Plant Hormones During the Responses of Plants to Pathogens." *Seminars in Cell & Developmental Biology* 56:174–189.

Shin, Y., and C.P. Brangwynne. 2017. "Liquid Phase Condensation in Cell Physiology and Disease." *Science* 357(6357):eaaf4382.

Shindo, C., G. Bernasconi, and C.S. Hardtke. 2007. "Natural Genetic Variation in Arabidopsis: Tools, Traits and Prospects for Evolutionary Ecology." *Annals of Botany* 99(6):1043–1054.

Sicker, R.J., W.M. Foster, B.J. Motil, W.V. Meyer, F.P. Chiaramonte, A. Abbott-Hearn, A. Atherton, et al. *Light Microscopy Module: International Space Station Premier Automated Microscope*. ID: 20170004565. Presentation at the Annual Meeting of the American Society for Gravitational and Space Research (ASGSR 2016).

Simmons, M. 2014. "Physics: A Fundamental Force for Future Security." *The Conversation*. https://theconversation.com/physics-a-fundamental-force-for-future-security-22121.

Simons, D.M., E.M. Gardner, and P.I. Lelkes. 2006. "Dynamic Culture in a Rotating-Wall Vessel Bioreactor Differentially Inhibits Murine T-Lymphocyte Activation by Mitogenic Stimuli Upon Return to Static Conditions in a Time-Dependent Manner." *Journal of Applied Physiology (1985)* 100(4):1287–1292.

Singh, N.K., D. Bezdan, A. Checinska Sielaff, K. Wheeler, C.E. Mason, and K. Venkateswaran. 2018. "Multi-Drug Resistant Enterobacter Bugandensis Species Isolated from the International Space Station and Comparative Genomic Analyses with Human Pathogenic Strains." *BMC Microbiology* 18(1):175.

Sirignano, W.A. 2014. "Advances in Droplet Array Combustion Theory and Modeling." *Progress in Energy and Combustion Science* 42:54–86. https://doi.org/10.1016/j.pecs.2014.01.002.

Smith, R.C., M.S. Cramer, P.J. Mitchell, J. Lucchesi, A.M. Ortega, E.W. Livingston, D. Ballard, et al. 2020. "Inhibition of Myostatin Prevents Microgravity-Induced Loss of Skeletal Muscle Mass and Strength." *PLOS One* 15(4):e0230818.

Smith, S.M., M.A. Heer, L.C. Shackelford, J.D. Sibonga, L. Ploutz-Snyder, and S.R. Zwart. 2012. "Benefits for Bone from Resistance Exercise and Nutrition in Long-Duration Spaceflight: Evidence from Biochemistry and Densitometry." *Journal of Bone and Mineral Research* 27(9). https://doi.org/10.1002/jbmr.1647.

Smooke, M.D., C.S. McEnally, L.D. Pfefferle, R.J. Hall, and M.B. Colket. 1999. "Computational and Experimental Study of Soot Formation in a Coflow, Laminar Diffusion Flame." *Combustion and Flame* 117(1):117–139.

Sng, N.J., B. Kolaczkowski, R.J. Ferl, and A.-L. Paul. 2018. "A Member of the CONSTANS-Like Protein Family Is a Putative Regulator of Reactive Oxygen Species Homeostasis and Spaceflight Physiological Adaptation." *AoB PLANTS* 11(1).

Soliman, H.M. 1986. "The Mist-Annular Transition During Condensation and Its Influence on the Heat Transfer Mechanism." *International Journal of Multiphase Flow* 12(2):277–288.

Solon, A.P., J. Stenhammar, R. Wittkowski, M. Kardar, Y. Kafri, M.E. Cates, and J. Tailleur. 2015. "Pressure and Phase Equilibria in Interacting Active Brownian Spheres." *Physical Review Letters* 114(19):198301.

Soni, V., E.S. Bililign, S. Magkiriadou, S. Sacanna, D. Bartolo, M.J. Shelley, and W.T.M. Irvine. 2019. "The Odd Free Surface Flows of a Colloidal Chiral Fluid." *Nature Physics* 15(11):1188–1194.

Soucy, K.G., H.K. Lim, J.H. Kim, Y. Oh, D.O. Attarzadeh, B. Sevinc, M.M. Kuo, A.A. Shoukas, M.E. Vazquez, and D.E. Berkowitz. 2011. "HZE ^{56}Fe-Ion Irradiation Induces Endothelial Dysfunction in Rat Aorta: Role of Xanthine Oxidase." *Radiation Research* 176(4):474–485.

Soulages, J.M., G.H. McKinley, N.R. Hall, K.S. Magee, G.E. Chamitoff, and E.M. Fincke. "Extensional Properties of a Dilute Polymer Solution Following Preshear in Microgravity." Pp. 2199–2206 in *Earth and Space 2010: Engineering, Science, Construction, and Operations in Challenging Environments*. American Society of Civil Engineers Conference Proceedings.

Souvestre, P.A., C.K. Landrock, and A.P. Blaber. 2008. "Reducing Incapacitating Symptoms During Space Flight: Is Postural Deficiency Syndrome an Applicable Model?" *Hippokratia* 12(Suppl 1):41–48.

Spaans, S., R. Weusthuis, J. Van Der Oost, and S. Kengen. 2015. "NADPH-Generating Systems in Bacteria and Archaea." Review. *Frontiers in Microbiology* 6.

Spagnolie, S.E. 2015. *Complex Fluids in Biological Systems: Experiment, Theory, and Computation*, S.E. Spagnolie, ed. Biological and Medical Physics, Biomedical Engineering. New York: Springer.

Sparrow, B.D., W. Edwards, S.E.M. Munroe, G.M. Wardle, G.R. Guerin, J.F. Bastin, B. Morris, R. Christensen, S. Phinn, and A.J. Lowe. 2020. "Effective Ecosystem Monitoring Requires a Multi-Scaled Approach." *Biological Reviews of the Cambridge Philosophical Society* 95(6):1706–1719.

Spellings, M., M. Engel, D. Klotsa, S. Sabrina, A.M. Drews, N.H.P. Nguyen, K.J.M. Bishop, and S.C. Glotzer. 2015. "Shape Control and Compartmentalization in Active Colloidal Cells." *Proceedings of the National Academy of Sciences* 112(34):E4642–E4650.

Spivey, R., Sheredy, W.A., and G. Flores. 2008. "An Overview of the Microgravity Science Glovebox (MSG) Facility, and the Gravity-Dependent Phenomena Research Performed in the MSG on the International Space Station (ISS)." Paper for the 46th AIAA Aerospace Sciences Meeting and Exhibit.

Spooner, B.S., and J.A. Guikema. 1992. "The NASA Specialized Center of Research and Training (NSCORT) in Gravitational Biology." *Transactions of the Kansas Academy of Science. Kansas Academy of Science 95*(1–2):1–3.

Squires, T.M., and T.G. Mason. 2010. "Fluid Mechanics of Microrheology." *Annual Review of Fluid Mechanics* 42(1): 413–438.

Srinidhi, D., and J.L. Turner. 2021. "Can Space Gardening Help Astronauts Cope with Isolation?" NASA Human Research Program Strategic Communications. https://www.nasa.gov/feature/can-space-gardening-help-astronauts-cope-with-isolation.

Steller, J.G., R.S. Blue, R. Burns, T.M. Bayuse, E.L. Antonsen, V. Jain, M.M. Blackwell, and R.T. Jennings. 2020. "Gynecologic Risk Mitigation Considerations for Long-Duration Spaceflight." *Aerospace Medicine and Human Performance* 91(7):543–564.

Steward, W.G., R.V. Smith, and J.A. Brennan. 1995. "Cooldown Transients in Cryogenic Transfer Lines." Conference paper. Pp. 354–363 in *Advances in Cryogenic Engineering*, K.D. Timmerhaus, ed. Boston, MA: Springer.

Stockslager, M.A., S. Malinowski, M. Touat, J.C. Yoon, J. Geduldig, M. Mirza, A.S. Kim, et al. 2021. "Functional Drug Susceptibility Testing Using Single-Cell Mass Predicts Treatment Outcome in Patient-Derived Cancer Neurosphere Models." *Cell Reports* 37(1):109788.

Strayer, R.F., and C.F. Atkinson. 1997. "An Overview: Recycling Nutrients from Crop Residues for Space Applications." *Compost Science & Utilization* 5(3):25–31.

Stroud, J.E., M.S. Gale, S.R. Zwart, M. Heer, S.M. Smith, T. Montina, and G.A.S. Metz. 2022. "Longitudinal Metabolomic Profiles Reveal Sex-Specific Adjustments to Long-Duration Spaceflight and Return to Earth." *Cellular and Molecular Life Sciences* 79(11):578.

Stutte, G.W. 1996. "Nitrogen Dynamics in the CELSS Breadboard Facility at Kennedy Space Center." *Life Support & Biosphere Science* 3(1–2):67–74.

Su, S.H., M.A. Keith, and P.H. Masson. 2020. "Gravity Signaling in Flowering Plant Roots." *Plants (Basel)* 9(10).

Su, Y., J.T. Yrastorza, M. Matis, J. Cusick, S. Zhao, G. Wang, and J. Xie. 2022. "Biofilms: Formation, Research Models, Potential Targets, and Methods for Prevention and Treatment." *Advanced Science* 9(29):2203291.

Sugimoto, M., Y. Oono, O. Gusev, T. Matsumoto, T. Yazawa, M.A. Levinskikh, V.N. Sychev, G.E. Bingham, R. Wheeler, and M. Hummerick. 2014. "Genome-Wide Expression Analysis of Reactive Oxygen Species Gene Network in Mizuna Plants Grown in Long-Term Spaceflight." *BMC Plant Biology* 14(1):4.

Sui, Y., H. Ding, and P.D.M. Spelt. 2014. "Numerical Simulations of Flows with Moving Contact Lines." *Annual Review of Fluid Mechanics* 46(1):97–119.

Sulzman, F.M., D. Ellman, C.A. Fuller, M.C. Moore-Ede, and G. Wassmer. 1984. "Neurospora Circadian Rhythms in Space: A Reexamination of the Endogenous-Exogenous Question." *Science* 225:232–234.

Sumner, I. 1978. *Liquid Propellant Reorientation in a Low-Gravity Environment*. Cleveland, OH: NASA Lewis Research Center. https://ntrs.nasa.gov/api/citations/19780021464/downloads/19780021464.pdf.

Sun, P., C. Wu, F. Zhu, S. Wang, and X. Huang. 2020. "Microgravity Combustion of Polyethylene Droplet in Drop Tower." *Combustion and Flame* 222:18–26.

Swain, P., M. Mortreux, J.M. Laws, H. Kyriacou, E. De Martino, A. Winnard, and N. Caplan. 2022. "Bone Deconditioning During Partial Weight-Bearing in Rodents—A Systematic Review and Meta-Analysis," *Life Sciences in Space Research* 34:87–103. https://doi.org/https://doi.org/10.1016/j.lssr.2022.07.003.

Tabor, A. 2022a. "NASA Space Biology Bootcamp Teaches the Teachers, Reaching More Students." https://www.nasa.gov/feature/ames/nasa-space-biology-bootcamp-teaches-the-teachers-reaching-more-students.

Tabor, A. 2022b. "What Is Bionutrients?" NASA Ames Research Center. https://www.nasa.gov/ames/bionutrients.

Takamoto, M., Y. Tanaka, and H. Katori. 2022. "A Perspective on the Future of Transportable Optical Lattice Clocks." *Applied Physics Letters* 120(14).

Tamaru, H., C. Koyama, H. Saruwatari, Y. Nakamura, T. Ishikawa, and T. Takada. 2018. "Status of the Electrostatic Levitation Furnace (ELF) in the ISS-KIBO." *Microgravity Science and Technology* 30(5):643–651.

Taulavuori, E., K. Taulavuori, J.K. Holopainen, R. Julkunen-Tiitto, C. Acar, and I. Dincer. 2017. "Targeted Use of LEDs in Improvement of Production Efficiency Through Phytochemical Enrichment." *Journal of Science Food and Agriculture* 97(15):5059–5064.

Taylor, L.A. 1992. "Resources for a Lunar Base: Rocks, Minerals, and Soil of the Moon." Paper presented at the Second Conference on Lunar Bases and Space Activities. https://space.nss.org/wp-content/uploads/Lunar-Bases-conference-2-501-Resources-For-Lunar-Base.pdf.

Taylor, P.W. 2015. "Impact of Space Flight on Bacterial Virulence and Antibiotic Susceptibility." *Infection and Drug Resistance* 8:249–262.

Teale, W.D., I.A. Paponov, and K. Palme. 2006. "Auxin in Action: Signalling, Transport and the Control of Plant Growth and Development." *Nature Reviews: Molecular Cell Biology* 7(11):847–859.

Teixido, C., P. Castillo, C. Martinez-Vila, A. Arance, and L. Alos. 2021. "Molecular Markers and Targets in Melanoma." *Cells* 10(9):2320.

Tesei, D., A. Jewczynko, A.M. Lynch, and C. Urbaniak. 2022. "Understanding the Complexities and Changes of the Astronaut Microbiome for Successful Long-Duration Space Missions." *Life (Basel)* 12(4).

Théry, M., and L. Blanchoin. 2021. "Microtubule Self-Repair." *Current Opinion in Cell Biology* 68:144–154.

Thomas, W. 2023. "FY23 Budget Outcomes: NASA." *FYI: Science Policy News*. American Institute of Physics. January 27.

Thompson, L.R., J.G. Sanders, D. McDonald, A. Amir, J. Ladau, K.J. Locey, R.J. Prill, et al. 2017. "A Communal Catalogue Reveals Earth's Multiscale Microbial Diversity." *Nature* 551(7681):457–463.

Thompson, S.W., L.P. Jordan, and G.N. Flores. 2014. *An Overview of EXPRESS Rack, Microgravity Science Glovebox, and Sub-rack Facilities for Materials Science Research*. Presentation at the Materials LAB Workshop. April 15. https://ntrs.nasa.gov/api/citations/20140010107/downloads/20140010107.pdf.

Tierney, B.T., N.K. Singh, A.C. Simpson, A.M. Hujer, R.A. Bonomo, C.E. Mason, and K. Venkateswaran. 2022. "Multidrug-Resistant Acinetobacter Pittii Is Adapting to and Exhibiting Potential Succession Aboard the International Space Station." *Microbiome* 10(1):210.

Times of India. 2021. "NASA Astronaut Successfully Harvests 2 Plants in Space." *Times of India*. https://timesofindia.indiatimes.com/home/science/nasa-astronaut-successfully-harvests-2-plants-in-space/articleshow/82292154.cms.

Tisserand, P., L. Le Guillou, C. Afonso, J.N. Albert, J. Andersen, R. Ansari, É. Aubourg, et al. 2007. "Limits on the Macho Content of the Galactic Halo from the EROS-2 Survey of the Magellanic Clouds." *Astronomy and Astrophysics* 469(2):387–404.

Tolsma, J.S., K.T. Ryan, J.J. Torres, J.T. Richards, Z. Richardson, E.S. Land, I.Y. Perera, and C.J. Doherty. 2021. "The Circadian-Clock Regulates the Gravitropic Response." *Gravitational and Space Research* 9(1):171–186.

Touboul, P., G. Métris, M. Rodrigues, Y. André, Q. Baghi, J. Bergé, D. Boulanger, et al. 2017. "Microscope Mission: First Results of a Space Test of the Equivalence Principle." *Physical Review Letters* 119(23):231101.

Toulon, A., L. Breton, K.R. Taylor, M. Tenenhaus, D. Bhavsar, C. Lanigan, R. Rudolph, J. Jameson, and W.L. Havran. 2009. "A Role for Human Skin-Resident T Cells in Wound Healing." *Journal of Experimental Medicine* 206(4):743–750.

Trappe, S., D. Costill, P. Gallagher, A. Creer, J.R. Peters, H. Evans, D.A. Riley, and R.H. Fitts. 2009. "Exercise in Space: Human Skeletal Muscle After 6 Months Aboard the International Space Station." *Journal of Applied Physiology (1985)* 106(4):1159–1168.

Tretkoff, E. 2006. "March 20, 1800: Volta Describes the Electric Battery." *APS News* 15(3).

Tunyasuvunakool, K., J. Adler, Z. Wu, T. Green, M. Zielinski, A. Žídek, A. Bridgland, et al. 2021. "Highly Accurate Protein Structure Prediction for the Human Proteome." *Nature* 596(7873):590–596.

Turner, R.T., A.J. Branscum, C.P. Wong, U.T. Iwaniec, and E. Morey-Holton. 2020. "Studies in Microgravity, Simulated Microgravity and Gravity Do Not Support a Gravitostat." *Journal of Endocrinology* 247(3):273–282.

Turns, S., and D.C. Haworth. 2021. *An Introduction to Combustion: Concepts and Applications*, 4th ed. New York: McGraw-Hill.

Turroni, S., M. Magnani, P. KC, P. Lesnik, H. Vidal, and M. Heer. 2020. "Gut Microbiome and Space Travelers' Health: State of the Art and Possible Pro/Prebiotic Strategies for Long-Term Space Missions." Review. *Frontiers in Physiology* 11.

U.S. Census Bureau. 2023. "Quick Facts: United States." https://www.census.gov/quickfacts/fact/table/US/PST045222.

United Nations. 2023. "Benefits of Space: Agriculture." Office for Outer Space Affairs. https://www.unoosa.org/oosa/en/benefits-of-space/agriculture.html.

University of California, Berkeley. 2023. "Evidence for Speciation." *Understanding Evolution: Evo 101*. https://evolution.berkeley.edu/evolution-101/speciation/evidence-for-speciation.

University of Guelph. 2022. "CESRF: Controlled Environment Systems Research Facility." https://www.plant.uoguelph.ca/people-and-places/research-centres/cesrf-controlled-environment-systems-research-facility.

Urbaniak, C., P. van Dam, A. Zaborin, O. Zaborina, J.A. Gilbert, T. Torok, C.C.C. Wang, and K. Venkateswaran. 2019. "Genomic Characterization and Virulence Potential of Two *Fusarium Oxysporum* Isolates Cultured from the International Space Station." *mSystems* 4(2):e00345-00318.

Usik, M.A., M.A. Golubkova, and I.V. Ogneva. 2021. "State of Drosophila Melanogaster Ovaries After a Full Cycle of Gametogenesis Under Microgravity Modeling: Cellular Respiration and the Content of Cytoskeletal Proteins." *International Journal of Molecular Sciences* 22(17).

van der Schuren, A., C. Voiniciuc, J. Bragg, K. Ljung, J. Vogel, M. Pauly, and C.S. Hardtke. 2018. "Broad Spectrum Developmental Role of Brachypodium Aux1." *New Phytologist* 219(4):1216–1223.

Van Houtven, G., A. Honeycutt, B. Gilman, N. McCall, W. Throneburg, and K. Sykes. 2008. *Costs of Illness Among Older Adults: An Analysis of Six Major Health Conditions with Significant Environmental Risk Factors*. Research Triangle Park, NC: RTI Press.

Van Ombergen, A., A. Demertzi, E. Tomilovskaya, B. Jeurissen, J. Sijbers, I.B. Kozlovskaya, P.M. Parizel, et al. 2017. "The Effect of Spaceflight and Microgravity on the Human Brain." *Journal of Neurology* 264(Suppl 1):18–22.

Vandenbrink, J.P., R. Herranz, F.J. Medina, R.E. Edelmann, and J.Z. Kiss. 2016. "A Novel Blue-Light Phototropic Response Is Revealed in Roots of Arabidopsis Thaliana in Microgravity." *Planta* 244(6):1201–1215.

Verbeelen, T., N. Leys, R. Ganigué, and F. Mastroleo. 2021. "Development of Nitrogen Recycling Strategies for Bioregenerative Life Support Systems in Space." *Frontiers in Microbiology* 12:700810.

Vialva, T. 2019. "3D Bioprinting Solutions Print Bone Tissue on Board the ISS." 3D Printing Industry, updated December 17. https://3dprintingindustry.com/news/3d-bioprinting-solutions-print-bone-tissue-on-board-the-iss-166458.

Vickers, J. 2021. "Advanced Manufacturing Technologies." NASA. https://www.nasa.gov/directorates/spacetech/game_changing_development/projects/archived/AMT.

Vico, L., and A. Hargens. 2018. "Skeletal Changes During and After Spaceflight." *Nature Reviews: Rheumatology* 14(4):229–245.

Vijayendran, C., T. Polen, V.F. Wendisch, K. Friehs, K. Niehaus, and E. Flaschel. 2007. "The Plasticity of Global Proteome and Genome Expression Analyzed in Closely Related W3110 and MG1655 Strains of a Well-Studied Model Organism, Escherichia Coli-K12." *Journal of Biotechnology* 128(4):747–761.

Villacampa, A., M. Ciska, A. Manzano, J.P. Vandenbrink, J.Z. Kiss, R. Herranz, and F.J. Medina. 2021. "From Spaceflight to Mars g-Levels: Adaptive Response of A. Thaliana Seedlings in a Reduced Gravity Environment Is Enhanced by Red-Light Photostimulation." *International Journal of Molecular Sciences* 22(2).

Villacampa, A., I. Fañanás-Pueyo, F.J. Medina, and M. Ciska. 2022. "Root Growth Direction in Simulated Microgravity Is Modulated by a Light Avoidance Mechanism Mediated by Flavonols." *Physiologia Plantarum* 174(3):e13722.

Vining, K.H., and D.J. Mooney. 2017. "Mechanical Forces Direct Stem Cell Behaviour in Development and Regeneration." *Nature Reviews: Molecular Cell Biology* 18(12):728–742.

Viswanathan, V., E. Mazarico, S. Merkowitz, J.G. Williams, S.G. Turyshev, D. Currie, A.I. Ermakov, et al. 2021. "Extending Science from Lunar Laser Ranging." *Bulletin of the American Astronomical Society* 53(4).

Vitry, G., R. Finch, G. McStay, A. Behesti, S. Déjean, T. Larose, V. Wotring, and W.A. da Silveira. 2022. "Muscle Atrophy Phenotype Gene Expression During Spaceflight Is Linked to a Metabolic Crosstalk in Both the Liver and the Muscle in Mice." *iScience* 25(10):105213.

Vogt, G. 2022. "Paradigm Shifts in Animal Epigenetics: Research on Non-Model Species Leads to New Insights into Dependencies, Functions and Inheritance of DNA Methylation." *Bioessays* 44(8):e2200040.

Voigt, C.A. 2020. "Synthetic Biology 2020–2030: Six Commercially-Available Products That Are Changing Our World." *Nature Communications* 11(1):6379.

Volland, S., J. Esteve-Rudd, J. Hoo, C. Yee, and D.S. Williams. 2015. "A Comparison of Some Organizational Characteristics of the Mouse Central Retina and the Human Macula." *PLOS One* 10(4):e0125631.

Voorhies, A.A., C. Mark Ott, S. Mehta, D.L. Pierson, B.E. Crucian, A. Feiveson, C.M. Oubre, et al. 2019. "Study of the Impact of Long-Duration Space Missions at the International Space Station on the Astronaut Microbiome." *Scientific Reports* 9(1):9911.

Vorobieff, P., M. Anderson, J. Conroy, R. White, C.R. Truman, and S. Kumar. 2011. "Analogues of Rayleigh-Taylor and Richtmyer-Meshkov Instabilities in Flows with Nonuniform Particle and Droplet Seeding." *WIT Transactions on Engineering Sciences* 70:17–28.

Walker, J., and C. Granjou. 2017. "MELiSSA the Minimal Biosphere: Human Life, Waste and Refuge in Deep Space." *Futures* 92:59–69.

Walther, I., P. Pippia, M.A. Meloni, F. Turrini, F. Mannu, and A. Cogoli. 1998. "Simulated Microgravity Inhibits the Genetic Expression of Interleukin-2 and Its Receptor in Mitogen-Activated T Lymphocytes." *FEBS Letters* 436(1):115–118.

Wang, L., J. Xie, C. Mou, Y. Jiao, Y. Dou, and H. Zheng. 2021. "Transcriptomic Analysis of the Interaction Between Flowering Locus T Induction and Photoperiodic Signaling in Response to Spaceflight." *Frontiers in Cell & Developmental Biology* 9:813246.

Wang, L., X. You, L. Zhang, C. Zhang, and W. Zou. 2022. "Mechanical Regulation of Bone Remodeling." *Bone Research* 10(1):16.

Wang, M., D. Duday, E. Scolan, S. Perbal, M. Prato, C. Lasseur, and M. Hołyńska. 2021. "Antimicrobial Surfaces for Applications on Confined Inhabited Space Stations." *Advanced Materials Interfaces* 8(13):2100118.

Waters, S.M., S.M. Ledford, A. Wacker, S. Verma, B. Serda, J. McKaig, J. Varelas, P.M. Nicoll, K. Venkateswaran, and D.J. Smith. 2021. "Long-Read Sequencing Reveals Increased Occurrence of Genomic Variants and Adenosine Methylation in *Bacillus Pumilus* SAFR-032 After Long-Duration Flight Exposure Onboard the International Space Station." *International Journal of Astrobiology* 20(6):435–444.

Watson, J.L., D. Juergens, N.R. Bennett, B.L. Trippe, J. Yim, H.E. Eisenach, W. Ahern, et al. 2022. "Broadly Applicable and Accurate Protein Design by Integrating Structure Prediction Networks and Diffusion Generative Models." *bioRxiv* 2022.12.09.519842, preprint.

Webber, M.J., and M.W. Tibbitt. 2022. "Dynamic and Reconfigurable Materials from Reversible Network Interactions." *Nature Reviews Materials* 7(7):541–556.

Weber, J.K.R. 2010. "The Containerless Synthesis of Glass." *International Journal of Applied Glass Science* 1(3):248–256. https://doi.org/10.1111/j.2041-1294.2010.00026.x.

Weiland, K.J., F.G. Gati, M.E. Hill, T.F. O'Malley, and R.L. Zurawski. 2005. *The Fluids and Combustion Facility: Enabling the Exploration of Space*. Technical Report TM-2005-213973. Washington, DC: NASA.

Weinhold, A. 2018. "Transgenerational Stress-Adaption: An Opportunity for Ecological Epigenetics." *Plant Cell Reports* 37(1):3–9.

Wheeler, R.M. 2017. "Agriculture for Space: People and Places Paving the Way." *Open Agriculture* 2(1):14–32.

Wheeler, R.M., B.V. Peterson, J.C. Sager, and W.M. Knott. 1996. "Ethylene Production by Plants in a Closed Environment." *Advances in Space Research* 18(4–5):193–196.

White House. 2023. "Fact Sheet: Biden-Harris Administration Announces New Actions to Advance Open and Equitable Research." Washington, DC: Office of Science and Technology Policy. https://www.whitehouse.gov/ostp/news-updates/2023/01/11/fact-sheet-biden-harris-administration-announces-new-actions-to-advance-open-and-equitable-research.

Wicaksono, I., P.G. Hwang, S. Droubi, F.X. Wu, A.N. Serio, W. Yan, and J.A. Paradiso. 2022. "3DKnITS: Three-Dimensional Digital Knitting of Intelligent Textile Sensor for Activity Recognition and Biomechanical Monitoring." Paper presented at the 2022 44th Annual International Conference of the IEEE Engineering in Medicine & Biology Society (EMBC), July 11–15.

Wilkinson, M.D., M. Dumontier, I.J. Aalbersberg, G. Appleton, M. Axton, A. Baak, N. Blomberg, et al. 2016. "The FAIR Guiding Principles for Scientific Data Management and Stewardship." *Scientific Data* 3(1):160018.

Willey, J.S., R.A. Britten, E. Blaber, C.G.T. Tahimic, J. Chancellor, M. Mortreux, L.D. Sanford, A.J. Kubik, M.D. Delp, and X.W. Mao. 2021. "The Individual and Combined Effects of Spaceflight Radiation and Microgravity on Biologic Systems and Functional Outcomes." *Journal of Environmental Science & Health, Part C: Toxicology and Carcinogenesis* 39(2):129–179.

Williams, J., S.-W. Chiow, N. Yu, and H. Müller. 2016. "Quantum Test of the Equivalence Principle and Space-Time Aboard the International Space Station." *New Journal of Physics* 18(2):025018.

Williamson Smith, A. 2019. "Published Results from Crystallization Experiments on the ISS Could Help Merck Improve Cancer Drug Delivery." ISS National Laboratory. https://www.issnationallab.org/merck-lab-publishes-pembrolizumab-results.

Wilson, J.W., C.M. Ott, L. Quick, R. Davis, K. Höner zu Bentrup, A. Crabbé, E. Richter, et al. 2008. "Media Ion Composition Controls Regulatory and Virulence Response of Salmonella in Spaceflight." *PLOS One* 3(12):e3923.

Wilson, J.W., C.M. Ott, K.H. zu Bentrup, R. Ramamurthy, L. Quick, S. Porwollik, P. Cheng, et al. 2007. "Space Flight Alters Bacterial Gene Expression and Virulence and Reveals a Role for Global Regulator Hfq." *Proceedings of the National Academy of Sciences* 104(41):16299–16304.

Wilson, J.W., J.L. Shinn, R.K. Tripathi, R.C. Singleterry, M.S. Clowdsley, S.A. Thibeault, F.M. Cheatwood, et al. 2001. "Issues in Deep Space Radiation Protection." *Acta Astronautica* 49(3):289–312.

Witze, A. 2020. "Astronauts Have Conducted Nearly 3,000 Science Experiments Aboard the ISS." *Nature News*. https://www.nature.com/articles/d41586-020-03085-8.

Wnorowski, A., A. Sharma, H. Chen, H. Wu, N.Y. Shao, N. Sayed, C. Liu, et al. 2019. "Effects of Spaceflight on Human Induced Pluripotent Stem Cell-Derived Cardiomyocyte Structure and Function." *Stem Cell Reports* 13(6):960–969.

Wojcik, P., A. Kini, B. Al Othman, L.A. Galdamez, and A.G. Lee. 2020. "Spaceflight Associated Neuro-ocular Syndrome." *Current Opinion in Neurology* 33(1):62–67.

Wollman, A., M. Weislogel, B. Wiles, D. Pettit, and T. Snyder. 2016. "More Investigations in Capillary Fluidics Using a Drop Tower." *Experiments in Fluids* 57(4):57.

Wolverton, C. 2022. *Characterizing Plant Gravity Perception Systems*. https://taskbook.nasaprs.com/Tbp/index.cfm?action=public_query_taskbook_content&TASKID=15274.

Wong, C.P., U.T. Iwaniec, and R.T. Turner. 2021. "Evidence for Increased Thermogenesis in Female C57BL/6J Mice Housed Aboard the International Space Station." *npj Microgravity* 7(1):23.

Wong, M.M., G.L. Chong, and P.E. Verslues. 2017. "Epigenetics and RNA Processing: Connections to Drought, Salt, and ABA?" *Methods in Molecular Biology* 1631:3–21.

Wright, P.F., M.E. Ackerman, and E.B. Brickley. 2019. "Mucosal Immunity: The Forgotten Arm of the Immune System." *Journal of the Pediatric Infectious Disease Society* 8(1):53–54.

Wu, P.-H., S.S. Gambhir, C.M. Hale, W.-C. Chen, D. Wirtz, and B.R. Smith. 2020. "Particle Tracking Microrheology of Cancer Cells in Living Subjects." *Materials Today* 39:98–109.

Wu, X.-T., X. Yang, R. Tian, Y.-H. Li, C.-Y. Wang, Y.-B. Fan, and L.-W. Sun. 2022. "Cells Respond to Space Microgravity Through Cytoskeleton Reorganization." *The FASEB Journal* 36(2):e22114.

Xu, D., Y. Gao, L. Guo, C. Lin, and Y. Sun. 2018. "Effect of Dys-1 Mutation on Gene Expression Profile in Space-Flown Caenorhabditis Elegans." *Muscle Nerve* 10.1002/mus.26076.

Xu, P., H. Chen, J. Hu, and W. Cai. 2021. "Potential Evidence for Transgenerational Epigenetic Memory in Arabidopsis Thaliana Following Spaceflight." *Communications Biology* 4(1):835.

Xu, V., M. Jaffe, C.D. Panda, S.L. Kristensen, L.W. Clark, and H. Müller. 2019. "Probing Gravity by Holding Atoms for 20 Seconds." arXiv:1907.03054 [physics.atom-ph]. https://doi.org/10.48550/arXiv.1907.03054.

Xu, Y., W. Pei, and W. Hu. 2022. "A Current Overview of the Biological Effects of Combined Space Environmental Factors in Mammals." Review. *Frontiers in Cell and Developmental Biology* 10.

Xue, Y.-F., J.-F. Zhao, J.-J. Wei, J. Li, D. Guo, and S.-X. Wan. 2011. "Experimental Study of Nucleate Pool Boiling of FC-72 on Smooth Surface Under Microgravity." *Microgravity Science and Technology* 23(1):75.

Yagi-Utsumi, M., S. Yanaka, C. Song, T. Satoh, C. Yamazaki, H. Kasahara, T. Shimazu, K. Murata, and K. Kato. 2020. "Characterization of Amyloid B Fibril Formation Under Microgravity Conditions." *npj Microgravity* 6(1):17.

Yan, X., S.P. Sasi, H. Gee, J. Lee, Y. Yang, R. Mehrzad, J. Onufrak, et al. 2014. "Cardiovascular Risks Associated with Low Dose Ionizing Particle Radiation." *PLOS One* 9(10):e110269.

Yang, J., J. Barrila, C.M. Ott, O. King, R. Bruce, R.J.C. McLean, and C.A. Nickerson. 2021. "Longitudinal Characterization of Multispecies Microbial Populations Recovered from Spaceflight Potable Water." *npj Biofilms and Microbiomes* 7(1):70.

Yang, J., J. Barrila, K.L. Roland, C.M. Ott, and C.A. Nickerson. 2016. "Physiological Fluid Shear Alters the Virulence Potential of Invasive Multidrug-Resistant Non-Typhoidal Salmonella Typhimurium D23580." *npj Microgravity* 2(1):16021.

Yang, J., Y. Fu, and H. Liu. 2022. "Microbiomes of Air Dust Collected During the Ground-Based Closed Bioregenerative Life Support Experiment 'Lunar Palace 365.'" *Environmental Microbiome* 17(1):4.

Yang, J., S.G. Thornhill, J. Barrila, C.A. Nickerson, C.M. Ott, and R.J.C. McLean. 2018. "Chapter 1—Microbiology of the Built Environment in Spacecraft Used for Human Flight." *Methods in Microbiology* 45:3–26, V. Gurtler and J.T. Trevors, eds. Academic Press, Reprint.

Yi, Y., W. Xu, Y. Fan, and H.-X. Wang. 2021. "Drosophila as an Emerging Model Organism for Studies of Food-Derived Antioxidants." *Food Research International* 143:110307.

Yin, J., Y. Cao, Y.-H. Li, S.-K. Liao, L. Zhang, J.-G. Ren, W.-Q. Cai, et al. 2017. "Satellite-Based Entanglement Distribution Over 1200 Kilometers." *Science* 356(6343):1140–1144.

Youssef, M., T. Hueckel, G.-R. Yi, and S. Sacanna. 2016. "Shape-Shifting Colloids via Stimulated Dewetting." *Nature Communications* 7(1):12216.

Yu, C.-N., K. Lazaridis, Y. Wu, E. Voroshilov, M.D. Krivilyov, S.D. Mesarovic, and D.P. Sekulic. 2021. "Filling a Hole by Capillary Flow of Liquid Metal–Equilibria and Instabilities." *Physics of Fluids* 33(3):034109.

Yu, T., B.W. Parks, S. Yu, R. Srivastava, K. Gupta, X. Wu, S. Khaled, P.Y. Chang, J.H. Kabarowski, and D.F. Kucik. 2011. "Iron-Ion Radiation Accelerates Atherosclerosis in Apolipoprotein E-Deficient Mice." *Radiation Research* 175(6):766–773.

Yuan, M., M.A. Custaud, Z. Xu, J. Wang, M. Yuan, C. Tafforin, L. Treffel, et al. 2019. "Multi-System Adaptation to Confinement During the 180-Day Controlled Ecological Life Support System (CELSS) Experiment." *Frontiers in Physiology* 10:575.

Zea, L., R.J.C. McLean, T.A. Rook, G. Angle, D.L. Carter, A. Delegard, A. Denvir, et al. 2020. "Potential Biofilm Control Strategies for Extended Spaceflight Missions." *Biofilm* 2:100026.

Zea, L., Z. Nisar, P. Rubin, M. Cortesão, J. Luo, S.A. McBride, R. Moeller, et al. 2018. "Design of a Spaceflight Biofilm Experiment." *Acta Astronautica* 148:294–300.

Zea, L., N. Prasad, S.E. Levy, L. Stodieck, A. Jones, S. Shrestha, and D. Klaus. 2016. "A Molecular Genetic Basis Explaining Altered Bacterial Behavior in Space." *PLOS One* 11(11):e0164359. https://doi.org/10.1371/journal.pone.0164359.

Zeidler, C., G. Woeckner, J. Schöning, V. Vrakking, P. Zabel, M. Dorn, D. Schubert, B. Steckelberg, and J. Stakemann. 2021. "Crew Time and Workload in the EDEN ISS Greenhouse in Antarctica." *Life Sciences in Space Research* 31:131–149.

Zeitlin, C., D.M. Hassler, F.A. Cucinotta, B. Ehresmann, R.F. Wimmer-Schweingruber, D.E. Brinza, S. Kang, et al. 2013. "Measurements of Energetic Particle Radiation in Transit to Mars on the Mars Science Laboratory." *Science* 340(6136): 1080–1084.

Zell, M., J. Straub, and A. Weinzierl. 1984. "Nucleate Pool Boiling in Subcooled Liquid Under Microgravity: Results of TEXUS Experimental Investigations." In *ESA Fifth European Symposium on Materials Science Under Microgravity. Results of Spacelab* 1:327–333.

Zhang, L., X. Guo, Z. Zhang, A. Wang, and J. Zhu. 2021. "Cold-Regulated Gene LeCOR413PM2 Confers Cold Stress Tolerance in Tomato Plants." *Gene* 764:145097.

Zhang, W., and R. Ramautar. 2021. "CE-MS for Metabolomics: Developments and Applications in the Period 2018–2020." *Electrophoresis* 42(4). https://doi.org/10.1002/elps.202000203.

Zhang, J., P.J. Santos, P.A. Gabrys, S. Lee, C. Liu, and R.J. Macfarlane. 2016. "Self-Assembling Nanocomposite Tectons." *Journal of the American Chemical Society* 138(50):16228–16231.

Zhang, Y., L. Wang, J. Xie, and H. Zheng. 2015. "Differential Protein Expression Profiling of Arabidopsis Thaliana Callus Under Microgravity on Board the Chinese SZ-8 Spacecraft." *Planta* 241(2):475–488.

Zhao, T., G. Liu, D. Liu, Y. Yi, B. Xie, and H. Liu. 2022. "Water Recycle System in an Artificial Closed Ecosystem—Lunar Palace 1: Treatment Performance and Microbial Evolution." *Science of the Total Environment* 806(Pt 3):151370.

Zheng, D., T. Liwinski, and E. Elinav. 2020. "Interaction Between Microbiota and Immunity in Health and Disease." *Cell Research* 30(6):492–506.

Zheng, X., J. Dolde, V. Lochab, B.N. Merriman, H. Li, and S. Kolkowitz. 2022. "Differential Clock Comparisons with a Multiplexed Optical Lattice Clock." *Nature* 602(7897):425–430.

Zhou, M., N.J. Sng, C.E. LeFrois, A.-L. Paul, and R.J. Ferl. 2019. "Epigenomics in an Extraterrestrial Environment: Organ-Specific Alteration of DNA Methylation and Gene Expression Elicited by Spaceflight in Arabidopsis Thaliana." *BMC Genomics* 20(1):205.

Zia, R.N. 2018. "Active and Passive Microrheology: Theory and Simulation." *Annual Review of Fluid Mechanics* 50(1):371–405.

Zimmermann, A. 2023. *U.S. R&D and Innovation in a Global Context: The 2023 Data Update*. Washington, DC: American Association for the Advancement of Science. https://www.aaas.org/sites/default/files/2023-05/AAAS%20Global%20RD%20Update%20April%202023.pdf.

Zohar, E., J.I. Cirac, and B. Reznik. 2012. "Simulating Compact Quantum Electrodynamics with Ultracold Atoms: Probing Confinement and Nonperturbative Effects." *Physical Review Letters* 109(12):125302.

Zudell, D. 2021. "Experiment Bound for Space Station Turns Down the Heat." NASA. https://www.nasa.gov/feature/glenn/2021/experiment-bound-for-space-station-turns-down-the-heat.

Zupanska, A.K., F.C. Denison, R.J. Ferl, and A.L. Paul. 2013. "Spaceflight Engages Heat Shock Protein and Other Molecular Chaperone Genes in Tissue Culture Cells of Arabidopsis Thaliana." *American Journal of Botany* 100(1):235–248.

Zupanska, A.K., E.R. Schultz, J. Yao, N.J. Sng, M. Zhou, J.B. Callaham, R.J. Ferl, and A.L. Paul. 2017. "ARG1 Functions in the Physiological Adaptation of Undifferentiated Plant Cells to Spaceflight." *Astrobiology* 17(11):1077–1111.

Appendixes

A

Statement of Task

The National Academies of Sciences, Engineering, and Medicine will appoint a decadal survey committee to carry out a decadal survey of biological and physical sciences research in space. The study will generate consensus recommendations to implement a comprehensive strategy and vision for a decade of transformative science at the frontiers of biological and physical sciences research in space. The results of the study will assist NASA in defining and aligning biological and physical sciences research to uniquely advance scientific knowledge, meet the needs of human and robotic exploration missions, and provide terrestrial benefits.

The committee, with input from study panels covering the breadth of biological and physical sciences in space, will carry out the following tasks:

1. Conduct a review of the current state of knowledge in the major and emerging areas of space-related biological and physical sciences research;
2. Identify the most compelling science challenges and frontiers to be pursued utilizing spaceflight environments and analogues of spaceflight conditions that will enable scientific discovery, address the needs of space exploration, and/or result in applications on Earth;
3. Develop a comprehensive research strategy to advance the frontiers of biological and physical sciences research in space that will include identifying, recommending, and ranking the highest priority research activities, taking into account for each activity the scientific case, international and commercial activities, and opportunities for partnerships. Where feasible and useful, such factors as timing, cost category and cost risk, technical readiness, and technical risk will also be considered. The strategy should:

   a. Recommend approaches to facilitate the development of a robust, resilient, and appropriately balanced program of biological and physical science space research that will enable scientific discovery, address the scientific and technological needs of space exploration, and/or result in applications on Earth.
   b. Identify facility and platform capabilities and environmental requirements for each of the recommended research activities as appropriate, including facilities or capabilities not currently available but which could be developed in the future.
   c. Assemble notional proof-of-concept research campaigns, where appropriate synergies may be achieved, that address identified prioritized research activities as part of complex or multi-disciplinary missions or mission sets including those that make use of cost-effective ground analogues and sub-orbital flights.

d. Utilize the Technical Risk and Cost Evaluation (TRACE) process on large recommended spaceflight projects (those costing more than $100 million) and on any other projects selected by the committee.
e. Organize the recommended research activities, associated facilities and platforms, and proof-of-concept research campaigns into broad cost categories in order to assist NASA's understanding of the top-level scientific performance and resource options.

4. Recommend broad decision rules, where appropriate, so that NASA can consider accommodating significant deviations in the projected budget or changes in priorities precipitated by new discoveries or new commercial achievements.

As part of its review the committee will consider and address relevant cross-cutting research needs and key aspects of the infrastructure—including NASA, commercial, and international programs and plans—that may affect the conduct of research. Potential infrastructure topics that will be considered include:

- Where NASA capabilities or ability to assume high risk enable it to uniquely support the research enterprise, and where support roles might feasibly be transitioned to commercial providers;
- Roles played by NASA's biological and physical sciences program in supporting the conduct of space research, particularly in light of the mission and capabilities of the International Space Station (ISS) National Laboratory, limited lifetime of the ISS, and the prospect of commercial platform(s) in low Earth orbit (LEO);
- Existing and potential new research synergies between NASA and other U.S. government agencies, opportunities for collaborative research that are relevant to science priorities between Science Mission Directorate's science divisions, as well as with commercial entities and international partners; and
- The current position and expected evolution of the United States relative to other countries in the areas covered by the study, including the uniqueness (or lack thereof) of U.S. efforts.

CONSIDERATIONS

In considering the emerging availability of commercial platforms, facilities, and services for research in LEO and beyond, the committee should include an awareness of the space-related programs within other agencies and non-government organizations. While the focus of the committee will be on the next decade, the committee may also identify potential research objectives that extend beyond that timeframe. This study should build upon the findings and recommendations of previous National Academies' studies. As part of its work, the committee will also:

- Describe how the identified research objectives could produce knowledge, enable exploration activities, or provide benefits to terrestrial and other applications;
- Recommend criteria for identifying and updating a high-value research portfolio that is enabled by exploration and/or enables exploration; and
- Apply these criteria to recommend a high-value research portfolio that is enabled by exploration and/or enables exploration.

SCOPE

When selecting biological and physical science disciplines that will be reviewed in this report, the committee will generally consider discipline areas covered in the previous decadal report, as well as emerging areas of interest. However, the committee will not review the discipline areas of NASA's program of risk identification and mitigation for astronauts except to the extent that biological research in microbiology, animal biology, or plant biology could inform that program. Translational research, innovative methods and procedures, and preclinical studies (particularly those involving understanding biological processes, normal or pathophysiological adaptation to microgravity, and mechanisms of action) may be included.

B

Acronyms and Abbreviations

AAAS	American Academy of Arts and Sciences
AAPI	Asian American and Pacific Islander
ABS	acrylonitrile butadiene styrene
ACE	advanced colloids experiment
ACME	advanced combustion microgravity experiment
AFM	atomic force microscopy
AFRL	Air Force Research Laboratory
AHA2	Arabidopsis plasma membrane proton APTase 2
AI	artificial intelligence
ALS	advanced life support
ALSDA	animal life sciences data archive
AM	additive manufacturing
AMF	ISS Additive Manufacturing Facility
AMO	atomic, molecular, and optical (physics)
AMR	anti-microbial resistance
AOX	alternative oxidase
APH	Advanced Plant Habitat
ASEP	advanced space experiment processor
ATP	adenosine triphosphate
AWG	Analysis Working Group (GeneLab)
BASS	Burning and Supression of Solids (NASA)
BBB	blood-brain barrier
BCAT	binary colloidal aggregation test
BEC	Bose-Einstein condensate
BFF	BioFabrication Facility
BGC	biosynthetic gene cluster
BiPS-Free	Biological and Physical Science Free Flyer
BLEO	beyond LEO

BLiSS	Bioregenerative Life Support Systems
BPAC	Biological and Physical Advisory Committee
BPS	biological and physical sciences
BPS Division	Division of Biological and Physical Sciences (NASA)
BRIC	Biological Research in Canister
BSM	beyond standard model
BSP	Biospecimen Sharing Program
CACES	Cold Atom Clock Experiment in Space (China)
CAL	Cold Atom Laboratory
CASIS	Center for the Advancement of Science in Space
CBEF	Cell Biology Equipment Facility (JAXA)
CBPSS	Committee on Biological and Physical Space Sciences
CERISS	Commercially Enabled Rapid Space Science Initiative
CESA1	cellulose synthase 1
CESI-MS	capillary electrospray ionization
CFD	computational fluid dynamics
CFM	cryogenic fluid management
CheX	confined helium experiment
CHF	critical heat flux
CI	computational infrastructure
CIR	Combustion Integrated Rack
CLD	coflow laminar diffusion
CLD	Commercial LEO Destination
CLPS	commercial lunar payload services (NASA)
CM-HT	flow condensation module for heat transfer measurements
CNS	central nervous system
COTS	commercial off-the-shelf
CPU	computer processing unit
CRISPR	Clustered Regularly Interspaced Short Palindromic Repeats
CSA	Canadian Space Agency
CSF	cerebrospinal fluid
CSK	cytoskeleton
CTE	coefficient of thermal expansion
CubeSat	cube-satellite
DARPA	Defense Advanced Research Projects Agency
DEIA	diversity, equity, inclusion, and accessibility
DLR	German Space Agency
DNA	deoxyribonucleic acid
DoD	Department of Defense
DOE	Department of Energy
DOTT	Deep Space OTT
DSAC	deep space atomic clock
DYNAMX	science requirements document for the critical dynamics in microgravity
ECLSS	environmental control and life support system (ISS)
ECM	extracellular matrix
EEA	Erasmus Experiment Archive
EEF	ecology experiment facility

EGF	epidermal growth factor
ELF	Electrostatic Levitation Furnace (JAXA)
EMCS	European Modular Cultivation System
EMILI	European Molecular Indicator of Life
EML	ISS Electromagnetic Levitator (ESA)
EMU	extravehicular mobility unit
EPS	extracellular polymeric substance
ESA	European Space Agency
EV	extracellular vesicle
EVA	extravehicular activity
EXPRESS	Expedite the Processing of Experiments to Space Station
FAIR	findable, accessible, interoperable, and reusable
FBCE	flow boiling and condensation experiment
FBM	flow boiling module
FFE	far-from-equilibrium
FFI	fractional frequency instability
FLEX	flame extinguishment experiment
FPS	fission power system
FTIR	Fourier transformed infrared spectroscopy
GC	gas chromatography
GCR	galactic cosmic ray
GEM	graduate education for minority students
GEO	geostationary orbit
GPI	glycosylphosphatidylinositol
GPS	Global Positioning System
GSE	ground support equipment
GW	gravitational wave
HBCU	Historically Black College and University
HCCI	homogeneous charge compression ignition
HCU	Habitat Cage Unit (JAXA)
HEO	Human Exploration and Operations Mission Directorate (NASA)
HLS	human landing system (NASA)
HRP	Human Research Program (NASA)
HSF	heat shock factor
HSI	Hispanic-serving institution
HSP	heat shock protein
ICP	intracranial pressure
ISRU	in situ resource utilization
ISS	International Space Station
ISSNL	ISS National Laboratory
ISSPO	ISS Program Office
JAXA	Japan Aerospace Exploration Agency
JWST	James Webb Space Telescope
KSQ	key scientific question

LC	liquid chromatography
LEO	low Earth orbit
LGBTQ+	lesbian, gay, bisexual, transgender, queer, +
LHC	Large Hadron Collider
LIGO	Laser Interferometer Gravitational-Wave Observatory
LINC	linker of nucleoskeleton and cytoskeleton
LISA	Laser Interferometer Space Antenna
LMM	Light Microscopy Module
LOI	limiting oxygen index
LOTT	LEO OTT
LPE	Lambda Point Experiment
LSDA	Life Sciences Data Archive
LTEE	long-term evolution experiment
M2TIC	micro-mercury trapped-ion clock
MAIUS	matter-wave interferometry in microgravity (DLR)
MATRICES	Manufacturing Materials and Processes for Sustainability in Space
MCS	multi-component search
MDCA	Multi-User Droplet Combustion Apparatus
ME-LIF	microchip electrophoresis laser induced fluorescence
MELiSSA	Micro-Ecological Life Support System Alternative (ESA)
MGCB	Microgravity Database
MI&O	ISSPO integration and operations costs
MICROSCOPE	Microsatellite à traînée compensée pour l'observation du principe d'equivalence
miRNA	micro-RNA
MISSE	materials international space station experiment
MISSE-FF	MISSE flight facility
MLE	mid-deck locker equivalent
MLOC	minimum limiting oxygen concentration
MoBE	microbes of the built environment
MOC	maximum oxygen concentration
mRNA	messenger RNA
MS	mass spectroscopy
MSG	microgravity science glovebox
MSI	minority-serving institution
mtDNA	mitochondrial DNA
MVP	Multi-Use Variable-G Platform
NADPH	nicotinamide adenine dinucleotide phosphate
NASA	National Aeronautics and Space Administration
NBISC	NASA Biological Institutional Scientific Collection
NCATS	National Center for Advancing Translational Sciences
NGO	non-government organization
NGS	next-generation sequencing
NIH	National Institutes of Health
NIST	National Institute of Standards and Technology
NMR	nuclear magnetic resonance
NO	nitric oxide
NSF	National Science Foundation
NSRL	NASA Space Radiation Laboratory

OECD	Organisation for Economic Co-operation and Development
OHALO	operational crop production system and prototype for a Mars transit vehicle
OLC	optical lattice clock
OTT	optical time transfer
PAESMEM	Presidential Award for Excellence in Science, Mathematics and Engineering Mentoring
PCE	Polymer and Composite Experiment
PCR	polymerase chain reaction
PE	polyethylene
PEC	Passive Experiment Container
PET	polyethylene terephthalate
PFaST	Probing the Fabric of Spacetime
PGP	Plant Gravity Perception
PGP	plant growth promoting
PHA	poly-hydroxy alkanoic acid
PHAB	plate habitat
PHB	poly-hydroxy butyrate
PRoMO	polar radiation of model organisms
PSI	Physical Science Informatics
QUESS	Quantum Experiments at Space Scale (China)
R&D	research and development
RCCI	reactivity-controlled compression ignition
RCLF	relevant civilian labor force
RNA	ribonucleic acid
ROS	reactive oxygen species
ROS	Russian Orbital Segment
ROSES	Research Opportunities in Earth and Space Program
RTI	Rayleigh-Taylor Instability
S&E	science and engineering
SABL	Space Automated Bioproduct Laboratory
SAMS	space acceleration measurement system
SANS	Spaceflight-Associated Neuro-Ocular Syndrome
SciAct	science activation
SEDI	socially or economically disadvantaged individual
SES	senior executive service
SHERE	shear history extensional rheology experiment
SL	senior level
SLPSRA	Division of Space Life and Physical Sciences Research and Application (NASA)
SM	standard model
SMD	Science Mission Directorate (NASA)
SMG	simulated microgravity
SOC	space optical clock
SoFIE	Solid Fuel Ignition and Extinction
SoT	statement of task
SOTT	space-to-space OTT
SPE	solar particle event
SSERVI	Solar System Exploration Research Institute (NASA)

STAR	spaceflight technology, applications, and research
STEM	science, technology, engineering, and mathematics
STMD	Space Technology Mission Directorate (NASA)
SWaP	size, weight, and power
SynCom	synthetic microbial community
TCA	tricarboxylic acid
TCS	thermal control system
TOAST	Test of Arabidopsis Space Transcriptome
TRACE	technical risk and cost evaluation
TRISH	Translational Institute for Space Health
TWOTFT	two-way optical time and frequency transfer
UHTC	ultra-high temperature ceramic
UNOOSA	United Nations Office for Outer Space Affairs
USDA	U.S. Department of Agriculture
USOS	U.S. Orbital Segment, portion of the International Space Station
UV	ultraviolet
VCC	vapor compression cycle
Veggie	ISS vegetable production system
VFB	Vented Fly Box
VLBI	very long-baseline interferometry
VOC	volatile organic compound
WIMP	weakly interacting massive particle
XRD	X-ray diffraction
ZBOT	zero boil-off tank

C

Steering Committee and Staff Biographical Information

STEERING COMMITTEE

ROBERT J. FERL, *Co-Chair*, is a distinguished professor and the assistant vice president for research at the University of Florida. His research includes analysis of the fundamental biological processes involved in plant physiological adaptations to environments, with an emphasis on the environmental opportunities presented by space exploration. He is an expert in plant gene responses to environmental stresses and the signal transduction processes that control those responses. His research has taken him to the high Canadian Arctic to study biological responses to remote terrestrial impact craters and to Antarctica to investigate advanced plant production in support of space exploration habitats. His honors include the NASA Medals for Exceptional Scientific Achievement and Exceptional Public Service. He previously served on the National Academies' Decadal Survey on Biological and Physical Sciences in Space: Plant and Microbial Biology Panel (2009–2011) and as the co-chair of the Committee on Biological and Physical Sciences in Space.

KRYSTYN J. VAN VLIET, *Co-Chair*, is a professor of engineering and the vice president for research and innovation at Cornell University, Cornell Tech, and Weill Cornell Medicine. At the outset of this study and survey, she was the Koerner Professor of Materials Science and Engineering and Biological Engineering at the Massachusetts Institute of Technology (MIT). Her research focuses on material chemomechanics: the material behavior at the interface of mechanics, chemistry, physics, and biology, and, in particular, thermodynamically metastable surfaces and interfaces. Van Vliet served previously as MIT associate provost and associate vice president for research, overseeing campus space management, technology licensing, and corporate relations, and co-directed the MIT Biomedical Engineering Minor Program. As MIT faculty, Van Vliet also conducted research in Singapore, where her interdisciplinary team invented and developed new technology platforms for diagnostics and treatment of cell and tissue disease, as well as cell therapy manufacturing solutions.

ADAM P. ARKIN is the Dean A. Richard Newton Memorial Professor in the Department of Bioengineering at the University of California, Berkeley, and a senior faculty scientist at the Lawrence Berkeley National Laboratory. Arkin is the director of the Center for Utilization of Biological Engineering in Space. The center seeks microbial and plant-based biological solutions for in situ resource utilization that reduce the launch mass and improve reliability and quality of food, pharmaceuticals, fuels, and materials for astronauts on a mission to Mars.

Arkin's laboratory for systems and synthetic biology seeks to uncover the evolutionary design principles of cellular networks and populations and to exploit them for applications in the areas of systems and synthetic biology, environmental microbiology of bacteria and viruses, bioenergy, biomedicine, and bioremediation. Arkin earned a PhD in physical chemistry from MIT.

SUSAN M. BAILEY is a professor and radiation cancer biologist in the Department of Environmental and Radiological Health Sciences at Colorado State University. Part of Bailey's current research program includes being one of 10 investigations selected for the NASA Twins Study, an integrated effort to launch human space life science research into a new era of molecular or "omics"-based studies. Bailey and her team assessed changes in telomere length and telomerase activity, DNA damage responses in the space- and Earth-bound twins and in a cohort of unrelated astronauts, and matched ground control subjects. Bailey previously served on the National Academies' Committee to Review NASA's Evidence Reports on Human Health Risks. She serves on the editorial board of two scientific journals, is an author on over 70 peer-reviewed publications and three book chapters, and is an inventor on one patent application. Bailey received a PhD in biomedical sciences from the University of New Mexico School of Medicine.

DEBJYOTI BANERJEE is a professor of mechanical engineering and petroleum engineering and the dean's fellow in the Inter-Collegiate School of Engineering Medicine at Texas A&M University. Previously, he was a hiring manager of the Fluidics and Device Engineering Group at Applied Biosystems (Life Technologies). Banerjee received 17 U.S. patents. His research interests are in thermal-fluids sciences—for example, multiphase flows with phase change (e.g., the "nanoFin Effect" in boiling and condensation), thermal management (chip cooling), thermal energy storage (phase change materials), micro/nano-fluidics (confined fluids), lab-on-chip technologies (neural organoids, diagnostics), bio-MEMS, bio/nano-technologies (Dip-Pen Nanolithography, nano-calorimetry/"nano-nose"), sustainability (desalination) and renewable energy (molten salt nanofluids for nuclear and concentrated-solar power), food-energy-water nexus—including smart-irrigation/precision-agriculture (e.g., for microgravity), and numerical simulations (lumped/network models, finite elements analyses, computational fluids/heat transfer, molecular dynamics, and machine learning).

PAUL M. CHAIKIN is a Silver Professor of Physics at New York University. He joined the physics faculty at the University of California, Los Angeles (UCLA), in 1972 and studied thermopower, density waves, and high field phenomena mostly in organic superconductors. The lure of actually seeing the microscopics of a system led him to soft matter. He helped develop techniques to measure elasticity and motion and understand colloidal interactions. His hard and soft matter interests continued after joining the faculty at the University of Pennsylvania (1983), the staff at Exxon Research (1983), and the faculty at Princeton University (1988). His interests in geometry/topology led to his founding contributions to diblock copolymer nanolithography and studies of defects, annealing, and pattern formation. He helped demonstrate and explain why ellipsoids pack more densely than spheres. In 2005 he helped found the Center for Soft Matter Research at New York University. His more recent research centers on artificial self-replication, self-assembly, active matter, DNA nanotechnology, topological defects on curved surfaces, and quantifying order far from equilibrium.

KATHLEEN E. CULLEN is a professor in the Departments of Biomedical Engineering, Neuroscience, and Otolaryngology, and the co-director of the Center for Hearing and Balance at Johns Hopkins University. Cullen founded and directs Johns Hopkins' Systems Neuroscience and Neuroengineering Laboratory, which spans the interdisciplinary fields of neural engineering and neuroscience, harnessing the power of innovative computational and neurophysiological methodologies. In addition to those research activities, Cullen currently serves as the program chair and vice president of the Society for the Neural Control of Movement. Cullen has long been committed to improving diversity in science, including the promotion, visibility, and representation of women and underrepresented minorities. Cullen has been an active member of the scientific advisory board of the National Space Biomedical Research Institute, which works with NASA to identify health risks in extended space flight. Cullen earned a PhD in biomedical engineering from the University of Chicago.

DANIEL H. GESCHWIND is the Gordon and Virginia MacDonald Distinguished Professor of Human Genetics, Neurology and Psychiatry at UCLA and the lead of the Geschwind Laboratory. Geschwind is also the founder, senior associate dean, and associate vice chancellor of the Institute for Precision Health at UCLA. Geschwind is a pioneer in the transcriptomic and functional genomic analyses of the healthy nervous system and the application of systems biology methods in brain disease. Geschwind defined the molecular pathology of autism and several other major psychiatric disorders and has made major contributions to discover new pathways involved in neurodegeneration and to facilitate neural regeneration. Geschwind has received several awards for his laboratory's work, including the Gold Medal from the Society of Biological Psychiatry and the Ruane Prize, and he is an elected member of the American Association of Physicians and the National Academy of Medicine. Geschwind earned an MD and a PhD in neurobiology from the Yale University School of Medicine.

ROBERT W. HYERS is the George I. Alden Professor and the head of the Department of Mechanical and Materials Engineering at Worcester Polytechnic Institute and a principal scientist and owner of RHA Materials, LLC, which offers contract research and consulting on materials processing and properties to the government and academic institutions as well as private businesses. Previously, Hyers was the president and chief technology officer of the Boston Electrometallurgical Corporation and a staff scientist at the NASA Marshall Space Flight Center/ Electrostatic Levitation Laboratory. Hyers's research includes high-temperature materials, condition monitoring and prognosis of structures, and physics-based modeling of materials processing and failure. Hyers holds a patent for porous hydroxyapatite networks for synthetic bone material. Hyers is a fellow at ASM International. Hyers earned a PhD in materials engineering from MIT.

YIGUANG JU is the Robert Porter Patterson Professor in Mechanical and Aerospace Engineering and serves as the director of the sustainable energy program at Princeton University. Ju's research interests include combustion, fuels, and low carbon energy conversion in the areas of near limit and supercritical combustion, plasma-assisted combustion and chemical manufacturing, alternative fuels, chemical kinetics, and energy materials. Ju is a fellow of the American Society of Mechanical Engineers, a founding fellow of the Combustion Institute, and serves on the Board of Directors of the Combustion Institute and the Institute for Dynamics of Explosions and Reactive Systems. Ju's awards include the Distinguished Paper Award from the International Symposium on Combustion, the NASA Director's Certificate of Appreciation award, the Friedrich Wilhelm Bessel Research Award from the Alexander von Humboldt Foundation, the International Prize from the Japanese Combustion Society, and the Propellants and Combustion Award from the American Institute of Aeronautics and Astronautics. Ju earned a PhD in mechanical and aerospace engineering from Tohoku University.

CHRISTOPHER E. MASON is a professor at Weill Cornell Medicine, with appointments at the Tri-Institutional Program in Computational Biology and Medicine, the Sandra and Edward Meyer Cancer Center, and the Feil Family Brain and Mind Research Institute. Mason's laboratory develops and deploys computational and experimental methodologies to identify the functional genetic elements of the human genome and metagenome. The laboratory performs research in three principal areas: (1) molecular profiling in patients with extreme phenotypes, including brain malformations, aggressive cancers, and astronauts; (2) creating new biochemical and computational techniques in DNA/RNA sequencing and DNA/RNA base modifications; and (3) the development of models for cellular and genome engineering. These systems-based methods can enable a better understanding of the functional elements of the human genome and metagenome, such that these genetic networks for ameliorating disease can be repaired or reengineered and the foundation can be laid for long-term human spaceflight and exploration. Mason completed a PhD in genetics at Yale University.

MICHAEL J. PECAUT is a research professor and the vice chair in the Division of Biomedical Engineering Sciences at Loma Linda University. In the past 30 years, Pecaut has focused on the gravitational and radiation components of the spaceflight environment. In addition to actual space flight opportunities involving rodent models (STS-77, -108, -118, -135 and RR-3, -9, -12, -18), flight conditions have been modeled using a variety of forcing functions, including anti-orthostatic tail suspension, centrifugation, and low dose/low dose rate radiation. Further research has focused on immune function. Pecaut earned a PhD in aerospace engineering from the University of Colorado Boulder.

WILLIE S. ROCKWARD is a chair and a professor of physics at Morgan State University. Research interests include micro/nano-optics lithography, extreme ultraviolet interferometry, metamaterials, and the spectroscopy of binary stars. As the department chair, Rockward investigated the barriers faced by the Department of Physics of Historically Black Colleges and Universities and launched the We C.A.R.E. approach meant to improve the overall number of African Americans in physics and science, technology, engineering, and mathematics (STEM). Rockward previously served as a tenured professor at Morehouse College, where he also served as the chair of the Department of Physics and Dual Degree Engineering Program and as the research director of the Materials and Optics Research and Engineering Laboratory. Rockward is also a former president of the National Society of Black Physicists. Rockward received a PhD in physics from the Georgia Institute of Technology.

ELBA E. SERRANO is a Regents Professor of Biology at New Mexico State University (NMSU). Serrano's research focuses on the sensory systems for hearing and balance, neurogenetics, and glial neurobiology. An advocate for interdisciplinary research and education, Serrano is an elected fellow of the American Association for the Advancement of Science (AAAS) and the recipient of a Presidential Award for Excellence in Science, Mathematics and Engineering Mentoring. Serrano serves as the NMSU lead principal investigator for the National Science Foundation's (NSF's) Hispanic Serving Institutions (HSI) National STEM Resource Hub, a capacity-building and faculty development initiative that aspires to increase STEM student success at over 550 HSI across all 50 states and the U.S. territories. Her honors include being named one of 100 inspiring Hispanic/Latinx scientists in America. Serrano received a PhD in biological sciences from Stanford University with an emphasis in neuroscience and biophysics.

PETER VOROBIEFF is a professor and the assistant chair of the Department of Mechanical Engineering at the University of New Mexico. Before joining the ranks of the academia, Vorobieff worked in a variety of places including the Russian Association of Space Explorers and Los Alamos National Laboratory. Primary research interests lie in the area of fundamental hydrodynamic instability studies. Distinctions include being named the Halliburton Professor at the University of New Mexico School of Engineering and 2015 Professor of the Year by Pi Tau Sigma International Mechanical Engineering Honor Society. Vorobieff earned a PhD in mechanical engineering and applied mathematics from Lehigh University.

RONALD L. WALSWORTH is the Minta Martin Professor in the Departments of Physics and Electrical and Computer Engineering and the founding director of the Quantum Technology Center at the University of Maryland, College Park. Walsworth leads an interdisciplinary research group with a focus on developing precision measurement tools and then applying them to problems in both the physical and life sciences. Honors include the Francis Pipkin Award in Precision Measurements, American Physical Society; a Distinguished Traveling Lecturer, American Physical Society; and a fellow of the American Physical Society. Walsworth earned a PhD in physics from Harvard University.

SARAH WYATT is a professor in the Department of Environmental and Plant Biology and the director of Interdisciplinary Molecular and Cellular Biology graduate program at Ohio University. Wyatt's research interests include transition of floral dimorphisms; plant signaling, especially as it relates to the response to gravity; and the use of genetic, genomic, proteomic, and bioinformatics strategies to identify novel signaling components. The Wyatt Laboratory uses molecular, genetic, and genomic tools to study plant growth and development and has run several experiments aboard the International Space Station. Wyatt is a fellow of the American Society of Plant Biology. Wyatt received a PhD in plant physiology and molecular biology from Purdue University.

ZHUOMIN ZHANG is the J. Erskine Love, Jr. Professor of Heat Transfer, Combustion, and Energy Systems at the Georgia Institute of Technology. Zhang engages in a broad spectrum of research in heat transfer and thermophysical engineering, including measurements of the radiative properties and optical response of high-temperature superconducting thin-film materials and devices; thermal modeling of absolute cryogenic radiometers, space-based solar radiometers, and pulsed-laser calorimeters; radiation temperature measurement for rapid thermal processing (RTP) and heat transfer modeling of RTP systems; and investigation of the bidirectional reflectance of rough silicon wafers. Zhang is a fellow of the American Physical Society and the AAAS. Zhang received a PhD in mechanical engineering from MIT.

STAFF

DWAYNE A. DAY is a senior program officer and a study director at the National Academies of Sciences, Engineering, and Medicine. He has served as the study director for the planetary sciences decadal survey midterms, as well as assisted on previous planetary science and astrophysics decadal surveys. He has also served as the study director for studies on NASA's near Earth objects efforts, advanced aerial mobility, and the size of the NASA astronaut corps. Prior to coming to the National Academies, he was an investigator on the Columbia Accident Investigation Board and the historian for the Congressional Budget Office, and worked for the U.S. Air Force and The George Washington University Space Policy Institute. He is a frequent author on civil, military, and intelligence space history and policy and a former Guggenheim Fellow and Verville Fellow at the Smithsonian's National Air and Space Museum.

ARUL MOZHI is a senior program officer and an associate director of the Aeronautics and Space Engineering Board (ASEB) and the Space Studies Board (SSB) at the National Academies. Since 1999, Mozhi has been directing projects in the areas of defense and broader science and technology carried out by numerous committees of ASEB, SSB, the Laboratory Assessments Board, the Army Research Laboratory Technical Assessment Board, the Naval Studies Board, and the National Materials and Manufacturing Board. Prior to joining the National Academies, Mozhi held technical and management positions in systems engineering and applied materials research and development (R&D) at several small- and mid-size high-tech R&D and consulting companies in the Washington, DC, and Boston areas—UTRON, Roy F. Weston, and Marko Materials. He received his MS and PhD (the latter in 1986) in materials engineering from The Ohio State University and then served as a postdoctoral research associate there for 2 years. He received his B.Tech. in metallurgical engineering from the Indian Institute of Technology, Kanpur, in 1982.

GREGORY MACK served as a senior program officer at the National Academies. Along with managing the Committee on Astronomy and Astrophysics and the Committee on Radio Frequencies, he was the study co-director for the Astro2020 decadal survey and is also the study co-director for the Biological and Physical Sciences Research in Space Decadal Survey. Prior to the National Academies, Mack was a government relations professional at the APS, a program director in the Division of Astronomical Sciences at NSF, an AAAS Science and Technology Policy Fellow placed in the Division of Physics at NSF, and a visiting assistant professor of physics and astronomy at Ohio Wesleyan University. Mack received his PhD in physics from The Ohio State University with a focus on theoretical astrophysics. His research concerned theoretical particle astrophysics and cosmology, specifically the particle properties of dark matter.

SANDRA J. GRAHAM has served as a senior program officer at the National Academies' SSB since 1994. During that time Graham has directed a large number of major studies, many of them focused on space research in biological and physical sciences and technology, including the comprehensive 2011 decadal survey, *Recapturing a Future for Space Exploration: Life and Microgravity Sciences Research for a New Era*. High-profile studies in other areas have included an assessment of servicing options for the Hubble Space Telescope, a study of the societal impacts of severe space weather, and the Pathways to Exploration report reviewing the U.S. human space program. Prior to joining the SSB, Graham held the position of senior scientist at the Bionetics Corporation, where she provided technical and science management support for NASA's Microgravity Science and Applications Division. Graham's honors include the National Research Council's (NRC's) Distinguished Service Award, the NRC Division on

Engineering and Physical Sciences Exceptional Achievement Award, and the Orr-Reynolds Distinguished Service Award from the American Society for Gravitational and Space Research. She is a member of the American Chemical Society and the Project Management Institute. She received her PhD in inorganic chemistry from Duke University, where her research focused on rate modeling and reaction chemistry for biological metal complexes and their analogs.

JULIE ANNE SCHUCK is a senior program officer at the National Academies. In this role, Schuck regularly manages projects and staff teams; supports ad hoc expert committees and facilitates their work and consensus building; provides analytical, administrative, and editorial support for studies, convening activities, and report development; and serves as a technical writer. Projects have covered a wide range of subjects, including this study; law and justice issues; challenges in the implementation of health care; national security; STEM education; the science of human-system integration; workforce development; and the evaluations of several federal research programs. Schuck received a BS in engineering physics from the University of California, San Diego, and an MS in education from Cornell University.

ERIK SVEDBERG is a National Materials and Manufacturing Board Scholar at the National Academies, where he has been employed since 2008. In this role, he works with experts from across the nation to develop, negotiate, and oversee scientific and technical advisory studies for federal agencies related to questions of materials science, manufacturing, and engineering design. His previous and current activities at the National Academies include work as a study director for *Materials Needs and R&D Strategy for Future Military Aerospace Propulsion Systems*; *Research Opportunities in Corrosion Science and Engineering*; *Opportunities in Protection Materials Science and Technology for Future Army Applications*; *Optics and Photonics: Essential Technologies for Our Nation*; *A Quadrennial Review of the National Nanotechnology Initiative: Nanoscience, Applications, and Commercialization*; *Airline Passenger Security Screening: New Technologies and Implementation Issues*; *Frontiers of Materials Research: A Decadal Survey*; *Strategic Long-Term Participation by DoD in Its Manufacturing USA Institutes*; and *High-Performance Bolting Technology for Offshore Oil and Natural Gas Operations*. His work also includes overseeing workshops such as *Big Data in Materials Research and Development*; *Limited Affordable Low-Volume Manufacturing*; *Materials and Manufacturing Capabilities for Sustaining Defense Systems*; *Novel Processes for Advanced Manufacturing: Summary of a Workshop*; and *Domestic Manufacturing Capabilities for Critical DoD Applications: Emerging Needs in Quantum-Enabled Systems: Proceedings of a Workshop*. Svedberg has a decade of industry experience with both small and large companies in the materials science area and has been a guest researcher at the National Institute of Standards and Technology for several years. He has been awarded and overseen five research grants and has published over 80 scientific articles, been granted two patents, and is cited over 1,700 times with an h-index of 22. His PhD and master's degree are in materials science and he is a fellow of the AAAS, the American Vacuum Society, and the Washington Academy of Science, including being its 2019 leadership in material science award recipient.

STEVEN MOSS served as a senior program officer with the Board on Life Sciences at the National Academies. During his tenure at the National Academies, Moss has focused on the future of the bioeconomy, biotechnology, and needs for the future of research in the life sciences. In addition to his role at the National Academies, Moss is the secretary of the Committee for Environmental Improvement of the American Chemical Society and a member of the Political Action Committee of the Biophysical Society. Prior to joining the National Academies, Moss received his PhD in chemistry and chemical biology from the University of California, San Francisco, where he focused on the signaling pathways of infectious bacteria.

GAYBRIELLE HOLBERT joined the SSB and the ASEB as a program assistant in 2019. In this role, she assists with ongoing projects and workshops by providing logistics, report creation, and project support including the Standing Committee on Earth Science and Applications from Space. She previously worked as a communication specialist for the United Black Fund Inc. a non-profit organization that helped inner-city youth by providing after-school programs, scholarships, and resources to engage and enhance their educational needs. Prior to that, she was the social media consultant for the Development Corporation of Columbia Heights, a non-profit advocacy platform that supports responsible, community-focused economic development and as a production assistant for a startup multimedia production company. She holds a BA in mass media communications from the University of the District of Columbia.

DIONNA WISE is a program coordinator with the SSB, having previously worked for the National Academies' Division of Behavioral and Social Sciences and Education. Recently, she was the lead study coordinator for the Astro2020 decadal survey. Wise has a long career in office administration, having worked as a supervisor in several capacities and fields. She attended the University of Colorado, Colorado Springs, and majored in psychology.

COLLEEN N. HARTMAN joined the National Academies in 2018, as the director for both the SSB and the ASEB. In 2021, she additionally became the director for the Board on Physics and Astronomy. After beginning her government career as a presidential management intern under Ronald Reagan, Hartman worked on Capitol Hill for House Science and Technology Committee Chairman Don Fuqua as a senior engineer building spacecraft at the NASA Goddard Space Flight Center and as a senior policy analyst at the White House. She has served as Planetary Division director, deputy associate administrator, and acting associate administrator at NASA's Science Mission Directorate, as deputy assistant administrator at the National Oceanic and Atmospheric Administration, and as deputy center director and director of science and exploration at NASA's Goddard Space Flight Center. Hartman has built and launched scientific balloon payloads, overseen the development of hardware for a variety of Earth-observing spacecraft, and served as NASA program manager for dozens of missions, the most successful of which was the Cosmic Background Explorer (COBE). Data from the COBE spacecraft gained two NASA-sponsored scientists the Nobel Prize in physics in 2006. She also played a pivotal role in developing innovative approaches to powering space probes destined for the solar system's farthest reaches. While at NASA Headquarters, she spearheaded the selection process for the New Horizons probe to Pluto. She helped gain administration and congressional approval for an entirely new class of funded missions that are competitively selected, called "New Frontiers," to explore the planets, asteroids, and comets in the solar system. She has several master's degrees and a PhD in physics. Hartman has received numerous awards, including two prestigious Presidential Rank Awards.

D

Complementarity of NASA's Division of Biological and Physical Sciences and Human Research Program

This appendix references the scope of Chapter 4, and specifically the recognition that *some* biological/life science research in the NASA Science Mission Directorate (SMD) Division of Biological and Physical Sciences (BPS Division) that enables space exploration has natural connections to the separate Human Research Program (HRP) that has focused on astronaut health and safety.

The biological and physical sciences (BPS) research community perceives a natural, beneficial overlap between the applied, human, risk mitigation–focused research that is the focus of HRP and the fundamental, discovery-focused research that is the focus of the BPS Division and extends beyond humans. That said, this decadal survey comes at a time of adjustment within NASA as a funding agency, especially with the move of the BPS Division from the Human Exploration and Operations (HEO) Mission Directorate to SMD. To emphasize, NASA's BPS Division rests within the *Science* Mission Directorate and sponsors scientific research to answer fundamental questions requiring access to space, among other important roles in science-focused missions. Separately, the HRP is a NASA program within the *Human Exploration and Operations* Mission Directorate, focused on discovering the best methods and technologies to support safe, productive human space travel. HRP is headquartered and administered at Johnson Space Center in Houston, Texas, and includes other elements at other NASA centers.

These foci are complementary; new methods and technologies for safe and impactful human space travel can benefit from the underlying science of mechanisms conferring or avoiding risks—in natural or engineered systems. The BPS research community's perspective and preference inferred from input papers, panel discussions, and committee discussions is that the BPS Division maintains a strong, positive interaction with the biological sciences at the HRP within NASA. That relationship continues to evolve, with HRP focusing ever more directly on the science of human health in space, with the BPS Division focusing more on the fundamental biology of model systems to inform the basic science that impacts human health and the biology of all systems that are part of traveling in space.

In identifying priority key scientific questions (KSQs) for the BPS decadal survey, it was necessary for the BPS research community to reflect on commonalities and distinctions of research areas supported by two different parts of NASA, the BPS Division and HRP. To be clear, several other government agencies and some companies fund research related to human health and safety, and the BPS Division's research focus is much broader than fundamental science related to mechanism associated with health or countermeasures in space environments. But NASA's unique focus on space science is key to momentum in BPS research and is of finite resources well below that of the National Institutes of Health (NIH). Thus, concerns about delineation between BPS Division priorities and HRP priorities draws attention of the BPS research community. This decadal survey is not tasked with advising NASA on its own internal organization and budget lines. However, from the BPS researcher community, the sense built over this decadal survey is that these are viewed appropriate as a Venn diagram of overlapping scientific interests buttressed by distinct missions to apply that understanding.

Recommendation 4-1 emphasized scientific exchange (e.g., of plans and data) and coordination (including of sponsoring scientific research of complementary benefit) between these two NASA units. This recommendation stems from strong preference by the science community that artificial siloes of funding and information exchange are minimized so that answers to critical scientific questions and U.S. leadership in finding and using those answers are not stalled by U.S. agency-internal administrative details. When the primary beneficiary of the answer is the astronaut of the coming decade, this is HRP-leaning research; when the primary beneficiary of the answer is the Earth-based population now and in the distant future, this is BPS-leaning research. *But rather than debates about whether the scientific question falls into a BPS or HRP bucket, debates among research administrators and planners and peer reviewers can be focused on the relative importance of the question and impact of answering it for society.* One way (among others) to consider the Venn diagram position of the research plan could be: If not for humans traveling to and exploring space safely and productively, would answering this research question, achieving this research aim, or testing this research hypothesis be prioritized? Or, If not for access to the more resource-intensive laboratory of the space environment, would less precise or accurate answers to this research question be a societal detriment on Earth?

Consider the following two science questions where one grapples with this real or perceived overlap, given awareness of finite resources (including both available funding and researchers' time):

1. How does reduced gravity affect aging processes in the brain?
2. How does fluid flow in reduced gravity through narrow channels?

Clearly, addressing either of these science questions *can* aid fundamental understanding and discovery unrelated to human health and safety, and *can also aid* astronaut human health and safety. A proposed research plan would inform the balance of each. In broad question 1, research project plans could be focused on durations and conditions and countermeasure technologies relevant to cognition during astronaut missions (HRP-leaning), or they could be focused on leveraging the space environment because accelerated effects of aging facilitate organoid-based assays of demyelination that would take decades on Earth but help address untreatable neurodegenerative diseases if understood (BPS-leaning). In broad question 2, research project plans could be focused on experiments that could improve fire safety associated with transfer of specific combustible liquids in the space environment for refueling (HRP-leaning), or they could be a computational modeling effort requiring validation of semi-empirical constants with potential applications of that validated general model to blood flow in astronauts on the way to Mars, to water flow in LEO-located plants, and to improved spatial tolerance when three-dimensional (3D) printing with complex fluids on Earth's moon or on Earth (BPS-leaning).

These examples show that while one can map the relative position of a research project plan on a Venn diagram, the more important exercise during finite resource allocation is understanding the relative impact of the knowledge gained from the research—including the impact of fundamental knowledge that may have no immediate utility in space or on Earth—and the relative quality of the research project plan. The "so what?" of the research outcomes should be compelling but need not be immediately useful to astronauts or to Earth-based communities to be compelling.

With this context specific to NASA organization, the research areas below are those that were discussed within the BPS decadal survey, with thoughtful input across the BPS research community. These research areas are not an exhaustive list of complementary research areas between HRP and the BPS Division, but they are those developed in the process of considering potential scientific questions of interest to the BPS community. These research areas were not included among the KSQs for BPS in the coming decade, not because they are not important avenues of study in the coming decade, but because they represent areas much closer to the HRP side of synergistic overlap in NASA between HRP and the BPS Division.

IMPACT OF THE SPACEFLIGHT ENVIRONMENT ON THE MUSCULOSKELETAL SYSTEM AND ITS FUNCTION

The response of the musculo-skeletal system to microgravity includes significant muscle wasting and bone loss (Backup et al. 1994; Fitts et al. 2001; Radugina et al. 2018; Shenkman et al. 2003; Trappe et al. 2009). Astronauts are estimated to lose ~1–3 percent of bone mass per month while in space. Over the course of an exploration class mission, this rate of bone loss will increase fracture risk for any astronaut transitioning to a partial or full gravity environment. Such an event could be mission-critical and potentially life threatening, depending on any complications. Indeed, post-flight fractures have occurred in long-duration astronauts, although not yet during space missions. There is also significant muscle loss in arms and hands of astronauts in response to long-duration spaceflight (Puglia et al. 2018), despite the fact that arms and hands are continuously engaged in daily activities. This has prompted the implementation of daily exercise to counteract muscle and bone remodeling. Despite such measures, astronauts still experience muscle weakness and bone loss when returned to Earth's gravity (Fitts et al. 2010; Gopalakrishnan et al. 2010; LeBlanc et al. 2000), suggesting that other factors, in addition to reduced activity, play roles in musculoskeletal remodeling.

Potential Research Areas

The main mechanosensory cells in bone are osteocytes. Osteocytes normally play an essential role in modulating bone formation and resorption by recruiting osteoblast (bone-forming) or osteoclast (bone-resorbing) cells. Bones exposed to microgravity have greater numbers of empty osteocyte lacunae than control bones, indicating the death of osteocytes. It has been proposed that this loss of viable osteocytes dramatically alters bone mechanosensation during spaceflight. Numerous studies have further shown altered expression in osteoclasts (Backup et al. 1994), which has relevance for bone loss in astronauts. Several studies have suggested that one factor is a decline in protein synthesis (Allen et al. 1996; Caiozzo et al. 1994; Esser and Hardeman 1995; Fitts et al. 2000; Gambara et al. 2017; Martin 1988). Following exposure to microgravity, osteocyte apoptosis occurs, which leads to further osteoclast recruitment. If decreases in viable osteocytes alter the mechanosensory systems of bone, this may also impact nearby skeletal muscle, given the degree of biochemical crosstalk between these two tissues. Understanding how to best mitigate these effects remains an open question.

Impact of the Spaceflight Environment on Aging

The cause of aging, an inevitable biological process that affects almost all living organisms, is still an area of significant controversy. Although numerous hypotheses, including increased oxidative stress, accumulation of damaged DNA molecules and mutations, and telomere shortening (da Costa et al. 2016; Horvath and Raj 2018), have been proposed, none of these hypotheses completely explains aging phenotypes or processes. Thus, it is difficult to determine if the space/planetary environment affects aging or not. However, all the hypothetical drivers of aging listed above have been shown to occur in response to the space environment, indicating the likelihood that accelerated aging does indeed occur. One unifying factor that supports the idea that space travel increases aging is the level of stress experienced. Exposure to chronic adverse conditions, and the resultant activation of the neurobiological stress response cascade, has been associated with an increased risk of early-onset, age-related disease as well as an older biological age. This body of research has led to the hypothesis that exposure to stressful life experiences, when occurring repeatedly or over a prolonged period, may accelerate the rate at which the body ages (Polsky et al. 2022). Astronauts undoubtedly experience chronic adverse conditions, whether in a spacecraft or on a planetary base, and are therefore likely to experience psychological and physiological stress leading to accelerated aging.

To address the premise that exposure to the spaceflight environment represents an environment of accelerated aging study, 3D cell models have recently been explored on the International Space Station (ISS) as part of an NIH initiative.[1] This initiative comprised nine separate projects that explored the effects of spaceflight on a spectrum of

[1] National Institutes of Health, "Tissue Chip for Drug Screening," updated May 10, 2023, https://ncats.nih.gov/tissuechip.

disease states and organ functions, including immune responses, kidney and cardiac function, osteoarthritis, and biological barrier responses. But, given the recency of these studies, results are yet to be published. In this context, it is noteworthy that the bone loss and muscle atrophy are hallmark aging phenotypes (Bettis et al. 2018). In addition, simulated microgravity has been shown to accelerate aging of cultured human skeletal muscle myoblasts, with these effects remaining upon return to Earth's gravity (Takahashi et al. 2021). Given these similarities, it is highly likely that studies of organisms and tissue interactions in altered gravities will shed light on fundamental aging processes occurring in humans on Earth.

What Are the Mechanisms and Impact of Spaceflight-Induced Changes in Telomere Length?

A key age-related marker is telomere length, which has been assessed in astronauts (Bailey et al. 2022). In the majority of human cells, telomeres shorten with time until reaching a critically short length, at which point a permanent cell cycle arrest, known as replicative senescence, is entered. This underlies a major theory of aging. Surprisingly, spaceflight-specific telomere elongation was confirmed in three unrelated astronauts during 1-year and 6-month missions aboard the ISS using multiple assays. For example, telomere length in humans was altered in the Twin study; although in this case, telomeres lengthened in space but returned to normal upon return to Earth (Garrett-Bakelman et al. 2019). However, on returning to Earth, astronauts showed a rapid shortening to a length below that seen before flight, indicating that increased telomere shortening, and therefore accelerated aging, is a feature of spaceflight.

What is the impact of spaceflight-induced changes in known hallmarks of aging (genomic instability, epigenetic alterations, mitochondrial dysfunction, cellular senescence, stem cell exhaustion, and altered intercellular communication)? Many changes in accepted hallmarks of aging have been shown to occur after exposure to space radiation, suggesting that aging is accelerated in the space environment. A caveat to this view is that the dose rates of charged-particle radiation in the extraterrestrial environments accessed to date are too low to affect these biological responses significantly. However, the nature of aging and age-related degenerative diseases (e.g., age-related Alzheimer's) is that they develop gradually; the process is incremental and cumulative, meaning that small amounts of insults over time can eventually lead to an overall large impact. This point is particularly relevant for future long-term missions in space where such cumulative damage could be expedited. To effectively answer these questions, future studies focused on improving long-term monitoring of aging markers in current and former space travelers are required.

Is it possible to alter and/or leverage the regenerative capacity of stem cells to develop targeted countermeasures for the physiological stress induced by space exploration? Stem cells are critically important for regeneration of muscle cells, bone cells, and nervous system and immune system function, but understanding of the effects of altered gravity is limited. A number of different studies have sent eukaryotic stem cell models to the ISS, including, but not limited to, iPSC-derived cardiomyocytes (iPSC-CM) (Acharya et al. 2022; Wnorowski et al. 2019). Experiments to date suggest that the ability of stem cells to differentiate along multiple phenotypic lineages is negatively affected by spaceflight, although cells returned to Earth's gravity subsequently exhibited increased stemness and ability to differentiate into (e.g., myocardial [Blaber et al. 2015]) cells. Similarly, human mesenchymal stem cells were reported to show an increased capacity to secrete cytokines and growth factors and were more potent in their immunosuppressive capacity compared to ground controls in one study (Huang et al. 2020), although other studies found no significant change (Ludtka et al. 2021). Nevertheless, the current literature suggests that individual stem cells have intrinsic gravity-sensing mechanisms and that understanding the cellular basis of these effects could unlock the regenerative capacity of stem cells to develop targeted countermeasures.

IMPACT OF THE SPACEFLIGHT ENVIRONMENT ON THE NERVOUS SYSTEM AND ITS FUNCTION

The abrupt transition to a reduced gravity environment during spaceflight, as well as the abrupt reappearance of gravity experiences during landing, has a marked impact on the nervous system. Astronauts experience significant impairments regarding their perceptual and motor performance. This is because the reduced gravity environment of space unloads the vestibular otolith organs of the inner ear, which normally detect the orientation of the head relative to gravity (Carriot et al. 2021). As a result, the otolith organs are no longer stimulated as they would be on Earth. This also affects vestibular nuclei and cortical projection, where different sensory inputs are integrated (Morita et al. 2016). There is emerging evidence that the brain uses two main strategies to adapt to such changes. The first involves the rapid updating of a cerebellum-based internal model of the sensory consequences of gravity (Mackrous et al. 2019). The second involves the reweighting of extra-vestibular information (Jamali et al. 2019) as the vestibular system becomes less (i.e., entering microgravity) and then again more reliable (i.e., return to Earth). Research in animal models and humans exposed to analogs of spaceflight have revealed substantial reorganization of the peripheral and central nervous systems. However, knowledge of environmentally induced changes remains largely based on ground-based studies that have focused on the influence of altered gravity (Carriot et al. 2021; Clément et al. 2020). Space-based experiments are required to advance knowledge of the neural mechanisms that mediate neural adaptation to the space environment, as well as establish how to best optimize these mechanisms to improve behavioral and cognitive performance.

In addition, although the central nervous system (CNS) is not traditionally considered to be the most radiosensitive organ, any damage to it can be particularly devastating to health and quality of life, and it is difficult to repair. Because all body systems are regulated to some degree by neural input, impaired CNS and/or neural damage can result in a cascade of negative functional outcomes. Recent research has revealed significant neural damage and concomitant cognitive deficits induced by simulated galactic cosmic radiation on Earth. For example, new functional dentate granule cells deriving from adult neural stem cells are formed during adult hippocampal neurogenesis and then integrated into the existing neural circuits. These cells are particularly vulnerable to galactic cosmic radiation. Damage to their structure will impact cognitive ability to perform pattern separation, specifically the brain's ability to temporally distinguish information in order to store memories independently of each other. Galactic cosmic radiation also decreases dendrite complexity and density, which has been shown to negatively impact decision-making and critical thinking functions. Epigenetic effects have been demonstrated in mouse retina, which is likely relevant to spaceflight-associated neuro-ocular syndrome (SANS). CNS responses to radiation, including repair potential, can be analyzed and mitigated at different levels—from molecular (DNA damage, reactive oxygen species), to cellular (cell membrane damage, cell death), to vascular leakage and disrupted electrochemical connections between neurons, to tissue and organ damage that eventually culminates in behavioral deficits (Greene-Schloesser et al. 2012). It is important to understand the impact of galactic cosmic radiation in the actual space environment at low Earth orbit (LEO) and interplanetary regions.

Potential Research Areas

How Do the Environmental Challenges Experienced During Space Exploration Alter the Peripheral and Central Nervous Systems?

Microgravity and radiation can alter and ultimately damage sensitive sensory organs, neurons, and the connectivity between structures. Additionally, ground-based studies have shown that central neural pathways are highly plastic and display rapid reorganization in response to altered environmental constraints. Importantly, there is now emerging evidence that central neural pathways likewise display significant reorganization following exposure to microgravity that produces anatomical and structural changes across multiple regions (e.g., brain stem, hippocampus, and sensorimotor cortex) (Carriot et al. 2021; Clément et al. 2020). Thus, research is required to establish the mechanisms underlying changes within the CNS during space exploration, as well as to understand the long-term functional consequences of the resulting central neural pathway reorganization.

How Does the Space Environment Impact the Blood-Brain Barrier and Lymphatic Vessels, and Do These Changes Impair Function?

The blood-brain barrier plays an important role in protecting the brain from many compounds circulating in blood supply and in preventing edema in the brain compartment. There is evidence that blood-brain barrier integrity is compromised after combined hindlimb unloading and chronic low-dose, low-dose-rate gamma radiation, which could lead to brain edema and cognitive impacts (Bellone et al. 2016). Similarly, lymphatic vessels form a network in parallel to the cardiovascular system and are important to the regulation of tissue fluid volume and immune responses. Lymph flow is enabled by local tissue deformation and by gravity-induced tissue fluid pressures, and hence is disrupted in the absence of gravity. Impaired drainage of cerebrospinal fluid (CSF) from the brain via lymphatics may contribute to elevated intracranial pressure and possibly blurring of vision while in LEO (Rasmussen et al. 2020). There is evidence that elements of the spaceflight environment impact the lymphatics in other systems. For example, combined proton radiation and hindlimb unloading in rodents reduces lymphocyte counts in the spleen (Mao et al. 2018) and impairs clearance of bacterial infections (Li et al. 2014), which implicates altered lymphatics given their important role in normal spleen and immune function. These data on blood-brain barrier integrity and lymphatics alterations suggest that a more thorough understanding of both is needed, especially with longer-term sojourns in the space environment.

How Does the Space Environment Modify Risks of Preexisting Abnormal Heart Rhythms (e.g., Atrial Fibrillation) and Progression of Underlying Atherosclerosis (Most Common Form of Heart Disease)?

Atrial fibrillation has been detected in ~5 percent of astronauts, similar to the rate in non-astronauts but presenting at an earlier age. Six months of spaceflight has induced transient changes in atrial electrophysiology; if this chronic condition is preexisting prior to spaceflight, longer-term complications may be possible. Sex-specific differences in atrial fibrillation and stroke put men at increased risk of serious complications in terrestrial environments; further, men may be asymptomatic in early stages. Any induction or aggravation of ventricular arrythymias can be life-threatening. Further, there is growing evidence that space-relevant radiation doses in mammals may induce (Vernice et al. 2020) and/or accelerate growth of preexisting atherosclerotic plaque (Boerma et al. 2015) and impair small arteries' dilation capabilities (Delp et al. 2016). Functional consequences of these effects and rate of progression could be better defined.

How Does Prevention of Pregnancy and Menstrual Suppression via Oral Contraceptives Used by Women Crew Interact with Risk of In-Flight Blood Clot (Thrombus) Formation and Cancer Risks?

Many hormonal-based oral contraceptives increase risks of venous thromboembolisms (Reyes et al. 2022); this is of increased relevance given recent documentation of jugular vein thrombi in crew members.

IMPACT OF THE SPACEFLIGHT ENVIRONMENT ON IMMUNE FUNCTION

Virtually all immune cell populations are reduced after spaceflight (Baqai et al. 2009). Studies in both animal models and humans have shown that the spaceflight environment can influence mitogen-induced proliferation, cytokine production and reactivity (Baqai et al. 2009; Chapes et al. 1999; Crucian et al. 2011, 2013; Hwang et al. 2015), and leukocyte subpopulation distributions (Crucian et al. 2008, 2013; Hwang et al. 2015; Pecaut et al. 2000; Stowe et al. 2011). The overall development of macrophage subpopulations is affected both in vivo (Ortega et al. 2009) and in vitro (Armstrong et al. 1995; Ortega et al. 2012). Blood monocyte counts increased in astronauts after 9 days and decreased after 16 days in space (Stowe et al. 2003). There also appear to be functional deficits in these populations. For example, monocytes isolated from astronauts exposed to the spaceflight environment had a diminished capacity to generate an oxidative burst and phagocytize *E. coli* (Kaur et al. 2005). Similar changes were noted in neutrophils (Kaur et al. 2004). However, the functional consequences of these changes are not well understood.

Potential Research Areas

Potential areas of research may seek to answer the following questions: Given impacts of the space environment on the immune system and exposure of humans on Earth to circulating viral infections, will reactivation of latent viruses become problematic during exploration missions? Will immune memory developed from vaccinations be effective in space? A review of ISS medical records indicates that prolonged rashes, allergies, and cold sores are linked to viral reactivation and immune system dysregulation (Crucian et al. 2016). What countermeasures (like those currently used aboard the ISS to mitigate reactivation of Epstein-Barr virus, cytomegalovirus, and varicella-zoster virus) (Crucian et al. 2020) will be needed to protect future space travelers, given the endemic nature of commonly circulating viruses like HIV and SARS-CoV-2, the virus causing the COVID-19 pandemic? Reactivation of either of these viruses during long-term space travel warrants consideration.

What Are the Negative Impacts of Lunar or Martian Dust on Inflammatory Processes in the Lung?

Once astronauts arrive and live on the surface, they will be exposed to elevated levels of fine dust generated from the surrounding regolith. On Earth, chronic exposure to fine particles or dust results in alveolitis, emphysema, and chronic obstructive pulmonary disease. In the Apollo era, astronauts coming back from the Moon complained about the pernicious dust from their moonwalks; it was fine and electrostatic, and it posed a respiratory hazard in a low gravity environment. Little is known about how martian or Moon dust will impact astronaut physiology, but it is likely that pulmonary function will be one of the first physiological systems to respond to extraterrestrial dust.

Will Asymptomatic, Preclinical Cancers (and Which Ones) Be Accelerated in the Space Environment and How Rapidly?

Ionizing radiation of the space environment increases risks of cancer generation, with a higher incidence in women. Sex-specific cancers such as breast, ovary, and uterus are highest risks terrestrially that could be monitored during spaceflight. During over-winter Antarctic missions, diagnoses of breast, lung, and testicular cancers have occurred, despite precautions and screening (Reyes et al. 2022).

From BPS Theme 1: Adapting to Space; Fluid Shifts

How Do Fluid Shifts Affect Vascular and Cardiac Remodeling, Baroreceptor Function, and Cardiovascular Subsystem Function?

Microgravity causes adaptive changes in cardiovascular physiology that pose deleterious consequences for astronauts. Fluid shifts alter the shape, size, and contractility of the heart. Increased cardiac filling and diuresis results in reduced blood volume and diastolic pressure. Microgravity also increases cytosolic calcium concentrations in cardiomyocytes, which enables Ca^{2+}/calmodulin remodeling. Transcriptomic analysis in *Drosophila* hearts demonstrated a reduction in extracellular matrix proteins and an increase in proteases owing to microgravity. Thus, remodeling of heart tissue is significant following exposure to microgravity, which plays a role in pathology. For example, orthostatic intolerance is a consequence to long-term exposure to microgravity. Identifying remodeling pathways may pinpoint therapeutic targets for prophylaxis.

Fluid Shift Impacts on Visual Acuity

SANS—a condition characterized by a loss of visual acuity—has been observed in human crew members. For example, exposure to microgravity reduces intraocular—intracranial pressure gradients and has been linked to impaired visual acuity (Eklund et al. 2016; Lawley et al. 2017; Mader et al. 2011). Notably, SANS generally does not resolve after return to 1 g. Furthermore, it is more prevalent in those exposed to the space environment for longer periods of time (e.g., >6 months). There is accumulating evidence that headward fluid shifts (and perhaps reduced lymphatic drainage from the cranial compartment) contributes to increased intracranial pressure, resulting in impaired function of the optic nerve and retinal ischemia (Ly et al. 2022). Future work is required to establish whether reversing headward fluid shifts for brief periods each day (e.g., by lower body negative pressure [LBNP] countermeasures) can effectively mitigate these visual deficits and if radiation exposure exacerbates these vision changes.

From BPS Theme 1, Question 4: Multigenerational Effects

If Sperm Development and/or Ovarian Function (and Related Hormonal Factors) in Sexually Mature Mammals Is Altered by the Space Environment, Does This Result in Functional Impacts on Fertility?

This question is of immediate concern to current crew members who spend at least several months in space, and it is also critically important to long-term settlement of planetary surfaces.

How Are Conception and Pregnancy (Including Implantation in the Placenta) Initiated in the Space Environment Affected?

There is nearly zero information on this in mammalian systems, and the impact on the next generation needs to be understood.

How Is Fetal Organ Development, Particularly the Ovaries and Testes, Affected by the Independent or Combined Factors of Microgravity and GCR Exposure?

Gonadal development and any impacts on the germline will be particularly important to understanding the impact on a second generation produced by adults conceived and raised in the space environment.

Theme 3, Question 8: Probing Phenomena Hidden by Gravity or Terrestrial Limitations

Bone Growth and Remodeling

Bone loss in response to microgravity is well documented (Moosavi et al. 2021). Osteocytes show intrinsic ability to sense changes in applied stress, specifically although not uniquely among tissue cell types (Yang et al. 2018). Apoptosis plays a critical role in bone and muscle loss in response to both real and simulated microgravity (Hughes-Fulford 2003; Prasad et al. 2020). The deleterious effects of microgravity on muscle structure and function are well documented (Moosavi et al. 2021). These effects can be highly specific; for example, changes in expression of oxidative enzymes differ depending on the specific muscle fiber type being analyzed within a muscle (Thomason et al. 1992). Apoptosis plays a critical role in bone and muscle loss in response to both real and simulated microgravity (Prasad et al. 2020).

REFERENCES

Acharya, A., H. Nemade, S. Papadopoulos, J. Hescheler, F. Neumaier, T. Schneider, K. Rajendra Prasad, et al. 2022. "Microgravity-Induced Stress Mechanisms in Human Stem Cell-Derived Cardiomyocytes." *iScience* 25(7):104577.

Allen, D.L., W. Yasui, T. Tanaka, Y. Ohira, S. Nagaoka, C. Sekiguchi, W.E. Hinds, R.R. Roy, and V.R. Edgerton. 1996. "Myonuclear Number and Myosin Heavy Chain Expression in Rat Soleus Single Muscle Fibers After Spaceflight." *Journal of Applied Physiology* 81(1):145–151.

Armstrong, J.W., R.A. Gerren, and S.K. Chapes. 1995. "The Effect of Space and Parabolic Flight on Macrophage Hematopoiesis and Function." *Experimental Cell Research* 216(1):160–168.

Backup, P., K. Westerlind, S. Harris, T. Spelsberg, B. Kline, and R. Turner. 1994. "Spaceflight Results in Reduced mRNA Levels for Tissue-Specific Proteins in the Musculoskeletal System." *American Journal of Physiology-Endocrinology and Metabolism* 266(4):E567–E573.

Bailey, S.M., J.J. Luxton, M.J. McKenna, L.E. Taylor, K.A. George, S.G. Jhavar, and G.P. Swanson. 2022. "Ad Astra—Telomeres in Space!" *International Journal of Radiation Biology* 98(3):395–403.

Baqai, F.P., D.S. Gridley, J.M. Slater, X. Luo-Owen, L.S. Stodieck, V. Ferguson, S.K. Chapes, and M.J. Pecaut. 2009. "Effects of Spaceflight on Innate Immune Function and Antioxidant Gene Expression." *Journal of Applied Physiology (1985)* 106(6):1935–1942.

Bellone, J.A., P.S. Gifford, N.C. Nishiyama, R.E. Hartman, and X.W. Mao. 2016. "Long-Term Effects of Simulated Microgravity and/or Chronic Exposure to Low-Dose Gamma Radiation on Behavior and Blood–Brain Barrier Integrity." *npj Microgravity* 2(1):16019.

Bettis, T., B.J. Kim, and M.W. Hamrick. 2018. "Impact of Muscle Atrophy on Bone Metabolism and Bone Strength: Implications for Muscle-Bone Crosstalk with Aging and Disuse." *Osteoporosis International* 29(8):1713–1720.

Blaber, E.A., H. Finkelstein, N. Dvorochkin, K.Y. Sato, R. Yousuf, B.P. Burns, R.K. Globus, and E.A. Almeida. 2015. "Microgravity Reduces the Differentiation and Regenerative Potential of Embryonic Stem Cells." *Stem Cells and Development* 24(22):2605–2621.

Boerma, M., G.A. Nelson, V. Sridharan, X.-W. Mao, I. Koturbash, and M. Hauer-Jensen. 2015. "Space Radiation and Cardiovascular Disease Risk." *World Journal of Cardiology* 7(12):882–888.

Caiozzo, V.J., M.J. Baker, R.E. Herrick, M. Tao, and K.M. Baldwin. 1994. "Effect of Spaceflight on Skeletal Muscle: Mechanical Properties and Myosin Isoform Content of a Slow Muscle." *Journal of Applied Physiology* 76(4):1764–1773.

Carriot, J., I. Mackrous, and K.E. Cullen. 2021. "Challenges to the Vestibular System in Space: How the Brain Responds and Adapts to Microgravity." Review. *Frontiers in Neural Circuits* 15.

Chapes, S.K., S.J. Simske, G. Sonnenfeld, E.S. Miller, and R.J. Zimmerman. 1999. "Effects of Spaceflight and PEG-IL-2 on Rat Physiological and Immunological Responses." *Journal of Applied Physiology* 86(6):2065–2076.

Clément, G.R., R.D. Boyle, K.A. George, G.A. Nelson, M.F. Reschke, T.J. Williams, and W.H. Paloski. 2020. "Challenges to the Central Nervous System During Human Spaceflight Missions to Mars." *Journal of Neurophysiology* 123(5):2037–2063.

Crucian, B., S. Johnston, S. Mehta, R. Stowe, P. Uchakin, H. Quiriarte, D. Pierson, M.L. Laudenslager, and C. Sams. 2016. "A Case of Persistent Skin Rash and Rhinitis with Immune System Dysregulation Onboard the International Space Station." *The Journal of Allergy and Clinical Immunology: In Practice* 4(4):759–762.

Crucian, B.E., G. Makedonas, C.F. Sams, D.L. Pierson, R. Simpson, R.P. Stowe, S.M. Smith, et al. 2020. "Countermeasures-Based Improvements in Stress, Immune System Dysregulation and Latent Herpesvirus Reactivation Onboard the International Space Station—Relevance for Deep Space Missions and Terrestrial Medicine." *Neuroscience & Biobehavioral Reviews* 115:68–76.

Crucian, B., R. Stowe, S. Mehta, P. Uchakin, H. Quiriarte, D. Pierson, and C. Sams. 2013. "Immune System Dysregulation Occurs During Short Duration Spaceflight on Board the Space Shuttle." *Journal of Clinical Immunology* 33(2):456–465.

Crucian, B.E., R.P. Stowe, D.L. Pierson, and C.F. Sams. 2008. "Immune System Dysregulation Following Short- vs Long-Duration Spaceflight." *Aviation, Space, and Environmental Medicine* 79(9):835–843.

Crucian, B., R. Stowe, H. Quiriarte, D. Pierson, and C. Sams. 2011. "Monocyte Phenotype and Cytokine Production Profiles Are Dysregulated by Short-Duration Spaceflight." *Aviation, Space, and Environmental Medicine* 82(9):857–862.

da Costa, J.P., R. Vitorino, G.M. Silva, C. Vogel, A.C. Duarte, and T. Rocha-Santos. 2016. "A Synopsis on Aging-Theories, Mechanisms and Future Prospects." *Ageing Research Reviews* 29:90–112.

Delp, M.D., J.M. Charvat, C.L. Limoli, R.K. Globus, and P. Ghosh. 2016. "Apollo Lunar Astronauts Show Higher Cardiovascular Disease Mortality: Possible Deep Space Radiation Effects on the Vascular Endothelium." *Scientific Reports* 6(1):29901.

Eklund, A., G. Jóhannesson, E. Johansson, P. Holmlund, S. Qvarlander, K. Ambarki, A. Wåhlin, L.-O.D. Koskinen, and J. Malm. 2016. "The Pressure Difference Between Eye and Brain Changes with Posture." *Annals of Neurology* 80(2):269–276.

Esser, K.A., and E.C. Hardeman. 1995. "Changes in Contractile Protein mRNA Accumulation in Response to Spaceflight." *American Journal of Physiology-Cell Physiology* 268(2):C466–C471.

Fitts, R.H., D.R. Riley, and J.J. Widrick. 2000. "Physiology of a Microgravity Environment Invited Review: Microgravity and Skeletal Muscle." *Journal of Applied Physiology* 89(2):823–839.

Fitts, R.H., D.R. Riley, and J.J. Widrick. 2001. "Functional and Structural Adaptations of Skeletal Muscle to Microgravity." *Journal of Experimental Biology* 204(18):3201–3208.

Fitts, R.H., S.W. Trappe, D.L. Costill, P.M. Gallagher, A.C. Creer, P.A. Colloton, J.R. Peters, J.G. Romatowski, J.L. Bain, and D.A. Riley. 2010. "Prolonged Space Flight-Induced Alterations in the Structure and Function of Human Skeletal Muscle Fibres." *The Journal of Physiology* 588(18):3567–3592.

Gambara, G., M. Salanova, S. Ciciliot, S. Furlan, M. Gutsmann, G. Schiffl, U. Ungethuem, P. Volpe, H.C. Gunga, and D. Blottner. 2017. "Gene Expression Profiling in Slow-Type Calf Soleus Muscle of 30 Days Space-Flown Mice." *PLOS One* 12(1):e0169314.

Garrett-Bakelman, F.E., M. Darshi, S.J. Green, R.C. Gur, L. Lin, B.R. Macias, M.J. McKenna, et al. 2019. "The NASA Twins Study: A Multidimensional Analysis of a Year-Long Human Spaceflight." *Science* 364(6436).

Gopalakrishnan, R., K.O. Genc, A.J. Rice, S.M.C. Lee, H.J. Evans, C.C. Maender, H. Ilaslan, and P.R. Cavanagh. 2010. "Muscle Volume, Strength, Endurance, and Exercise Loads During 6-Month Missions in Space." *Aviation, Space, and Environmental Medicine* 81(2):91–104.

Greene-Schloesser, D., M.E. Robbins, A.M. Peiffer, E.G. Shaw, K.T. Wheeler, and M.D. Chan. 2012. "Radiation-Induced Brain Injury: A Review." *Frontiers in Oncology* 2:73.

Horvath, S., and K. Raj. 2018. "DNA Methylation-Based Biomarkers and the Epigenetic Clock Theory of Ageing." *Nature Reviews Genetics* 19(6):371–384.

Huang, P., A.L. Russell, R. Lefavor, N.C. Durand, E. James, L. Harvey, C. Zhang, S. Countryman, L. Stodieck, and A.C. Zubair. 2020. "Feasibility, Potency, and Safety of Growing Human Mesenchymal Stem Cells in Space for Clinical Application." *npj Microgravity* 6:16.

Hughes-Fulford, M. 2003. "Function of the Cytoskeleton in Gravisensing During Spaceflight." *Advances in Space Research* 32(8):1585–1593.

Hwang, S.A., B. Crucian, C. Sams, and J.K. Actor. 2015. "Post-Spaceflight (STS-135) Mouse Splenocytes Demonstrate Altered Activation Properties and Surface Molecule Expression." *PLOS One* 10(5):e0124380.

Jamali, M., J. Carriot, M.J. Chacron, and K.E. Cullen. 2019. "Coding Strategies in the Otolith System Differ for Translational Head Motion vs. Static Orientation Relative to Gravity." *eLife* 8.

Kaur, I., E.R. Simons, V.A. Castro, C.M. Ott, and D.L. Pierson. 2004. "Changes in Neutrophil Functions in Astronauts." *Brain, Behavior, and Immunity* 18(5):443–450.

Kaur, I., E.R. Simons, V.A. Castro, C.M. Ott, and D.L. Pierson. 2005. "Changes in Monocyte Functions of Astronauts." *Brain, Behavior, and Immunity* 19(6):547–554.

Lawley, J.S., L.G. Petersen, E.J. Howden, S. Sarma, W.K. Cornwell, R. Zhang, L.A. Whitworth, M.A. Williams, and B.D. Levine. 2017. "Effect of Gravity and Microgravity on Intracranial Pressure." *Journal of Physiology* 595(6):2115–2127.

LeBlanc, A., C. Lin, L. Shackelford, V. Sinitsyn, H. Evans, O. Belichenko, B. Schenkman, et al. 2000. "Muscle Volume, MRI Relaxation Times (T2), and Body Composition After Spaceflight." *Journal of Applied Physiology (1985)* 89(6):2158–2164.

Li, M., V. Holmes, Y. Zhou, H. Ni, J.K. Sanzari, A.R. Kennedy, and D. Weissman. 2014. "Hindlimb Suspension and SPE-Like Radiation Impairs Clearance of Bacterial Infection." *PLOS One* 9(1):E85665.

Ludtka, C., J. Silberman, E. Moore, and J.B. Allen. 2021. "Macrophages in Microgravity: The Impact of Space on Immune Cells." *npj Microgravity* 7(1):13.

Ly, V., S.R. Velichala, and A.R. Hargens. 2022. "Cardiovascular, Lymphatic, and Ocular Health in Space." *Life (Basel)* 12(2).

Mackrous, I., J. Carriot, M. Jamali, and K.E. Cullen. 2019. "Cerebellar Prediction of the Dynamic Sensory Consequences of Gravity." *Current Biology* 29(16):2698–2710.

Mader, T.H., C.R. Gibson, A.F. Pass, L.A. Kramer, A.G. Lee, J. Fogarty, W.J. Tarver, et al. 2011. "Optic Disc Edema, Globe Flattening, Choroidal Folds, and Hyperopic Shifts Observed in Astronauts After Long-Duration Space Flight." *Ophthalmology* 118(10):2058–2069.

Mao, X.W. 2018. "Impact of Spaceflight and Artificial Gravity on the Mouse Retina: Biochemical and Proteomic Analysis." *International Journal of Molecular Sciences* 19(9).

Martin, T.P. 1988. "Protein and Collagen Content of Rat Skeletal Muscle Following Space Flight." *Cell and Tissue Research* 254(1):251–253.

Moosavi, D., D. Wolovsky, A. Depompeis, D. Uher, D. Lennington, R. Bodden, and C.E. Garber. 2021. "The Effects of Spaceflight Microgravity on the Musculoskeletal System of Humans and Animals with an Emphasis on Exercise as a Countermeasure: A Systematic Scoping Review." *Physiological Research* 70:119–151.

Morita, S., E. Furube, T. Mannari, H. Okuda, K. Tatsumi, A. Wanaka, and S. Miyata. 2016. "Heterogeneous Vascular Permeability and Alternative Diffusion Barrier in Sensory Circumventricular Organs of Adult Mouse Brain." *Cell and Tissue Research* 363(2):497–511.

Ortega, M.T., N. Lu, and S.K. Chapes. 2012. "Evaluation of in Vitro Macrophage Differentiation During Space Flight." *Advances in Space Research* 49(10):1441–1455.

Ortega, M.T., M.J. Pecaut, D.S. Gridley, L.S. Stodieck, V. Ferguson, and S.K. Chapes. 2009. "Shifts in Bone Marrow Cell Phenotypes Caused by Spaceflight." *Journal of Applied Physiology (1985)* 106(2):548–555.

Pecaut, M.J., S.J. Simske, and M. Fleshner. 2000. "Spaceflight Induces Changes in Splenocyte Subpopulations: Effectiveness of Ground-Based Models." *American Journal of Physiology Regulatory, Integrative and Comparative Physiology* 279(6):R2072–R2078.

Polsky, L.R., K.E. Rentscher, and J.E. Carroll. 2022. "Stress-Induced Biological Aging: A Review and Guide for Research Priorities." *Brain, Behavior, and Immunity* 104:97–109.

Prasad, B., D. Grimm, S.M. Strauch, G.S. Erzinger, T.J. Corydon, M. Lebert, N.E. Magnusson, M. Infanger, P. Richter, and M. Krüger. 2020. "Influence of Microgravity on Apoptosis in Cells, Tissues, and Other Systems in Vivo and in Vitro." *International Journal of Molecular Sciences* 21(24).

Puglia, I., M. Balsamo, M. Vukich, and V. Zolesi. 2018. "Long-Term Microgravity Effects on Isometric Handgrip and Precision Pinch Force with Visual and Proprioceptive Feedback." *International Journal of Aerospace Engineering* 2018:1952630.

Radugina, E.A., E.A.C. Almeida, E. Blaber, V.A. Poplinskaya, Y.V. Markitantova, and E.N. Grigoryan. 2018. "Exposure to Microgravity for 30 Days Onboard Bion M1 Caused Muscle Atrophy and Impaired Regeneration in Murine Femoral Quadriceps." *Life Sciences in Space Research* 16:18–25.

Rasmussen, J.C., S. Kwon, A. Pinal, A. Bareis, F.C. Velasquez, C.F. Janssen, J.R. Morrow, C.E. Fife, R.J. Karni, and E.M. Sevick-Muraca. 2020. "Assessing Lymphatic Route of CSF Outflow and Peripheral Lymphatic Contractile Activity During Head-Down Tilt Using Near-Infrared Fluorescence Imaging." *Physiological Reports* 8(4):e14375.

Reyes, D.P., K.S. Masterova, M. Walton, E.L. Kerstman, and E.L. Antonsen. 2022. "Assessment of Sex-Dependent Medical Outcomes During Spaceflight." *Journal of Women's Health (Larchmt)* 31(8):1145–1155.

Shenkman, B.S., I.N. Belozerova, P. Lee, T.L. Nemirovskaya, and I.B. Kozlovskaya. 2003. "Effects of Weightlessness and Movement Restriction on the Structure and Metabolism of the Soleus Muscle in Monkeys After Space Flight." *Neuroscience and Behavioral Physiology* 33(7):717–722.

Stowe, R.P., C.F. Sams, and D.L. Pierson. 2003. "Effects of Mission Duration on Neuroimmune Responses in Astronauts." *Aviation, Space, and Environmental Medicine* 74(12):1281–1284.

Stowe, R.P., C.F. Sams, and D.L. Pierson. 2011. "Adrenocortical and Immune Responses Following Short- and Long-Duration Spaceflight." *Aviation, Space, and Environmental Medicine* 82(6):627–634.

Takahashi, H., A. Nakamura, and T. Shimizu. 2021. "Simulated Microgravity Accelerates Aging of Human Skeletal Muscle Myoblasts at the Single Cell Level." *Biochemical and Biophysical Research Communications* 578:115–121.

Thomason, D.B., P.R. Morrison, V. Oganov, E. Ilyina-Kakueva, F.W. Booth, and K.M. Baldwin. 1992. "Altered Actin and Myosin Expression in Muscle During Exposure to Microgravity." *Journal of Applied Physiology (1985)* 73(2 Suppl):90s–93s.

Trappe, S., D. Costill, P. Gallagher, A. Creer, J.R. Peters, H. Evans, D.A. Riley, and R.H. Fitts. 2009. "Exercise in Space: Human Skeletal Muscle After 6 Months Aboard the International Space Station." *Journal of Applied Physiology (1985)* 106(4):1159–1168.

Vernice, N.A., C. Meydan, E. Afshinnekoo, and C.E. Mason. 2020. "Long-Term Spaceflight and the Cardiovascular System." *Precision Clinical Medicine* 3(4):284–291.

Wnorowski, A., A. Sharma, H. Chen, H. Wu, N.Y. Shao, N. Sayed, C. Liu, et al. 2019. "Effects of Spaceflight on Human Induced Pluripotent Stem Cell-Derived Cardiomyocyte Structure and Function." *Stem Cell Reports* 13(6):960–969.

Yang, X., L.-W. Sun, C.-F. Du, X.-T. Wu, and Y.-B. Fan. 2018. "Finite Element Analysis of Osteocytes Mechanosensitivity Under Simulated Microgravity." *Microgravity Science and Technology* 30(4):469–481.

E

Technical Risk and Cost Evaluation Related to the Chapter 6 Research Campaigns

WHAT IS THE TRACE PROCESS AND HOW WAS IT APPLIED TO THE CONCEPT OF RESEARCH CAMPAIGNS?

The technical risk and cost evaluation (TRACE) process is a specific application of risk-advised programmatic evaluation. It has evolved over more than a decade of work with the National Academies of Sciences, Engineering, and Medicine and The Aerospace Corporation to provide a common framework for the programmatic evaluation of concepts under consideration for decadal surveys and other studies by the National Academies. Previous works include two decadal surveys for both astronomy and astrophysics and planetary sciences, and one each for Earth sciences and applications from space and solar and space physics. The TRACE process evolved from cost and schedule modeling methodologies that have been used for evaluating NASA mission options for many decades. It continues to evolve for the decadal surveys, and each decadal survey applies the TRACE process differently, both in how they conduct it, and how the steering committees use the results in their decision-making. TRACE is an input to the decision-making but does not determine the decisions.

The approach used for this assessment was to identify (1) a compelling need, (2) a set of research questions and knowledge goals required to meet that need, and (3) set(s) of experiments that if executed could be expected to provide answers to those questions with successive experiments building the body of knowledge. From the experiment plan, the required support requirements for the number and type of experiments (e.g., up/down mass, facilities needed, crew time, etc.) were identified. Once this work was accomplished, the Panel on Engineering and Science Interface iterated with the Aerospace team to identify the complete set of elements needed to execute the campaign, refine the analogies and costed elements, and execute the process that is diagrammed below.

For example, the Bioregenerative Life Support Systems (BLiSS) research campaign was developed around the compelling need to develop self-sustainable biological life support systems that produce food, clean water, and renew air to enable long-duration space missions. From this, four research thrusts were developed and mapped into the key scientific question from Chapters 4 and 5. Five sets of experiment groups were developed, each with secondary experiment/measurements goals. Approximately 500 experiments and a similar number of inhabitable environment samples were identified. For each of those, the needed facility (e.g., Veggie, Ohalo, bioreactor) and the type of experiment was identified to allow for calculation of crew time, up/down/mass, and so on.

Leaning heavily on International Space Station (ISS) experiences and lessons learned, the operations concepts were described to include items such as the ground facilities to pre- and post-process experimental materials,

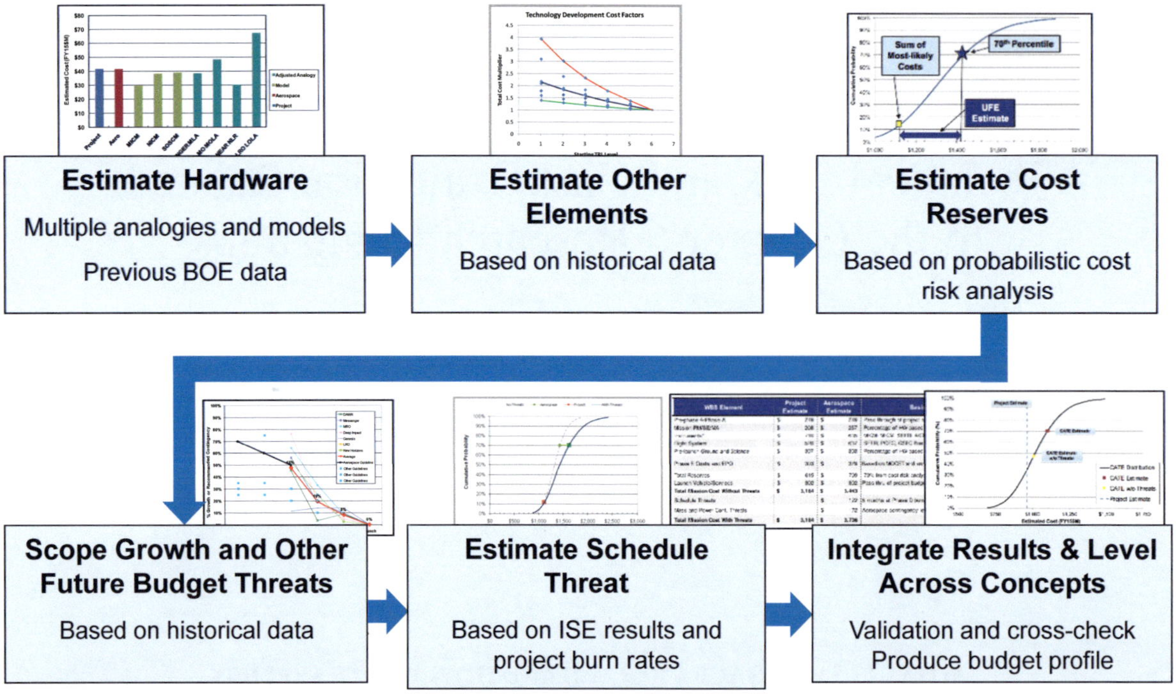

FIGURE E-1 Technical risk and cost evaluation (TRACE) cost estimating approach overview. NOTE: BOE, Basis of Estimate; ISE, Independent Schedule Estimate. SOURCE: Provided by The Aerospace Corporation under contract with the National Academies of Sciences, Engineering, and Medicine.

perform house control experiments, and to begin assembling BLiSS composite systems. Also included were the terrestrial research programs to prepare for and exploit the knowledge from the space-based experiments, the needed technology developments, and technical and programmatic management.

The majority of the key development and operations data was provided by NASA. The operations concepts for the Commercial LEO Destinations have been published, but there is of course no performance history, so ISS experiences were used to modify and inform those estimates as appropriate.

Once this work was accomplished, the panel iterated with the Aerospace team to identify the complete set of elements needed to execute the research campaign, refine the analogies and costed elements, and execute the process diagrammed below. (See Figure E-1.)

Key elements of this work are (1) identifying and rating risks associated with technology development requirements, flight systems, instrumentation, facility design, and operations; (2) estimating budget scope for proposed activities; (3) developing scenarios for phasing and implementing activities consistent with the predicted budget, including future cost threats, complexity growth, or campaign or scope creep or schedule realism; and (4) using a Monte Carlo analysis to capture appropriate reserves based on concept maturity and technology readiness.

The results were captured into tables and spreadsheets with the mean cost by element and aggregate cost risk captured as reserves. The latter items were captured as explicit budget reserves for development items and funded schedule extension risk for the operations phase.

F

Key Scientific Questions and Recommendations for BPS Space Research Over the Decade 2023–2032

This appendix collects in a single location the key scientific questions for biological and physical sciences (BPS) space research over the decade 2023–2032 (Table F-1) and the recommendations for BPS space research over the decade 2023–2032 (Table F-2) that is found out at various locations in this report.

TABLE F-1 Key Scientific Questions for Biological and Physical Sciences Space Research Over the Decade 2023–2032

Themes	Key Scientific Questions
Adapting to Space (Chapter 4)	• How does the space environment influence biological mechanisms required for organisms to survive the transitions to and from space, and thrive while off Earth? • How do genetic diversity and life history influence physiological adaptation to the space environment? • How does the space environment alter interactions between organisms?
Living and Traveling in Space (Chapter 4)	• What are the important multi-generational effects of the space environment on growth, development, and reproduction? • What principles guide the integration of biological and abiotic systems to create sustainable and functional extraterrestrial habitats? • What principles enable identification, extraction, processing, and use of materials found in extraterrestrial environments to enable long-term, sustained human and robotic space exploration? • What are the relevant chemical and physical properties and phenomena that govern the behavior of fluids in space environments?
Probing Phenomena Hidden by Gravity or Terrestrial Limitations (Chapter 5)	• What are the mechanisms by which organisms sense and respond to physical properties of surroundings and to applied mechanical forces, including gravitational force? • What are the fundamental principles that organize the structure and functionality of materials, including but not limited to soft and active matter? • What are the fundamental laws that govern the behavior of systems that are far from equilibrium? • What new physics, including particle physics, general relativity, and quantum mechanics, can be discovered with experiments that can only be carried out in space?

TABLE F-2 Recommendations for Biological and Physical Sciences Space Research Over the Decade 2023–2032

Chapter	Recommendation
Framework for Thriving in Space (Chapter 3)	Recommendation 3-1: NASA should direct its research resources toward the key scientific questions identified in this study. Recommendation 3-2: NASA should work with other U.S. government agencies and other nations' space agencies to coordinate research resources toward the key scientific questions, as relevant to multiple-agency missions. Recommendation 3-3: As activity in low Earth orbit increases, and lunar and martian missions are increasingly likely, NASA should increase resources dedicated to understanding the answers to these key scientific questions.
Science to Enable Space Exploration (Chapter 4)	Recommendation 4-1: NASA should continue to strengthen the science exchange between the Biological and Physical Sciences Program and the Human Research Program. Such effort may include establishing a coordinating body and shared research initiatives as well as the two-way exchange of technologies, data, mission science, specimen banking, and plans. Recommendation 4-2: NASA should increase resources dedicated to producing and understanding the answers to the key scientific questions that address the transitions to and from space. The committee sees potential for significant advances in space exploration if a biological and physical sciences portfolio in the coming decade is aimed at understanding the biological responses that occur during transitions between the Earth and space environments over extended duration and distance to fundamentally enable space exploration; genetic diversity to understand positive and negative responses and long-term adaptations to spaceflight to accelerate the identification of risks, mechanisms of adaptation, and potential positive adaptations that could improve life in space; and how cells, systems, and organisms concurrently adapt to the spaceflight environment and develop mechanisms for encouraging positive and countering negative communicated responses. Recommendation 4-3: To ensure the long-term survival of life in the spaceflight environment, NASA should ramp up investigations into space impacts on sustained human presence in space by investigating • Reproduction, development, and evolution within all relevant biological systems; • The relationships between biology and space hardware to ensure structural integrity, optimize recycling, and utilization of local resources; • Effective chemical, physical, and biological methods for locating, extracting, and processing local resources, especially from the Moon, for use in local habitation and downstream production; and • Fluid physics, combustion, and related sciences to enable sustainable space exploration and habitation.
Science Enabled by the Space Environment (Chapter 5)	Recommendation 5-1: NASA should substantially increase resources dedicated to producing and understanding the answers to the key scientific questions detailed in this report. This investment recognizes the potential for significant societal impacts utilizing the space environment for the biological and physical sciences portfolio in the coming decade, aimed at • Identifying the mechanisms by which organisms sense and respond to the surrounding environment, including gravitational force; • Advancing knowledge of material structure, self-assembly, and stability of materials, including but not limited to soft/active matter, in space environments, cognizant of but distinct from the applications of that knowledge to space exploration and habitation (e.g., manufacturing in space); • Supporting ground-based and microgravity research on understanding the fundamental laws of systems far from equilibrium, especially those that underlie the existence of life; and • Identifying new principles of physics that can only be discovered through experiments in space, including those governing particle physics, general relativity, and quantum mechanics.

TABLE F-2 Continued

Chapter	Recommendation
	Recommendation 5-2: For fundamental physics in space, NASA should facilitate durable formation of collaborations and efficient knowledge transfer between researchers working in multi-disciplinary teams. This scope to address these multi-decadal key scientific questions should include ground-based infrastructure, theoretical and experimental physics, precision measurement and technology development with private sector participants, and should be coordinated with missions in which biological and physical sciences research is one among several whole-of-government objectives.
	Recommendation 5-3: In all of the space-enabled research areas, NASA should allocate funding with an anticipation that new directions of research may arise.
Research Campaigns (Chapter 6)	Recommendation 6-1: NASA should pursue dedicated research campaigns that, through the coming decade, will drive resolution to specific groups of key scientific questions. Coordination beyond NASA, including other federal agencies and the private sector as well as public–private partnerships, should be considered for the dedicated new funding and materials to support these research campaigns. • BLiSS (Bioregenerative Life Support Systems) to build and understand the systems that would provide high-quality food, refresh air and water, process wastes, and enable the creation of space environments sustainable for long periods of time independent of Earth; and • MATRICES (Manufacturing Materials and Processes for Sustainability in Space) to understand and harness the physical processes by which materials and complex fluids can be repeatably utilized in space, to enable sustainable exploration and circular lifecycles for the built environment on Earth and in space.
	Recommendation 6-2: NASA should pursue development of the Probing the Fabric of Space Time (PFaST) initiative in this decade only if it can obtain substantial (greater than 75 percent) funding from external (i.e., other than NASA) sources.
Strategy and Challenges (Chapter 7)	Recommendation 7-1: Because the nation benefits from global leadership in space science and technology, and given the emergence of commercial platforms that can be tasked to the nation's science, NASA should • Seek significant funding increases for biological and physical sciences with new monies or through rebalancing the portfolio across the Science Mission Directorate, and in coordination with other U.S. government agencies, as the community needs to grow significantly in size to reach the science goals of the nation; • Actively engage commercial spaceflight firms, using science funding as a driver and with all due haste, to ensure that science needs are met with clear priority, guaranteeing that national science needs are enabled along with those of potential commercial customers using those platforms; and • Ensure that the funded science community fully engages diversity and inclusivity in the pursuit of the nation's space exploration science priorities.
	Recommendation 7-2: To maintain research campaign momentum, NASA should require external advisory committees to evaluate research campaign team progress and emergent technologies annually.
	Recommendation 7-3: Because key questions identified in this study benefit from access to multiple spaceflight-related platforms, the Biological and Physical Sciences Program should • Coordinate funding opportunities with the Space Technology Mission Directorate such that access to the range of spaceflight and spaceflight-related platforms is efficiently employed to answer key science questions, especially those questions that inform technology development for space exploration; and • Maintain a foundational approach to science, building through a strong, vibrant program of ground-based, suborbital, orbital, lunar, martian, and beyond missions.
	Recommendation 7-4: Because key scientific questions identified in this study support the effective utilization of, and benefit from access to, deep-space exploration platforms, NASA should ensure that scientific opportunities are maximized within the range of spaceflight and spaceflight-related platforms intended for lunar, cislunar, and Mars transit solutions.

continued

TABLE F-2 Continued

Chapter	Recommendation
	Recommendation 7-5: The U.S. government, including NASA, should develop and maintain sufficient ground-based infrastructure to validate and support biological and physical sciences missions. Some of these facilities already exist and simply need to be upgraded, while others have yet to be conceived and built.
	Recommendation 7-6: NASA should continue to expand the investment in open and shared computational infrastructure (CI) to support storage, analysis, and dissemination of its biological and physical data, while ensuring linkage to the original and archived samples. • For biological sciences, GeneLab should be continued and efforts made to ensure findable, accessible, interoperable, and reusable access from other critical international biological resource CIs. • NASA should recognize the need for long-term investment to maintain, update, and improve such community-serving CI and physical repositories over time.
	Recommendation 7-7: NASA should work with the other appropriate U.S. government agencies with a goal to establish an office or a mechanism for commercial sponsorship and collaboration with nonprofit organizations, including academia and government research agencies. That office/mechanism should have the primary focus of • Coordinating the work between these commercial sectors and government agencies; • Providing guidance on or facilitating research compliance, data security, and material transfer agreements, including prototype agreements; • Representing multiple space environments and destinations (e.g., not only the International Space Station in low Earth orbit); and • Communicating these opportunities to the research community.
	Recommendation 7-8: NASA should work with appropriate government agencies to establish clear guidelines for international collaborations within the biological and physical sciences—in particular, for support of non-U.S. students and scholars, to balance two goals: • Sustain and advance the U.S. leadership in the relevant areas of research, possibly by attracting the best and brightest globally; and • Support a robust global research community and information exchange, fostering partnerships with other space programs and U.S. access to other nations' ground-based and space assets.
	Recommendation 7-9: To retire many of the key scientific questions by the end of the decade, NASA should establish support for the Biological and Physical Sciences Program to levels that reflect the current national need and to build the science community in size, diversity of technical expertise and lived experience, and capability to reach the science goals of the nation, toward levels that are an order of magnitude above the current funding and well before the end of the decade.
	Recommendation 7-10: To maintain a viable scientific community, the numerical majority of supported principal investigators (i.e., fraction of research team leaders) should be extramural (i.e., not NASA employees), and funding levels should be commensurate with addressing the key scientific questions.
	Recommendation 7-11: NASA should establish periodic reviews of selected research campaigns to ensure coordinated access to the space environment, publicly communicated progress on research milestones, and facilitation of collaborations and public–private partnerships as required to meet these ambitious goals.
	Recommendation 7-12: NASA should identify mechanisms to compete new or additional research campaigns within 5 years, in light of anticipated changes to access to low Earth orbit and the inevitable but unknown changes in research, technology, funding, and space mission directives that will ensue after this report is issued.

TABLE F-2 Continued

Chapter	Recommendation
	Recommendation 7-13: NASA should ensure diversity, equity, inclusivity, and accessibility in the pursuit of the nation's space exploration science priorities, including instituting a requirement of documented progress in diversity among NASA-sponsored research teams seeking multi-year funding or multiple sponsorship requests over the coming decade. This inclusivity should be intentionally broad in concept, with respect to visible and less visible characteristics of historically underrepresented groups in BPS research and leadership.
	Recommendation 7-14: Project grants should be funded at levels and duration consistent with the project aims with full support for trainees (postdoctorates, graduate students, and undergraduates), including travel for trainees and principal investigators to support the mission and participate in scientific meetings. Full funding representing the total costs of research (direct and indirect) is imperative to be inclusive of participation by all trainees.